Biogenic Nanoparticles

Interplay with Climate Change and Implications for Human Health

Biogenic Nanoparticles

Interplay with Climate Change and Implications for Human Health

Edited by

Vishnu D. Rajput
Academy of Biology and Biotechnology, Southern Federal University, Rostov-on-Don, Russia

Abhishek Singh
Faculty of Biology, Yerevan State University, Yerevan, Armenia

Karen Ghazaryan
Faculty of Biology, Yerevan State University, Yerevan, Armenia

Athanasios Alexiou
Novel Global Community Educational Foundation, Hebersham, Australia

Tatiana Minkina
Soil Science and Land Evaluation Department, Southern Federal University, Russia

ELSEVIER

Elsevier
Radarweg 29, PO Box 211, 1000 AE Amsterdam, Netherlands
125 London Wall, London EC2Y 5AS, United Kingdom
50 Hampshire Street, 5th Floor, Cambridge, MA 02139, United States

ISBN: 978-0-443-33819-9

For Information on all Elsevier publications
visit our website at https://www.elsevier.com/books-and-journals

Publisher: Candice Janco
Acquisitions Editor: Jessica Mack
Editorial Project Manager: Helena Beauchamp
Production Project Manager: Paul Prasad Chandramohan
Cover Designer: Greg Harris

Typeset by MPS Limited, Chennai, India

Contents

Section II Interaction of biogenic nanomaterials and the environment

Section III Human health effects of biogenic nanomaterials

List of contributors

Diptikanta Acharya
Department of Biotechnology, GIET University, Gunupur, Rayagada, Odisha, India

Shreni Agrawal
Department of Life Sciences, Parul Institute of Applied Sciences, Parul University, Waghodia, Gujarat, India

Elsayed Fathi Abd Allah
Plant Production Department, College of Food and Agricultural Sciences, King Saud University, Riyadh, Saudi Arabia

Iqra Arooj
Department of Microbiology and Molecular Genetics, The Women University Multan, Multan, Punjab, Pakistan

Laraib Aslam
Department of Microbiology and Molecular Genetics, The Women University Multan, Multan, Punjab, Pakistan

Kousik Atta
Faculty of Agricultural Sciences, Institute of Applied Sciences and Humanities, GLA University, Mathura, Uttar Pradesh, India

Diptanu Banik
Department of Agronomy, School of Agriculture, Lovely Professional University, Phagwara, Punjab, India

Indrani Barman
Program of Biochemistry, Faculty of Science, Assam Down Town University, Guwahati, Assam, India

Upasana Bhagawati
Amity Institute of Microbial Technology, Amity University, Noida, Uttar Pradesh, India

Rupjyoti Bharali
Department of Botany, Biswanath College, Biswanath Chariali, Assam, India

Ayushi Bhatnagar
School of Life and Basic Sciences, Jaipur National University, Jaipur, Rajasthan, India

Indrani Bhattacharya
Department of Life Sciences, Parul Institute of Applied Sciences, Parul University, Waghodia, Gujarat, India

Jutishna Bora
Amity Institute of Biotechnology, Amity University Jharkhand, Ranchi, Jharkhand, India

Marina Burachevskaya
Laboratory of Soil Chemistry and Ecology, Tula State Lev Tolstoy Pedagogical University (Tolstoy University), Tula, Russia

Armine Chakhmakhchyan
Faculty of Biology, Yerevan State University, Yerevan, Armenia

Ranju Chetri
Department of Zoology, Tezpur College, Tezpur, Assam, India

Sanjay Chetry
Department of Fruit Science, Tamil Nadu Agricultural University, Coimbatore, Tamil Nadu, India

Simran Choudhury
School of Life Sciences, Central University of Gujarat, Vadodara, Gujarat, India

Utpal Das
Department of Horticulture and Food Science, VIT School of Agricultural Innovations and Advanced Learning, Vellore Institute of Technology, Vellore, Tamil Nadu, India

Hassan El-Ramady
Soil and Water Department, Faculty of Agriculture, Kafrelsheikh University, Kafr El-Sheikh, Egypt

Mohamed S. Elshikh
Department of Botany and Microbiology, College of Science, King Saud University, Riyadh, Saudi Arabia

Morishwala Mohammad Farhan
Department of Life Sciences, Parul Institute of Applied Sciences, Parul University, Waghodia, Gujarat, India

Karen Ghazaryan
Faculty of Biology, Yerevan State University, Yerevan, Armenia

Tuhina Ghosh
Division of Seed Science & Technology, ICAR-Indian Agricultural Research Institute, Pusa Campus, New Delhi, India

Anna Harutyunyan
Faculty of Biology, Yerevan State University, Yerevan, Armenia

Abeer Hashem
Botany and Microbiology Department, College of Science, King Saud University, Riyadh, Saudi Arabia

Ani Hayrapetyan
Faculty of Biology, Yerevan State University, Yerevan, Armenia

Abid Hussain
Department of Biological and Chemical Engineering, Aarhus University, Aarhus, Denmark

Mohd Zakir Hussain
Department of Agronomy, Bihar Agricultural University, Sabour, Bhagalpur, Bihar, India

Sumit Jangra
Tropical Research and Education Center, Homestead, Florida

Anita Jaswal
Department of Agronomy, School of Agriculture, Lovely Professional University, Phagwara, Punjab, India

Roshni Jha
Amity Institute of Biotechnology, Amity University Jharkhand, Ranchi, Jharkhand, India

Dharm Nath Kamat
Department of Plant Breeding and Genetics, Sugarcane Research Institute, Dr. Rajendra Prasad Central Agricultural University, Samastipur, Bihar, India

Bal Krishna
Department of Plant Breeding & Genetics, Bihar Agricultural University, Sabour, Bhagalpur, Bihar, India

Ashwini Kumar
Department of Molecular Biology and Biotechnology, Acharya Narendra Deva University of Agriculture and Technology, Kumarganj, Ayodhya, Uttar Pradesh, India

Rakesh Kumar
Division of Crop Research, ICAR Research Complex for Eastern Region, Patna, Bihar, India

Ravi Ranjan Kumar
Department of Molecular Biology and Genetic Engineering, Bihar Agricultural University, Sabour, Bhagalpur, Bihar, India

Vandna Kushwaha
Department of Molecular Biology and Biotechnology, Acharya Narendra Deva University of Agriculture and Technology, Kumarganj, Ayodhya, Uttar Pradesh, India

Smita Lata
Amity Institute of Biotechnology, Amity University Jharkhand, Ranchi, Jharkhand, India

Dilip Kumar Mahto
Bhola Paswan Shastri Agricultural College, Purnea, Bihar, India

Sumira Malik
Amity Institute of Biotechnology, Amity University Jharkhand, Ranchi, Jharkhand, India

Gohar Margaryan
Faculty of Biology, Yerevan State University, Yerevan, Armenia

Rahul Kumar Maurya
Department of Molecular Biology and Biotechnology, Acharya Narendra Deva University of Agriculture and Technology, Kumarganj, Ayodhya, Uttar Pradesh, India

Chandra Mohan Mehta
Department of Agronomy, School of Agriculture, Lovely Professional University, Phagwara, Punjab, India

Ankita Mitra
School of Medicine Stanford University, Stanford, CA, United States

Sagar Mondal
Amity Institute of Biotechnology, Amity University Jharkhand, Ranchi, Jharkhand, India

Hasmik Movsesyan
Faculty of Biology, Yerevan State University, Yerevan, Armenia

Darbinyan Nare
Faculty of Biology, Yerevan State University, Yerevan, Armenia

Nishi Patel
Department of Life Sciences, Parul Institute of Applied Sciences, Parul University, Waghodia, Gujarat, India

Kratika Pathak
School of Life and Basic Sciences, Jaipur National University, Jaipur, Rajasthan, India

Arunava Pradhan
Department of Biology, Centre of Molecular and Environmental Biology (CBMA), Aquatic Research Network (ARNET), Associate Laboratory, University of Minho, Campus of Gualtar, Braga, Portugal; Institute for Science and Innovation for Bio-Sustainability (IB-S), University of Minho, Campus of Gualtar, Braga, Portugal

Shambhoo Prasad
Department of Molecular Biology and Biotechnology, Acharya Narendra Deva University of Agriculture and Technology, Kumarganj, Ayodhya, Uttar Pradesh, India

Yogesh Pratap
Department of Molecular Biology and Biotechnology, Acharya Narendra Deva University of Agriculture and Technology, Kumarganj, Ayodhya, Uttar Pradesh, India

Swati Priya
Amity Institute of Biotechnology, Amity University Jharkhand, Ranchi, Jharkhand, India

Francisco Roberto Quiroz-Figueroa
Laboratorio de Fitomejoramiento Molecular, Centro Interdisciplinario de Investigación para el Desarrollo Integral Regional-Unidad Sinaloa, Instituto Politécnico Nacional, Guasave, Sinaloa, Mexico

Priyadarshani Rajput
Academy of Biology and Biotechnology, Southern Federal University, Rostov-on-Don, Russia

Vishnu D. Rajput
Academy of Biology and Biotechnology, Southern Federal University, Rostov-on-Don, Russia

Rakesh Kumar Rout
Department of Biotechnology, GIET University, Gunupur, Rayagada, Odisha, India

Sarvesh Rustagi
School of Applied and Life Sciences, Uttaranchal University, Dehradun, Uttarakhand, India

Saba Saba
Department of Microbiology and Molecular Genetics, The Women University Multan, Multan, Punjab, Pakistan

Ramesh K. Sadawarti
Department of Horticulture, School of Agriculture, Lovely Professional University, Phagwara, Punjab, India

Mani Salaria
Department of Horticulture, School of Agriculture, Lovely Professional University, Phagwara, Punjab, India

Nikita Sangwan
School of Life and Basic Sciences, Jaipur National University, Jaipur, Rajasthan, India

Parismita Sarma
Department of Botany, Rangia College, Rangia, India

Juhi Sharma
School of Life and Basic Sciences, Jaipur National University, Jaipur, Rajasthan, India

Riya Sharma
Amity Institute of Microbial Technology, Amity University, Noida, Uttar Pradesh, India

Tarun Sharma
Department of Agronomy, School of Agriculture, Lovely Professional University, Phagwara, Punjab, India

Shobhna
Department of Horticulture, School of Agriculture, Lovely Professional University, Phagwara, Punjab, India

Abhishek Singh
Faculty of Biology, Yerevan State University, Yerevan, Armenia

Aditya Pratap Singh
Department of Genetics and Plant Breeding, Bidhan Chandra Krishi Viswavidyalaya, Mohanpur, Nadia, West Bengal, India

Anil Kumar Singh
Bihar Agricultural University, Sabour, Bhagalpur, Bihar, India

Harshit Singh
College of Post Graduate Studies in Agricultural Sciences, Central Agricultural University (Imphal), Umiam, Meghalaya, India

Rupesh Kumar Singh
Centro de Investigação e Tecnologias Agroambientais e Biológicas (CITAB), UTAD, Vila Real, Portugal

Sakshi Singh
Faculty of Biology, Yerevan State University, Yerevan, Armenia

M.S. Smitha
Amity Institute of Microbial Technology, Amity University, Noida, Uttar Pradesh, India

Shivangee Solanki
Department of Life Sciences, Parul Institute of Applied Sciences, Parul University, Waghodia, Gujarat, India

João Ricardo Sousa
Centro de Investigação e Tecnologias Agroambientais e Biológicas (CITAB), UTAD, Vila Real, Portugal; Academy of Biology and Biotechnology, Southern Federal University, Rostov-on-Don, Russia

Nayan Talukdar
Program of Biotechnology, Faculty of Science, Assam Downtown University, Guwahati, Assam, India

Henrique Trindade
Centro de Investigação e Tecnologias Agroambientais e Biológicas (CITAB), UTAD, Vila Real, Portugal; Academy of Biology and Biotechnology, Southern Federal University, Rostov-on-Don, Russia

Jaagriti Tyagi
Amity Institute of Microbial Technology, Amity University, Noida, Uttar Pradesh, India; Biotechnology Department, Dr. MPS Group of Institutions, Agra, Uttar Pradesh, India

Priyanka Upadhyay
Department of Genetics and Plant Breeding, School of Agriculture, Lovely Professional University, Jalandhar, Phagwara, Punjab, India

Praveen Kumar Yadav
Division of Seed Science and Technology, ICAR-Indian Agricultural Research Institute, New Delhi, India

About the editors

Dr. Vishnu D. Rajput is working as the Head of International Laboratory on Nanobiotechnologies at Southern Federal University, Rostov-on-Don, Russia. His ongoing research is based on the investigation of effective remediation approaches using biochar-/nano-biochar-based sorbents and nanomaterials. With long experience and experimental work, Dr. Rajput comprehensively detailed the state of research in environmental science in regard to "how nanoparticles/heavy metals interact with plants, soil, microbial community, and the larger environment as well as possible remediation technology using nanoparticles/nano-biochar. Recently, he is working on nano-enhanced remediation of soil and water systems. He has published 350 scientific publications, 88 chapters, and 28 books, and has an H-Index: 49 (i10-index 145) since December 2023. He is an internationally recognized reviewer, received an Outstanding Reviewing Certificate from Elsevier and Springer. He is an editorial board member of various high-impact journals such as *Biochar*. He is holding national and international projects including Mega-Grant, BRICS. Dr. Rajput has received "Certificate for Appreciation 2019," "Certificate of Honor 2020," Diploma Award 2021 and 2022, and Letter of Gratitude 2022 and 2023 from the Southern Federal University, Russia, for outstanding contributions in academic, creative research, and publication activities. He has also received "Highly Qualified Specialist" status from Russian government. In the year of 2023, he is included in the list of 2% of the worlds' potential scientists by Elsevier BV publishing house.

Dr. Abhishek Singh is a dedicated postdoctoral scientist currently conducting cutting-edge research at Yerevan State University in the Republic of Armenia. His academic journey is firmly rooted in the fields of agriculture and biotechnology, and he has earned his doctoral degree from the esteemed Sardar Vallabh Bhai Patel University of Agriculture and Technology in Meerut, UP, India. Dr. Singh's educational journey commenced with an Integrated MSc in Biotechnology, with a specialization in Plant Studies, from the prestigious Sam Higginbottom University of Agriculture, Technology, and Sciences in Allahabad, India. At present, Dr. Singh's research efforts are primarily focused on unraveling the intricate effects of salinity stress on various aspects of plant biology. His work spans a wide spectrum, encompassing physio-biochemical characteristics, morphological transformations, intricate anatomical variations, and detailed examinations of cellular and subcellular organelles' ultrastructure. Moreover, his research delves into the complexities of DNA, RNA, and protein damage, seeking to comprehensively understand the multifaceted impacts of salinity stress on plant life. In addition to his work on salinity stress, Dr. Singh is deeply engaged in exploring the fascinating interactions between metallic nanoparticles and plants. His research touches upon the entire lifecycle of these nanoparticles, including their bioaccumulation, uptake, translocation, and the potential toxic effects they may induce. His contributions in this area are crucial for advancing sustainable agriculture by shedding light on the interactions between nanoparticles, plants, and soil, as well as their physiological and genomic responses. Dr. Singh's dedication to scholarship is evident through his extensive publication record, which includes 25 full-length research articles, 10 insightful books, and 30 chapters, all of which are indexed in the prestigious Scopus database. Furthermore, he has presented his research findings through six articles at scientific conferences. Beyond academia, he has shared his expertise through over 30 articles related to agriculture in various magazines, further contributing to the dissemination of knowledge in his field. Recognized for his expertise and commitment to rigorous scientific evaluation, Dr. Singh actively participates in peer review activities for reputable publications on a global scale. His contributions have earned him accolades, including a certificate as a peer reviewer from Springer and recognition for his valuable insights. In addition to his research and review activities, Dr. Singh generously lends his expertise to the editorial boards of numerous journals and societies, contributing to the advancement of knowledge in the fields of agriculture and biotechnology. In recognition of his outstanding contributions, the state government of Uttar Pradesh, India, honored Dr. Abhishek Singh with the prestigious "Appreciation Award 2021." This award underscores his significant achievements and underscores his position as a distinguished researcher in the fields of agriculture and biotechnology.

Dr. Karen Ghazaryan stands as a prominent figure in academia, currently holding the esteemed position of Dean at the Faculty of Biology, Yerevan State University. With an impressive 16-year journey in the realm of scientific research, he has left an indelible mark across multiple domains, making substantial contributions to various critical areas such as soil ecology, water resources monitoring and management, climate change and its environmental consequences, ecological management of the environment, ecological evaluation of the environment, and biodiversity protection. Dr. Ghazaryan's profound expertise is vividly evident through his substantial portfolio of more than 96 scientific research articles, showcasing his unwavering dedication to advancing knowledge and understanding in his areas of specialization. His scholarly accomplishments extend to the authorship of influential books, further solidifying his position as a thought leader and authoritative voice in his field. Beyond his scholarly achievements, Dr. Ghazaryan actively contributes to the progress of scientific publishing as a distinguished member of various editorial boards for peer-reviewed journals. His commitment to advancing research and fostering academic excellence has had a lasting impact on the scientific community. His invaluable insights continue to shape the ever-evolving landscape of environmental sciences. Dr. Ghazaryan's exemplary career, unswerving dedication, and remarkable contributions to academia position him as an outstanding and respected figure in the realm of soil ecology and environmental sciences. His profound commitment to advancing knowledge and preserving the delicate ecological balance of our planet serves as an inspiration to aspiring researchers and scientists worldwide, making him a true luminary in his field.

Dr. Athanasios Alexiou holds a BSc in Applied Mathematics from the National and Kapodistrian University of Athens, a PhD in Bioinformatics Algorithms and Mathematical Modeling in Biology, and completed a 2-year Postdoc Research Fellowship in Neuroinformatics Applications from the Ionian University. He currently collaborates with research teams from King Abdulaziz University (Saudi Arabia) and the University of Sharjah (United Arab Emirates). Additionally, he serves as a Faculty Member at the Department of Science and Engineering at the Novel Global Community Education Foundation (an educational platform registered in Australia) and as a Senior Researcher at the AFNP Med Company in Austria. Dr. Alexiou's primary research interests lie in the field of computational biomedicine. His work involves developing innovative mathematical models and algorithms for neurodegenerative diseases, creating prognostic Bayesian tools for Alzheimer's disease, identifying molecular mechanisms of infectious diseases, particularly the SARS-CoV-2 virus, and exploring the ethical implications of scientific research.

He holds prestigious roles in academia, including serving as an Associate Editor in esteemed journals like *Frontiers in Genetics* and *Pharmacology* in the Section of ELSI, as well as *Frontiers in Neuroscience* and *Psychology* journals. Dr. Alexiou is also a Section Editor in the *Current Alzheimer Research Bentham Science* journal, an Associate Editor in *The Open Public Health Journal* and the *Infectious Medicine* Elsevier journal, and a Guest Editor in the *Biomedicines* MDPI journal and the *Oxidative Medicine and Cellular Longevity* Hindawi journal. With approximately 200 peer-reviewed scientific papers, chapters, and books published in high-impact publications, along with four patents to his name, he is recognized as an accomplished researcher. Moreover, Dr. Alexiou has edited several books published by reputable publishing houses, delivered numerous talks and presentations in international conferences and workshops, and actively participates in various internationally funded programs. He is also a valued member of several scientific and program committees for conferences and serves as a Reviewer and Editorial Board Member for esteemed international journals. With over 20 years of experience in mentoring and education, he has taught computer science and theory of algorithms for the Hellenic Ministry of Education. His biography was featured in the 33rd Marquis Who's Who International Edition in 2016.

Professor Dr. Tatiana Minkina is an eminent figure in the academic realm, holding the prestigious position of Head of the Soil Science and Land Evaluation Department at the Southern Federal University. Her impressive career also encompasses leadership as the Head of the International Master's Degree Educational Program, "Management and Estimation of Land Resources," from 2015 to 2022. Notably, this program received accreditation by ACQUIN, highlighting its academic excellence and rigor. Dr. Minkina's extensive expertise lies in the fields of soil science, biogeochemistry of trace elements, and environmental soil chemistry. Her research endeavors encompass diverse areas, including soil monitoring, assessment, modeling, and remediation, with a particular emphasis on physicochemical treatment methods. In recognition of her outstanding contributions to education and the cultivation of highly skilled professionals, Dr. Minkina received the esteemed Diploma of the Ministry of Education and Science of the Russian Federation in 2015. Her dedication to academic excellence has left an indelible mark on the educational landscape. Currently, Prof. Dr. Tatiana Minkina is actively engaged in various research projects funded by prestigious organizations, including the Russian Scientific Foundation, Ministry of Education and Science of the Russian Federation, and the Russian Foundation of Basic Research. These projects underscore her commitment to advancing scientific knowledge and addressing critical environmental challenges. Dr. Minkina is a distinguished member of several esteemed organizations, including the Expert Group of the Russian Academy of Science, the International Committee on Contaminated Land, Eurasian Soil Science Societies, the International Committee on Protection of the Environment, and the International Scientific Committee of the International Conferences on Biogeochemistry of Trace Elements. Her active participation in these organizations reflects her dedication to collaborative research and global scientific advancement. Prof. Dr. Tatiana Minkina's remarkable research portfolio boasts an impressive total of 757 scientific publications, with 389 published in English, emphasizing her global reach and influence. She also serves as an invited editor for the esteemed Open Access Journal by MDPI, *Water*, renowned for its Impact Factor of 2.524. Additionally, she holds editorial board memberships in esteemed journals such as

Geochemistry: Environment, Exploration, Analysis, noted for its Impact Factor of 1.109, and *Eurasian Journal of Soil Science*, further solidifying her role as a leading authority in her field. Dr. Tatiana Minkina's unwavering dedication and significant contributions to soil science and environmental chemistry have earned her a well-deserved reputation as a respected and accomplished researcher. Her work continues to shape the landscape of her domain, making her an influential figure in the academic community.

Preface

The 21st century presents humanity with unparalleled challenges in the form of climate change and environmental deterioration. These global issues not only endanger ecosystems but also significantly impact human health. As we search for sustainable answers, biogenic nanoparticles—naturally occurring nanomaterials—emerge as a promising solution for addressing environmental stressors and improving human well-being. The book *Biogenic Nanoparticles: Interplay with Climate Change and Implications for Human Health* offers a thorough examination of this rapidly growing field. It serves as an essential resource for researchers, scientists, policymakers, and health professionals, providing state-of-the-art insights into the interactions between biogenic nanoparticles, climate systems, and human health outcomes. The complex relationship between these nanomaterials and the environment is increasingly recognized as a crucial area of study, offering new avenues for climate resilience, pollution control, and disease prevention. This book is structured into several sections, each carefully designed to guide readers through the fundamental principles, recent advancements, and future implications of biogenic nanoparticles. It covers topics ranging from the synthesis mechanisms and behavior of these nanoparticles in various environmental conditions to their role in combating pollutants and mitigating climate change effects. A substantial portion of the work focuses on the impact of biogenic nanoparticles on human health. As climate change intensifies environmental stressors—such as increased exposure to pollutants, rising temperatures, and altered ecosystems—the potential for biogenic nanoparticles to alleviate or complicate these effects becomes a critical research area. This book explores both positive and negative implications, highlighting how these nanoparticles could offer solutions for improving air and water quality, reducing toxic exposure, and contributing to medical advancements. However, the utilization of biogenic nanoparticles also raises ethical and regulatory concerns. As new technologies develop, it is vital to consider potential risks and ensure their deployment is safe, equitable, and sustainable. This volume addresses these pressing issues, offering thoughtful discussions on the governance and societal implications of biogenic nanoparticles in the context of climate and health policies. This book benefits from collaborations with leading experts in nanotechnology, environmental science, and public health. Their collective expertise and diverse perspectives have created a resource that not only deepens our understanding of the interplay between biogenic nanoparticles, climate change, and human health but also provides practical insights for addressing these urgent global challenges. The editors express their gratitude to the contributors whose work has shaped this volume, and hope that it will inspire further innovation and discourse. This book aims to serve as a catalyst for new research, informed policies, and sustainable solutions that can help protect both the planet and human health in the coming years.

Vishnu D. Rajput
Abhishek Singh
Karen Ghazaryan
Athanasios Alexiou
Tatiana Minkina

Acknowledgments

Vishnu D. Rajput acknowledge the financial support by the Ministry of Science and Higher Education of the Russian Federation (no. FENW-2023-0008). Abhishek Singh and Karen Ghazaryan acknowledge the financial support by higher education and science committee MESCS Republic of Armenia provided the 23PostDoc-4D007 grant to support AS. Higher education and science committee MESCS Republic of Armenia has provided grant numbers 21AG-4C075 to support KG. The financial support of above agencies has been essential in realizing the edited volume, Biogenic Nanoparticles: Interplay with Climate Change and Implications for Human Health, set to be released by Elsevier in the United States. Their dedication to promoting research and understanding in green nanotechnology and sustainable agriculture has been vital in bringing this project to fruition. Their backing has allowed us to compile and share important insights, fostering advancements and innovation in these crucial fields.

We convey our profound appreciation for their foresight and commitment, which have not only enhanced the substance of this book but also contributed to the worldwide conversation on sustainable agriculture. We anticipate ongoing collaboration and the positive influence it will undoubtedly exert on the future of agricultural practices and scientific inquiry.

SECTION I

Biogenic nanomaterials - formation, characterization, and fate

CHAPTER

1 A comprehensive introduction on biogenic nanoparticles: fabrication, implementation, and environmental concern

Saba Saba[1], Laraib Aslam[1], Abid Hussain[2] and Iqra Arooj[1]

[1]*Department of Microbiology and Molecular Genetics, The Women University Multan, Multan, Punjab, Pakistan*

[2]*Department of Biological and Chemical Engineering, Aarhus University, Aarhus, Denmark*

1.1 Introduction

"Nano" is a Greek term which means dwarf or really little. While Tokyo Science University Professor Norio Taniguchi invented the term nanotechnology, it was physicist Professor Richard Feynman who first presented the notion of nanotechnology in his seminal address "there's plenty of room at the bottom."[1] They include materials with at least one dimension on the nanometre scale, that is, 1 μm down to 10 Å.[2,3] A new era in materials science began in the 1960s with the development of nanotechnology. They are categorized according to the maximum number of dimensions that an electron can occupy, such as 0D/0-dimensional, 1D/1-dimensional, and so on. Thin films (2D), wires and rods (1D), and spheres (0D) are a few examples of frequent geometries. Because of their special electrical, magnetic, and physiochemical characteristics—such as size control, an improved surface area to volume ratio, and functionalitie, the confinement of electrons within these particles and the modification of bandgap energies by changing particle size, are the primary causes of these special characteristics of nanoparticles (NPs).[4–6]

Previous research has mostly relied on chemical and physical techniques. However, in addition to high prices, a major worry was the use of harmful and potentially hazardous compounds for synthesis.[7] For example, (1) chemical techniques like thermolysis, microemulsion, sol–gel, and photoreduction and (2) physical techniques like melt mixing, physical vapour deposition, sputtering, and laser ablation. These techniques frequently cause dangerous byproducts to be produced, unstable NPs, and harmful compounds to adhere to the surface of the nanoparticles.[7–9] Controlling NP geometries to produce desired electrical,

Biogenic Nanoparticles. DOI: https://doi.org/10.1016/B978-0-443-33819-9.00005-6

optoelectronic, and physicochemical properties is therefore crucial. For the synthesis of NPs with varying morphologies and sizes, many synthetic techniques have been used. Even if these techniques have produced better NPs, there is still a need for a better knowledge of the production process, with the primary goal being to create NPs that are ecologically-friendly.[10,11]

As a result, a variety of biological synthesis techniques that produce nontoxic, ecologically safe, more biocompatible, and environmentally benign nanostructures have been documented. Numerous microorganisms exist, and they mostly create nanoparticles by reacting in diverse ways with metal precursors. Nanostructures with the required composition and geometries can be created by managing the biological synthesis.[12–14] One of the primary characteristics of NPs is the ability to manipulate particle size, shape, and crystalline nature, which could be used in a variety of fields.[15] They use a biomimetic technique to produce pure metal/M and metal oxide/MO particles on the nanometre scale. Aside from their distinct physiochemical properties, NPs serve as a link between bulk materials and atomic or molecular structures. As a result, they are suitable for use in a variety of disciplines.[16–18]

Nanotechnology has employed two primary methodologies. These are called the top-down and bottom-up approaches. The former entails transforming atoms into molecules in order to create nanostructures, materials, and gadgets. The second method includes constructing structures and devices from larger entities that lack specialized atomic control.[19] *Biologenic NPs* synthesis is a bottom-up, green strategy that is gaining prominence these days. The green synthesis uses biological resources such as plant extract, biodegradable waste, and microorganisms such as bacteria, algae, fungi, and yeast as a reducing agent, and it is known as "green or bionanofactories," Biogenic nanomaterial synthesis, such as NPs, nanorods, nanowires, and nanotubes, is considered an economically viable alternative to producing nanomaterials for a variety of applications.[20–22] *Nanotechnology is a megatrend* that has emerged as a *general-purpose technology*. The National Nanotechnology Initiative, established by executive action in 2000, was formalized in 2003 with the 21st Century Nanotechnology Research and Development Act.[23] It has the power to shift our ideas and expectations, giving us the opportunity to resolve global issues.[24] Following are stated the realistic applications of biogenic NPs (Fig. 1.1).

1.2 Biomedical applications

1.2.1 Antibacterial applications

Metallic nanoparticles (such as iron, copper, silver, titanium, and zinc) have been a primary focus for numerous biomedical applications, including antibacterial against infections caused by multidrug-resistant pathogens.[25] Nanoparticles exhibit antibacterial characteristics due to release of metal ion as well as oxidative

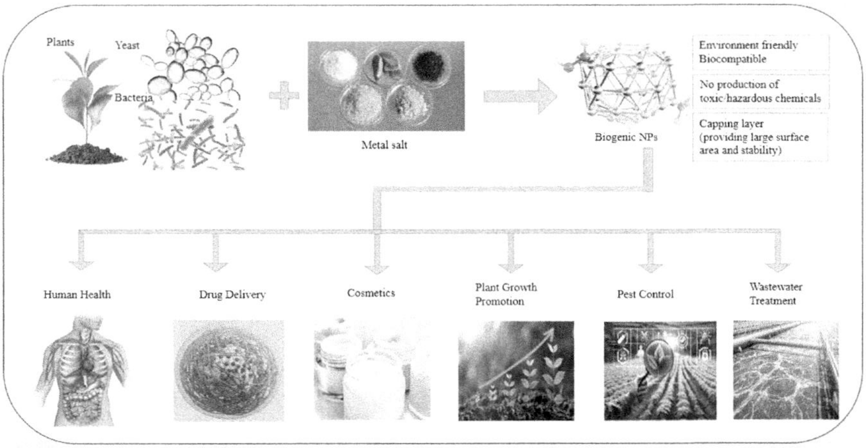

FIGURE 1.1

Realistic applications of biogenic NPs.

and nonoxidative stress, all happening at the same time. Biogenic AgNPs from *Sporosarcina koreensis* DC4, *Bhargavaea indica* DC1, and *Brevibacterium frigoritolerans* DC2 show antibacterial efficacy against a variety of microorganisms. Copper nanoparticles from *Sida acuta* demonstrated antibacterial efficacy against *Proteus vulgaris*, *S. aureus*, and *Escherichia coli*. When used with conventional antibiotics, these nanoparticles increased their antibacterial activity. Zinc oxide has also been shown to have antibacterial activity against *P. aeruginosa, S. aureus*, and *E. coli*. A comparison of biological and chemical nanoparticles revealed that biological nanoparticles have stronger antibacterial properties than chemically synthesized nanoparticles. Nickel NPs derived from *Desmodium gangeticum* shown greater antioxidant, biocompatible, antibacterial activity in (epithelial) cell lines, that is, LLC PK1 than chemically synthesized NPs.[26–28] For many years, antibacterial nanoparticles, especially those coated in nanosilver, have been used in surgical instruments, bone prostheses, and dental implants. They have also been used as a covering for wound dressings to fight wound microbes. Furthermore, they specifically target the bacteria and cause the impairment of vital membrane processes including permeability and respiration. Moreover, they hinder gene transfer and cell division by interacting with intracellular elements such as proteins and nucleic acids.[29] For the treatment of lesion infections, magnetite nanoparticles made from *Mirabilis jalapa* roots are useful in a variety of ways. By acting as antioxidants, these NPs lessen the oxidative damage/stress which is caused by free radicals.[30] Because of their small size and large surface area, gold NPs have demonstrated strong antibacterial activity. They cause death or stagnant growth by inhibiting respiratory chain enzymes, which are essential for the formation of bacterial cell walls. It has been demonstrated that spherical

Table 1.1 Advantages of biogenic NPs.

	Properties	NP formulation by chemical methods	NP formulation by biological/greener methods
1.	Nature	Toxic and expensive	Nontoxic and cost-effective
2.	Reducing agent	Sodium citrate, sodium borohydride, *N,N*-dimethylformamide, dimethylformamide, hydrazine hydrate, ethylene glycol	Polysaccharides, amino acids, proteins, flavones, phenolics, terpenoids, nitrate reductase enzyme
3.	Methodology	Surfactant/stabilizer is required to prevent the NPs agglomeration	Stabilizer is not required
4.	Environmental impact	Energy intensive and also pollute the environment	Safe, performed in natural condition with no energy requirement
5.	Antibacterial activity	Less activity against pathogens	High activity against pathogens

Au NPs from the blue green alga *S. platensis* inhibit both *B. subtilis* and *S. aureus*. NPs attach themselves to the membranes of gram-positive bacteria, break their connections, and into the microbe. Because of their high lignin, cellulose, pectin, and hemicellulose content, banana peels make ideal templates for the production of NP and have strong antibacterial and yeast action (Table 1.1).

1.2.2 Generalized applications

In the last few years, there has been a lot of research done on nanomaterials because of their potential to fight cancer and reduce inflammation. For these objectives, plant and microbial-based methods have been favored due to the cytotoxicity and detrimental by-product formation linked with chemical reagents. Since no capping agents or surfactants are used, the characteristics of these NPs differ from those synthesized using other traditional and chemical procedures. NPs produced from waste materials and algae show a wide range of applications. Iron and iron oxide NPs are highly paramagnetic and are widely used in biomedical applications, including medication administration, tissue healing, magnetic resonance imaging (MRI), and cell labelling.[31] AgNPs are a powerful antibacterial and biocompatible agent that has garnered a lot of interest. Better antibacterial activity requires them to connect with the cell wall, which they can do successfully.[32,33] Because AgNPs can mimic the synthesis of molecules that produce free radicals, they function as antioxidants. AgNPs have a function in antioxidant agent absorption and distribution in addition to their free radical scavenging action. AgNPs with possible anticancer properties have been used in cancer treatment as drug delivery systems, therapies, diagnostic tools, and theranostics. For instance, Ag-based nanoparticles conjugated with doxorubicin are employed to suppress the development of B16F10 cells in place of free doxorubicin. AgNPs' fibrinolytic and antiplatelet aggregation characteristics are being investigated.[34]

AuNPs have shown to be a valuable tool in a variety of possible biomedical applications, such as DNA modelling and as a promising substitute for serious illnesses. Different sized AuNPs exhibit the optical characteristics required for biosensor applications, particularly in cancer nanotechnology. When compared to tumor necrosis factor-alpha (TNF-α), a cytokine with anticancer efficaciousness but limited therapeutic uses, PEG/plyethlene glycol coated AuNPs maximise tumor damage.[35–37] Because of their small size and large surface area, gold NPs have demonstrated strong antibacterial activity. They cause death or stagnant growth by inhibiting respiratory chain enzymes, which are essential for the formation of bacterial cell walls. It has been demonstrated that spherical AuNPs from the blue green alga *S. platensis* inhibit both *B. subtilis* and *S. aureu*. NPs attach themselves to the membranes of gram-positive bacteria, break their connections, and into the microbe. Because of their high lignin, cellulose, pectin, and hemicellulose content, banana peels make ideal templates for the production of NP and have strong antibacterial and yeast action.[38,39]

According to a study, reducing elements including flavonoids and polyphenols, which function as capping agents, are present in *Cassia fistula* extract. When ZnO NPs were examined for scavenging action, they showed exceptional antibacterial activity against *Plasmodium desmolyticum, S. aureus*, and *E. coli*. Using floral extract from *Trifolium pratense*, the biological production of ZnO NPs from zinc acetate was demonstrated. With more antibacterial activity than gentamicin against *Pseudomonas aeruginosa*, the nanomaterial demonstrated antimicrobial efficacy against *S. aureus* and *P. aeruginosa*. An extract from the leaves of *Boerhavia diffusa* shown significant antibacterial and antibiofilm properties against strains of Methicillin-resistant *S. aureus* (MRSA). To create multifunctional ZnO NPs, *Bacillus megaterium* extracts were utilized as capping and reducing agents. The scientists discovered that reactive oxygen species (ROS) generated on the surface of the NPs bind to the surface of bacteria and destroy them by electrostatic forces. The antibacterial action of free Zn^{+2} ions was facilitated by the mechanical interaction between bacteria and the ZnO rod surface.[40] *Tinospora cordifolia* endophytic fungal extract was used to create biogenic AgNPs, which had stronger antibacterial, antioxidant, and antiinflammatory properties. Additionally, it was shown to have antiproliferative effect in breast and cervical cancer cells. AgNPs ultimately discovered hemocompatibility and cytocompatibility. The results show that biogenic silver nanoparticles, which were synthesized as a potential multifunctional therapeutic material with low systemic toxicity, exhibit numerous and obvious biological activities.[41]

1.2.3 Disease therapy and drug delivery

One of the leading causes of mortality worldwide is cancer. Numerous medications are available to treat various forms of cancer. Chemotherapy in conjunction with cytotoxic medications is the most generally prescribed course of treatment. However, it has several unfavourable side effects that can occur whether it is taken alone, in conjunction with radiation therapy or cytotoxic drug therapy. For this reason, studies investigating plant extracts' potential anticancer effects have received funding from the National Cancer Institute (USA). Natural compounds obtained from therapeutic plants are thought to be excellent candidates for the creation of anticancer medications.[42] MNPs (metal oxide) and MONPs (metal oxide NPs) biogenic nanoparticles are thought to be a useful tool for cancer detection and preferred medicine delivery to tumor sites. These biosynthesized nanoparticles selectively eliminate cancer cells while causing no harm to normal/healthy cells by leveraging the unique features of nanoparticles as well as the antioxidant and antitumor capabilities of plants. The specific method by which various MNPs and MONPs cause cytotoxicity in diverse cancer cell lines is unknown. Apoptosis has long been thought to be the primary mechanism of cell death. The 3-(4, 5-dimethylthiazolyl-2)-2, 5-diphenyltetrazolium bromide/MTT test is widely used in cancer cells to assess the cytotoxicity of MNPs/MONPs in cancer cells.[43] Metal/metal oxide nanosystems (M-NSs) are being studied for cancer treatment due to their properties like early-stage diagnosis, immune development, nanotherapeutics, tailored drug delivery systems. However, lack of biocompatibility, off-target effects, toxicity, and high cost limits their acceptability. Biomass and green chemistry-based M-NSs are developing as popular biogenic M-NSs due to their affordability, scalability, and acceptance in a variety of tissues and organs[44] (Fig. 1.2).

1.2.3.1 Criteria of nanoparticles for anticancer application

Size and shape are important factors for NPs' mobility inside the tumor vasculature and bloodstream. For drug delivery applications, spherical nanoparticles are preferred because of their larger surface volume ratio, which enables more medication to be loaded for better efficacy. Lastly, conjugating different targeted ligands can increase the delivery mechanism's effectiveness.[45–47]

1.2.3.2 Drug delivery system

The innovative drug delivery methods may increase the effectiveness of the pharmaceutical compounds while also allowing the reintroduction of components that had been previously eliminated due to formulation inefficiencies. This approach is made even more appealing by reducing side effects, improving efficacy and selectivity, guarding against photo/thermal degradation, and regulating the release of active constituents.[48–50] To minimize harmful compounds and enable lower medication dosages, green chemistry focuses on creating biogenic nanoparticles. Physicochemical and biological barriers have been addressed with the development of controlled drug-delivery systems (DDS). Because of their large surface area, NPs can bind, adsorb, or trap therapeutic substances. These NPs enhance loading capacity, target penetration, bioavailability, and sustained drug release. Additionally, they might postpone drug resistance and improve the efficacy of

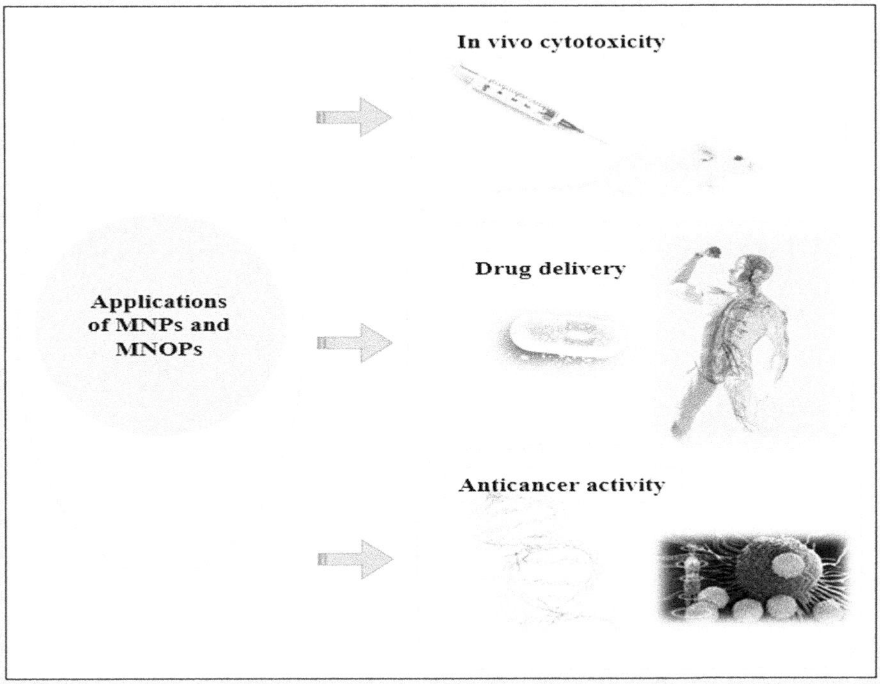

FIGURE 1.2

Application of MNPs and MNOPs.

traditional therapy.[51–54] Certain drugs, such as doxorubicin and vancomycin, are also made from natural substances. Drug targeting can either be active or passive where antibodies and/or peptides and (drug delivery) vectors are employed in association, to make them attached with receptor structures. Physical and chemical interactions can cause drug release, and the physicochemical characteristics of drugs are exploited to choose a particular metallic nanovector. By creating covalent bond, hydrogen bond or electrostatic interactions between the polarity of the drug and capping or nanoparticles, natural materials can help with pharmaceutical loading issues.[55]

1.2.3.3 Mechanism of antitumor activity

It has recently been demonstrated that by conventional cancer drugs, not only the growth of tumor cells is inhibited, but healthy cells are also damaged. High dosages consequently cause systemic tolerance against the conventional therapeutic medicines/drugs such as paclitaxel, adriamycin, epiamicyn, docetaxel, and so forth develops. Therefore, there is a clear and pressing need for the creation of innovative medicines. Due to their high drug-loading capacity, large surface area,

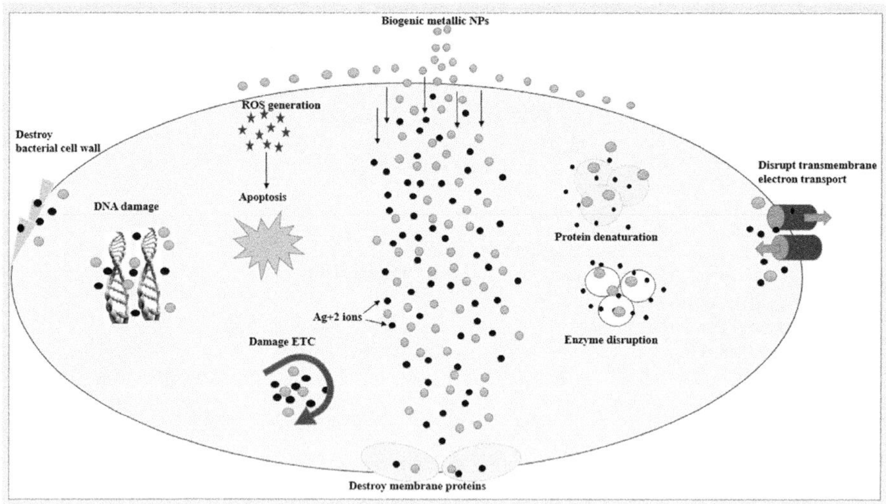

FIGURE 1.3

General mechanism of cytotoxicity by biogenic NPs.

and ability to target specific tumors, nano-drugs have drawn a lot of attention as the best option for treating tumors. Biogenic NPs have been shown in numerous studies to have significant chemotherapeutic effects on a variety of cancer cells. The primary processes that underlie the antitumor activity of biogenic NPs are as follows: (1) apoptotic pathways, which are set off by elevated ROS levels and lead to oxidative damage and DNA fragmentation; (2) disruption of cell function caused by interference with macromolecules like proteins and DNA; and (3) association with cell membranes, which modifies permeability and causes mitochondrial dysfunction. The main reason why scientists have an interest in biogenic NPs made from plants is that phytoconstituents have the ability to inhibit cancer cells through phagocytosis, the production of ROS, intercellular signalling, and cell proliferation regulation. Additionally, it was shown that phenols and polyphenolic compounds inhibit cancer and are linked to atherosclerosis[56] (Fig. 1.3).

1.2.3.4 Antitumor drugs

Some nanomaterials can be employed as medicines due to their inherent anticancer properties. They have certain surface features and physical-chemical properties which enable them to have cytotoxic action. NPs also function as nanocarriers, either passively or actively by (1) targeting the tumor through EPR/enhanced permeability and retention and (2) targeting solid tumors through interactions of ligands with receptors respectively. Active targeting refers to the conjugation of various compounds, such as peptides, antibodies, and DNA to target specific cells.[57,58] Targeting cytotoxic medications specifically to malignant cells appears to be a viable technique for improving anticancer therapy efficacy and lowering severe side effects. Nanocarrier delivery are highly efficient in the treatment of various tumors due to the following properties: (1) they increase the drug solubility, (2) they reduce interaction with nontarget tissue, and (3) they avoiding drug degradation and hence minimize unfavorable responses.[59]

1.3 Petroleum industry

In the oil and gas sector, nanoparticles are being utilised more frequently for well stimulation, oil recovery, and drilling fluids. They are easier to enter formation pore spaces and more reactive because of their small size and high volume-to-surface area ratio. Viscous liquids known as drilling fluids help in drilling by managing formation pressure, forcing, removing cuttings, and maintaining bore stability. Conventional additives need to take into account a number of variables, including temperature and particle size. It has been demonstrated that nanoparticles enhance filtration, borehole cleaning, and rheological characteristics. Iron oxide nanoparticles enhanced the potential of xanthan gum drilling fluid, while copper (Cu), aluminum (Al), and magnesium (Mg) oxide NPs boosted its rheological qualities as well. Other NPs, including titanium oxide, zinc titanate, and silicon dioxide, have been investigated for their ability to improve wellbore stability, lubrication, and filtration.[59–61] Recently there has also been a lot of interest in the increased oil recovery systems based on nanoparticles. Gas, chemical, and thermal injection are three popular tertial oil recovery techniques that are enhanced by the introduction of nanoparticles. Various nanoparticles including aluminum oxide, silicon dioxide, and graphene oxide are employed for reduced oil viscosity, increased foam stability and mobility, and enhanced oil recovery respectively.[62,63]

1.4 Pest management

Silicon (Si) induces antibiotic and antixenotic tolerance when employed to manage insect pests in a variety of crops. Due to their excellent efficiency in controlled release, silica (SiO_2) nanoparticles, or SiNPs, have been used recently for targeted pesticide treatment, as well as soil and foliar applications. SiNPs can be readily transported into plant systems by roots and lodged in a variety of tissues, bolstering plant defences, because of their tiny size and mesoporous structure. With the use of nanocapsules, which are dubbed "magic bullets" because they release active compounds into tissues and cuticles on a regular basis, nanotechnology also aids plants in their battle against illnesses and pests.[64] In a different study, 3 nm mesoporous Si-NPs were employed as a carrier to introduce chemicals and DNA into plant cells. The chemicals are subsequently triggered to fix the particles across cell walls without causing any harm to the plants. This strategy has been effectively implemented in tobacco and maize plants.[65] Novel developments in agriculture have been made possible by recent advancements in nanotechnology. Because of their distinct physiological and structural characteristics, SiNPs, among other nanoparticles, provide significant benefits to agriculture as nanofertilizers, nanopesticides, nanozeolites, and customized delivery systems. Their ability to stimulate plant development in both typical and stressful environments is well established. It has been shown that nanosilicon increases a plant's ability to withstand a variety of environmental challenges, and it is thought to be a safe and effective substitute for preventing plant diseases.[66]

1.5 Eradication of salinity and drought stress

Salinity and drought inhibit plant growth by limiting root growth. Silicon nanoparticles (SiNPs) can enhance water intake, root growth, and biomass even in dry situations by increasing root area and regulating solute levels. This lowers the osmotic shock induced by sodium chloride toxicity. Amorphous silicon precipitation acts as a mechanical barrier for plants, and the sort of Si deposited varies depending on the leaf type. Polymerized Si in plant cell walls and apoplasts inhibits pathogen penetration and damaging infections. Biogenic SiNPs were discovered to reduce cadmium stress in rapeseed by acting on cellular (oxidative repair) pathways. A greenhouse experiment found that 250 mg/kg of SiNPs considerably improved the growth metrics, for example, fresh weight, dry weight, and photosynthetic efficiency as well. Antioxidant enzymes including ascorbate peroxidase, catalase, peroxidase, and superoxide dismutase all greatly increased in activity after receiving SiNP therapy. These enzymes are crucial for scavenging ROS and reducing oxidative stress caused by cadmium (Cd). Silicon NPs treatment decreased Cd-induced damage to leaves ultrastructure, according to TEM research, whereas SiNPs treatment changed the expression of genes linked to stress and antioxidant defense, according to qPCR research.[67] Though Si for plants does not appear necessary, but it does encourage their growth and development. Among the quasi-essential elements are selenium, cobalt, aluminum, sodium, and zinc. Si increases antioxidant activity and reduces the uptake of soil pollutants, hence mitigating the adverse consequences of physiological stress on plants. The application of silicon-based NPs in food production and agriculture has gained attention. NP-based Si fertilizers have high bioavailability and thus seem essential for sustainable agricultural output, as conventional fertilisers have low bioavailability. SiNPs can modify microbes in the rhizosphere and biomass content, change microbial communities and biomass, and increase soil fertility.[68] Biogenic AgNPs are increasingly being used in agriculture due of their stress tolerance. These nanomaterials improve crop stress tolerance by increasing nutrient uptake, enzyme responses, and the presence of growth-promoting microorganisms. Silver-based nanomaterials have been proven in studies to preserve water balance in lentils under water stress, promote germination during drought, and boost photosynthetic pigments in. The use of these nanoparticles increases growth parameters, photosynthetic potential, and plant hormone balance. It is suggested that these nanomaterials may help plants adapt to harsh circumstances.[69,70]

1.6 Wastewater treatment

The T&EPs, that is, treatment of toxic and emerging pollutants are posing new challenges to conventional wastewater treatment technologies. The development of dependable and affordable methods for managing T&EPs has become the focus of study due to the expensive installation and operating costs of current treatment systems.[71]

1.6.1 Heavy metal removal

Fe_2O_3 nanoparticles produced using *Aloe vera* leaf extract can effectively remove arsenic (As), a hazardous pollutant. These nanoparticles can be recycled up to four times, with a maximum ability to adsorb 38.47 mg/g at 20°C. Trivalent and pentavalent arsenic were eliminated with the use of iron oxide nanowires that were obtained from the chemoautotrophic bacteria *M. ferrooxydans.* To remove Cr(VI) from aqueous solutions, biocompatible FeNPs were produced from *Y. lipolytica* using *P. granatum* extract. Cr(VI) was effectively separated with 98.1% purity using zero-valent iron nanoparticles made from *Eucalyptus globulus* leaves extract in under 30 minutes. Additionally, it was demonstrated that *S. jambos* could separate Cr(VI), having a removal capacity of 0.69 g/g. Pb(II) and Cr(III) were extracted from water-based matrices using magnetic iron nanoparticles made from *Cnidium monnieri* seeds extract. To extract Cd from simulated wastewater, *P. aeruginosa* JP-11 biogenic CdS nanoparticles were used. To extract Cu ions from groundwater, FeNPs have been produced from *S. thermolineatus*, while selenite nanoparticles were extracted from bacterial PS.[72] A study[73] discovered that hazardous heavy metals may be specifically eliminated from a 20% fly ash aqueous solution by using maghemite (γ-Fe2O3) NPs sized 20–40 nm. Furthermore, the leaf extracts of a medicinal plant called *Spilanthes acmella* was used to create IO-NPs. IO-NPs were utilized in an optimized pH (7.0) procedure to remove As^{3+}, Cd^{2+}, Cu^{2+}, and Co^{2+}, ions via adsorptive process from aqueous solutions. 0.8 g/L for Cd^{2+} and Cu^{2+} and 0.6 g/L for As^{3+} and Co^{2+} were the IO-NP dose. The biosynthesised IO-NPs shown comparable efficacy in concurrently adsorbing different heavy metal ions, suggesting that they have the potential to be useful nanoadsorbents.[74]

1.6.2 Removal of dyes

The bulbous extract of the herb *Syzygium aromaticum* (clove) was used to create Fe_2O_3 NPs and c-Fe_2O_3 NPs, which were examined for their adsorption capacity and usage in removing the dye called methylene blue (MB) from aqueous solution. A study that was published in the literature defined this process as nontoxic and straightforward. When it came to MB, the biosynthesized Fe_2O_3 NPs demonstrated a greater clearance rate compared to -Fe_2O_3 which were 92% and 88% respectively.[75] In another study, MB, rhodamine-B/RhB, and p-nitrophenol/p-NP were reduced using *Thymbra spicata* extract-mediated AgNPs as a catalyst in an aqueous medium. $NaBH_4$ was used as a reducing agent in this process. The investigation showed that MB, RhB, and p-NP underwent catalytic degradation in 60, 50, and 60 minutes, respectively, with kapp values of 0.057, 0.086, and 0.065/s. For every reaction, 1 mg of catalyst was needed.[76–79]

1.7 Promotion of plant growth

One study found that CuO NPs from *Pseudanabaena foetida* RJ1 (cyanobacteria) stimulates plant development. CuO NPs mediated by cyanobacteria and an inexpensive zeolite adsorbent were used to characterize and create nano-formulations from the NPs. When the nano-formulations were added to paddy seeds that had germinated, they significantly increased the promotion of plant development in hydroponic settings. According to this study, micronutrient NPs may be used as an innovative, reasonably priced nanofertilizer to aid in the growth of plants.[80] Thyme leaves and lavender flowers were processed to produce Ag, ZnO, and Fe_3O_4 nanoparticles using inexpensive, nontoxic, and ecologically friendly chemicals. On basil seeds, the effects of various NP concentrations were contrasted. Extracts from lavender flowers served as capping agents and reduced Ag-NPs, while extracts from thyme leaves acted on ZnO-NPs and Fe_3O_4-NPs. For root length, stem length, and seed germination, the greatest values were discovered at 25, 50, and 100 mg/L of Fe_3O_4-NP. By breaking the NP's seed dormancy, the germination-promoting effects of NPs were discovered to be correlated with size, surface area, shape, and concentration. This resulted in early and rapid germination.[81] By boosting water intake, carbon nanotubes—a useful element of nanotechnology—play a significant role in improving the development of plants and seed germination. In addition, it has made a name for being a pioneer in the field of nano-fertilizers to advance the agricultural industry. Additionally, it greatly reduces the variety of microorganisms used in agricultural areas and increases the bioavailability of hazardous metals in agricultural products by inducing apoptosis through the creation of ROS.[82] In another work, TiO_2 nanoparticles promoted spinach growth by photoreducing ambient N_2 to NH_3.[83] Furthermore, TiO_2 NPs effectively boosted cucumber photosynthetic activity and spinach seed germination, both of which had a positive influence on the plants. Interestingly, rutile and TiO_2 enhanced chlorophyll production in spinach by 28% and 19%, respectively.[84]

1.8 Formulation of cosmetics

Nowadays, people engage in cosmetics solely for the purpose of improving their appearance, thus issues such as wrinkles, ageing, hair loss, and dandruff must be addressed effectively. Nanocarriers not only improve the efficacy of cosmeceutical products by offering better and longer-lasting benefits, but they also help to improve the goods' esthetic look. This dual benefit not only improves the product's overall quality and efficacy but also boosts user happiness. Nanocarriers also provide UV ray protection, which adds to the cosmeceutical product's overall benefits. Nanotechnology is also gaining traction because to its great penetration of actives into the deeper layers of skin. It might be claimed that nanotechnology is taking over all the disadvantages of traditional products.[85] In traditional medicine, the nonrhizomatous plant *Iris tuberosa* has been used to treat dandruff, acne, and skin disorders. A biogenic AgNP was created in a study to be a preservative in moisturising creams. *Iris tuberosa* leaves extract was prepared and utilized for the synthesis of biogenic Ag NPs, which were then examined under a UV–Vis spectrum, DLS, and TEM. It was determined that the characteristics were antibacterial and antioxidant. An effective moisturising cream containing biogeninc-AgNPs was developed and put to the test. The biogenic AgNP preservative was cheap to create, soluble in water, effective against all microorganisms, and exhibited broad antibacterial action.[86]

1.9 Future perspectives and myths

See Table 1.2.

Finding impartiality among (1) process, (2) cost, and (3) scalability of the NPs, remains a significant challenge. (1) In order to maintain, create, and purify the particles, sophisticated instruments are required at each stage of the process, especially for microorganisms that make biogenic nanoparticles. For instance, the production of NPs requires incubators with controlled temperature and shaking, and the removal of contaminants of nanoparticles requires centrifuges. A significant amount of heavy machinery is needed for the entire process of creating and purifying nanoparticles, which raises the relative expense of the technology. (2) When it comes to plants management of natural resources, for example, plants and culture, is necessity and requires attention. (3) The majority of biosynthesis techniques are still being developed, and problems such as NP separation and purification need to be solved in more detail. (4) In addition, there are gaps as well in our knowledge regarding the metabolic routes and mechanisms underlying the bio-mediated synthesis of nanoparticles. (5) Understanding about the moieties which actively bind and take part in synthesis and offer surface stability and biocompatibility is essential when working towards biomedical goals. Creating biocompatible nanostructures on a large scale with a narrow size distribution is

Table 1.2 Properties of biogenic NPs.[87–91]

1.	Better penetration potential
	Nanoparticles offer numerous benefits to the pharmaceutical industry, including enhanced therapeutic activity, in vivo protection, efficient delivery, selectivity, and longer release profiles. Their nano size allows for greater penetration, reducing dosage and adverse effects.
2.	Capping
	NPs interact with biomolecules (due to their ability to cross blood brain barrier), to change their physicochemical, biochemical, and cellular characteristics. The majority of biological entities function as capping agents to stop aggregation and coalescence and help with stabilization.
3.	Biocompatible and abdication of agglomeration
	When NPs decrease surface energy, they increase particle size and decrease surface area, causing agglomeration. Biopolymers have an impact on bio-dispersity by enhancing NP biocompatibility and preventing cluster aggregation. Biological reactions, bioavailability, cellular absorption, and toxicity can all be impacted by NP stability. Composition, surface characteristics, and aggregation state are among the variables influencing stability. The confluence of biological molecules implicated in capping and NP reduction enhances stability, prevents agglomeration, and influences the treatment outcomes.
4.	Bioreduction
	The bioreduction process, which microorganisms typically display, entails the build-up of metallic ions to lessen their toxicity. Microorganism bioreduce extracellularly through diverse metabolites, or intracellularly through a variety of reducing species found not only on the cell wall but inside the cell as well. Because of a variety of flavonoids, proteins, and water-soluble macromolecules, plants can also reduce.
5.	Stability
	The maintenance of particular attributes of nanostructures, such as size, crystallinity aggregation, form, composition, and surface chemistry, is defined as stability of NPs. The capping layer of green nanoparticles, which is formed during biogenic nanoparticle synthesis and comes from biological extracts, affects the stability and biocompatibility of these particles. This layer is helpful for long-term stability and influences the biological activity of nanoparticles. According to a study, the extra surface can trap microorganism-formed nanoparticles, which demonstrate exceptional stability.

another problem facing modern biotechnology techniques. Their biomedical applications demand precise size and shape and may be hampered by their heterogeneity. To create exact and reproducible biosynthesis processes, it is necessary to precisely determine the reducing components. (6) In addition, it is important to thoroughly examine the structure, makeup, and coatings of biogenic NPs in addition to assessing any possible interactions with the biological system. (7) To prevent unanticipated harmful consequences, it is also important to evaluate the potential for aggregation and the creation of agglomerates, which may have an impact on the characteristics and bioavailability of the individual NPs. (8) Their distribution profile, release kinetics, and organismic clearance present additional challenges for the biological use of these NPs. Comparing the conjugates produced by biogenic processes to their equivalents synthesized physiochemically, the majority of them are biomaterials with lower toxicity. As such, biogenic NPs are useful in avoiding nano-cytotoxicity, the primary problem with contemporary drug delivery technologies.[92–94] The development of biogenic nanocomposites for drug delivery shows promise for significantly advancing the science of biomedicine by employing an even more strategic and intelligent methodology.[95,96]

It can also be used to create multifunctional nanomaterials that have the potential to treat illnesses or deliver medications in a way that does not disrupt the body or have negative side effects. The use of genetically engineered microbes for the synthesis of such NPs with the required form and size is a recent and promising technique in this sector.

Future research should primarily focus on laying the groundwork for nanobiomonitoring, identifying the biological consequences of nanoparticle in the surrounding environment, and developing new tools for nanotoxicology.[97] These situations make it clear that integrating nanotechnology and phytochemicals would open up new possibilities in the realm of biomedicine. Because metallic nanoparticles are antimicrobial, their uses are not just restricted to the biomedical field; they may also be found in textiles, food packaging, computer keyboards, washing machines and self-cleaning coatings on mobile phones. The biogenic nanoparticles haven't been made available for purchase for these uses, though.[98] Future research and commercialization in antimicrobial applications must emphasise the advantages of biogenic nanoparticles compared to physiochemically synthesised nanoparticles. Finding the ideal balance between biogenic nanoparticles' applications, scalability, and manufacturing expenses is actually the challenging part of using them. Thus a great deal of research is required to predict low-cost options of creating biogenic nanoparticles so that they may be easily obtained for a range of future uses connected to the antimicrobial period or otherwise.

1.10 Conclusion

From the relevant studies, biogenesis seems to be a cost-effective and environmentally beneficial substitute for the production of conventional nanoparticles. Synergy between NPs and materials from the organism used for the manufacturing, is typically made possible via biogenic synthesis. Thus it is concluded that it's a multifaceted field with diverse applications. Moreover, certain aspects are needed to addressed in order to pursue its use in different fields for the betterment of humans and entire globe.

References

1. Rai, M.; Gade, A.; Yadav, A. *Biogenic Nanoparticles: An Introduction to What They Are, How They Are Synthesized and Their Applications. Metal Nanoparticles in Microbiology;* Springer, 2011, 1–14
2. Bruus, H. *Introduction to Nanotechnology;* Department of Micro and Nanotechnology, Technical University of Denmark, 2004.
3. Silva, G. A. Introduction to Nanotechnology and Its Applications to Medicine. *Surg. Neurol.* **2004**, *61* (3), 216–220.
4. Sharma, V. K.; Yngard, R. A.; Lin, Y. Silver Nanoparticles: Green Synthesis and Their Antimicrobial Activities. *Adv. Colloid interface Sci.* **2009**, *145* (1-2), 83–96.
5. Wu, Q.; Miao, W.-S.; Zhang, Y.-D.; Gao, H.-J.; Hui, D. Mechanical Properties of Nanomaterials: A Review. *Nanotechnol. Rev.* **2020**, *9* (1), 259–273.
6. Iravani, S. Green Synthesis of Metal Nanoparticles Using Plants. *Green Chem.* **2011**, *13* (10), 2638–2650.
7. Govindaraju, K.; Tamilselvan, S.; Kiruthiga, V.; Singaravelu, G. Biogenic Silver Nanoparticles by Solanum torvum and Their Promising Antimicrobial Activity. *J. Biopesticides* **2010**, *3* (Special Issue), 394.
8. Singh, P.; Kim, Y. J.; Wang, C.; Mathiyalagan, R.; Yang, D. C. TheDevelopment of a Green Approach for the Biosynthesis of Silver and Gold Nanoparticles by Using Panax ginsengRoot Extract,and Their Biological Applications. *Artif. Cells, Nanomedicine, Biotechnol.* **2016**, *44* (4), 1150–1157.
9. Singh, P., et al. Green Synthesis of Gold and Silver Nanoparticles from Cannabis sativa (Industrial Hemp) and Their Capacity for Biofilm Inhibition. *Int. J. Nanomed.* **2018**, 3571–3591.
10. Xia, Y.; Yang, H.; Campbell, C. T. *Nanoparticles for Catalysis*, Vol. 46. ACS Publications, 2013, 1671–1672
11. Charitidis, C. A.; Georgiou, P.; Koklioti, M. A.; Trompeta, A.-F.; Markakis, V. Manufacturing Nanomaterials: From Research to Industry. *Manuf. Rev.* **2014**, *1*, 11.
12. Grasso, G.; Zane, D.; Dragone, R. Microbial Nanotechnology: Challenges and Prospects for Green Biocatalytic Synthesis of Nanoscale Materials for Sensoristic and Biomedical Applications. *Nanomaterials* **2019**, *10* (1), 11.
13. Moitra, P.; Alafeef, M.; Dighe, K.; Frieman, M. B.; Pan, D. Selective Naked-Eye Detection of SARS-CoV-2 Mediated by N Gene Targeted Antisense Oligonucleotide Capped Plasmonic Nanoparticles. *ACS Nano* **2020**, *14* (6), 7617–7627.
14. Fang, X.; Wang, Y.; Wang, Z.; Jiang, Z.; Dong, M. Microorganism Assisted Synthesized Nanoparticles for Catalytic Applications. *Energies* **2019**, *12* (1), 190.
15. Antonyraj, C. A., et al. Selective Oxidation of HMF to DFF using Ru/γ-Alumina Catalyst in Moderate Boiling Solvents Toward Industrial Production. *J. Ind. Eng. Chem.* **2013**, *19* (3), 1056–1059.
16. Suhag, D.; Thakur, P.; Thakur, A. Introduction to Nanotechnology. *Integr. Nanomater. their Appl.* **2023**, 1–17.
17. Qidwai, A.; Pandey, A.; Kumar, R.; Shukla, S.; Dikshit, A. Advances in Biogenic Nanoparticles and the Mechanisms of Antimicrobial Effects. *Indian J. Pharm. Sci.* **2018**, *80* (4).
18. Sharma, D.; Kanchi, S.; Bisetty, K. Biogenic Synthesis of Nanoparticles: A Review. *Arab. J. Chem.* **2019**, *12* (8), 3576–3600.

19. Subramani, K.; Elhissi, A.; Subbiah, U.; Ahmed, W. *Introduction to Nanotechnology. Nanobiomaterials in Clinical Dentistry;* Elsevier, 2019, 3–18
20. Yang, P.; Lipowsky, R.; Dimova, R. Nanoparticle Formation in Giant Vesicles: Synthesis in BiOmimetic Compartments. *Small* **2009**, *5* (18), 2033–2037.
21. Honary, S.; Gharaei-Fathabad, E.; Paji, Z. K.; Eslamifar, M. A Novel Biological Synthesis of Gold Nanoparticle by Enterobacteriaceae Family. *Tropical J. Pharm. Res.* **2012**, *11* (6), 887–891.
22. Fu, H.; Yang, X.; Jiang, X.; Yu, A. Bimetallic Ag–Au Nanowires: Synthesis, Growth Mechanism, and Catalytic Properties. *Langmuir* **2013**, *29* (23), 7134–7142.
23. Bhushan, B. Introduction to anotechnology. *Springer Handb. Nanotechnol.* **2017**, 1–19.
24. Nasrollahzadeh, M.; Sajadi, S. M.; Sajjadi, M.; Issaabadi, Z. *An Introduction to anotechnology. Interface Science and Technology*, 28. Elsevier, 2019, 1–27
25. Durán, N.; Nakazato, G.; Seabra, A. B. Antimicrobial Activity of Biogenic Silver Nanoparticles, and Silver Chloride Nanoparticles: An Overview and Comments. *Appl. Microbiology Biotechnol.* **2016**, *100*, 6555–6570.
26. Zaidi, S.; Misba, L.; Khan, A. U. Nano-Therapeutics: A Revolution in Infection Control in Post Antibiotic Era. *Nanomedicine: Nanotechnology, Biol. Med.* **2017**, *13* (7), 2281–2301.
27. Singh, P.; Singh, H.; Kim, Y. J.; Mathiyalagan, R.; Wang, C.; Yang, D. C. Extracellular Synthesis of Silver and Gold Nanoparticles by Sporosarcina koreensis DC4 and Their Biological Applications. *Enzyme Microb. Technol.* **2016**, *86*, 75–83.
28. Singh, P.; Kim, Y.-J.; Zhang, D.; Yang, D.-C. Biological Synthesis of Nanoparticles from Plants and Microorganisms. *Trends Biotechnol.* **2016**, *34* (7), 588–599.
29. Singh, P.; Garg, A.; Pandit, S.; Mokkapati, V.; Mijakovic, I. Antimicrobial Effects of Biogenic Nanoparticles. *Nanomaterials* **2018**, *8* (12), 1009.
30. Madhavi, B.; Prameeladevi, S.; Annapurna, N.; Padma, M. Biogenic Synthesis of Magnetite Nanoparticles using the Roots of Mirabilis Jalapa for Efficient Removal of Eluent, Anti-Bacterial, and Antioxidant: A Sustainable Approach. *J. Physics: Conf. Ser.* **2024**, *2765* (1), 012004. IOP Publishing.
31. Pankhurst, Q. A.; Connolly, J.; Jones, S. K.; Dobson, J. Applications of Magnetic Nanoparticles in Biomedicine. *J. Phys. D: Appl. Phys.* **2003**, *36* (13), R167.
32. Jung, W. K.; Koo, H. C.; Kim, K. W.; Shin, S.; Kim, S. H.; Park, Y. H. Antibacterial Activity and Mechanism of Action of the Silver Ion in Staphylococcus aureus and Escherichia coli. *Appl. Environ. Microbiology* **2008**, *74* (7), 2171–2178.
33. Eckhardt, S.; Brunetto, P. S.; Gagnon, J.; Priebe, M.; Giese, B.; Fromm, K. M. Nanobio Silver: Its Interactions With Peptides And Bacteria, And Its Uses In Medicine. *Chem. Rev.* **2013**, *113* (7), 4708–4754.
34. Das, S.; Bhardwaj, A.; Pandey, L. M. Functionalized Biogenic Nanoparticles for Use in Emerging Biomedical Applications: A Review. *Curr. Nanomaterials* **2021**, *6* (2), 119–139.
35. Khan, M. S.; Vishakante, G. D.; Siddaramaiah, H. Gold Nanoparticles: A Paradigm Shift in Biomedical Applications. *Adv. Colloid Interface Sci.* **2013**, *199*, 44–58.
36. Gupta, A. K.; Gupta, M. Synthesis and Surface Engineering of Iron Oxide Nanoparticles for Biomedical Applications. *Biomaterials* **2005**, *26* (18), 3995–4021.
37. Cai, W.; Gao, T.; Hong, H.; Sun, J. Applications of Gold Nanoparticles in Cancer Nanotechnology. *Nanotechnology, Sci. Appl.* **2008**, 17–32.

38. Ibrahim, H. M. Green Synthesis and Characterization of Silver Nanoparticles using Banana Peel Extract and Their Antimicrobial Activity Against Representative Microorganisms. *J. Radiat. Res. Appl. Sci.* **2015**, *8* (3), 265–275.
39. Suganya, K. U., et al. Blue Green Alga Mediated Synthesis of Gold Nanoparticles and Its Antibacterial Efficacy against Gram Positive Organisms. *Mater. Sci. Engineering: C.* **2015**, *47*, 351–356.
40. Jha, S.; Rani, R.; Singh, S. Biogenic Zinc Oxide Nanoparticles and Their Biomedical Applications: A Review. *J. Inorg. Organomet. Polym. Mater.* **2023**, *33* (6), 1437–1452.
41. Bagur, H., et al. Endophyte Fungal Isolate Mediated Biogenic Synthesis and Evaluation of Biomedical Applications of Silver Nanoparticles. *Mater. Technol.* **2022**, *37* (3), 167–178.
42. Rao, P. V.; Nallappan, D.; Madhavi, K.; Rahman, S.; Jun Wei, L.; Gan, S. H. Phytochemicals and Biogenic Metallic Nanoparticles as Anticancer Agents. *Oxid. Med. Cell. Longev.* **2016**, *2016* (1), 3685671.
43. Rani, N.; Saini, K. Biogenic Metal and Metal Oxides Nanoparticles as Anticancer Agent: A eview. *IOP Conf. Series: Mater. Sci. Eng.* **2022**, *1225* (1), 012043 . IOP Publishing.
44. Chaudhary, V., et al. Biogenic Green Metal Nano Systems as Efficient Anti-cancer Agents. *Environ. Res.* **2023**, *229*, 115933.
45. Gnanavel, V.; Palanichamy, V.; Roopan, S. M. Biosynthesis and Characterization of Copper Oxide Nanoparticles and Its Anticancer Activity on Human Colon Cancer Cell Lines (HCT-116). *J. Photochemistry Photobiology B: Biol.* **2017**, *171*, 133–138.
46. Ramkumar, V. S., et al. Biofabrication and Characterization of Silver Nanoparticles using Aqueous Extract of Seaweed Enteromorpha compressa and Its Biomedical Properties. *Biotechnol. Rep.* **2017**, *14*, 1–7.
47. Wu, W.; Wu, Z.; Yu, T.; Jiang, C.; Kim, W.-S. Recent Progress on Magnetic Iron Oxide Nanoparticles: Synthesis, Surface Functional Strategies and Biomedical Applications. *Sci. Technol. Adv. Mater.* **2015**, *16* (2), 023501.
48. Mainardes, R. M.; Urban, M. C.; Cinto, P. O.; Chaud, M. V.; Evangelista, R. C.; Gremiao, M. D. Liposomes and Micro/nanoparticles as Colloidal Carriers for Nasal Drug Delivery. *Curr. Drug delivery* **2006**, *3* (3), 275.
49. Chorilli, M.; Brizante, A. C.; Rodrigues, C. A.; Salgado, H. R. N.; Pombo-Nascimento, E.; Ventura, D. Aspectos gerais em sistemas transdérmicos de liberação de fármacos. *Rev. bras. farm.* **2007**, *88* (1), 7–13.
50. Venugopal, J., et al. Continuous Nanostructures for the Controlled Release of rugs. *Curr. Pharm. Des.* **2009**, *15* (15), 1799–1808.
51. Awan, S. S., et al. Ailanthus altissima Leaf Extract Mediated Green Production of ZINC Oxide (ZnO) Nanoparticles for Antibacterial and Antioxidant Activity. *Saudi J. Biol. Sci.* **2023**, *30* (1), 103487.
52. Vergheese, M.; Vishal, S. K. Green Synthesis of Magnesium Oxide Nanoparticles using Trigonella foenum-graecum Leaf Extract and Its Antibacterial Activity. *J. Pharmacognosy Phytochemistry* **2018**, *7* (3), 1193–1200.
53. Abdelmigid, H. M.; Hussien, N. A.; Alyamani, A. A.; Morsi, M. M.; AlSufyani, N. M.; Kadi, H. A. Green Synthesis of Zinc Oxide Nanoparticles using Pomegranate Fruit Peel and Solid Coffee Grounds vs. Chemical Method of Synthesis, with Their

Biocompatibility and Antibacterial Properties Investigation. *Molecules* **2022**, *27* (4), 1236.

54. Devi, H. S.; Boda, M. A.; Shah, M. A.; Parveen, S.; Wani, A. H. Green Synthesis of Iron Oxide Nanoparticles using Platanus orientalis Leaf Extract for Antifungal Activity. *Green. Process. Synth.* **2019**, *8* (1), 38–45.
55. Patra, J. K., et al. Nano Based Drug Delivery Systems: Recent Developments and Future Prospects. *J. Nanobiotechnology* **2018**, *16*, 1–33.
56. Nikolova, M. P.; Joshi, P. B.; Chavali, M. S. Updates on Biogenic Metallic and Metal Oxide Nanoparticles: Therapy, Drug Delivery and Cytotoxicity. *Pharmaceutics* **2023**, *15* (6), 1650.
57. Zhang, D.; Ma, X.-L.; Gu, Y.; Huang, H.; Zhang, G.-W. RETRACTED: Green Synthesis of Metallic Nanoparticles and Their Potential Applications to Treat Cancer. *Front. Chem.* **2020**, *8*, 799.
58. Yang, G., et al. GSH-Activatable NIR Nanoplatform with Mitochondria Targeting for Enhancing Tumor-specific Therapy. *ACS Appl. Mater. & Interfaces* **2019**, *11* (48), 44961–44969.
59. Martínez-Carmona, M.; Gun'Ko, Y.; Vallet-Regí, M. ZnO Nanostructures for Drug Delivery and Theranostic Applications. *Nanomaterials* **2018**, *8* (4), 268.
60. Fakoya, M. F.; Shah, S. N. Emergence of Nanotechnology in the Oil and Gas Industry: Emphasis on the Application of Silica Nanoparticles. *Petroleum* **2017**, *3* (4), 391–405.
61. Fakoya, M. F.; Shah, S. N. Effect of Silica Nanoparticles on the Rheological Properties and Filtration Performance of Surfactant-Based and Polymeric Fracturing Fluids and Their Blends. *SPE Drill. & Completion* **2018**, *33* (02), 100–114.
62. Parizad, A.; Shahbazi, K.; Tanha, A. A. Enhancement of Polymeric Water-based Drilling Fluid Properties using Nanoparticles. *J. Pet. Sci. Eng.* **2018**, *170*, 813–828.
63. Elshawaf, M. Consequence of Graphene Oxide Nanoparticles on Heavy Oil Recovery. In *SPE Kingdom of Saudi Arabia Annual Technical Symposium and Exhibition*, 2018, pp. SPE-192245-MS: SPE.
64. Pérez-de-Luque, A.; Rubiales, D. Nanotechnology for Parasitic Plant Control. *Pest. Manag. Sci.* **2009**, *65* (5), 540–545.
65. Jeer, M. Recent Developments in Silica-Nanoparticles Mediated Insect Pest Management in Agricultural Crops. *Silicon Nano-silicon Environ. Stress. Manag. Crop. Qual. Improvement* **2022**, 229–240.
66. Mahawar, L., et al. Silicon Nanoparticles: Comprehensive Review on Biogenic Synthesis and Applications in Agriculture. *Environ. Res.* **2023**, *232*, 116292.
67. Ahmed, T., et al. Biogenic Silicon Nanoparticles Mitigate Cadmium (Cd) Toxicity in Rapeseed (Brassica napus L.) by Modulating the Cellular Oxidative Stress Metabolism and Reducing Cd Translocation. *J. Hazard. Mater.* **2023**, *459*, 132070.
68. Rajput, V. D., et al. Effects of Silicon and Silicon-Based Nanoparticles on Rhizosphere Microbiome, Plant Stress and Growth. *Biology* **2021**, *10* (8), 791.
69. Abasi, F., et al. Biogenic Silver Nanoparticles as a Stress Alleviator in Plants: A Mechanistic Overview. *Molecules* **2022**, *27* (11), 3378.
70. Chakraborty, S.; Singh, A.; Roychoudhury, A. Biogenic Nanoparticles and Generation of Abiotic Stress-Resilient Plants: A New Approach for Sustainable Agriculture. *Plant. Stress.* **2022**, *6*, 100117.

71. Ali, I., et al. Overview of Microbes Based Fabricated Biogenic Nanoparticles for Water and Wastewater Treatment. *J. Environ. Manag.* **2019**, *230*, 128–150.
72. Bano, A.; Gupta, A.; Sharma, S.; Sharma, R. Recent Developments in Nanocarrier-Based Nutraceuticals for Therapeutic Purposes. *Biogenic Nano-Particles their Use Agro-ecosystems* **2020**, 371–391.
73. Yadav, V. K.; Fulekar, M. Biogenic Synthesis of Maghemite Nanoparticles (γ-Fe2O3) using Tridax Leaf Extract and Its Application for Removal of Fly Ash Heavy Metals (Pb, Cd). *Mater. Today: Proc.* **2018**, *5* (9), 20704–20710.
74. Sharma, R.; Garg, R.; Bali, M.; Eddy, N. O. Biogenic Synthesis of Iron Oxide Nanoparticles using Leaf Extract of Spilanthes acmella: Antioxidation Potential and Adsorptive Removal of Heavy Metal Ions. *Environ. Monit. Assess.* **2023**, *195* (11), 1345.
75. Jain, A.; Wadhawan, S.; Mehta, S. Biogenic Synthesis of Non-Toxic Iron Oxide NPs via Syzygium aromaticum for the Removal of Methylene Blue. *Environ. Nanotechnology, Monit. & Manag.* **2021**, *16*, 100464.
76. Alshamsi, H. A.; Jaffer, A. A. New Hibiscus Sabdariffa L Petals Extract Based Green Synthesis of Zinc Oxide Nanoparticles for Photocatalytic Degradation of Rhodamine B Dye Under Solar Light. *AIP Conf. Proc.* **2022**, *2394* (1). AIP Publishing.
77. Veisi, H.; Azizi, S.; Mohammadi, P. Green Synthesis of the Silver Nanoparticles Mediated by Thymbra spicata Extract and Its Application as a Heterogeneous and Recyclable Nanocatalyst for Catalytic Reduction of a Variety of Dyes in Water. *J. Clean. Prod.* **2018**, *170*, 1536–1543.
78. Shayegan Mehr, E.; Sorbiun, M.; Ramazani, A.; Taghavi Fardood, S. Plant-Mediated Synthesis of Zinc Oxide and Copper Oxide Nanoparticles by using Ferulago angulata (schlecht) boiss Extract and Comparison of Their Photocatalytic Degradation of Rhodamine B (RhB) under Visible Light Irradiation. *J. Mater. Science: Mater. Electron.* **2018**, *29*, 1333–1340.
79. Naseem, K., et al. Biogenic Synthesized Nanocatalysts and Their Potential for the Treatment of Toxic Pollutants: Environmental Remediation. A Review. *Int. J. Environ. Sci. Technol.* **2024**, *21* (2), 2163–2194.
80. Yadav, R.; Kumar, M.; Tomar, R. S. Cyanobacteria Based Nanoformulation of Biogenic CuO Nanoparticles for Plant Growth Promotion of Rice Under Hydroponics Conditions. *Curr. Microbiology* **2024**, *81* (5), 118.
81. Sencan, A.; Kilic, S.; Kaya, H. Stimulating Effect of Biogenic Nanoparticles on the Germination of Basil (Ocimum basilicum L.) Seeds. *Sci. Rep.* **2024**, *14* (1), 1715.
82. Vithanage, M.; Seneviratne, M.; Ahmad, M.; Sarkar, B.; Ok, Y. S. Contrasting Effects of Engineered Carbon Nanotubes on Plants: A Review. *Environ. Geochem. Health* **2017**, *39*, 1421–1439.
83. Yang, N.; WeiHong, L.; Hao, L. Biosynthesis of Au Nanoparticles using Agricultural Waste Mango Peel Extract and Its In Vitro Cytotoxic Effect on Two Normal Cells. *Mater. Lett.* **2014**, *134*, 67–70.
84. Zheng, L.; Hong, F.; Lu, S.; Liu, C. Effect of Nano-TiO2 on Strength of Naturally Aged Seeds and Growth of Spinach. *Biol. Trace Elem. Res.* **2005**, *104*, 83–91.
85. Sethi, M.; Rana, R.; Sambhakar, S.; Chourasia, M. K. Nanocosmeceuticals: Trends and Recent Advancements in Self Care. *AAPS PharmSciTech* **2024**, *25* (3), 51.
86. Mondéjar-López, M., et al. Biogenic Silver Nanoparticles from Iris tuberosa as Potential Preservative in Cosmetic Products. *Molecules* **2021**, *26* (15), 4696.

87. Singh, P.; Kim, Y. J.; Wang, C.; Mathiyalagan, R.; Yang, D. C. Weissella oryzae DC6-Facilitated Green Synthesis of Silver Nanoparticles and Their Antimicrobial Potential. *Artif. Cells, Nanomedicine, Biotechnol.* **2016**, *44* (6), 1569−1575.
88. Singh, P., et al. In Vitro Anti-Inflammatory Activity of Spherical Silver Nanoparticles and Monodisperse Hexagonal Gold Nanoparticles by Fruit Extract of Prunus serrulata: A Green Synthetic Approach. *Artif. Cells, Nanomedicine, Biotechnol.* **2018**, *46* (8), 2022−2032.
89. Singh, H.; Du, J.; Singh, P.; Yi, T. H. Ecofriendly Synthesis of Silver and Gold Nanoparticles by Euphrasia officinalis Leaf Extract and Its Biomedical Applications. *Artif. Cells, Nanomedicine, Biotechnol.* **2018**, *46* (6), 1163−1170.
90. Abbai, R., et al. Green Synthesis of Multifunctional Silver and Gold Nanoparticles from the Oriental Herbal Adaptogen: Siberian ginseng. *Int. J. Nanomed.* **2016**, 3131−3143.
91. Phan, H. T.; Haes, A. J. What Does Nanoparticle Stability Mean? *J. Phys. Chem. C.* **2019**, *123* (27), 16495−16507.
92. Singh, A.; Rajput, V. D.; Agrawal, S.; Ghazaryan, K.; Minkina, T.; Tawaha, A. R. M. A., et al. Nanoparticles Mediated Salt Stress Resilience: A Holistic Exploration of Physiological, Biochemical, and Nano-Omics Approaches. *Rev. Environ. Contamination Toxicol.* **2024**, *262*. Available from: https://doi.org/10.1007/s44169-024-00070-4.
93. Verma, A.; Singh, A.; Rawat, S.; Rajput, P.; Minkina, T.; Mandzhieva, S., et al. Unveiling Nanomaterial-Induced Toxicity. *Adv. Environ. Eng. Green. Technol. book. Ser.* **2024**, 236−253. Available from: https://doi.org/10.4018/979-8-3693-1471-5.ch011.
94. Singh, A.; Sharma, A.; Singh, O.; Rajput, V. D.; Movsesyan, H. S.; Minkina, T., et al. In-Depth Exploration of Nanoparticles for Enhanced Nutrient Use Efficiency and Abiotic Stresses Management: Present Insights and Future Horizons. *Plant. Stress.* **2024**, 100576. Available from: https://doi.org/10.1016/j.stress.2024.100576.
95. Margaryan, G., et al. Unveiling the Salinity Tolerance Potential of Armenian Dandur (*Portulaca oleracea* L.) Genotypes: Enhancing Sustainable Agriculture and Food Security. *J. King Saud. Univ. - Sci.* **2024**, *36*, 103332. Available from: https://doi.org/10.1016/j.jksus.2024.103332.
96. Sharma, A., et al. Revolutionizing Agricultural Sustainability and Food Security and Management to Achieve SDGs Goals via Nanotechnology. *Adv. Environ. Eng. Green. Technol. Book. Ser.* **2024**, 276−288. Available from: https://doi.org/10.4018/979-8-3693-1471-5.ch013.
97. Ghazaryan, K.; Verma, A.; Rawat, S.; Rajput, P.; Minkina, T.; Mandzhieva, S., et al. Implementing Sustainable Nano-Bioremediation for Emerging Pollutants. *Adv. Environ. Eng. Green. Technol. Book. Ser.* **2024**, 316−332. Available from: https://doi.org/10.4018/979-8-3693-1471-5.ch016.
98. Sharma, A.; Singh, A.; Rajput, V. D.; Minkina, T.; Mandzhieva, S.; Elshikh, M. S., et al. Addressing Hidden Hunger via Improving Soil Health and Crop Nutrients Through Nanofortification. *Adv. Environ. Eng. Green. Technol. Book. Ser.* **2024**, 90−108. Available from: https://doi.org/10.4018/979-8-3693-1471-5.ch006.

CHAPTER

2 Realistic application of biogenic nanoparticles: future perspectives and myth

Aditya Pratap Singh[1], Sanjay Chetry[2], Harshit Singh[3], Tuhina Ghosh[4], Kousik Atta[5], Rakesh Kumar Rout[6], Utpal Das[7] and Diptikanta Acharya[6]

[1]*Department of Genetics and Plant Breeding, Bidhan Chandra Krishi Viswavidyalaya, Mohanpur, Nadia, West Bengal, India*

[2]*Department of Fruit Science, Tamil Nadu Agricultural University, Coimbatore, Tamil Nadu, India*

[3]*College of Post Graduate Studies in Agricultural Sciences, Central Agricultural University (Imphal), Umiam, Meghalaya, India*

[4]*Division of Seed Science & Technology, ICAR-Indian Agricultural Research Institute, Pusa Campus, New Delhi, India*

[5]*Faculty of Agricultural Sciences, Institute of Applied Sciences and Humanities, GLA University, Mathura, Uttar Pradesh, India*

[6]*Department of Biotechnology, GIET University, Gunupur, Rayagada, Odisha, India*

[7]*Department of Horticulture and Food Science, VIT School of Agricultural Innovations and Advanced Learning, Vellore Institute of Technology, Vellore, Tamil Nadu, India*

2.1 Introduction

Nanotechnology is ever evolving and has a prominent impact on material science,[1] with a notable impact on life science, particularly biotechnology and biomedical science. A unique property exhibited by nanoparticles contain specific characteristics such as shape, size, and distribution. Nature serves as a big "biolaboratory" with its rich diversity of biomolecules found in plants, algae, fungi, yeast, and other microbes. It has shown to play an active role in production of a variety of nanoparticle sizes and morphologies, thereby facilitating the development of safer, greener, and more ecologically friendly protocols for nanoparticles synthesis.[2]

Biogenic nanoparticles are produced through the use of biological entities like plants, bacteria, fungi, algae, or yeast. This approach is environmentally friendly and sustainable, minimizing the need for harmful chemicals commonly employed in traditional nanoparticle production methods. These nanoparticles have attracted

Biogenic Nanoparticles. DOI: https://doi.org/10.1016/B978-0-443-33819-9.00006-8

significant interest because of their distinctive characteristics and diverse applications. Key elements related to this process are summarized in Table 2.1.

Extracts from latex, seeds, stems, leaves, and roots of plants have also been used to create nanoparticles as stabilizing or reducing agents. Stable nanoparticles of Au, Ag, and Au–Ag alloy were synthesized using extracts from the leaves of *Jasminum sambac*.[3] For morphological control, spherical, hexagonal, and triangular-shaped Au NPs were created at a higher temperature with hot water and olive leaf extract than NPs manufactured with chemical techniques.[4] Iron–polyphenol (Fe–P) nanoparticles (NPs) which display photocatalytic activity against Acid Black dye, were synthesized using native Australian plants such as *Rosemarinus officinalis*, *Eucalyptus tereticornis*, and *Melaleuca nesophila*.[5] The ability of different types of algae to bioremediate hazardous metals and transform them into more malleable forms has led to their usage as model systems among the numerous biological systems utilized for NP synthesis. Furthermore, according to Patel et al.,[6] these are capable of producing a variety of metal and metal oxide nanoparticles. As a result, designing NPs with environmentally friendly methods should be the main priority. These offer answers to the escalating problems associated with environmental concerns. In contrast to chemical methods, the biological approach, which uses a variety of microorganisms, has the advantage of being more economical, environmentally friendly, and energy-efficient. The NPs produced by biological methods are also more biocompatible due to the coating of biological molecules on their surface.[7,8]

Table 2.1 Key components of biogenic nanoparticles.

Content	Key components		
Biogenic nanoparticles	**Characteristics**	Ecofriendly synthesis, surface functionalization and biocompatibility	
	Uses of biogenic nanoparticles	Medical applications	1. Drug delivery 2. Antimicrobial agents 3. Imaging and diagnostics
		Agricultural applications	1. Pest control 2. Soil fertility enhancement
		Environmental applications	1. Water treatment 2. Pollution remediation
		Industrial applications	1. Catalysis 2. Cosmetics
	Advantages of biogenic nanoparticles	1. Cost-effectiveness 2. Scalability 3. Sustainability	

Biomedicine and allied fields can benefit greatly from the biocompatibility of bio-inspired nanoparticles.[9] According to Riddin et al.[10] the use of biogenic methods allows for the creation of nanoparticles (NPs) with unique morphologies and a range of sizes. Platinum NPs with distinct sizes and shapes have been created using the sulfate-reducing bacteria's cell-soluble protein extract.[10] These NPs produced via a biogenic enzymatic approach were better than those made chemically because they required less costly ingredients and had greater catalytic activity. *Penicillium rugulosum*, a significant industrial fungus, was utilized to create uniformly sized Au NPs, which are simpler to work with than other bacteria or yeast.[11]

2.2 Synthesis process of BNPs

Going back to history, nanotechnology is not a new scientific age; rather, it is a part of the history of the arts and living things. In the art of sculpture, copper and gold are combined with other materials and reduced to their corresponding metal ions at specific temperatures. These naturally occurring nanomaterials have the unique ability to self-assemble by recognizing specific molecules.[12] A prime example of a natural nanoparticle is the papillae, nanoscopic wax crystal found on the upper side of each epidermal cell of a lotus leaf which reduce the surface area in contact with water. his scientific principle has led to the development of products with superhydrophobic surfaces, such as Lotusan®, a self-cleaning paint, and slippery liquid-infused porous surfaces (SLIPS), used in refrigeration. Additionally, Bhasma, a distinctive herbal–mineral–metallic compound used in Ayurveda, typically ranges in size from 5 to 50 nm.

In the modern period, nano-based concepts and applications have been flourishing across all scientific fields since the 1990s. In particular, the concept of nanobiotechnology was initially presented at the turn of the 20th century when scientists started looking into a variety of cutting-edge methods for improving nanomedicine and developing sustainable ecosystems. Thanks to bioinspired techniques, researchers have created viable applications and controlled NP form and size.[13] It's interesting to note that there are two conventional methods for fabricating NPs with the necessary form and size, depending on how they are assembled: top-down and bottom-up[14] (Fig. 2.1). These two approaches provide different NPs with different properties in the end, although they share the same synthesis principles. The top-down approach divided bulk materials into minute fragments, producing fine generations of NPs. Each method has a unique capacity for producing high-quality NPs.

The biogenic synthesis approaches are also known to be green synthesis of nanoparticle. it differs from other methods of nanoparticle synthesis as it uses biological resources—plant, bacteria, algae, fungi etc. It is always the bottom-up approach of nanoparticle synthesis as this method uses ions and form bulkier biogenic nanomaterial. The biogenic nanoparticle synthesis has two discrete process mythologies: extracellular and intracellular methods (Figs. 2.2 and 2.3).

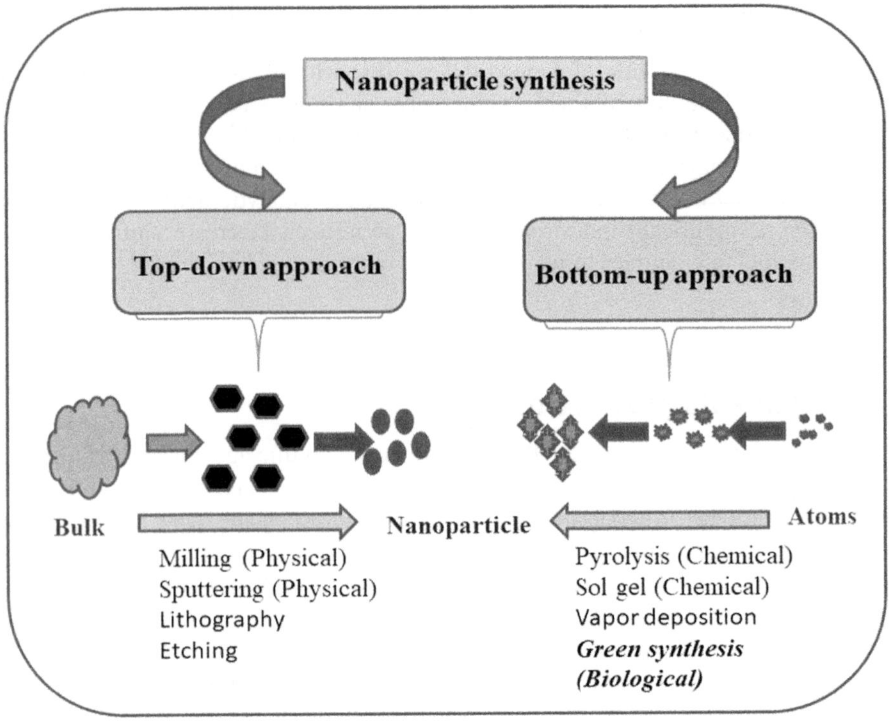

FIGURE 2.1

Top-down and bottom-up approaches of NP synthesis.

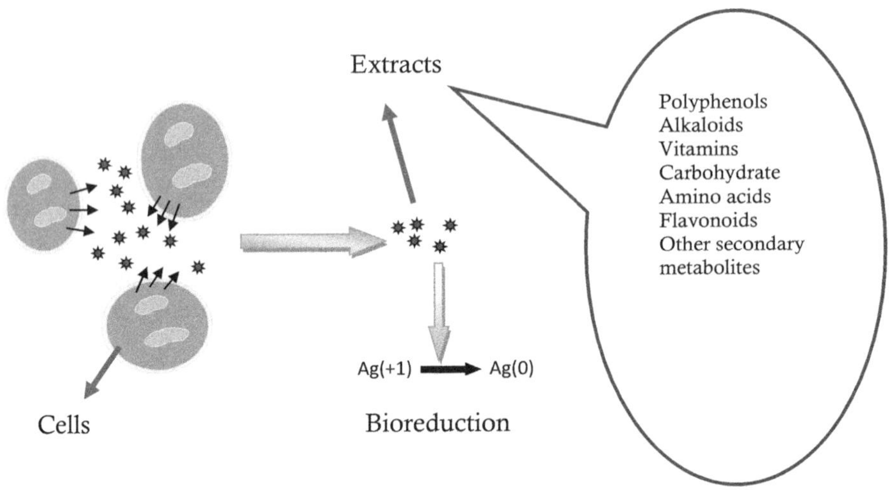

FIGURE 2.2

Extracellular method of BNP synthesis.

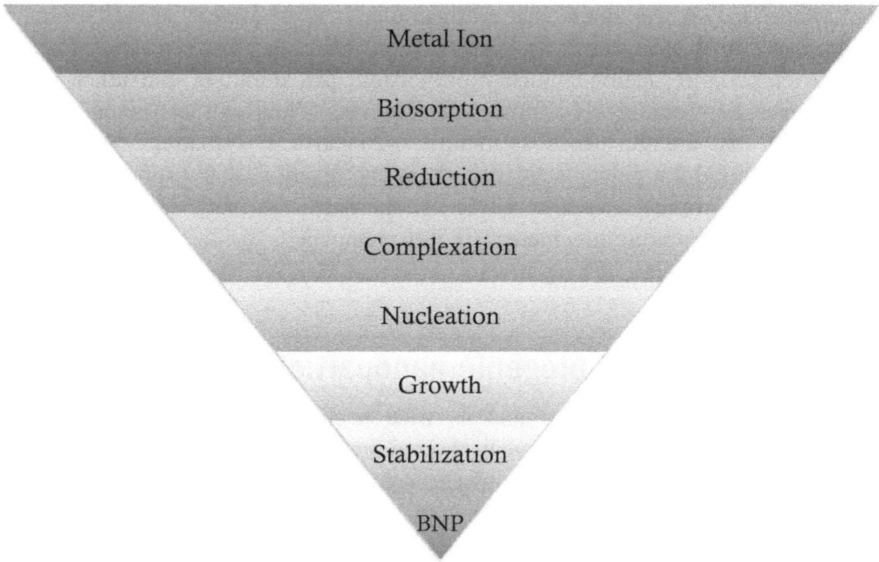

FIGURE 2.3

Intracellular method of BNP synthesis.

In the case of extracellular, the reducing agent and capping agents containing extract is prepared from biological resource secretions and these prepared extracts are used for synthesis of biogenic nanoparticle. contrarily, extract which are generated or present within the cell are obtained and used for synthesis of biogenic nanoparticle. The most significant feature of all is that the method does not uses any hazardous toxic chemical.

In biogenic synthesis, bioreduction and bioprecipitation using extract is the main feature that converts toxic ions to nontoxic biogenic nanoparticles. In case of intracellular, the metal ion in the surrounding is taken up by the microbial cell or plant cell. The alkaloids, polyphenols and other secondary metabolite aid in bioreduction. The enzymes present inside also help in bioreduction and bioprecipitation. The enzymes that are present in such extract are mostly of oxidoreductase class and NAD(P)H/NADH dependent. The biogenic method of synthesis can be further classified on the basis of source of bio extract:

1. Plant mediated biogenic synthesis
2. Algae mediated biogenic synthesis
3. Bacteria mediated biogenic synthesis
4. Fungi mediated biogenic synthesis.

The time this method takes to synthesize biogenic nanoparticle is greater as compared to other methods. Additionally, there is no evidence of top-down approach of biogenic synthesis. If extracellular and intracellular methods are put into comparison, it was found that extracellular is most preferable as the method requires no purification step.

The temperature, stereochemistry proportion, pH, duration, and other factors all influence how quickly NP forms and degrades. Plant extracts contain phytoconstituents that are the cause of metallic nanoparticle formation. Many active ingredients found in plants start the chemical reaction that produces metallic nanoparticles. Similarly, metallic oxide nanoparticles are made by the same process, however the precise mechanism is still unknown.

2.3 Applications of biogenic nanoparticle

The biogenic nanoparticle has found its wide application in various fields which includes agricultural technology, medical, environmental remediation, biosensor and imaging technology etc. Metals such as Cu, Ag, Fe, Au, and Mg, long used in traditional medicine for their inherent antimicrobial properties, have become a focal point in nanotechnology. Researchers have leveraged these properties to develop metallic and metal oxide nanoparticles (NPs) as potential antimicrobial agents. Metallic nanoparticles synthesized from plant-based sources have demonstrated a broad spectrum of biocidal activities against both Gram-positive and Gram-negative bacteria, as well as eukaryotic organisms.[15] Furthermore, Baptista et al.[16] highlighted that metallic nanoparticles (NPs) exhibited potent inhibitory effects against various resistant strains, including ampicillin-resistant *Escherichia coli*, vancomycin-resistant *Staphylococcus aureus* (VRSA), methicillin-resistant *Staphylococcus aureus* (MRSA), erythromycin-resistant *Streptococcus pyogenes*, and *Pseudomonas aeruginosa.* The antibacterial action of the NPs varies based on the type of metals used, as well as the genus, and the species of the targeted microorganisms. Due to their larger surface area-to-volume ratio compared to bulk metals, NPs are generally believed to interact more effectively with cell membranes.[17]

Generally speaking, metal ions that interact with the cell membrane are released by BNPs, leading to damage and the subsequent formation of pits/gaps that cause the membrane to fragment.[18] Therefore, when BNPs interact with phosphorus and sulfur/thiol in proteins or DNA, they cause metabolic disruptions that ultimately lead to cell death. These disruptions include inhibition of protein synthesis, DNA replication, and the respiratory chain. Similar to this, BNPs combat bacteria by generating reactive oxygen species (ROS), which in turn causes damage to the integrity of the cell wall and internal protein, DNA, and RNA degradation due to phospholipid oxidation.[19]

2.3.1 Agricultural applications of BNPs

The world has endured many significant shifts lately, the prime examples are climate change and global warming. This brings repercussions to many sectors. The sector that suffers the most is the agricultural sector. The climate irregularities cause crop failure and eventual economic loss. Moreover, the crop yield and quality has depreciated in recent decades. The prevailing environmental condition induces stress like condition to crops which makes difficult for them to survive. Conventional methods seem to be inefficient to deal with the complication. The synthesis of biogenic nanoparticle has showed new ways to handle the current situation. Nanotechnology has the potential to bring a revolution in the field of agriculture.

2.3.2 Biomedical applications

Advent of nanotechnology has opened numerous paths across various fields, with significant advancement observed in medicine in recent years. Incorporation of nanoparticle with drug and its application is one of the breakthroughs. It has created new therapies and associated technologies.

2.3.2.1 Role of biogenic nanoparticles in drug delivery systems

The drug that are used in present times to deal with the diseases are not efficient. The main reason behind is the inefficient drug delivery mechanism due to which the drug that reaches the target site in not enough. For this reason, the drug is unable to curb the disease progression or cure it completely. To overcome this problem, the drugs dose is increased so that enough drug could reach the target site. Increasing dose of administered drug raises a serious issue of drug toxicity, which clearly imparts deleterious effect on the body.

Researchers have unraveled properties of nanoparticles which are useful in drug delivery mechanism. The nanosize of nanoparticle enables it to get assimilated in the body and escape from any bodily barrier. Hence, nanoparticles are used as nanocarriers of drug molecules to the site of action. Advancement in the field of nanomedicine has boosted the development of personalized medicine which deals with patient specific treatment. The drug dose for a particular patient is not same as other. The drug dose depends upon the severity and lethality of a disease. Employing nanoparticles reduces the extra doses, by administering required amount of drug which avoids the chances of drug toxicity.

2.3.2.2 Therapeutic applications in cancer treatment and targeted therapy

In recent times, nanomedicines have been used in cancer therapy and it is found that it could successfully deal with the cancer proliferation and progression with high efficacy. The precise cellular targeting is the main feature of biogenic nanoparticles. Cellular targeting follows two mechanisms active and passive targeting. In the case of active targeting, the BNPs (in the form of nanocapsule or nanosphere) are fixed with a ligand which binds to specific cancer receptor that is present inside the body. However, in passive targeting the therapy is not receptor specific. The passive approach utilizes EPR effect of tumor which represents leakiness of tumor. Cancer treatment via biogenic nanoparticles mainly includes cancer immunotherapy and cryotherapy. Cancer immunotherapy uses antibody fixed BNPs that binds to neoplastic antigen and mounts strong immunotherapy. This approach is somewhat similar to active targeting. For instance, adding monoclonal antibodies like Herceptin or Rituxan on the surface of drug loaded BNP to treat breast cancer. The Herceptin targets Human epidermal growth factor receptor 2 (HER 2).

The cryotherapy is quite innovative and intriguing. This method utilizes BNPs to introduce extreme temperature below 0°C to destroy neoplasm. Such freezing temperature creates stress conditions and induces apoptotic pathway. Ultimately necrosis is seen in such case. Formation of ice crystals damages cell and causes cell death.

2.3.3 Environmental remediation

The soaring anthropogenic activities lead to environmental deterioration. The persistent accumulation of pollutant could possibly enter the food chain. Once any contaminant infiltrates food chain, it raises serious health issues like carcinogenicity. Though the primary reason of all is pollution, environmental remediation is the solution to bring things to normal. To address environmental challenges without causing further harm, there is a need to either develop novel methods or improve existing ones. The biogenic nanoparticles have found its way to the environmental remediation by its remarkable properties (Table 2.2). It was also found that biogenic nanoparticle can also be employed to remove heavy metal and radioactive substances from water. Biogenic nanoparticle has the potential to transform wastewater treatment by making the process more efficient and cost effective in future.

BNPs possess the ability to carry out photodegradation. Photocatalytic degradation of organic pollutant clears pollutants from water. The BNPs interact with photons and aqueous surrounding. They form oxygen free radicals or reactive oxygen species (ROS), then these ROS react with pollutants and completely disintegrate them.

Additionally, these BNP assisted ROS formed could also interact with pathogens present in water and triggers cell death. For this purpose, we don't even need hefty amount of BNP. Only a minute amount of BNP can remove enormous range of pollutant.

If it becomes possible to synthesize biogenic magnetic nanoparticles/nanomaterials and surface functionalizing them with specific functional group that specifically bind to pollutant only. In later part, magnetic field is applied to separate out the pollutant-BNP complex from wastewater. This could be one of the possible innovative approaches.

2.3.4 Bioimaging and sensors

Bioimaging and sensor play a vital role in the field of medicine. It allows real-time imagery of biochemical processes that are happening inside a biological system. It also aids in early detection of any sort of biochemical anomalies that

Table 2.2 Biogenic nanoparticles (BNPs) and their target pollutant.

S. no.	BNPs	Produced from	Target pollutant
1	AgNPs	*C. japonicum* extracts	3′,3″,5′,5″-tetrabromophenolsulfonphthalein/ bromophenol blue (BPB)
2	SeNP	*E. harbicola* biomass	Methylene blue dye, Eriochrome black T dye, methyl orange dye
3	FeNP	Oolang tea extract Green tea extract	Malachite green dye (carcinogenic)
4	AgNP	*A. agallocha* leaves juice	Victoria blue dye
5	Bimetallic Fe/Pd NP	Tea extract	Azo dyes
6	Ag NP	Mentha leaves	Congo red
7	Ag NP	*Trigonella foenum-graecum* seed extract	Eosin Y Methylene blue Methyl orange
8	AgNP	*Cicer arietinum* leaves	Congo red Methylene blue 4-nitrophenol
9	AgNP	*A. millefolium* extract	Methylene blue 4-nitrophenol Methyl orange
10	AgNP	*Palm shell* extract	Copper phthalocyanine-based dye azo dye xanthene dye
11	AgNP	*S. acuminata* fruit extract	4-nitrophenol Methylene blue Methyl orange Phenol red Direct blue 24 (DB24)
12	FeNP	*L. speciosa* leaves extract	Methylene blue Methyl orange Allura red Brilliant blue Green S dye
13	Fe/Zn NP	*A. dubius* aqueous leaf extract	Naphthalene
14	Au NP	*T. viride and H. lixii* fungal aqueous extract	4-nitrophenol
15	AuNP	*B. merisflavi* cellular extract	Congo red and methylene blue
16	MnO_x NP	*Desmodesmus* sp. WR1	Bisphenol
17	MnNP	*Pseudomonas* sp. G7	Chlorophenols
18	Fe_2O_3 NP	Aloe vera leaf extract	Arsenic (As)
19	Fe NP	*Punica granatum* extract Surface modified with *Y. lipolytica yeast*	Chromium (Cr)
20	CdS NPs	*P. aeruginosa* JP-11	Cadmium (Cd)

might turn miserable. For instance, early detection of tumorous growth via bioimaging would possibly avoid cancer progression in the body. It enables doctors to identify the line of treatment and measures that should to be adapted to eliminate the disease from the body at a very early stage. Bioimaging technology includes ultrasound, X-rays, magnetic resonance imaging (MRI), computerized tomography (CT), positron emission tomography (PET), etc. Despite of numerous bioimaging instruments, the imaging results lacks well characterization of traumatized area or lesion. It is fails to clearly demarcate the affected tissues from healthy tissue. So, there is need to improvement in the imaging technology.

2.3.4.1 Biogenic nanoparticles as contrast agents in biomedical imaging

In order to obtain clear and accurate information contrasting agents are employed. Conventional contrasting agents possess the risk of potential toxicities despite of high metabolism in the body. Additionally, they have nonspecific distribution. Recent development has enabled the use of biogenic nanomaterials in the bioimaging. The reason behind is the remarkable signal amplification, enhanced image contrast and even distribution. Most importantly, they have high quantum yield and absorbency, resistant to photobleaching, stable in in vivo conditions etc.

In addition, biogenic nanoparticles are biocompatible, highly bioavailable and remains in blood stream for prolonged period. Magnetostatic bacteria produce magnetosome, which is nanoparticle which size ranges between 40 and 100 nm. A magnetosome is made of magnetite (Fe_2O_3) or pyrite (Fe_3S_4) and it possess magnetic property, which can be useful in bioimaging techniques. The magnetosome can be used as good contrasting agent for MRI. For instance, the magnetosome can be targeted for HER2 peptide which helps detecting regions for breast cancer.

2.4 Case studies of successful use of biogenic nanoparticles

To contextualize realistic application of BNPs, we will discuss enhanced bactericidal mechanisms of biogenic silver nanoparticles compared to chemical counterparts, a study conducted by Kumari et al.[20] in detail.

Silver has been renowned for its antimicrobial properties, and the advent of nanotechnology has amplified its efficacy through the development of silver nanoparticles (AgNPs). Kumari et al.[20] examined the enhanced bactericidal mechanisms of biogenically synthesized silver nanoparticles (BSNPs) using *Trichoderma viride* metabolites, compared to chemically synthesized silver nanoparticles (CSNPs).

The study investigated how the biological synthesis process involving *T. viride* not only forms AgNPs but also endows them with enhanced antimicrobial properties due to the unique secondary metabolites from the fungus. These metabolites are believed to provide superior capping, stability, and antimicrobial efficiency compared to the chemically synthesized counterparts. The synthesis of BSNPs involved the reduction of silver nitrate using cell-free filtrates from *T. viride*. The filtrate, rich in proteins and secondary metabolites, facilitated the reduction and stabilization of silver ions, resulting in the formation of nanoparticles. CSNPs, on the other hand, were synthesized using a standard chemical reduction process with trisodium citrate as the reducing agent. UV-visible spectroscopy confirmed the formation of nanoparticles through surface plasmon resonance, with BSNPs showing a peak at 422 nm and CSNPs at 431 nm. Dynamic light scattering (DLS) analysis revealed that both BSNPs and CSNPs were monodispersed with a similar hydrodynamic size. However, BSNPs demonstrated better stability and less aggregation over time. Transmission electron microscopy (TEM) further confirmed the spherical shape and uniform distribution of both types of nanoparticles, with sizes ranging from 10 to 20 nm.

2.4.1 Enhanced antimicrobial activity of BSNPs

The antimicrobial efficacy of BSNPs and CSNPs was evaluated against three bacterial strains: *Shigella sonnei* (Gram negative), *Pseudomonas aeruginosa* (Gram negative), and *Staphylococcus aureus* (Gram positive). The study utilized growth inhibition assays, colony-forming unit (CFU) counts, and disk diffusion tests to assess the bactericidal properties of the nanoparticles.

BSNPs exhibited significantly higher antimicrobial activity than CSNPs across all bacterial strains. At a concentration of 1 μg/mL, BSNPs were as effective as 5 μg/mL of CSNPs, indicating a more potent bactericidal effect. The superior efficacy of BSNPs is attributed to the unique coating of antimicrobial metabolites from *T. viride*, which enhanced the nanoparticles' ability to disrupt bacterial cell membranes.

2.4.2 Mechanisms of bacterial cell disintegration

The study explored the mechanisms behind the enhanced bactericidal action of BSNPs. Flow cytometry analysis showed that BSNPs had higher cellular internalization rates compared to CSNPs, particularly in *S. sonnei* and *S. aureus*. This increased internalization is believed to be due to the stronger capping provided by the fungal metabolites, which prevents aggregation and maintains a higher surface area for interaction with bacterial cells. The generation of reactive oxygen species (ROS) was also significantly higher in BSNP-treated bacteria. ROS plays a crucial role in oxidative stress, leading to severe damage to bacterial cell membranes and intracellular components. The study found that BSNPs induced a more substantial increase in ROS levels, correlating with higher internalization and resulting in enhanced bacterial disintegration. Transmission electron microscopy provided visual evidence of the extensive morphological damage caused by BSNPs. Treated bacteria exhibited disrupted cell membranes, leakage of cytoplasmic contents, and overall structural disintegration, which were more pronounced in BSNP-treated cells than in those treated with CSNPs. The findings suggest that BSNPs hold significant potential for use in various antimicrobial applications, including medical devices, wound dressings, and coatings for surfaces prone to bacterial contamination. The green synthesis approach also aligns with the growing demand for environmentally friendly and sustainable technologies in combating antibiotic-resistant pathogens. Some other successful use cases of BNPs have been enlisted in Table 2.3.

2.5 Challenges and limitations

2.5.1 Regulatory challenges and safety considerations

The utilization of biogenic nanoparticles (BNPs) has sparked widespread interest due to its potential applications across various sectors, including agriculture, foods and healthcare. These applications frequently face regulatory scrutiny across various countries to ensure their safety and, in some instances, their efficacy. Nonetheless, a primary barrier to the commercial advancement of nanotechnology is the limited geographical reach of legislative frameworks, regulatory guidelines, and prevailing negative public perceptions.[39] The pressing issues of global food security and climate change further underscore the necessity for the commercialization of nanotechnology-based products.[40] Consequently, there is an urgent requirement for risk assessments concerning the use of nanoparticles, as advocated by the European Parliament and several nongovernmental organizations (NGOs), leading to the establishment of various regulations concerning foods and healthcare (Table 2.4).

In the European Union, the safety of various substances and products, including nanoparticles, is governed by a comprehensive legal framework that encompasses both general and sector-specific regulations.[42] Important legislations include the Food Information to Consumers Regulation (1169/2011), the Regulation on Plastic Materials and Articles Intended for Food Contact (10/2011), the Regulation on Active and Intelligent Materials (450/2009), the Biocidal Products Regulation (528/2012) and the Cosmetic Products Regulation (1223/2009). The principal regulation is REACH (Registration, Evaluation, Authorisation, and Restriction of Chemicals), which specifically addresses the application of nanoparticles in areas such as plant protection products, food additives and materials that come into contact with food (European Commission, 2013). Moreover, the EU Scientific Committee on Emerging and Newly Identified Health Risks (SCENIHR) has advocated for a case-by-case risk assessment approach for nanoparticles (SCENIHR, 2007). The European Commission has also issued a recommendation defining nanoparticles, emphasizing a size range of 1 to 100 nm as a key criterion.[42] This definition serves as a foundational guideline that can be modified for specific legislative contexts.

Table 2.3 Some successful case studies on the use of biogenic nanoparticles.

Nanoparticles	Origin/source	Applications	Mode of action/ concentration	References
Ag NPs	*Bacillus subtilis* KMS2-2 strain	Antibacterial activity was observed against pathogenic bacteria, including *Pseudomonas fluorescens*, *Proteus mirabilis*, *Escherichia coli*, *Bacillus cereus,* and *Staphylococcus aureus.*	At 100 μL concentrations of NPs, the maximum inhibition zone (16 mm) was detected against *Pseudomonas fluorescens*, whereas the minimum inhibition zone (08 mm) was noted against *Proteus mirabilis*.	21
Ag NPs	Plant extract of *Melia dubia*	In vitro anticancer effect against human breast cancer cell lines MCF-7.	IC50 value of NPs for the cancerous cell line was 31.2 μg/mL (IC50: the concentration, which cause the death of 50% cancer cell).	22
Ag NPs	*Solanum torvum*	Antimicrobial effect against pathogenic bacteriae *Pseudomonas aeruginosa*, and *Staphylococcus aureus*; as well as pathogenic fungi *Aspergillus flavus* and *Aspergillus niger*.	In 100 μL of NPs, the zone of inhibition (mm) was as follows: *Pseudomonas aureginosa* 16.9, *Staphylococcus aureus* 17.6, *Aspergillus flavus* 15.2, *Aspergillus niger* 14.8.	23
Ag/zeolite nanocomposite	Leaf extract of *Euphorbia prolifera*	Cause catalytic reduction of detrimental dyes like 4-nitrophenol (4-NP), methyl orange (MO), congo red (CR), methylene blue (MB) and rhodamine B (RhB) in the presence of excess NaBH4 at room temperature.	Ag/zeolite nanocomposite required for catalytic reduction: 10 mg (Rh B), 3 mg (MB), 7 mg (MO, CR).	24
Ag@AgCl-NPs	Aqueous extract green marine alga (*Chaetomorpha* sp).	Nanosensors useful for detection of Hg^{2+} in two different types of water samples such as mineral and tap waters.	Minimum detectable Hg^{2+} concentration was found to be 4.19 nM by using 5 mL of crude extract.	25

(Continued)

Table 2.3 Some successful case studies on the use of biogenic nanoparticles. *Continued*

Nanoparticles	Origin/source	Applications	Mode of action/ concentration	References
SeNPs	*Bacillus* sp. B2 strain	Antibacterial effects against *Staphylococcus aureus*, *Pseudomonas aeruginosa*, *Escherichia coli*, and *Klebsiella pneumoniae.*	By applying 100 μg of SeNPs, the maximum diameter of Zone of Inhibition (mm) was 13 ± 0.62 (*S. aureus*), 22 ± 1.04 (*P. aeruginosa*), 18 ± 0.22 (*E. coli*), 16 ± 0.63 (*K. pneumonia*).	26
Fe NPs	Manglicolous (mangrove) fungus *Aspergillus niger* BSC-1	Elimination of detrimental hexavalent chromium from aqueous solution/ water.	Highest Cr (VI) exclusion was observed at 40°C and pH 3 with 2.5 g/L of Fe NPs dose.	27
CuO NPs	Seed extract of *Cucurbita maxima, Cucurbita pepo*, and *Cucurbita moschata*	Anticancer effect against human colorectal cancer cell line (HCT-116).	CuO NPs displayed 50% inhibitory concentration (IC50) at 25 μg/mL against the HCT-116 cell line.	28
Cu_2O nanoparticles	*Camellia sinensis* leaf extract	Exhibiting major cell death and antihuman ovarian cancer properties against Caov-3, SW-626, and SK-OV-3 cell lines.	IC50 value (μg/mL) perceived for the NPs was Caov-3 (263), SW-626 (208) and SK-OV-3(315).	29
CuNPs	*Ilex paraguariensis*	Inhibiting multiplication and survivability of carcinoma lung cells (A549).	At 100 μg/mL concentrations of NPs, the highest rate of inhibition of A549 cell line was 36.89% after 72 h time exposure.	30
PdNPs	Leaf extracts from *Allium fistulosum, Basella alba* and *Tabernaemontana divaricate*	Antifungal activity against *Aspergillus flavus, Penicillium* sp., and *Candida albicans*.	For 150 nm NPs, the zone of inhibition (mm) observed was, *A. flavus (14), Penicillium sps (18)*, and *C. albicans* (10).	31

LyAgNP, LyFeNP, and LyGNP	Lycopene from *Lycopersicon esculentum*	Nanoparticles showed good cytotoxic effect against colon cancer cells COLO320DM, HT29 and cervical cancer Hella cell line.	100 μg/mL concentration of each nanoparticle was effective for cytotoxic activity.	32
MgO NPs	Endobacterium *Burkholderia rinojensis*	Antifungal and antibiofilm effect against *Fusarium oxysporum f.* sp. lycopersici.	At 15.36 μg/mL concentration of MgO, NPs mycelial growth of the fungus was entirely repressed, whereas the biofilm formation of the pathogen was totally inhibited by 1.92 μg/mL concentrations of MgO NPs.	33
GNPs (Gold NPs)	Methanolic extract of the endophytic isolate *Aspergillus terreus*	Antimicrobial effect against various food-borne pathogens and phytopathogens as well as antioxidant activity against the ABTS radical.	At a concentration of 400 μg/mL, the nanoparticles exhibited the greatest inhibition zones against *S. aureus* and *V. cholera*. At a lower concentration of 200 μg/mL, they inhibited the radial mycelial growth of *Fusarium oxysporum* and *Rhizoctonia solani* by 52.5% and 65.46%, respectively. Additionally, the gold nanoparticles demonstrated significant antioxidant activity against the ABTS radical, with an IC50 value of 38.61 μg/mL.	34
ZnNPs	Leaf extract of *Sorghum bicolor* L	Significant enhancement of the fresh and dry weight (g) of shoot and root. Moreover, increased chlorophyll contents were detected under normal and saline conditions in okra. Upsurge in the antioxidative activity was found under saline stress conditions.	The foliar application of 0.3% Zn-NPs was effective in the regulation of the antioxidant defense mechanism under a saline stress condition.	35

(Continued)

Table 2.3 Some successful case studies on the use of biogenic nanoparticles. *Continued*

Nanoparticles	Origin/source	Applications	Mode of action/ concentration	References
Neodymium oxide (Nd_2O_3-IL) NPs	Leaves extract of *Andrographis paniculata*	Antibacterial activity against the bacterial strains of *Staphylococcus aureus* and *Escherichia coli*. As well as antioxidant activity.	For the Gram-negative bacteria (*E. coli*), the maximum zone of inhibition of Nd_2O_3-IL is found to be 4 mm at 10 μg/mL concentration, whereas the zone of inhibition of the Nd_2O_3-IL NPs against *S.aureus* was about 7 mm. At the highest concentration (500 μg/ mL), the inhibition potential of Nd_2O_3-IL NPs was found to be 52.21%.	36
Titanium nanoparticles	*Trichoderma harzianum*	Biological control agent, inhibiting mycelial growth and the development of new sclerotia of *S. sclerotiorum.*	At the concentrations of 108 NPs/ mL, the nanoparticles were effective against *Sclerotium sp*. Whereas, no harmful effects were found at the concentration of 2.1 × 109 NPs/mL against the agriculturally important microorganisms including *Bacillus thuringiensis, Pseudomonas aeruginosa, Bradyrhizobium japonicum and Beauveria bassiana*.	37
Magnetite nanoparticles (MNPs)	*Anthemis pseudocotula* (AP)	Protecting environment through assortment of oil spills in the presence of an external magnetic field.	The analysis determined that the optimal ratio of APH-MNPs to crude oil for achieving a 90% removal of crude oil was 1:10.	38

Table 2.4 Country and their organization(s) involved in regulatory and safety of foods and medical products.

Country	Responsible organization(s) Foods	Healthcare
Australia	Foods Standards Australia New Zealand	–
Canada	Health Canada	Health Canada
European Union	European Commission (EC), European Food Safety Authority (EFSA)	European Commission, European Medicines Agency
France	Ministry of Agriculture/ General Directorate on Food, European Commission, European Food Safety Authority	Ministry of Environment/General Directorate for Risk Prevention/ Department of chemical products, diffuse pollution and agriculture, European Commission, European Medicines Agency
Germany	Federal Ministry of Food, Agriculture and Consumer Protection, European Commission, European Food Safety Authority	Ministry of Health, Labor, and Welfare
Korea	Korea Food and Drug Administration	–
Japan	Ministry of Health, Labor and Welfare	Ministry of Health, Labor and Welfare
Netherlands	European Commission, European Food Safety Authority	Medicines Evaluation Board, European Commission, European Medicines Agency
Norway	European Commission, Norwegian Food Safety Authority, European Food Safety Authority	–
Poland	European Commission, European Food Safety Authority	European Commission, European Medicines Agency
Russian Federation	Federal Service for Surveillance on Consumer Rights Protection and Human Well-being, Ministry of Health and Social Development of the Russian Federation	Federal Service for Surveillance on Consumer Rights Protection and Human Well-being, Ministry of Health and Social Development of the Russian Federation
United States	Food and Drug Administration (FDA), Department of Health and Human Services (HHS); and Food Safety and Inspection Service (FSIS), Department of Agriculture (USDA)	Food and Drug Administration (FDA), Department of Health and Human Services (HHS)

Source: OECD.[41]

Regulatory strategies for nanotechnology differ widely across the globe. While the European Union and Switzerland have implemented specific laws addressing nanotechnology, other nations, such as the United States, have taken a more adaptable stance with nonmandatory frameworks and guidance.[41,43] In the United States, for instance, the Food and Drug Administration (FDA) and the Environmental Protection Agency (EPA) have yet to establish an official definition for nanoparticles, although they do utilize comprehensive methods when evaluating products that incorporate nanotechnology.[44] This disparity in regulatory approaches, along with varying public perceptions, has significantly obstructed the broader commercialization of nanotechnology in the agricultural sector, as these products encounter substantial hurdles in entering the market.[40]

The successful governance of nanoparticles necessitates international collaboration and open dialogue to devise effective regulatory frameworks that mitigate biosafety risks. As nanotechnology rapidly advances, it is imperative that regulatory entities, including the FDA, OECD, and ISO, work in concert to establish clear and consistent guidelines that enable the safe and beneficial application of nanoparticles while maintaining public confidence.[42,45] Public perception of biogenic nanoparticles (BNPs) is a critical determinant of their acceptance and adoption. A disconnect frequently exists between scientific knowledge and public perception, resulting in resistance against the use of nanomaterials in consumer goods. Consequently, transparency in research, regulation, and communication is essential for building public trust.[46–48]

2.5.2 Scalability and reproducibility issues in nanoparticle synthesis

Utilizing biological resources for nanoparticle synthesis provides several benefits, such as being environmentally friendly and optimizing material utilization.[49] Microorganisms, plants, and marine organisms can be employed for intracellular and extracellular synthesis through bio reduction processes.[50] However, scaling up production and ensuring consistent outcomes pose significant challenges in this field. These issues become particularly critical when transitioning from small-scale studies to large-scale manufacturing, where maintaining uniformity, cost-effectiveness, and quality assurance are paramount. While nanoparticles and nanodevices have been successfully produced in laboratory settings, scaling up their production remains a substantial technological challenge. Large-scale preparation of nanoparticles should prioritize developing processes that comply with safety standards and environmental regulations, enabling the integration of nanoparticles into commercially viable products for everyday use. Therefore, scalability and reproducibility are crucial yet demanding aspects of nanoparticle synthesis.

Scaling up the production of nanoparticles involves overcoming significant challenges. Although numerous synthetic methods have been proposed worldwide, the conditions necessary for large-scale industrial production often differ from those used in laboratory settings. Currently, many technologies that could benefit from nanoparticles lack access to commercially available equipment and the capacity to produce large quantities of the desired materials. Key obstacles to large-scale nanoparticle fabrication include synthesis conditions, material quality, and the cost-benefit ratio. Additionally, our understanding of the potential adverse effects of nanoparticles on human health and the environment is limited, particularly since these particles can easily penetrate biological barriers.[51] Another frequently overlooked challenge is the monitoring methods, which are essential because any changes to synthesis conditions can lead to products that fail to meet required quality standards, resulting in economic losses for the industry. Effective monitoring is crucial for maintaining consistency and ensuring that nanoparticles meet safety and performance criteria.[13,52] Moreover, the transition from laboratory-scale synthesis to industrial production requires the development of scalable processes that adhere to safety and environmental regulations. This includes ensuring the reproducibility of nanoparticle characteristics, such as size and shape, which are critical for their intended applications.

Scaling up green synthesis processes presents its own set of challenges, particularly when using plant materials. Several factors influence the scalability of this approach, including the restricted availability of raw materials, challenges in managing the molecular composition and concentration, the absence of adequate equipment for handling large quantities, and the inherent variability of plant-derived materials. Furthermore, the metabolites formed due to the presence of enhancers or inhibitors in subsequent experiments may not be suitable for the eco-friendly synthesis of metallic nanoparticles. Large-scale production of metabolites from wild plants could introduce significant variation in their metabolic profiles, which could result in nanoparticles with varying physicochemical properties.[53] Plants experience various types of stress throughout their life cycle and in order to achieve consistent and reliable outcomes, it is essential to obtain plant extracts from the same growth cycle and under similar environmental conditions. However, these requirements can complicate the process, as different species may have varying growth durations and may not thrive under controlled conditions. To facilitate scaling up, tissue culture emerges as an optimal method, as it allows for the production of uniform and independent biochemical products by growing plants in a controlled environment.[54] In large-scale green synthesis, substantial volumes of plant extracts—ranging from a few milliliters to several liters—can be utilized to ensure a rapid, reproducible, and scalable process. However, currently, only a limited number of batches are available at a commercial scale. It is anticipated that innovative instruments will emerge in the coming years, potentially transforming the green synthesis of metallic nanoparticles.[53] Additionally, fostering strong interdisciplinary collaboration is essential to bridge the gap between industry and academic institutions.[55]

Reproducibility issues are a significant challenge in the field of nanoparticle synthesis, particularly when scaling up from laboratory settings to large-scale production. The establishment of standardized protocols and advancements in analytical techniques are crucial to guarantee consistent results and allow for effective comparisons across studies.[49] Advanced tools are often required to accurately measure nanoparticle properties, but variability in the calibration and sensitivity of these tools can affect the reproducibility of results. Even when a synthesis process has been optimized at the lab scale, reproducing the same level of optimization at a larger scale is challenging, often necessitating reoptimization, which is often both time-intensive and expensive. Despite these hurdles, the wide-ranging use of biogenic nanoparticles in areas such as biomedicine, environmental remediation, targeted drug delivery and biosensing underscore the importance of further research to better understand biologically dependent synthesis processes and to develop green manufacturing methods for metallic nanoparticles.[50,56]

2.6 Public perception and myths

2.6.1 Common misconceptions about biogenic nanoparticles

Biogenic nanoparticles (BNPs) are synthesized through biological processes using microorganisms, plants, marine sources etc.[50] Common misconceptions about BNPs include their perceived inferiority to chemically synthesized nanoparticles. The understanding of nanoparticles, which have been in existence for over fifty years, remains somewhat unclear among researchers. The principles of green chemistry advocate for the development of nanoparticles as a sustainable and environmentally friendly alternative to traditional synthesis methods. This green chemistry framework emphasizes that the synthesis of nanoparticles should be a safe and nontoxic process, utilizing natural resources while minimizing the use of hazardous substances and keeping costs low. Another major misconception is that biogenic nanoparticles are inherently safe and environmentally friendly. While it is true that biogenic nanoparticle synthesis often relies on green methodologies, which can be more ecofriendly than traditional chemical methods, the safety and environmental impact of these nanoparticles are not always straightforward.[50,57] The specific characteristics of the nanoparticles, such as their size, shape, and surface properties, can have a significant impact on their toxicity and environmental behavior, and these factors can vary depending on the synthesis methodology and the source organism. Another common misconception is that biogenic nanoparticles are always more stable and homogeneous than their chemically synthesized counterparts. While it is true that biological synthesis can produce nanoparticles with a narrow size range and good stability, this is not always the case. The stability and homogeneity of biogenic nanoparticles can be influenced by factors such as the growth conditions, the specific biomolecules involved in the synthesis process, and the post-synthesis processing.[57,58] However, BNPs offer advantages such as ecofriendliness, sustainability, and biocompatibility. They have diverse applications in biomedicine, drug delivery, analytical sensing, and food preservation.[59] Fungi are particularly favored for BNP synthesis due to their high biomass production and metal tolerance. Despite their potential, challenges remain in optimizing production, controlling size and shape, and improving stability.[59] Additionally, further research is needed to elucidate molecular mechanisms, degradation pathways, and biodistribution of BNPs to facilitate their transition from laboratory to clinical applications.[50]

2.6.2 Addressing concerns regarding safety and environmental impact

Biogenic nanoparticles (NPs) can raise environmental and safety concerns due to their potential to bind, transport pollutants and accumulate in organisms. Xu et al.[60] highlighted that that exposure to silver nanoparticles (AgNPs) at relevant environmental concentrations can reduce the abundance and activity of soil microorganisms, potentially impairing functions crucial for plant growth and the degradation of xenobiotic compounds. Rastogi et al.[61] further emphasizes that the concentration and size of nanoparticles can have distinct effects on photosynthesis, either positively or negatively. According to Saha et al.,[62] AgNPs that accumulate in plant tissues can be transferred through the food chain to consumers, potentially causing lethal effects in species that are not tolerant to these nanoparticles.

While biosynthesis of NPs using fungi and plant extracts is considered eco-friendly compared to chemical methods,[63,64] the potential for nanopollution and bioaccumulation in soil and water environments is a significant issue. The unique properties of biogenic NPs, such as enhanced bioavailability and dispersibility, may lead to adverse biological effects and toxicity.[65] While these NPs show promise in various applications, including crop stress management and pollutant removal, further research is needed to fully characterize their potential toxicity and ensure their safe use. Balancing the benefits of biogenic NPs with their potential environmental risks remains a critical challenge in the field of nanotechnology.

2.7 Future perspectives

Biogenic nanoparticles have shown potential in various fields, including pharmaceuticals, medicine, environmental remediation, and biosensing.[59,66] Their low cytotoxicity and biocompatibility make them particularly attractive for biomedical applications, such as disease detection and cancer treatment. As research progresses, addressing challenges in synthesis optimization and exploring new applications will be crucial for advancing the field of biogenic nanoparticles. The biosynthesis of BNPs offers researchers the opportunity to develop safer nanomaterials while enhancing the understanding of their health and safety implications. This approach facilitates the production of valuable materials on a practical scale by avoiding the use of toxic chemicals. Significant research is focused on various capping agents, such as biomolecules and polysaccharides, which serve dual functions as chelating/reducing and capping agents in nanoparticle synthesis. Consequently, these particles are shielded from further reactions and aggregation, thereby improving their stability and longevity. Plant extracts rich in biomolecules can effectively reduce metal ions to generate nanoparticles with unique properties.[67] These nanoparticles exhibit promising antibacterial, antioxidant, and cytotoxic activities, making them potential alternatives to conventional treatments for multidrug-resistant bacteria.[68] Microorganisms like bacteria, fungi, and algae also offer a distinct environment for nanoparticle synthesis, although challenges remain in optimizing production and stability.[59]

Biogenic nanoparticles have gained significant attention due to their ecofriendly synthesis, low cytotoxicity, and biocompatibility.[66] These nanomaterials offer potential applications in various fields, in biomedicine, agriculture, food packaging, and environmental management including pharmaceuticals, medicine, and environmental remediation.[64] The synthesis of biogenic nanoparticles involves green chemistry principles, reducing environmental pollutants and conserving natural resources. Compared to conventional methods, biogenic synthesis offers advantages such as mild reaction conditions, reasonable energy demands, and the ability to produce nanoparticles with tailored properties.[69] Applications of biogenic nanoparticles include disease detection, cancer treatment, wastewater treatment, pollutant removal, and solar energy conversion.[64,66] Fruit and vegetable waste, rich in bioactive compounds, serve as sustainable sources for nanoparticle synthesis, potentially transforming biodegradable waste into valuable products. The green synthesis of nanomaterials presents advantages such as low cost, better reducing and stabilizing potential, and near-zero contamination.[70–75] Future research should focus on addressing safety concerns, toxicity aspects, and developing optimal commercialization strategies for biogenic nanomaterials.[75–79] Despite their potential, challenges remain in the large-scale production and commercialization of biogenic nanoparticles, necessitating further research to address knowledge gaps and improve feasibility for industrial adoption.[69] Overall, biogenic nanoparticles offer a sustainable approach to nanotechnology with potential for wide-ranging applications across multiple industries.

2.8 Conclusion

Biogenic nanoparticles (BNPs) represent a significant advancement in the field of nanotechnology, merging the principles of green chemistry with cutting-edge applications across multiple sectors, including agriculture, medicine, and environmental remediation. This chapter has provided an overview of the potential uses of BNPs from improving crop tolerance to their applications in drug delivery and pollution remediation. Nevertheless, even though these applications seem so promising, there are several factors that prevent biomass nanoparticles from widespread practical use, such as problems with their scale-up, certain regulations, and lack of more detailed studies on their toxicity. Such new developments of BNP technology will be based on the solutions to these obstacles targeted into interdisciplinary studies, technology development, and establishment of the acceptable regulatory framework. In the course of development in this area, it will be imperative to manage perceptions and misperceptions, and ensure that the added advantages of using BNPs are proportionate to the costs involved. Therefore, it is clear that industries will be transformed through the application of BNPs owing to their prospective value, but this should be tempered by the need to harness their potential in a prudent and efficient manner.

References

1. Astruc, D. Introduction: nanoparticles in catalysis. *Chem. Rev.* **2020**, *120*, 461–463. Available from: https://doi.org/10.1021/acs.chemrev.8b00696.
2. Sharma, D.; Kanchi, S.; Bisetty, K. Biogenic synthesis of nanoparticles: a review. *Arab. J. Chem.* **2019**, *12*, 3576–3600. Available from: https://doi.org/10.1016/j.arabjc.2015.11.002.
3. Yallappa, S.; Manjanna, J.; Dhananjaya, B. Phytosynthesis of stable Au, Ag and Au–Ag alloy nanoparticles using J. sambac leaves extract, and their enhanced antimicrobial activity in presence of organic antimicrobials. *Spectrochimica Acta Part. A: Mol. Biomolecular Spectrosc.* **2015**, *137*, 236–243. Available from: https://doi.org/10.1016/j.saa.2014.08.030.
4. Khalil, M. M.; Ismail, E. H.; El-Magdoub, F. Biosynthesis of Au nanoparticles using olive leaf extract: 1st nano updates. *Arab. J. Chem.* **2012**, *5*, 431–437. Available from: https://doi.org/10.1016/j.arabjc.2010.11.011.
5. Wang, Z.; Fang, C.; Megharaj, M. Characterization of iron–polyphenol nanoparticles synthesized by three plant extracts and their fenton oxidation of azo dye. *ACS Sustain. Chem. & Eng.* **2014**, *2*, 1022–1025. Available from: https://doi.org/10.1021/sc500021n.
6. Patel, V.; Berthold, D.; Puranik, P.; Gantar, M. Screening of cyanobacteria and microalgae for their ability to synthesize silver nanoparticles with antibacterial activity. *Biotechnol. Rep.* **2015**, *5*, 112–119. Available from: https://doi.org/10.1016/j.btre.2014.12.001.
7. Hakim, L. F.; Portman, J. L.; Casper, M. D.; Weimer, A. W. Aggregation behavior of nanoparticles in fluidized beds. *Powder Technol.* **2005**, *160*, 149–160. Available from: https://doi.org/10.1016/j.powtec.2005.08.019.
8. Tripp, S. L.; Pusztay, S. V.; Ribbe, A. E.; Wei, A. Self-assembly of cobalt nanoparticle rings. *J. Am. Chem. Soc.* **2002**, *124*, 7914–7915. Available from: https://doi.org/10.1021/ja0263285.
9. Huang, J.; Lin, L.; Sun, D.; Chen, H.; Yang, D.; Li, Q. Bio-inspired synthesis of metal nanomaterials and applications. *Chem. Soc. Rev.* **2015**, *44*, 6330–6374. Available from: https://doi.org/10.1039/C5CS00133A.
10. Riddin, T.; Gericke, M.; Whiteley, C. Biological synthesis of platinum nanoparticles: effect of initial metal concentration. *Enzyme Microb. Technol.* **2010**, *46*, 501–505. Available from: https://doi.org/10.1016/j.enzmictec.2010.02.006.
11. Mishra, A.; Tripathy, S. K.; Yun, S.-I. Fungus mediated synthesis of gold nanoparticles and their conjugation with genomic DNA isolated from *Escherichia coli* and *Staphylococcus aureus*. *Process. Biochem.* **2012**, *47*, 701–711. Available from: https://doi.org/10.1016/j.procbio.2012.01.017.

12. Griffin, S.; Masood, M. I.; Nasim, M. J.; Sarfraz, M.; Ebokaiwe, A. P.; Schäfer, K.-H.; Keck, C. M.; Jacob, C. Natural nanoparticles: a particular matter inspired by nature. *Antioxidants* **2017**, *7*, 3. Available from: https://doi.org/10.3390/antiox7010003.
13. Patil, S.; Chandrasekaran, R. Biogenic nanoparticles: a comprehensive perspective in synthesis, characterization, application and its challenges. *J. Genet. Eng. Biotechnol.* **2020**, *18*, 67. Available from: https://doi.org/10.1186/s43141-020-00081-3.
14. Gwo, S.; Chen, H.-Y.; Lin, M.-H.; Sun, L.; Li, X. Nanomanipulation and controlled self-assembly of metal nanoparticles and nanocrystals for plasmonics. *Chem. Soc. Rev.* **2016**, *45*, 5672–5716. Available from: https://doi.org/10.1039/C6CS00450D.
15. Eltarahony, M.; Zaki, S.; ElKady, M.; Abd-El-Haleem, D. Biosynthesis, characterization of some combined nanoparticles, and its biocide potency against a broad spectrum of pathogens. *J. Nanomaterials* **2018**, *2018*, 5263814. Available from: https://doi.org/10.1155/2018/5263814.
16. Baptista, P. V.; McCusker, M. P.; Carvalho, A.; Ferreira, D. A.; Mohan, N. M.; Martins, M.; Fernandes, A. R. Nano-strategies to fight multidrug resistant bacteria—"A Battle of the Titans. *Front. Microbiol.* **2018**, *9*, 1441. Available from: https://doi.org/10.3389/fmicb.2018.01441.
17. Slavin, Y. N.; Asnis, J.; Hńfeli, U. O.; Bach, H. Metal nanoparticles: understanding the mechanisms behind antibacterial activity. *J. Nanobiotechnology* **2017**, *15*, 1–20. Available from: https://doi.org/10.1186/s12951-017-0308-z.
18. Rajkuberan, C.; Sudha, K.; Sathishkumar, G.; Sivaramakrishnan, S. Antibacterial and cytotoxic potential of silver nanoparticles synthesized using latex of *Calotropis gigantea* L. *Spectrochimica Acta Part. A: Mol. Biomolecular Spectrosc.* **2015**, *136*, 924–930. Available from: https://doi.org/10.1016/j.saa.2014.09.115.
19. Karakoti, A. S.; Hench, L. L.; Seal, S. The potential toxicity of nanomaterials—the role of surfaces. *Jom* **2006**, *58*, 77–82. Available from: https://doi.org/10.1007/s11837-006-0147-0.
20. Kumari, M.; Shukla, S.; Pandey, S.; Giri, V. P.; Bhatia, A.; Tripathi, T.; Kakkar, P.; Nautiyal, C. S.; Mishra, A. Enhanced cellular internalization: a bactericidal mechanism more relative to biogenic nanoparticles than chemical counterparts. *ACS Appl. Mater. & Interfaces* **2017**, *9*, 4519–4533. Available from: https://doi.org/10.1021/acsami.6b15473.
21. Mathivanan, K.; Selva, R.; Chandirika, J. U.; Govindarajan, R.; Srinivasan, R.; Annadurai, G.; Duc, P. A. Biologically synthesized silver nanoparticles against pathogenic bacteria: synthesis, calcination and characterization. *Biocatalysis Agric. Biotechnol.* **2019**, *22*, 101373. Available from: https://doi.org/10.1016/j.bcab.2019.101373.
22. Kathiravan, V.; Ravi, S.; Ashokkumar, S. Synthesis of silver nanoparticles from *Melia dubia* leaf extract and their in vitro anticancer activity. *Spectrochimica Acta Part. A: Mol. Biomolecular Spectrosc.* **2014**, *130*, 116–121. Available from: https://doi.org/10.1016/j.saa.2014.03.107.
23. Govindaraju, K.; Tamilselvan, S.; Kiruthiga, V.; Singaravelu, G. Biogenic silver nanoparticles by *Solanum torvum* and their promising antimicrobial activity. *J. Biopesticides* **2010**, *3*, 394.
24. Hatamifard, A.; Nasrollahzadeh, M.; Sajadi, S. M. Biosynthesis, characterization and catalytic activity of an Ag/zeolite nanocomposite for base-and ligand-free oxidative hydroxylation of phenylboronic acid and reduction of a variety of dyes at room

temperature. *N. J. Chem.* **2016**, *40*, 2501–2513. Available from: https://doi.org/10.1039/C5NJ02909K.

25. Karimi, S.; Samimi, T. Green and simple synthesis route of Ag@ AgCl nanomaterial using green marine crude extract and its application for sensitive and selective determination of mercury. *Spectrochimica Acta Part. A: Mol. Biomolecular Spectrosc.* **2019**, *222*, 117216. Available from: https://doi.org/10.1016/j.saa.2019.117216.
26. Bharathi, S.; Kumaran, S.; Suresh, G.; Ramesh, M.; Thangamani, V.; Pugazhvendan, S.; Sathiyamurthy, K. Extracellular synthesis of nanoselenium from fresh water bacteria *Bacillus* sp., and its validation of antibacterial and cytotoxic potential. *Biocatalysis Agric. Biotechnol.* **2020**, *27*, 101655. Available from: https://doi.org/10.1016/j.bcab.2020.101655.
27. Chatterjee, S.; Mahanty, S.; Das, P.; Chaudhuri, P.; Das, S. Biofabrication of iron oxide nanoparticles using manglicolous fungus *Aspergillus niger* BSC-1 and removal of Cr (VI) from aqueous solution. *Chem. Eng. J.* **2020**, *385*, 123790. Available from: https://doi.org/10.1016/j.cej.2019.123790.
28. Tabrez, S.; Khan, A. U.; Mirza, A. A.; Suhail, M.; Jabir, N. R.; Zughaibi, T. A.; Alam, M. Biosynthesis of copper oxide nanoparticles and its therapeutic efficacy against colon cancer. *Nanotechnol. Rev.* **2022**, *11*, 1322–1331. Available from: https://doi.org/10.1515/ntrev-2022-0081.
29. Dou, L.; Zhang, X.; Zangeneh, M. M.; Zhang, Y. Efficient biogenesis of Cu2O nanoparticles using extract of *Camellia sinensis* leaf: evaluation of catalytic, cytotoxicity, antioxidant, and anti-human ovarian cancer properties. *Bioorganic Chem.* **2021**, *106*, 104468. Available from: https://doi.org/10.1016/j.bioorg.2020.104468.
30. Majid, A.; Faraj, H. R. Green Synthesis of copper nanoparticles using aqueous extract of yerba mate (Ilex Paraguarients St. Hill) and its anticancer activity. *Int. J. Nanosci. Nanotechnol.* **2022**, *18*, 99–108.
31. Vinodhini, S.; Vithiya, B. S. M.; Prasad, T. A. A. Green synthesis of palladium nanoparticles using aqueous plant extracts and its biomedical applications. *J. King Saud. University-Science* **2022**, *34*, 102017. Available from: https://doi.org/10.1016/j.jksus.2022.102017.
32. Shejawal, K. P.; Randive, D. S.; Bhinge, S. D.; Bhutkar, M. A.; Todkar, S. S.; Mulla, A. S.; Jadhav, N. R. Green synthesis of silver, iron and gold nanoparticles of lycopene extracted from tomato: their characterization and cytotoxicity against COLO320DM, HT29 and Hella cell. *J. Mater. Science: Mater. Med.* **2021**, *32*, 1–12. Available from: https://doi.org/10.1007/s10856-021-06489-8.
33. Abdel-Aziz, M. M.; Emam, T. M.; Elsherbiny, E. A. Bioactivity of magnesium oxide nanoparticles synthesized from cell filtrate of endobacterium *Burkholderia rinojensis* against *Fusarium oxysporum*. *Mater. Sci. Engineering: C.* **2020**, *109*, 110617. Available from: https://doi.org/10.1016/j.msec.2019.110617.
34. Mishra, R. C.; Kalra, R.; Dilawari, R.; Goel, M.; Barrow, C. J. Bio-synthesis of Aspergillus terreus mediated gold nanoparticle: antimicrobial, antioxidant, antifungal and in vitro cytotoxicity studies. *Materials* **2022**, *15*, 3877. Available from: https://doi.org/10.3390/ma15113877.
35. Zafar, S.; Hasnain, Z.; Aslam, N.; Mumtaz, S.; Jaafar, H. Z.; Wahab, P. E. M.; Qayum, M.; Ormenisan, A. N. Impact of Zn nanoparticles synthesized via green and chemical approach on okra (*Abelmoschus esculentus* L.) growth under salt stress. *Sustainability* **2021**, *13*, 3694. Available from: https://doi.org/10.3390/su13073694.

36. Sundrarajan, M.; Muthulakshmi, V. Green synthesis of ionic liquid mediated neodymium oxide nanoparticles by *Andrographis paniculata* leaves extract for effective biomedical applications. *J. Environ. Chem. Eng.* **2021**, *9*, 104716. Available from: https://doi.org/10.1016/j.jece.2020.104716.
37. Pasquoto-Stigliani, T.; Guilger-Casagrande, M.; Campos, E. V.; Germano-Costa, T.; Bilesky-José, N.; Migliorini, B. B.; Feitosa, L. O.; Sousa, B. T.; De Oliveira, H. C.; Fraceto, L. F. Titanium biogenic nanoparticles to help the growth of *Trichoderma harzianum* to be used in biological control. *J. Nanobiotechnology* **2023**, *21*, 166. Available from: https://doi.org/10.1186/s12951-023-01918-y.
38. Abdullah, M. M.; Atta, A. M.; Allohedan, H. A.; Alkhathlan, H. Z.; Khan, M.; Ezzat, A. O. Green synthesis of hydrophobic magnetite nanoparticles coated with plant extract and their application as petroleum oil spill collectors. *Nanomaterials* **2018**, *8*, 855. Available from: https://doi.org/10.3390/nano8100855.
39. Arts, J. H.; Hadi, M.; Keene, A. M.; Kreiling, R.; Lyon, D.; Maier, M.; Michel, K.; Petry, T.; Sauer, U. G.; Warheit, D. A critical appraisal of existing concepts for the grouping of nanomaterials. *Regulatory Toxicol. Pharmacology* **2014**, *70*, 492–506. Available from: https://doi.org/10.1016/j.yrtph.2014.07.025.
40. Rossi, M.; Cubadda, F.; Dini, L.; Terranova, M.; Aureli, F.; Sorbo, A.; Passeri, D. Scientific basis of nanotechnology, implications for the food sector and future trends. *Trends Food Sci. & Technol.* **2014**, *40*, 127–148. Available from: https://doi.org/10.1016/j.tifs.2014.09.004.
41. OECD, Regulatory Frameworks for Nanotechnology in Foods and Medical Products: Summary Results of a Survey Activity, in: T.a.I.P.P. OECD Science, No. 4 (Ed.), OECD Publishing, Paris, 2013. Available from: https://doi.org/10.1787/5k47w4vsb4s4-en.
42. Amenta, V.; Aschberger, K.; Arena, M.; Bouwmeester, H.; Moniz, F. B.; Brandhoff, P.; Gottardo, S.; Marvin, H. J.; Mech, A.; Pesudo, L. Q. Regulatory aspects of nanotechnology in the agri/feed/food sector in EU and non-EU countries. *Regulatory Toxicol. Pharmacology* **2015**, *73*, 463–476. Available from: https://doi.org/10.1016/j.yrtph.2015.06.016.
43. Rauscher, G.R.H.; Amenta, V.; Boix Sanfeliu, A.; Calzolai, L.; Emons, H.; Gaillard, C.; Gibson, N.; Linsinger, T.; Mech, A.; Quiros Pesudo, L.; Rasmussen, K.; Riego Sintes, J.; Sokull-Kluttgen, B.; Stamm, H. Towards a review of the EC Recommendation for a definition of the term "nanomaterial". Part 1: Compilation of information concerning the experience with the definition, JRC Scientific and Policy Report EUR 26567 EN, European Commission, Joint Research Centre, Institute for Health and Consumer Protection, Luxembourg, 2014. Available from: https://doi.org/10.2788/36237.
44. Hamburg, M. A. FDA's approach to regulation of products of nanotechnology. *Science* **2012**, *336*, 299–300. Available from: https://doi.org/10.1126/science.1205441.
45. FAO/WHO, State of the art on the initiatives and activities relevant to risk assessment and risk management of nanotechnologies in the food and agriculture sectors. FAO/WHO Technical Paper. 2013. http://www.fao.org/docrep/018/i3281e/i3281e.pdf.
46. Bleeker, E. A.; Swart, E.; Braakhuis, H.; Cruz, M. L. F.; Friedrichs, S.; Gosens, I.; Herzberg, F.; Jensen, K. A.; von der Kammer, F.; Kettelarij, J. A. Towards harmonisation of testing of nanomaterials for EU regulatory requirements on chemical safety—a proposal for further actions. *Regulatory Toxicol. Pharmacology* **2023**, *139*, 105360. Available from: https://doi.org/10.1016/j.yrtph.2023.105360.

47. Falkner, R.; Jaspers, N. Regulating nanotechnologies: risk, uncertainty and the global governance gap. *Glob. Environ. Politics* **2012**, *12*, 30–55. Available from: https://doi.org/10.1162/GLEP_a_00096.
48. Picecchi, D. Tiny things with a huge impact: the international regulation of nanomaterials. *Mich. J. Envtl & Admin L* **2017**, *7*, 447. Available from: https://doi.org/10.36640/mjeal.7.2.tiny.
49. Abegunde, S. M.; Afolayan, B. O.; Ilesanmi, T. M. Ensuring sustainable plant-assisted nanoparticles synthesis through process standardization and reproducibility: challenges and future directions—a review. *Sustain. Chem. One World* **2024**, 100014. Available from: https://doi.org/10.1016/j.scowo.2024.100014.
50. Kulkarni, D.; Sherkar, R.; Shirsathe, C.; Sonwane, R.; Varpe, N.; Shelke, S.; More, M. P.; Pardeshi, S. R.; Dhaneshwar, G.; Junnuthula, V. Biofabrication of nanoparticles: sources, synthesis, and biomedical applications. *Front. Bioeng. Biotechnol.* **2023**, *11*, 1159193. Available from: https://doi.org/10.3389/fbioe.2023.1159193.
51. Pini, M.; Rosa, R.; Neri, P.; Bondioli, F.; Ferrari, A. M. Environmental assessment of a bottom-up hydrolytic synthesis of TiO2 nanoparticles. *Green. Chem.* **2015**, *17*, 518–531. Available from: https://doi.org/10.1039/c4gc00919c.
52. de Oliveira, P. F.; Torresi, R. M.; Emmerling, F.; Camargo, P. H. Challenges and opportunities in the bottom-up mechanochemical synthesis of noble metal nanoparticles. *J. Mater. Chem. A* **2020**, *8*, 16114–16141. Available from: https://doi.org/10.1039/D0TA05183G.
53. Silva, L. P.; Reis, I. G.; Bonatto, C. C. Green synthesis of metal nanoparticles by plants: current trends and challenges. *Green Processes for Nanotechnology* **2015**, 259–275. Available from: https://doi.org/10.1007/978-3-319-15461-9_9.
54. Saleh, N.; Yousaf, Z. *Tools and techniques for the optimized synthesis, reproducibility and scale up of desired nanoparticles from plant derived material and their role in pharmaceutical properties. Nanoscale Fabrication, Optimization, Scale-Up and Biological Aspects of Pharmaceutical Nanotechnology;* Elsevier, 2018, 85–131. Available from: https://doi.org/10.1016/B978-0-12-813629-4.00003-6.
55. Barberio, M.; Antici, P. In situ study of nucleation and aggregation phases for nanoparticles grown by laser-driven methods. *Sci. Rep.* **2017**, *7*, 41372. Available from: https://doi.org/10.1038/srep41372.
56. Chopra, H.; Bibi, S.; Singh, I.; Hasan, M. M.; Khan, M. S.; Yousafi, Q.; Baig, A. A.; Rahman, M. M.; Islam, F.; Emran, T. B. Green metallic nanoparticles: biosynthesis to applications. *Front. Bioeng. Biotechnol.* **2022**, *10*, 874742. Available from: https://doi.org/10.3389/fbioe.2022.874742.
57. Mousavi-Kouhi, S. M.; Beyk-Khormizi, A.; Amiri, M. S.; Mashreghi, M.; Hashemzadeh, A.; Mohammadzadeh, V.; Alavi, F.; Mottaghipisheh, J.; Sarafraz Ardakani, M. R.; Taghavizadeh Yazdi, M. E. Plant gel-mediated synthesis of gold-coated nanoceria using ferula gummosa: characterization and estimation of its cellular toxicity toward breast cancer cell lines. *J. Funct. Biomater.* **2023**, *14*, 332. Available from: https://doi.org/10.3390/jfb14070332.
58. Herlekar, M.; Barve, S.; Kumar, R. Plant-mediated green synthesis of iron nanoparticles. *J. Nanopart.* **2014**, *2014*, 140614. Available from: https://doi.org/10.1155/2014/140614.
59. Mughal, B.; Zaidi, S. Z. J.; Zhang, X.; Hassan, S. U. Biogenic nanoparticles: synthesis, characterisation and applications. *Appl. Sci.* **2021**, *11*, 2598. Available from: https://doi.org/10.3390/app11062598.

60. Xu, J. Y.; Zhang, H. Long-term effects of silver nanoparticles on the abundance and activity of soil microbiome. *J. Env. Sci(China)* **2018**, *69*, 3–4. Available from: https://doi.org/10.1016/j.jes.2018.06.009.
61. Rastogi, A.; Zivcak, M.; Tripathi, D.; Yadav, S.; Kalaji, H. Phytotoxic effect of silver nanoparticles in *Triticum aestivum*: improper regulation of photosystem I activity as the reason for oxidative damage in the chloroplast. *Photosynthetica* **2019**, *57*. Available from: https://doi.org/10.32615/ps.2019.019.
62. Saha, N.; Gupta, S. D. Low-dose toxicity of biogenic silver nanoparticles fabricated by Swertia chirata on root tips and flower buds of *Allium cepa*. *J. Hazard. Mater.* **2017**, *330*, 18–28. Available from: https://doi.org/10.1016/j.jhazmat.2017.01.021.
63. Ahmed, S.; Chaudhry, S. A.; Ikram, S. A review on biogenic synthesis of ZnO nanoparticles using plant extracts and microbes: a prospect towards green chemistry. *J. Photochemistry Photobiology B: Biol.* **2017**, *166*, 272–284. Available from: https://doi.org/10.1016/j.jphotobiol.2016.12.011.
64. Kakkar, S.; Dash, S.; Vohra, N.; Saha, S. Engaging employees through effective performance management: an empirical examination. *Benchmarking: An. Int. J.* **2020**, *27*, 1843–1860. Available from: https://doi.org/10.1108/BIJ-10-2019-0440.
65. Sonawane, H.; Shelke, D.; Chambhare, M.; Dixit, N.; Math, S.; Sen, S.; Borah, S. N.; Islam, N. F.; Joshi, S. J.; Yousaf, B. Fungi-derived agriculturally important nanoparticles and their application in crop stress management–Prospects and environmental risks. *Environ. Res.* **2022**, *212*, 113543. Available from: https://doi.org/10.1016/j.envres.2022.113543.
66. Bhagat, D. S.; Gurnule, W. B.; Bumbrah, G. S.; Koinkar, P.; Chawla, P. A. Recent advances in biomedical applications of biogenic nanomaterials. *Curr. Pharm. Biotechnol.* **2023**, *24*, 86–100. Available from: https://doi.org/10.2174/1389201023666220513101628.
67. Soni, V.; Raizada, P.; Singh, P.; Cuong, H. N.; Rangabhashiyam, S.; Saini, A.; Saini, R. V.; Van Le, Q.; Nadda, A. K.; Le, T.-T. Sustainable and green trends in using plant extracts for the synthesis of biogenic metal nanoparticles toward environmental and pharmaceutical advances: a review. *Environ. Res.* **2021**, *202*, 111622. Available from: https://doi.org/10.1016/j.envres.2021.111622.
68. Qidwai, A.; Pandey, A.; Kumar, R.; Shukla, S.; Dikshit, A. Advances in biogenic nanoparticles and the mechanisms of antimicrobial effects. *Indian. J. Pharm. Sci.* **2018**, *80*. Available from: https://doi.org/10.4172/pharmaceutical-sciences.1000398.
69. Bandala, E. R.; Stanisic, D.; Tasic, L. Biogenic nanomaterials for photocatalytic degradation and water disinfection: a review. *Environ. Science: Water Res. & Technol.* **2020**, *6*, 3195–3213. Available from: https://doi.org/10.1039/d0ew00705f.
70. Omran, B. A. Versatile applications of biosynthesized nanoparticles, global safety issues, grand challenges, and future perspectives regarding nanobiotechnology. *Nanobiotechnology: A Multidiscip. Field Sci.* **2020**, 185–221. Available from: https://doi.org/10.1007/978-3-030-46071-6_6.
71. Sharma, A.; Singh, A.; Rajput, V. D.; Minkina, T.; Mandzhieva, S.; Elshikh, M. S., et al. Addressing hidden hunger via improving soil health and crop nutrients through nanofortification. *Nanotechnology Applications and Innovations for Improved Soil Health.* **2024**, 90–108. Available from: https://doi.org/10.4018/979-8-3693-1471-5.ch006.

72. Singh, A.; Rajput, V. D.; Agrawal, S.; Ghazaryan, K.; Minkina, T.; Tawaha, A. R. M. A., et al. Nanoparticles mediated salt stress resilience: a holistic exploration of physiological, biochemical, and nano-omics approaches. *Rev. Environ. Contamination Toxicol.* **2024**, *262*. Available from: https://doi.org/10.1007/s44169-024-00070-4.
73. Margaryan, G., et al. Unveiling the salinity tolerance potential of Armenian Dandur (*Portulaca oleracea* L.) genotypes: enhancing sustainable agriculture and food security. *J. King Saud. Univ.—Sci.* **2024**, *36*, 103332. Available from: https://doi.org/10.1016/j.jksus.2024.103332.
74. Sharma, A., et al. Revolutionizing agricultural sustainability and food security and management to achieve SDGs goals via nanotechnology. *Nanotechnology Applications and Innovations for Improved Soil Health.* **2024**, 276–288. Available from: https://doi.org/10.4018/979-8-3693-1471-5.ch013.
75. Ghazaryan, K.; Verma, A.; Rawat, S.; Rajput, P.; Minkina, T.; Mandzhieva, S., et al. Implementing sustainable nano-bioremediation for emerging pollutants. *Adv. Environ. Eng. green. Technol. book. Ser.* **2024**, 316–332. Available from: https://doi.org/10.4018/979-8-3693-1471-5.ch016.
76. Verma, A.; Singh, A.; Rawat, S.; Rajput, P.; Minkina, T.; Mandzhieva, S., et al. Unveiling nanomaterial-induced toxicity. *Adv. Environ. Eng. green. Technol. book. Ser.* **2024**, 236–253. Available from: https://doi.org/10.4018/979-8-3693-1471-5.ch011.
77. Singh, A.; Sharma, A.; Singh, O.; Rajput, V. D.; Movsesyan, H. S.; Minkina, T., et al. In-depth exploration of nanoparticles for enhanced nutrient use efficiency and abiotic stresses management: present insights and future horizons. *Plant. Stress.* **2024**, 100576. Available from: https://doi.org/10.1016/j.stress.2024.100576.
78. Upadhyay, P., et al. Nano-omics-based abiotic and biotic stresses management. *Harnessing Nano-Omics and Nanozymes for Sustainable Agriculture.* **2024**, 348–371. Available from: https://doi.org/10.4018/979-8-3693-1890-4.ch017.
79. Fathi, A.; Shiade, S. R. G.; Parmoon, G.; Yaghoubian, Y.; Pirdashti, H.; Rajput, V. D., et al. *Bioremediation of Heavy Metals Contaminated Soils Using Nanotechnology;* Elsevier eBooks, 2024611–628. Available from: https://doi.org/10.1016/b978-0-443-21610-7.00004-5.

CHAPTER

3 Beyond science: ethical and societal considerations in the era of biogenic nanoparticles

Simran Choudhury[1], Indrani Bhattacharya[2], Shreni Agrawal[2], Abhishek Singh[3], Vishnu D. Rajput[4], Sakshi Singh[3], Gohar Margaryan[3], Rupesh Kumar Singh[5], Francisco Roberto Quiroz-Figueroa[6], João Ricardo Sousa[4], Henrique Trindade[4], Hassan El-Ramady[7] and Karen Ghazaryan[3]

[1]*School of Life Sciences, Central University of Gujarat, Vadodara, Gujarat, India*

[2]*Department of Life Sciences, Parul Institute of Applied Sciences, Parul University, Waghodia, Gujarat, India*

[3]*Faculty of Biology, Yerevan State University, Yerevan, Armenia*

[4]*Academy of Biology and Biotechnology, Southern Federal University, Rostov-on-Don, Russia*

[5]*Centro de Investigação e Tecnologias Agroambientais e Biológicas (CITAB), UTAD, Vila Real, Portugal*

[6]*Laboratorio de Fitomejoramiento Molecular, Centro Interdisciplinario de Investigación para el Desarrollo Integral Regional-Unidad Sinaloa, Instituto Politécnico Nacional, Guasave, Sinaloa, Mexico*

[7]*Soil and Water Department, Faculty of Agriculture, Kafrelsheikh University, Kafr El-Sheikh, Egypt*

3.1 Introduction

The world has been witnessing the venturing of nanotechnology into the various major sectors of the world. These sectors include food and agriculture, energy, environment, medicine, technology, and even the textiles. Nanotechnology talks about quantities at the scale of 10^{-9} m, which is quite a minute quantity, but this field holds immense potential when studied and analyzed extensively and used judiciously. Nanoparticles, since their advent, have been an interesting science and have been conventionally been synthesized by physical and chemical processes. However, these methods are known to produce toxic and hazardous by-products, while the method itself involves the use of hazardous reagents (in wet chemistry) and is a highly energy consuming process (physical method). The newer synthesis method is much greener, and involves the use of biological substrates and agents. This method constitutes of green chemistry, and the nanoparticles hence produced are known as biogenic nanoparticles. This method is known for its less toxic nature, more biocompatibility, better nanoparticle production,

Biogenic Nanoparticles. DOI: https://doi.org/10.1016/B978-0-443-33819-9.00007-X

and less energy consumption, thus, lowering the production costs. Biogenic nanoparticle synthesis can be done using a range of biological components: plant extracts, fungi, algae, bacteria, etc.[1–3]; there are mentions of examples of such products in this article.

Furthermore, we discuss the applications of biogenic nanoparticles in various fields, such as environment, agriculture, packaging, medicine, etc., which would bring up the reason behind their employment in the same, while subsequently bringing in the need for innovations in the various fields. Biogenic nanoparticles are being immensely researched in the two fields of agriculture and biomedical sciences. The field of agriculture has already seen the incorporation and commercialization of nanoparticles. However, there still remain potential toxicity concerns, and above all there is a need of making agriculture much more sustainable, which is a major goal across the globe. Biogenic nanoparticle-based agricultural components are much safer and better for agricultural applications. In case of biogenic nanomedicines, the antimicrobial and antioxidant properties predominant in biogenic nanoparticles forms the basis of their application.[4–7] Similarly, another innovative avenue is that of sensing and analysis. The utilization of biogenic nanoparticles would improve the functioning and performance of the bioreceptor-nanomatrix system. A particularly unheard nexus is that of nanotechnology-based textile industry, where the biogenic nanoparticles can once again be utilized due to their antimicrobial, strengthening, and biodegradable properties. The applications of biogenic nanoparticles are, thus, multifaceted.

The next subject of utmost importance that has to be discussed in context of this article is that of potential toxicity of biogenic nanoparticles. Biogenic nanoparticles are indeed beneficial and sustainable, yet there is lack of toxicity studies which are essential for using them commercially. Toxicity studies are an essential prerequisite of production and consumption of any product. The interactions of nanoparticles as a whole have not been deciphered fully. Therefore, there is a need of analyzing the interactions of such elements in the environment.[8,9] There also exist inequity in laboratory and production processes, and lack of awareness among the population. Also, certain production processes can be expensive to the general population, especially among populations in low-income regions of the world. Hence, it is essential to take into account the various societal aspects and people's perspective about the same.

3.2 Defining biogenic nanoparticles: what and how in a nutshell

Nanoparticles, as is known, are materials in the size range of 10^{-9} m, which have several superior features, like high surface-to-volume ratio, precision targeting, and novel properties, that are difficult to achieve in their macromolecular counterparts (Fig. 3.1). Nanoparticles can be altered at the level of their atomic and

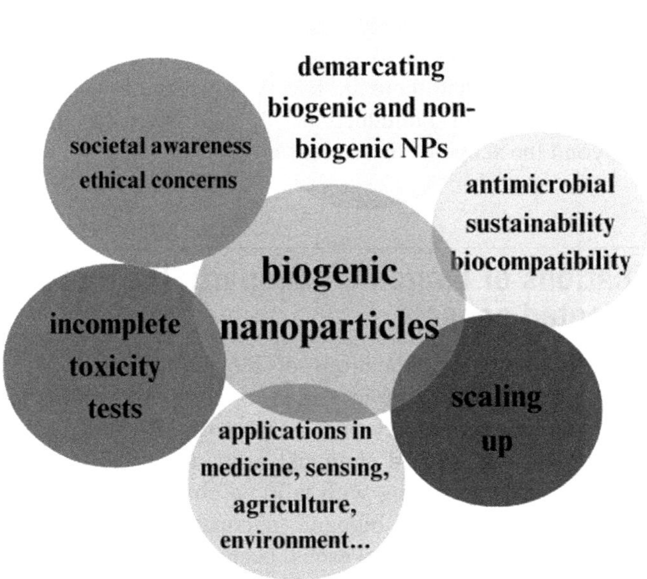

FIGURE 3.1

Application of biogenic nanoparticles in different fields of applied sciences.

molecular structures, thus, inspiring a sense of customization, especially in creating personalized medical agents and, sophisticated and dedicated strategies in environmental monitoring and restoration. Among nanoparticles, there stands a classification which categorizes them into broad classes depending on the synthesis method that has been employed in their production. One such production method is that of biogenic synthesis, where biological components are being employed to synthesize nanoparticles. These biological components can range from plant extracts to microorganisms such as bacteria, yeast, fungi, algae, etc.,[10] and such synthesis processes are considered to be the greener, that is, more eco-friendly and environment sensitive, compared to others.

These nanoparticles that have been synthesized using biological components, or biochemicals, are known as biogenic nanoparticles. The reason for the need of such production processes can be well demonstrated by the synthesis process of selenium nanoparticles. Selenium nanoparticles can be synthesized by either chemical or biogenic method. However, the chemical method employs the reduction of sodium selenite by GSH or glucose, whereas, the biogenic method employs microbes like *Sulfurospirillum* sp. and *Selenihalanaerobacter* sp. to reduce selenate or selenite to amorphous elemental selenium. The chemical synthesis process results in nanoparticles with a varied and large range of size of nanoparticles, while the biogenic synthesis process has much more uniform size range. Also, the silver nanoparticles produced by biogenic processes have unique antimicrobial processes, proving to be pioneers in the field of nanomedicines. The lack of uniformity in the size of the nanoparticles produced can prove to be a limitation in scaling up the production process. It should be noted that, there are two major methods of producing biogenic nanoparticles-biosorption and bioreduction.[11–15] However, such discussions are vast and also beyond the scope of the current article.

3.3 Applications of biogenic nanoparticles and their multifaceted aspects

There exists a multitude of applications of biogenic nanoparticles, which can range from those in the fields of biomedicine to environmental technologies. The reason lies in their unique chemistry and morphology. Biogenic nanoparticles are known to have novel physical and chemical properties. They are also known to have antimicrobial properties, along with tailored characteristics for their specific purpose. Most significantly, these biogenic nanoparticles are biocompatible and nontoxic, with enhanced catalytic properties. Nanoparticles, in general, are known for their precise and smart carrier roles, which can prove to be innovative in the field of medicine as well as agriculture. Drug delivery, imaging, sensing, agricultural products (such as nanofertilizers, nanopesticides, etc.), biomaterials, etc., can all utilize biogenic nanoparticles as their basis. Ecofriendly synthesis is another feather in the cap for biogenic nanoparticles.[16]

3.3.1 Biomedical sciences

In the field of biomedical science, biogenic nanoparticles are seen as innovative means to give a boost to the already established techniques, and also pioneer into new avenues. This ranges from drug delivery systems to tissue engineering. Nanoparticles such as silver nanoparticles (Ag-NPs) are widely known for their antimicrobial properties. When synthesized using biogenic processes, these Ag-NPs are known to have enhanced medicinal effects against pathogens. Some such examples include, *Azadarichta indica* leaf extract–AgNPs having antimicrobial effect on multidrug-resistant pathogen, due to its increased zeta potential, *Leptolyngbya* cell extract–AuNPs (gold nanoparticles) showing specific antibacterial activity against Gram-positive bacteria, CuNPs (copper nanoparticles) exhibiting antibacterial (against *Botrytis subtilis* and *E. coli*) and antifungal activities. Thus, biogenic nanoparticles exhibit a wide range of medicinal effects, including antibacterial, antifungal, and even antiviral activities. AgNPs have also shown antiviral activity against HIV. These AgNPs cause cell lysis, and hence, they prevent viral proliferation. These nanoparticles having biogenic origin also showcase enhanced stability, as is the case with *Mangifera indica* leaf extract fabricated AgNPs. The biogenic nanoparticles are known to increase the surface for interaction thus increasing antimicrobial efficacy. The nanoparticles surround the cell and form agglomerations to prevent cell division and proliferation; gradually, they encapsulate the pathogenic cell, thus, ultimately eliminating the pathogen. The biogenic nanoparticles are much more effective on viruses, compared to their counterparts from other methods. Apart from all such benefits, AgNPs also have antioxidant functions. Biogenic nanoparticles have further been used in cancer therapy. For example, selenium, titanium dioxide, and calcium carbonate nanoparticles have properties like antioxidant and antiinflammatory, ROS generation, apoptosis generating activities in cancer cells. These nanoparticles can specifically target and eliminate tumor cells by inhibiting optimum conditions, having inhibitory impacts, and inducing apoptosis. Such activities prevent angiogenesis. Such innovations are hindered by lack of standardization and lack of studies showing interaction between various nanoparticles and various types of cancer. Apart from this, the nanoparticles are also known to be excellent drug carriers, due to their biocompatibility and precise nature. The nanoparticles indeed gained quite a superiority due to their role as smart drug delivery system, owing to their ability to target specific sites. These nanoparticles have shown enhanced bioavailability, and slow and targeted drug release to enhance treatment efficacy. It especially notable to see their nexus with cancer treatment drugs, where, there are huge possibilities of side effects, due to non-targeted drug administration, that can be improved by the use of such biogenic nanoparticles, and targeting would decrease the side-effects of the same. However, nanoparticles are not just useful in medication, but also in arena of diagnostics. Biogenic nanoparticles, owing to their specificity and biocompatibility have come off to be a safer choice to carry out sensing and monitoring of diseases, conditions, and pathogens. Such analytical techniques are leading the field of medicine into a new era. These are often achieved by the use of nanobiosensors possessing a biomolecule as a bioreceptor, on a nanomatrix, which is connected to a transducer, amplifier, and monitoring system. These nanomatrix component with a bioreceptor is the site of interaction for the target analyte; these interactions induce biochemical energy or potential differences, which, in turn, are converted into electrical signals by the transducer to be further conveyed to the amplifier, which, subsequently, amplifies the signal and sends it to the display unit to display the same into readable form. Such complex networks create a space for precise sensing. In case of diagnostics, it is quite obvious to witness the significance of gold and silver nanoparticles, as these elements have long been known for their association with photothermal processes, since time immemorial, and thus, possess the quality of being used in imaging processes. Further, these biogenic nanoparticles can be used as scaffolds in tissue engineering, thus, lending additional strength to the otherwise fragile tissues, simultaneously bestowing them with cellular growth factors, and hastening the process of tissue regeneration. These biogenic nanoparticles, again owing to their biocompatible nature, would prove to provide an edge over the conventional tissue healing process, by hastening postreconstructive surgery processes or the impacts of regenerative medicine.[16–21]

3.3.2 Agricultural sciences

Another major field of use for biogenic nanoparticles is definitely precision agriculture or precision farming, which not just accounts for food security but also sustainability. Precision agriculture is focused on the use of substances that are known for their ability to integrate modern techniques, technologies, and even data science to create a spatial roadmap in farming, and its environmental impacts. Nanotechnology has definitely been one such science aiding in precision agriculture. Nanopesticides, nanofertilizers, nanoherbicides, nanobiosensors, etc., are all widely known for their novel characteristics, and less hazardous nature compared to their conventional counterparts. The conventional fertilizers are very much prone to leaching, and cause losses in soil nitrogen by 50%–60%, and their volatility increases the cost of production. On the contrary, the use of nanomaterial-based chemicals increases soil fertility, has controlled nutrient release, lesser water use among others.[22] All such advantages make nanoparticles a very tempting choice to integrate them in the agricultural field. Simultaneously, nanobiosensors in the field of agriculture are also a significant innovation, as they work more on the precautionary principles than on the cure principles. The nanobiosensors in agriculture are able to alert the agriculturists about soil health, pH, nutrient content, etc. in real time, and not just report the changes made after chemical application.[23] However, pesticides with biological origin are seen to be much more prospective for agriculture. Thus, nanomaterial-based agricultural chemicals having biological origin are much more favorable ad beneficial. These products are much more stable, show enhanced bioavailability and biocompatibility, and are much more ecofriendly than the chemical synthesized ones. As the world population rises and prepares itself to cross its 9 billion mark, there are significant concerns regarding food security and sustainability of the agricultural process. Sustainable agriculture would go a long way and induce perseverance. Green synthesis of such agricultural components has brought a promise of sustainability. These green alternatives have higher biocompatibility, are more soluble, better transported within the plants, have controlled and prolonged nutrient release aiding in higher crop yield, as compared to the same amount of conventional agricultural chemicals used. The use of biogenic components in agricultural practices improves the overall health of the plant (by enhancing its nutrient absorption efficiency) and soil, as well as keep the ecosystem in consideration. Another problem is that of fungal and other microbial diseases in plants. As has been discussed earlier, various nanoparticles have various antifungal and antimicrobial properties. There also exist nanoparticles used as pesticides. All such formulations are mostly in the form of either foliar or soil spray. Some such examples include nCuO used for foliar application in *Solanum lycopersicum* cultivation, nano-graphene oxide used against *Gryllotalpa orientalis*, ε-caprolactone encapsulated atrazine acting as nanoherbicides. Green synthesis of nanofertilizers is a promising approach not just due to the fact that they are more biocompatible and ecofriendlier but also due to the fact that the resources needed for biogenic synthesis are widely available and those too add novel properties to the formulations, while accounting for their stability, for example, the synthesis of phytochemical-capped nAu from galangal rhizome extract. Apart from these postsowing applications, there also exists nanopriming activities that increase a plants stress tolerance and its germination capacity; examples of such applications include the application of biogenic AgNPs on *Echinops macrochaetus* for saline stress tolerance and priming of *Glycine max* seedlings using AgNPs. Apart from these, the biofortification of crops using biogenic nanoparticles has found its way into agricultural improvement practices, so as to overcome the nutrition deficiency in various sections of the population, who largely depend on the staple food crops for their overall nutritional fulfillment. The utilization of biogenic nanoparticle encapsulated fertilizers and pesticides show slow and targeted release, thus, having prolonged effect. The ecotoxicity is also known to be low, thus, avoiding problems of leaching, eutrophication, and environmental damage.[24–34]

3.3.3 Cosmetics industry

The cosmetic products industry is very high grossing, with varied qualities of products available, with some of them having highly toxic components to others with extremely organic and natural ones. There is a rising concern among the consumers regarding the various chemicals present in the products. These toxic substances can be the actual major component or a preservative for the product. These chemicals can include parabens, triclosan, imidazolidinyl urea, etc., which are used as preservatives in the cosmetic. However, such chemicals can cause problems such as skin allergy, asthma, to DNA damages, to trigger a variety of immunogenic reactions. The use of more natural contents is much more acceptable, with clear declaration on the packaging. Such biomolecules with antimicrobial, bactericidal, and antifungal activity are well desired components in cosmetics. One such nanoparticle is AgNP which is predominantly known for all such properties. AgNPs are preferred for their novel physical and chemical properties. They are used in sunscreens for their unique properties. Biogenic AgNPs are synthesized from a variety of sources which include plant extracts of green tea, *Nelumbo nucifera*, *Iris tuberosa*, etc., which are known for their effective actions against acne, and other skin damages. Integration of such products becomes beneficial to the cosmetics. The common skin microorganisms (both pathogenic and opportunistic pathogenic) include *E. coli*, *Staphylococcus aureus*, *Candida albicans*, etc., which can be acted upon by such biogenic AgNPs. Such antibacterial agents are of broad-spectrum nature. These work by a variety of pathways, such as cell membrane or cell wall damage, DNA damage, or triggering ROS production in the bacterial cell. However, depending on the biogenic origin, the nanoparticles can have varied specificity in their antimicrobial action. For example, the nanoparticles synthesized from *Berberis vulgaris* are much more impactful on Gram-negative bacteria, than Gram-positive bacteria. The antioxidant properties of biogenic nanoparticles act as a plus point in their use as components of various cosmetics, such as skin care products and eye makeup.[35]

3.3.4 Food preservation and packaging

Biogenic nanoparticles are also an innovation in the food preservation and packaging industry. There are two major ways of utilizing biogenic nanoparticles in these industries. They can be either used as preservatives or be used as biomaterials for packaging purposes. Recently, the concept of using nanomaterials as packaging components are being analyzed and seen as a better strategy then the conventional ones. The utilization of biogenic AgNPs is definitely due to its bactericidal and bacteriostatic properties; however, the source of synthesis also matters a lot. The biogenic AgNPs from *Citrus limon* are much more effective than those from *Calendula offcinalis* and *Tagetes patula*, as was observed in preservation of berries (*Rubus fruticosus* and *Vaccinium myrtillus*). Such biogenic AgNPs hasten rotting and increase shelf-life, while preventing postharvest physiological diseases. The ecofriendly and biocompatible nature of biogenic nanoparticles make them better suited to current consumerism trends and regulations. The rise in the awareness among consumers for fresh ready to eat products and lesser use of harmful chemicals and preservatives have made biogenic nanoparticles a suitable candidate in this industry. Although various physical preservation methods (e.g., canning, drying, etc.) are in place, these methods can alter the taste and nutritional value of the produce, which is not desirable at all times. Also, these methods are energy-consuming strategies and, hence, are much pricier. Biogenic nanoparticles solve this problem in a very easy way. Apart from the above-mentioned extracts, AgNPs from *Citrus sinensis, Vitis vinifera, Ocimum tenuiflorum, Centella asiatica*, etc., are also very useful. The use of nanomaterials in the packaging industry is a very recent venture, and a very innovative strategy. Nanomaterials having antimicrobial and antioxidant properties are ideal for such purposes, as they inhibit microbial activity and prolong the shelf life of the produce. This is furthered by the formulation of nanocomposites for use as packaging materials. The polymers, like polyethylene and polypropylene, are used by lining them with a film of biogenic nanoparticles. This results in the food products having increased shelf-life, stability, lessened spoilage and oxidation, while enhancing the structural integrity of the packaging. Although, it is well-known that, AgNPs are the most widely used nanomaterials, yet, in the packaging industry, there exists a variety of other nanomaterials to facilitate the purpose, having better suitability. Such nanomaterials include SiNPs (Silica nanoparticles), ZnONPs (zinc oxide nanoparticles), carbon nanotubes, etc. The prime feature of such nanomaterials is their functioning as a barrier to various gases like carbon dioxide, oxygen, and moisture, and ethanol, which are known to cause rapid spoilage. Furthermore, these nanomaterials can provide protection against microbes such as *Saccharomyces cerevisiae, E. coli, Staphylococcus aureus*, etc., which are commonly responsible for food spoilage. The use of AgNPs in packaging materials for rice and potatoes has shown potential against *Aspergillus flavus*, while the use of ZnONPs has been observed in showing better UV blocking and better sturdy qualities. Additionally, these nanomaterials are biodegradable and ecofriendly, along with being proven to be smart and intelligent packaging system.[16,36] In the current scenario, where plastic packaging has become a nuisance and a prominent environmental contaminant, the utilization of such biodegradable alternatives is a positive step forward and is very much welcome in order to achieve the proposed SDGs (Sustainable Development Goals).

3.3.5 Environmental sciences

The next prominent field to be discussed would definitely be that of environmental monitoring and conservation. The most prominent feature of biogenic nanoparticles is undoubtedly their ecofriendly and biocompatible nature. There is indeed the presence of conventional techniques to achieve the same goal; however, biogenic nanomaterials have emerged to be the much more environmentally concerned strategy. The utilization of biogenic nanomaterials is far and wide in case of environment monitoring and preservation. First and foremost, the use of nanomaterials in sensing technology to detect various environmental contaminants and xenobiotics is a novel strategy. Biogenic nanoparticles are very much suitable to be used in optical, colorimetric, DNA-based, piezometric, thermometric, and other nanobiosensors. The low size dispersion of biogenically synthesized nanoparticles renders them with high surface to volume ratio, which is an essential criteria for the nanomatrix to be used in the development of nanobiosensors. These biogenic nanomatrices are much better suited due to their bioactivity. Such biogenic nanomaterials can easily be used to immobilize enzymes and catalysts on their surface, which is prime to the functioning of nanobiosensors. Also, the development of such nanobiosensors is undergoing, which can not only analyze indoor air quality in real time, but also remove the contaminants from the same. The integration of such innovative strategies with AI and IoT (Internet of Things) would be the future of environmental conservation and pollution control. Apart from these, it is also known that biogenic nanoparticles have high surface to volume ratio, and therefore, enhanced adsorption capability. This adsorption mechanism would prove to be useful for the removal of contaminants, especially from water and soil. Carbon nanotubes and metal oxide nanoparticles have such abilities to adsorb both organic and inorganic contaminants on their surface. These contaminants can include dyes, heavy metals, organic molecules, hazardous elements, etc., which can cause harm to the plants and animals alike, while damaging the ecosystem. Such mechanisms can effectively be used in wastewater treatment, especially before releasing the same into natural water bodies, thus, reducing their toxic impacts on the environment. Often, the waste water is dumped into freshwater bodies, which are significant sources of potable water; utilization of such purification methods would rejuvenate freshwater bodies and increase the availability of potable water to the mankind. Apart from adsorption, biogenic nanoparticles also possess catalytic activity and can perform oxidation of organic molecules, resulting in their further elimination, by their conversion into less toxic forms. These biogenic nanoparticles increase the immobility of the contaminants on their surface, owing to their high surface to volume ratio and catalytic activity and, hence, increase their potential of being used by microbes. Also, many of the nanoparticles do not exhaust their functionality after a single use and, hence, can be reused multiple times, adding to their environment friendly nature and their feasibility.[37–46]

3.3.6 Textile industry

The next field of interest is that of textile industry where there has been observed an integration of biogenic nanoparticles. Biogenic nanoparticles, especially AgNPs, have antimicrobial properties, which is considered as crucial and highly desirable property these days. Fungal invasions are definitely one of the biggest challenges in textile maintenance. There are genera of fungi known for their deterioration of textiles, such as *Chaetomium, Stachybotrys, Rhizopus, Mucor, Aspergillus*, etc.; all such fungi are common nuisance, especially in humid regions. They cause inconvenience in storage. The integration of biogenic nanoparticles would inhibit their growth, due to their antifungal activity. In some cases, such innovations can lead to elimination of over 90% of the pathogens, while ensuring the longevity of the textile. Additionally, these clothes are safer. Also, we know that these nanoparticles can add to the strength of the material. These modified textiles can better withstand the wear and tear, compared to their counterparts. Certain biogenic nanoparticles, like silica nanoparticles, can also confer the fabrics with oil and water resistant properties. Oil resistance can be used for clothes that are usually hard to clean, while water repellent properties are much better suited for clothes that are required to be kept dry for maximum period of time. The unique chemical and physical properties of these biogenic nanoparticles can be used to constitute fabrics that may be pH sensitive and color changing clothes or have special thermal properties for special purposes. Such adaptive and functional clothes are much in need nowadays. These specialized fabrics can also be modified to produce masks and other protective gears. Such fabrics can enhance the resources available to specialized professionals such as doctors, healthcare workers, laboratory personnels, and even students and patients.[47–49] Hazmat suits made with such preventive functions would increase the protection available to the personnels. As has been discussed, some biogenic nanoparticles have capability of radiation blocking; such innovations are going to bring in new functionalities into these avenues. Apart from this, the biodegradability of such gears makes them easier to dispose of, while being environmentally concerned. The disposability of such one-time gears is enhanced by these modulations. The property of antimicrobial and bacteriostatic activity would contain the spread of microbes, or at least deteriorate the same, from such gears. Such beneficial properties are very much beneficial.

Apart from the above-mentioned fields, there also exist several other fields where nanotechnology has been integrated, starting from renewable resources to advanced nanobots. Such discussions can be very much extensive and nonexhaustive and are beyond the scope of this article.

3.4 Getting an edge over others: advantages of biogenic synthesis

Nanoparticles are broadly synthesized by three methods: physical, chemical, and biological. The biological method is also alternatively termed as the green synthesis method, owing to the fact that it generates minimal toxic by-products and impacts, along with a host of other benefits. However, the chemical synthesis method is known to generate a myriad of toxic by-products. And also the biocompatibility of the nanoparticles is reduced. This creates an edge for biogenic synthesis method over its counterparts.

Chemical method of nanoparticle synthesis are wet methods, as they employ a huge range of solutions and reagents in their aqueous phase. As is known, there are various components involved in any chemical reaction such as substrates, stabilizers, and solvents. Here, most of these components are toxic chemicals and are labeled as hazardous. Till date most of these wet methods use toxic organic chemicals. However, it is to be noted that ambient room temperature synthesis is preferred over the high temperature ones. But, those room temperature synthesis methods are known to employ hazardous chemicals such as hydrazine, dimethylformamide, and sodium borohydride, thus, producing toxic by-products, and rendering the production process hazardous. Also, such toxic by-products may remain on the surface of the nanoparticle produced.

Furthermore, the employment of such synthesis methods requires sorption (both chemical and physical) of ligands on the surface of the nanoparticle, thus, making them harder to precipitate and lesser water soluble. This in turn reduces their biocompatibility. The utilization of nanoparticles in living systems requires them to be water soluble, to have properties for easy fine tuning in terms of surface chemistry, and to be biocompatible. These aspects are compromised in chemical methods of synthesis.

Owing to the above discussion, it may seem that the alternative to such production methods can be the use of physical methods for the production of these nanoparticles. However, the employment of physical methods require conditions like high voltage electric fields, high temperatures, etc.; such conditions lead to higher energy consumption and, hence, higher cost of production involved.[50] Thus, causing issues with the feasibility of the process, especially in their commercialization and industrialization phase. Scaling up of the production process thus, becomes harder, impacting their affordability in the market.

Biogenic synthesis of nanoparticles overcomes most of these limitations. Apart from biocompatibility and low toxicity, another important aspect is that of low dispersity or low size dispersion. This property is essential in the synthesis process due to the fact that it aids in low energy and reagent consumption, thus, increasing the overall yield of the process. In a much more precise way, low size dispersion refers to an increased "surface to volume ratio," which in turn increases the surface area available for interaction on the nanoparticle, hence, facilitating the purpose of their creation. Smaller particle size has also shown to be related to better catalytic property of the nanoparticle, and lower chances of aggregation, thus, maintaining the efficiency of the nanoparticles synthesized. If the nanoparticles formed aggregate, they would exhibit properties that are different from their originally desired physical and chemical properties at the nanoscale level. Therefore, to avoid such undesirable results, it is essential to achieve a low size dispersion. Furthermore, if the nanoparticles are being synthesized for nanobiosensor formulation, it is much more desirable to achieve a higher surface to volume ratio, so that the nanoparticles can be used in the formation of the nanomatrix holding the bioreceptor, and hence, providing better surface area for interaction and biochemical reactions to take place. Nanoparticles synthesized with greater surface to volume ratio are essential for the efficient functioning of nanosensors, especially electrical and optical.[29,50–52]

Green synthesis or biogenic synthesis is a way of eliminating the hazardous elements from the synthesis process of nanoparticles. These methods are also termed as "biosynthetic methods or biogenic synthetic method." These are proven to be useful in the creation of bespoke nanoparticles that are designed for specific purposes. These nanoparticles, produced using green chemistry or green synthetic methods, have greater biodegradability, lower toxicity, and higher water solubility, rendering them much more suitability, biocompatibility, and ecofriendly characteristics for working inside living systems and the environment. Apart from this, the use of biochemicals act as capping agents for nanoparticles, enhancing their catalytic activity.

Capping of nanoparticle refers to the adsorption of certain chemicals on the surface of the nanoparticle. This process induces stability to the nanoparticle. Hence, they are also known as stabilizers or capping agents. These capping agents can include anything, from biochemicals to surfactants to polymers. All these components shape the molecule and maintain its stability, integrity, and catalytic properties. The capping agent used also determines the final morphology of the nanoparticle. These agents aid in maintaining the colloidal properties of the nanoparticles. If the capping agent used is a biological molecule, such as proteins, polysaccharide, or any other biochemical, then the biocompatibility of the nanoparticle increases. Also, such biomolecule capped nanoparticle is better suited for biomedical purposes.[53–55]

It is very well known, that, nanoparticles are precise and smart molecules. These properties can be very well exploited in the biomedical and pharmaceutical industry, and utilized for drug delivery mechanisms. These nanoparticles when synthesized biosynthetically would possess better biocompatibility and less toxicity, thus, making them cut-out for such applications. These can also aid in diagnosis by performing sensing and imaging. The modulations in factors such as pH, temperature, substrates, etc., can result in specificity and desired characteristics in the nanoparticles.

Due to abundance and variation in the biological resources, and their easier modulation, it is much more suitable to synthesize nanoparticles using biogenic synthetic methods, not to mention that biogenic processes of synthesis are much simpler in nature. Thus, such methods have an edge over their other counterparts.[56–58]

3.5 Challenges of biogenic nanoparticles: potential toxicity, ethics, and regulations

Although there are so many advantages of the synthesis process and the biogenic nanoparticle itself, yet there exists so many challenges to the entire concept. Some of these challenges are related to the synthesis process, its scale up, some related to their applications, while the rest of them are related to their acceptance, demand, and feasibility in the market. Of all such challenges, some are being discussed in this article.

Nanoparticles, as is known, can be synthesized by three broad methods: physical, chemical, and biological methods. Among all these methods, the former two have been analyzed and studied much more extensively than those nanoparticles that are being produced biogenically. The lack of toxicity studies for green nanoparticles has posed to be a significant hurdle in commercialization and wide spread use. One issue is that the biological sources being utilized for synthesizing the nanoparticles may contain harmful toxic substances. An example is that of biogenic nanoparticle synthesis using algae. Algae contain a host of pigments and proteins, some of which are toxic to humans and other organisms. If nanoparticles are synthesized using the extract of such algae, then the toxins would also be incorporated. Upon using such nanoparticles, there would be possibility of toxicity in the consumers. Also, there is a lack of standardized processes for toxicity testing and replication of the experiments to check their accuracy and significance. The lack of statistical data regarding the toxicity testing is surely a major issue. However, in the case of nanomedicines, there do exist certain guidelines. This is attributed to the fact that, the major parameter in testing the toxicity of the nanomedicines is their dosage. The dosage quantity of nanomedicines determines their toxicity, in general terms. As is necessary with any other therapeutic drugs, nanomedicines must also undergo clinical trials and standardized processes for understanding their pros and cons on the patient. In order to determine the same, the patients are administered with the required therapeutic doses and then undergo toxicokinetic and pharmacokinetic profiling. A relief is that there are standardized and well-established immunological tests available for therapeutic drugs, as prescribed by WHO. However, there still remains uncertainties and questions regarding various aspects such as the fate of the nanocarrier and the presence of endotoxins in the nanocarrier that may generate immunogenic responses (the synthesis of biogenic nanoparticles from microbial sources may cause the incorporation of endotoxins in the nanoparticle; the content of such endotoxins in the nanoparticle has to be determined prior to the trial and administration procedure). All such loopholes must be filled at first, before commercialization. Many other such shortcomings and many solutions have been discussed in the REFINE project. The Refine Nanomed Regulatory Science Framework has been specifically formulated for nanomedicines and their toxicity, along with issues in their commercialization.

The hurdles in biological source determination can be very easily solved by the use of fruit and vegetable waste. Such wastes have a variety of antioxidants and beneficial biochemicals that can be utilized in biogenic nanoparticle synthesis. Another benefit is the lack of need of processing these sources, while reducing issues of presence of toxic substances, and maximizing the various beneficial properties, as can be seen in the aforementioned example of nanoparticle synthesis from *Citrus limon*.

However, an issue that remains is that, there are evidences of accumulation of nanoparticles in certain specific organs in model organisms. The most prominent system being impacted is the respiratory system, and the organ with maximum nanoparticle accumulation are the lungs. Some other organs, where major concentrations of nanoparticles were found, are the kidneys, the spleen, and the liver. Such high concentrations can lead to organ damage and failure.

The next challenge at hand is that of release of unwanted nanoparticles in the environment. One such example is that of silver nanoparticles being undesirably and unknowingly released into the aquatic environment. As has been discussed previously, AgNPs possess antimicrobial functions. Upon being released into the aquatic environment, these AgNPs become the source of unwanted dissolved silver particles that can cause damage to the microflora of the aquatic environment and, hence, deteriorate the ecosystem. Also, as is the case of biogenic nanoparticle containing textiles, over prolonged use, it is very much possible that the nanoparticles may start getting released from the apparel, especially during the regular washing process and, hence, contaminate the environment. Not just apparels but any product concerning nanoparticles may lose their stability as they age and lead into developing toxic properties. Consumption of such products would lead to problems such as cell and tissue damage, ROS activation, and DNA damage among the consumers. The intentional and unintentional release of nanoparticles in the environment is certainly a cause of concern. The nanoparticles are known to form aggregates in hardwater and seawater, owing to their chemical nature. The reactions that can take place in such conditions are not contemplated completely, as of yet, thus, determining the ecotoxicity in such conditions is also crucial.

Nanoparticles are extremely minute in size; hence, can easily cross living system components such as blood and brain barrier, and lymphatic tissues; and also, pass through the circulatory system. Thus, it is very much advisable to handle nanoparticles with utmost precaution. The nanoparticles can very conveniently enter the living system through various administration processes, such as through inhalation, ingestion, parenteral administration, and absorption. These nanoparticles upon entering the living body can cause oxidative stress, inflammation, apoptosis, oncogenesis, cytotoxicity, etc., along with causing changes to the cell transduction pathways and biosynthetic pathways.

As we talk about the process of biogenic synthesis, it is very well understood that the use of green and much more biocompatible formulations would lead to better health and sustainability. Therefore, the utilization of organic biomolecules, instead of metal ions, would be much more suitable. The interaction of biogenic nanomaterials with reducing agents would cause the nanoparticle formulation to become unstable.

The studies of toxicity in nanoparticles have been conducted quite adequately for nanoparticles but not for biogenic nanoparticles. There is lack of holistic knowledge. Moreover, there is a need to proportionate the experimental and laboratory production to that of the industrial level, by devising scale-up methods, which is a major challenge as of now. Also, as we talk about increasing the scale of utilization of nanoparticles in various sectors, it is essential to take into account that there do exist performance limits and manufacturing limits. Thus, estimating the scalability of the manufacturing process is crucial. These are further pushed down due to the lack of proper guidelines. While considering the use of any product, it is necessary to have proper demarcated and well-established rules and regulations. This is because of the fact that otherwise there would be chances of exploitation and abuse of the same.[18,19,25,59–65]

3.6 Conclusion: the perception of biogenic nanoparticles and future prospects

The changing times bring with them new technology as well as new perspectives. Few decades back, nanotechnology was just a thought for researchers, while today it has become a reality that is expanding its horizons by leaps and bounds.[66] It has transformed from a period where it only employed physical and chemical methods of nanoparticle synthesis to an era where nanoparticles are being synthesized from biological sources. These changes have also brought along themselves a myriad of path breaking changes.[67] These changes can be the employment of biogenic nanoparticles in sectors like medicine and textiles to their employment in sensing technology to their employment in removing environmental contaminants and environment conservation.[68,69] Such dynamics have definitely been noticed by the various sections of the population, resulting in enriching their knowledge regarding the field of nanotechnology. Nanotechnology, however, is not a detached field, rather it is very much integrated into biological sciences. One such nexus is that of nanobiotechnology, which is definitely a very much sought after field nowadays. The biogenic synthesis of nanoparticles with the help of technology, definitely does fall into this school.

Many people across the various sections of the society may not be aware of such developments, and it is very safe to say that large sections of the society are still in the dark. Most of such people belong to certain specific classes, such as these belonging to underprivileged classes and those who simply do not deal with such issues on a day-to-day basis. Thus, if nanotechnology is to be brought into the daily lives of the people, then the first step would definitely be sensitization.[70] Sensitization would not only create awareness regarding the existence and benefits of nanotechnology-based products but would also create a demand–supply and feedback–improvement system for the same products, opening the path for a new industry in the open market.

It Is necessary that we admit to the fact that there is lack of consumer awareness regarding this field and that, in order to educate people regarding the same, we have to explain them the origins, benefits, and limitations of this technology. There are so many beneficial biogenic nanoparticle-based products, but there is rarely awareness regarding the same due to lack of knowledge. Apparels with integrated biogenic nanoparticles are one such unknown wonder. These textiles can however be much more expensive those that are actually be used by the people. Economic disparity is also a cause behind the same.

Lastly, it is not necessary that sensitization will create acceptance. Nanoparticles might not still be completely accepted, due to various bioethics concerns and people's disbelief. A very important step in ensuring the commercialization of nanoparticles is that there has to be the existence of extensive data and results regarding the toxicological studies and environment interaction of the nanoparticle. This would create chances of easier acceptance, of this technology, and hence, practical eradication of various issues, where this tech would not just be a laboratory science but a practical production method. Thus, the analysis of Ecotoxicology and the fate of the product after prolonged use are the two aspects that need to be studied with urgency.

Acknowledgments

Higher education and science committee MESCS Republic of Armenia provided the 24FP-4C013 grant to support KG. VDR acknowledge the support by the Strategic Academic Leadership Program of the Southern Federal University ("Priority 2030").

References

1. Ingale, A. G. Biogenic synthesis of nanoparticles and potential applications: an eco-friendly approach. *J. Nanomedicine Nanotechnol.* **2013**, *04* (02). Available from: https://doi.org/10.4172/2157-7439.1000165.
2. Kumari, S.; Verma, R.; Chauhan, A.; Raja, V.; Kumari, S.; Kulshrestha, S. Biogenic approach for synthesis of nanoparticles via plants for biomedical applications: a review. *Mater. Today Proc.* **2023**. Available from: https://doi.org/10.1016/j.matpr.2023.04.242. pp. S2214785323021387.

3. Pooja, B.; Joginder, S. D.; Suresh, K. G. Biogenesis of nanoparticles: a review. *Afr. J. Biotechnol.* **2014**, *13* (28), 2778–2785. Available from: https://doi.org/10.5897/AJB2013.13458.
4. Basiuk, V. A.; Basiuk, E. V., Eds. *Green Processes for Nanotechnology: From Inorganic to Bioinspired Nanomaterials;* Springer International Publishing: Cham, 2015. Available from: https://doi.org/10.1007/978-3-319-15461-9.
5. Khandel, P.; Yadaw, R. K.; Soni, D. K.; Kanwar, L.; Shahi, S. K. Biogenesis of metal nanoparticles and their pharmacological applications: present status and application prospects. *J. Nanostructure Chem.* **2018**, *8* (3), 217–254. Available from: https://doi.org/10.1007/s40097-018-0267-4.
6. Qidwai, A.; Pandey, A.; Kumar, R.; Shukla, S. K.; Dikshit, A. Advances in biogenic nanoparticles and the mechanisms of antimicrobial effects. *Indian J. Pharm. Sci.* **2018**, *80* (4). Available from: https://doi.org/10.4172/pharmaceutical-sciences.1000398.
7. Singh, P.; Garg, A.; Pandit, S.; Mokkapati, V.; Mijakovic, I. Antimicrobial effects of biogenic nanoparticles. *Nanomaterials* **2018**, *8* (12), 1009. Available from: https://doi.org/10.3390/nano8121009.
8. Al-Hakkani, M. F. Biogenic copper nanoparticles and their applications: a review. *SN Appl. Sci.* **2020**, *2* (3), 505. Available from: https://doi.org/10.1007/s42452-020-2279-1.
9. Popescu, M.; and Velea, A. Biogenic production of nanoparticles. Digest J. Nanomat. Biostruct. 2010, 5, 1035–1040.
10. Dikshit, P., et al. Green Synthesis of metallic nanoparticles: applications and limitations. *Catalysts* **2021**, *11* (8), 902. Available from: https://doi.org/10.3390/catal11080902.
11. Jain, R.; Gonzalez-Gil, G.; Singh, V.; Hullebusch, E.D. V.; Farges, E.D. V.; Lens, P. N. L. Biogenic Selenium nanoparticles: production, characterization and challenges, 2014, Unpublished. Available from: https://doi.org/10.13140/2.1.4914.8489.
12. Olubukola Ajoke, A. R. Biogenic synthesis of silver nanoparticles and its synergistic antimicrobial potency: an overview. *J. Appl. Biotechnol. Bioeng.* **2019**, *6* (1), 22–28. Available from: https://doi.org/10.15406/jabb.2019.06.00170.
13. Simões, M.; Ottoni, C.; Antunes, A. Biogenic metal nanoparticles: a new approach to detect life on mars? *Life* **2020**, *10* (3), 28. Available from: https://doi.org/10.3390/life10030028.
14. Ying, S., et al. Green synthesis of nanoparticles: current developments and limitations. *Environ. Technol. Innov.* **2022**, *26*, 102336. Available from: https://doi.org/10.1016/j.eti.2022.102336.
15. Zambonino, M. C.; Quizhpe, E. M.; Mouheb, L.; Rahman, A.; Agathos, S. N.; Dahoumane, S. A. Biogenic selenium nanoparticles in biomedical sciences: properties, current trends, novel opportunities and emerging challenges in theranostic nanomedicine. *Nanomaterials* **2023**, *13* (3), 424. Available from: https://doi.org/10.3390/nano13030424.
16. Kumar, A., et al. Biogenic metallic nanoparticles: biomedical, analytical, food preservation, and applications in other consumable products. *Front. Nanotechnol.* **2023**, *5*, 1175149. Available from: https://doi.org/10.3389/fnano.2023.1175149.
17. Ambrin, G.; Kausar, H.; Naz, R.; Ahmad, A. *Current Status of Nanosensors in Biological Sciences.* In *Nanobiosensors for Agricultural, Medical and Environmental Applications;* Mohsin, M., Naz, R., Ahmad, A., Eds.; Springer Singapore: Singapore, 2020; pp 15–41. Available from: https://doi.org/10.1007/978-981-15-8346-9_2.
18. Chaturvedi, V. K., et al. Biosynthesized nanoparticles: a novel approach for cancer therapeutics. *Front. Med. Technol.* **2023**, *5*, 1236107. Available from: https://doi.org/10.3389/fmedt.2023.1236107.

19. Patil, S.; Chandrasekaran, R. Biogenic nanoparticles: a comprehensive perspective in synthesis, characterization, application and its challenges. *J. Genet. Eng. Biotechnol.* **2020**, *18* (1), 67. Available from: https://doi.org/10.1186/s43141-020-00081-3.
20. Puri, A., et al. From nature to nanotechnology: the interplay of traditional medicine, green chemistry, and biogenic metallic phytonanoparticles in modern healthcare innovation and sustainability. *Biomed. Pharmacother.* **2024**, *170*, 116083. Available from: https://doi.org/10.1016/j.biopha.2023.116083.
21. Uthra, C., et al. Exploring biogenic microbial synthesis and biomedical applications of silver nanoparticles: a comprehensive review. *Results Chem.* **2024**, *8*, 101555. Available from: https://doi.org/10.1016/j.rechem.2024.101555.
22. Vijayakumar, M. D.; Surendhar, G. J.; Natrayan, L.; Patil, P. P.; Ram, P. M. B.; Paramasivam, P. Evolution and recent scenario of nanotechnology in agriculture and food industries. *J. Nanomater.* **2022**, *2022*, 1–17. Available from: https://doi.org/10.1155/2022/1280411.
23. Singh, R. P. *Recent trends, prospects, and challenges of nanobiosensors in agriculture*. In Concepts and Strategies in Plant SciencesIn *Biosensors in Agriculture: Recent Trends and Future Perspectives;* Pudake, R. N., Jain, U., Kole, C., Eds.; Springer International Publishing: Cham, 2021; pp 3–13. Available from: https://doi.org/10.1007/978-3-030-66165-6_1.
24. Ahmad, S., et al. Rice seeds biofortification using biogenic ıron oxide nanoparticles synthesized by using Glycyrrhiza glabra: a study on growth and yield ımprovement. *Sci. Rep.* **2024**, *14* (1), 12368. Available from: https://doi.org/10.1038/s41598-024-62907-1.
25. Bano, A.; Gupta, A.; Sharma, S.; Sharma, R. *Recent developments in nanocarrier-based nutraceuticals for therapeutic purposes*. In *Biogenic Nano-Particles and their Use in Agro-ecosystems;* Ghorbanpour, M., Bhargava, P., Varma, A., Choudhary, D. K., Eds.; Springer Singapore: Singapore, 2020; pp 371–391. Available from: https://doi.org/10.1007/978-981-15-2985-6_20.
26. Biogenic. Synthesis of nanoparticles for sustainable crop production: a review. *Lett. Appl. NanoBioScience* **2022**, *12* (1), 10. Available from: https://doi.org/10.33263/LIANBS121.010.
27. Chauhan, A.; Anand, J.; Parkash, V.; Rai, N. Biogenic synthesis: a sustainable approach for nanoparticles synthesis mediated by fungi. *Inorg. Nano-Met. Chem.* **2023**, *53* (5), 460–473. Available from: https://doi.org/10.1080/24701556.2021.2025078.
28. Forini, M. M. L., et al. Nano-enabled weed management in agriculture: from strategic design to enhanced herbicidal activity. *Plant Nano Biol* **2022**, *1*, 100008. Available from: https://doi.org/10.1016/j.plana.2022.100008.
29. Gözde, K. A.; Dulta, K.; Chauhan, P.; Chauhan, P. K. *Biomimetic material-based biosensor for environmental monitoring*. In *Nanobiosensors for Environmental Monitoring;* Singh, R. P., Ukhurebor, K. E., Singh, J., Adetunji, C. O., Singh, K. R., Eds.; Springer International Publishing: Cham, 2022; pp 191–202. Available from: https://doi.org/10.1007/978-3-031-16106-3_10.
30. Guilger-Casagrande, M.; Bilesky-José, N.; Sousa, B. T.; Oliveira, H. C.; Fraceto, L. F.; Lima, R. Effects of biogenic silver and iron nanoparticles on soybean seedlings (Glycine max). *BMC Plant Biol.* **2022**, *22* (1), 255. Available from: https://doi.org/10.1186/s12870-022-03638-1.

31. Khan, S., et al. Promoting effect of biogenic AgNPs on *Echinops macrochaetus* for improving growth parameters and mitigating the toxic effect caused by salinity stress. *Biotechnol. Biotechnol. Equip.* **2024**, *38* (1), 2354710. Available from: https://doi.org/10.1080/13102818.2024.2354710.
32. Nanopriming Approach to Sustainable Agriculture*;* In *Advances in Environmental Engineering and Green Technologies;* Singh, A.; Rajput, V. D.; Ghazaryan, K.; Gupta, S. K.; Minkina, T.; Singh, A.; Rajput, V. D.; Ghazaryan, K.; Gupta, S. K.; Minkina, T., Eds. *;* IGI Global, 202310.4018/978-1-6684-7232-3.
33. Verma, K. K., et al. Recent Trends in nano-fertilizers for sustainable agriculture under climate change for global food security. *Nanomaterials* **2022**, *12* (1), 173. Available from: https://doi.org/10.3390/nano12010173.
34. Yin, J.; Su, X.; Yan, S.; Shen, J. Multifunctional nanoparticles and nanopesticides in agricultural application. *Nanomaterials* **2023**, *13* (7), 1255. Available from: https://doi.org/10.3390/nano13071255.
35. Mondéjar-López, M., et al. Biogenic silver nanoparticles from *Iris tuberosa* as potential preservative in cosmetic products. *Molecules* **2021**, *26* (15), 4696. Available from: https://doi.org/10.3390/molecules26154696.
36. Rodino, S.; Butu, M.; Butu, A. Application of biogenic silver nanoparticles for berries preservation. *Digest J. Nanomat. Biostruct.* 2019, 14, 601–606.
37. Azeez, L.; Lateef, A.; Olabode, O. An overview of biogenic metallic nanoparticles for water treatment and purification: the state of the art. *Water Sci. Technol.* **2023**, *88* (4), 851–873. Available from: https://doi.org/10.2166/wst.2023.255.
38. Goutam, S. P.; Saxena, G. *Biogenic nanoparticles for removal of heavy metals and organic pollutants from water and wastewater: advances, challenges, and future prospects. Bioremediation for Environmental Sustainability;* Elsevier, 2021623–-636*https://doi.org/10.1016/B978-0-12-820524-2.00025-0.*
39. Kulkarni, M. B.; Ayachit, N. H.; Aminabhavi, T. M. Recent advancements in nanobiosensors: current trends, challenges, applications, and future scope. *Biosensors* **2022**, *12* (10), 892. Available from: https://doi.org/10.3390/bios12100892.
40. Kumari, S.; Tyagi, M.; Jagadevan, S. Mechanistic removal of environmental contaminants using biogenic nano-materials. *Int. J. Environ. Sci. Technol.* **2019**, *16* (11), 7591–7606. Available from: https://doi.org/10.1007/s13762-019-02468-3.
41. Mandeep.; Shukla, P. Microbial nanotechnology for bioremediation of industrial wastewater. *Front. Microbiol.* **2020**, *11*, 590631. Available from: https://doi.org/10.3389/fmicb.2020.590631.
42. Mohamed, E. F.; Awad, G. *Development of nano-sensor and biosensor as an air pollution detection technique for the foreseeable future. Comprehensive Analytical Chemistry*, 99. Elsevier, 2022163–188*https://doi.org/10.1016/bs.coac.2021.11.003.*
43. Mughal, B.; Zaidi, S. Z. J.; Zhang, X.; Hassan, S. U. Biogenic nanoparticles: synthesis, characterisation and applications. *Appl. Sci.* **2021**, *11* (6), 2598. Available from: https://doi.org/10.3390/app11062598.
44. Pramanik, P.; Krishnan, P.; Maity, A.; Mridha, N.; Mukherjee, A.; Rai, V. *Application of nanotechnology in agriculture, in* environmental nanotechnology. Volume 4 32In *Environmental Chemistry for a Sustainable World;* Dasgupta, N., Ranjan, S., Lichtfouse, E., Eds.; 32; Springer International Publishing: Cham, 2020; pp 317–348. Available from: https://doi.org/10.1007/978-3-030-26668-4_9.

45. Rai, M.; Duran, N., Eds. *Metal Nanoparticles in Microbiology;* Springer Berlin Heidelberg: Berlin, Heidelberg, 2011. Available from: https://doi.org/10.1007/978-3-642-18312-6.
46. Riaz, M., et al. Synthesis of Biogenic silver nanocatalyst and their antibacterial and organic pollutants reduction ability. *ACS Omega* **2022**, *7* (17), 14723–14734. Available from: https://doi.org/10.1021/acsomega.1c07365.
47. Konsa, K.; Kormpaki, T.; Turu, J. *Biological damage to textiles and prevention methods*. In *Handbook of Museum Textiles;* Jose, S., Thomas, S., Pandit, P., Pandey, R., Eds.; . 1st ed. Wiley, 2022; pp 23–43. Available from: https://doi.org/10.1002/9781119983439.ch2.
48. Saleem, H.; Zaidi, S. Sustainable use of nanomaterials in textiles and their environmental impact. *Materials* **2020**, *13* (22), 5134. Available from: https://doi.org/10.3390/ma13225134.
49. Shah, M. A.; Pirzada, B. M.; Price, G.; Shibiru, A. L.; Qurashi, A. Applications of nanotechnology in smart textile industry: a critical review. *J. Adv. Res.* **2022**, *38*, 55–75. Available from: https://doi.org/10.1016/j.jare.2022.01.008.
50. Mondal, S.; Basu, S.; Begum, N. A.; Mandal, D. A brief introduction to the development of biogenic synthesis of metal nanoparticles. *J. Nano Res.* **2014**, *27*, 41–52. Available from: https://doi.org/10.4028/www.scientific.net/JNanoR.27.41.
51. Malik, P.; Katyal, V.; Malik, V.; Asatkar, A.; Inwati, G.; Mukherjee, T. K. Nanobiosensors: concepts and variations. *ISRN Nanomater.* **2013**, 1–9. Available from: https://doi.org/10.1155/2013/327435.
52. Woo Cho, H.; Nam, S.; Lim, S.; Kim, D.; Kim, H.; June Sung, B. Effects of size and interparticle interaction of silica nanoparticles on dispersion and electrical conductivity of silver/epoxy nanocomposites. *J. Appl. Phys.* **2014**, *115* (15), 154307. Available from: https://doi.org/10.1063/1.4871669.
53. Duval, R. E.; Gouyau, J.; Lamouroux, E. Limitations of recent studies dealing with the antibacterial properties of silver nanoparticles: fact and opinion. *Nanomaterials* **2019**, *9* (12), 1775. Available from: https://doi.org/10.3390/nano9121775.
54. Javed, R.; Zia, M.; Naz, S.; Aisida, S. O.; Ain, N. U.; Ao, Q. Role of capping agents in the application of nanoparticles in biomedicine and environmental remediation: recent trends and future prospects. *J. Nanobiotechnology* **2020**, *18* (1), 172. Available from: https://doi.org/10.1186/s12951-020-00704-4.
55. Javed, R.; Sajjad, A.; Naz, S.; Sajjad, H.; Ao, Q. Significance of capping agents of colloidal nanoparticles from the perspective of drug and gene delivery, bioimaging, and biosensing: an insight. *Int. J. Mol. Sci.* **2022**, *23* (18), 10521. Available from: https://doi.org/10.3390/ijms231810521.
56. Álvarez-Chimal, R.; Arenas-Alatorre, J. Á. *Green synthesis of nanoparticles: a biological approach.* In *Green Chemistry for Environmental Sustainability - Prevention-Assurance-Sustainability (P-A-S) Approach;* Shah, K. J., Ed.; IntechOpen, 2023. Available from: https://doi.org/10.5772/intechopen.1002203.
57. Karunakaran, G.; Sudha, K. G.; Ali, S.; Cho, E.-B. Biosynthesis of nanoparticles from various biological sources and its biomedical applications. *Molecules* **2023**, *28* (11), 4527. Available from: https://doi.org/10.3390/molecules28114527.
58. Pandit, C., et al. Biological agents for synthesis of nanoparticles and their applications. *J. King Saud Univ.—Sci.* **2022**, *34* (3), 101869. Available from: https://doi.org/10.1016/j.jksus.2022.101869.

59. De Jong, W. H.; Geertsma, R. E.; Borchard, G. Regulatory safety evaluation of nanomedical products: key issues to refine. *Drug Deliv. Transl. Res.* **2022**, *12* (9), 2042–2047. Available from: https://doi.org/10.1007/s13346-022-01208-4.
60. Egbuna, C., et al. Toxicity of nanoparticles in biomedical application: nanotoxicology. *J. Toxicol.* **2021**, *2021*, 1–21. Available from: https://doi.org/10.1155/2021/9954443.
61. Jagadeesh, P.; Rangappa, S. M.; Siengchin, S. Advanced characterization techniques for nanostructured materials in biomedical applications. *Adv. Ind. Eng. Polym. Res.* **2024**, *7* (1), 122–143. Available from: https://doi.org/10.1016/j.aiepr.2023.03.002.
62. Khan, F.; Karimi, M. N.; Khan, O. Exploring the scalability and commercial viability of biosynthesized nanoparticles for cooling panels with the help of Artificial Intelligence and solar energy systems. *Green Technol. Sustain.* **2023**, *1* (3), 100036. Available from: https://doi.org/10.1016/j.grets.2023.100036.
63. Khan, I.; Saeed, K.; Khan, I. Nanoparticles: properties, applications and toxicities. *Arab. J. Chem.* **2019**, *12* (7), 908–931. Available from: https://doi.org/10.1016/j.arabjc.2017.05.011.
64. Ottoni, C. A.; Lima Neto, M. C.; Léo, P.; Ortolan, B. D.; Barbieri, E.; De Souza, A. O. Environmental impact of biogenic silver nanoparticles in soil and aquatic organisms. *Chemosphere* **2020**, *239*, 124698. Available from: https://doi.org/10.1016/j.chemosphere.2019.124698.
65. Sharma, D.; Kanchi, S.; Bisetty, K. Biogenic synthesis of nanoparticles: a review. *Arab. J. Chem.* **2019**, *12* (8), 3576–3600. Available from: https://doi.org/10.1016/j.arabjc.2015.11.002.
66. Fathi, A.; Shiade, S.R.G.; Parmoon, G., et al. *Bioremediation of Heavy Metals Contaminated Soils Using Nanotechnology*; Elsevier eBooks, 2024; pp. 611–628. Available from: https://doi.org/10.1016/b978-0-443-21610-7.00004-5.
67. Upadhyay, P., et al. Nano-omics-based abiotic and biotic stresses management. *Advances in Environmental Engineering and Green Technologies Book Series* **2024**, 348–371. Available from: https://doi.org/10.4018/979-8-3693-1890-4.ch017.
68. Verma, A.; Singh, A.; Rawat, S.; Rajput, P.; Minkina, T.; Mandzhieva, S., et al. Unveiling nanomaterial-induced toxicity. *Advances in Environmental Engineering and Green Technologies Book Series* **2024**, 236–253. Available from: https://doi.org/10.4018/979-8-3693-1471-5.ch011.
69. Singh, A.; Sharma, A.; Singh, O.; Rajput, V. D.; Movsesyan, H. S.; Minkina, T., et al. In-depth Exploration of nanoparticles for enhanced nutrient use efficiency and abiotic stresses management: present insights and future horizons. *Plant Stress* **2024**., 100576. Available from: https://doi.org/10.1016/j.stress.2024.100576.
70. Sharma, A., et al. Revolutionizing agricultural sustainability and food security and management to achieve SDGs goals via nanotechnology. *Advances in Environmental Engineering and Green Technologies Book Series* **2024**, 276–288. Available from: https://doi.org/10.4018/979-8-3693-1471-5.ch013.

SECTION

II

Interaction of biogenic nanomaterials and the environment

CHAPTER

Nanotechnology for improved crop resilience in challenging environments

4

Bal Krishna[1], Priyanka Upadhyay[2], Anil Kumar Singh[3], Rakesh Kumar[4], Dilip Kumar Mahto[5], Ravi Ranjan Kumar[6], Dharm Nath Kamat[7], Mohd Zakir Hussain[8] and Praveen Kumar Yadav[9]

[1]*Department of Plant Breeding & Genetics, Bihar Agricultural University, Sabour, Bhagalpur, Bihar, India*

[2]*Department of Genetics and Plant Breeding, School of Agriculture, Lovely Professional University, Jalandhar, Phagwara, Punjab, India*

[3]*Bihar Agricultural University, Sabour, Bhagalpur, Bihar, India*

[4]*Division of Crop Research, ICAR Research Complex for Eastern Region, Patna, Bihar, India*

[5]*Bhola Paswan Shastri Agricultural College, Purnea, Bihar, India*

[6]*Department of Molecular Biology and Genetic Engineering, Bihar Agricultural University, Sabour, Bhagalpur, Bihar, India*

[7]*Department of Plant Breeding and Genetics, Sugarcane Research Institute, Dr. Rajendra Prasad Central Agricultural University, Samastipur, Bihar, India*

[8]*Department of Agronomy, Bihar Agricultural University, Sabour, Bhagalpur, Bihar, India*

[9]*Division of Seed Science and Technology, ICAR-Indian Agricultural Research Institute, New Delhi, India*

4.1 Introduction

The modern agricultural landscape is rife with challenges stemming from escalating human population growth, resource depletion, and unsustainable practices.[1,2] The Food and Agriculture Organization (FAO) predicts a daunting 50% increase in global food demand by 2050, driven by population growth and exacerbated by factors such as climate change, water scarcity, and soil degradation.[3] Such pressures are further compounded by the inherent vulnerability of plants to environmental stresses such as salinity, drought, and temperature extremes, which curtail yield of crops.[4–8] To address these challenges and ensure global food security, there's a pressing need for more resilient agricultural systems.[9] However, the intensification of biotic and abiotic stressors due to changing climate events has hampered plant growth and development and jeopardized for food security.[7,10–13] Agricultural productivity remains vulnerable to climate instability and associated factors like rising temperatures and changing rainfall patterns, which impede efforts to increase crop yields.[14,15] The detrimental impacts of synthetic fertilizers

Biogenic Nanoparticles. DOI: https://doi.org/10.1016/B978-0-443-33819-9.00008-1

on plant sustainability, coupled with the time-consuming nature of traditional breeding methods, underscore the urgency for more sustainable and efficient agricultural practices.[9,13,16,17] In response, researchers are increasingly turning to nanotechnology as a promising avenue to enhance agricultural productivity and address food security concerns.[17,18] By leveraging nanotechnology, agricultural systems can be bolstered against the challenges posed by climate change, with the potential to increase crop value and resilience.[2] This highlights the vital role that nanotechnology plays in securing sustainable agricultural development and accomplishing the objectives of global food security. Nanotechnology has emerged as a pivotal tool in agriculture, revolutionizing crop productivity and sustainability by leveraging nano-principles to address environmental challenges while ensuring food safety and security.[19,20] Agri-nanotechnology focuses on optimizing crop productivity through the efficient management of inputs such as fertilizers, pesticides, and agrochemicals, facilitated by real-time monitoring systems.[21] With applications spanning from vegetative growth to post-harvest processes, nano-based agriculture aims to enhance crop resilience to diverse environmental stresses, including climate variability and water scarcity, ultimately boosting productivity.[22] Nanotechnology aids plant protection by enabling rapid responses to climate change-induced threats, such as water shortages and high temperatures, thereby promoting crop growth and yield.[15] Engineered nanomaterials ($<$100 nm), key players in the agriculture technology revolution, facilitate efficient delivery of nutrients and pesticides through root or foliar applications, overcoming biological barriers and enhancing crop health.[10,23] Furthermore, nanocapsules and other nanostructures enable controlled release of agrochemicals, contributing to improved crop safety and productivity.[18,24] The versatility and efficacy of engineered nanomaterials position them as potential candidates for sustainable agriculture, offering enhanced nutrient distribution and reduced adverse environmental impacts.[25] This chapter underscores the critical role of nanotechnology in improving crop productivity, environmental sustainability, and food security, emphasizing the importance of understanding engineered nanomaterials as a systemic approach to agricultural enhancement.

4.2 Nanotechnology?

Nanotechnology, an interdisciplinary field encompassing life science, material science, and information technology, has gained prominence as a burgeoning industry, introducing engineered nanoparticles (NPs) into various sectors including agriculture.[3,26,27] This multidisciplinary research area, drawing from biology, physics, chemistry, medical science, and engineering, focuses on designing and characterizing materials and devices with functional entities at the nanometer scale.[21,28] While nano-based approaches in agriculture encompass a wide array of applications such as nanopesticides, nanofertilizers, and nanosensors, their development stages are still evolving.[29,30] Engineered NPs like aluminum (Al), zinc oxide (ZnO), titanium oxides (TiO_2), and others have been integrated into agricultural practices to enhance crop productivity and sustainability, with applications ranging from biosensors fabrication to genetic transformation of plants.[5,31,32] These nanoparticles, typically sized between 1 and 100 nm, exhibit enhanced adsorption properties and targeted delivery capabilities, owing to their small size, shape, and surface area.[3,31] Moreover, nanotools and devices find utility in various agricultural processes such as food processing, packaging, transportation, and shelf-life enhancement, contributing to food safety and quality maintenance.[19,33] Overall, nanotechnology offers promising avenues to address agricultural challenges and improve productivity through innovative nano-based solutions. Recent studies have highlighted the potential of engineered nanomaterials (ENMs) in agriculture to adapt to changing environmental conditions, demonstrating their ability to enhance targeted delivery of nutrients through molecular recognition elements like aptamers or antibodies.[34] By minimizing adverse environmental impacts and promoting sustainable nutrient use, ENMs coated with such elements have shown promise in improving crop yields.[35] Moreover, ENMs facilitate targeted delivery of agrochemicals, efficient utilization, and management in fields, overcoming challenges associated with conventional agricultural practices.[36] Utilizing ENMs for nutrient delivery can optimize resource utilization, develop smart sensors in plants to predict stress, create resilient cultivars via genetic modifications, enhance crop management to mitigate losses, and improve soil microbiome to bolster plant growth.[9] Various rare earth metal-based nanoparticles, including zinc oxide, silver, iron sulfide, cerium oxide, silica, gold, titanium dioxide, copper oxide, and carbon-based nanoparticles, have been deployed in controlled and natural field conditions to help plants thrive in stressed environments.[3,37,38] Additionally, nanotechnology holds promise in revolutionizing disease diagnosis and treatment, enhancing plant nutrient uptake capacity, and improving the efficacy of herbicides and insecticides through nano-based crystals.[15] This collective advancement in nanotechnology applications has the potential to significantly impact agriculture and the food industry, paving the way for more sustainable and efficient practices.

4.3 Nano-material-mediated crop resilience against changing environments

Nanotechnology has facilitated the development of novel fertilizers to address nutrient deficiencies in crops, thereby enhancing agricultural yield and ensuring food security.[39] Engineered nanomaterials (ENMs) play a crucial role in promoting crop sustainability by conferring resistance or resilience to various biotic and abiotic stresses, namely, temperature fluctuations, salinity, drought, flooding, and microbial infections, thus enabling crops to thrive in adverse conditions without compromising yield or nutritional value.[40] These ENMs offer versatile applications, serving as nanobiosensors for seed coating, nanofertilizers to enhance soil microbiome and soil health, nanopesticides for efficient control of pathogen, and nanocarriers for genetic engineering in plants.[9] Additionally, nanotechnology-mediated products like nanofertilizers have demonstrated the ability to increase agricultural productivity, particularly in challenging environmental conditions, while also reducing fertilizer and labor costs by minimizing nutrient loss from soil and optimizing nutrient uptake efficiency.[41,42] Tailored nanomaterials, including various metals and metal oxides, have been developed to meet the specific requirements of plant-related processes, enhancing crop growth, development, and quality across different species.[43–45] Examples include hydroxyl fullerenes, Zn, Fe, and Cu oxides, which have been shown to improve germination and physiological parameters in various crops like rice, maize, and barley.[15,46–48] Ashfaq et al. (2017) demonstrated the efficient translocation of copper carbon nano fertilizers (Cu-CNFs) in chickpea, enhancing nutrient uptake and plant growth.[49] Siddiqui and Al-Whaibi (2014) observed elevated germination in tomato seeds with SiO_2 nanoparticles.[50] Kumar et al. encapsulated Cu–Zn-CNFs in polymeric composites, promoting plant growth.[51] Nanomaterials have shown potential in minimizing fertilizer loss through slow and sustained release, as noted by Ijaz et al. (2023).[9] Liu et al.(2020) found that synthetic apatite NPs sustained soil phosphorus bioavailability, resulting in enhanced soybean yield and growth rate.[52] Plant nano-sensors, discussed by Khandelwal et al. (2019), offer a promising tool for detecting and overcoming pathogen attacks and nutrient deficiencies in plants.[21] They can detect a wide range of targets and monitor soil health, as outlined by Ijaz et al. (2023).[9] Wu et al. (2020) developed near-infrared fluorescent single-walled carbon nanotubes for monitoring stress-related signaling molecules in plants.[53] Ye et al. (2020) investigated MnNPs as a nano-priming agent to mitigate salinity stress in *Capsicum annum* during germination.[54] Despite their potential, the practical application and accuracy of nano-sensors in agriculture need optimization.[19]

4.4 Potential of nanotechnology for biotic stress management

To ensure sustainable crop production and minimize losses from pests, pathogens and environmental stresses, the utilization of nano-enabled pesticides has emerged as a promising solution.[55,56] Traditionally, farmers have heavily relied on chemical pesticides to mitigate crop losses, but these pose significant risks to human health and environmental sustainability.[15] Phytopathogens such as fungi, bacteria, viruses, and nematodes present formidable challenges to agriculture, threatening global food security.[57] The complex interaction between plants and phytopathogens involves intricate physiological, biochemical, and molecular responses, often managed using chemical pesticides that contribute to pesticide resistance.[58,59] Nanoparticles (NPs) offer a transformative approach to crop disease management by modulating plant immunity and combatting phytopathogen attacks.[60,61] NPs' unique properties enable them to penetrate plant tissues, interact with cell membranes, disrupt pathogen growth, and modulate plant defense pathways, enhancing resilience against diseases.[62–64] However, the success of NP-based approaches may be influenced by climatic factors, soil type, pH levels, and compatibility with other substances.[44,61] Ag nanoparticles, for instance, exhibit potent antimicrobial properties, with colloidal nano-Ag solutions demonstrating efficacy against fungal infections like powdery mildew.[65] Similarly, aqueous silicate solutions have been effective against powdery and downy mildew, enhancing crop growth, physiological activities, and stress resistance.[56,66–68] The antimicrobial activity of Ag nanoparticles made from *Glycosmis pentaphylla* fruit extract was investigated in recent studies by Dutta et al. (2022) against a variety of crop plant pathogens, including *Salmonella enterica*, *Bacillus subtilis*, *Escherichia coli*, *Alternaria alterata*, *Fusarium moniliforme*, and *Colletotrichum lindemuthianum*.[60] Savi et al. (2015) showed that nano-formulations provide more precise pesticide administration when they examined the effectiveness of zinc compounds against the development of deoxynivalenol and wheat *Fusarium* head blight.[69] Chitosan-coated iron and magnesium nanocomposites, for instance, showed assure in inhibiting the formation of biofilms by eliminating the rice bacterial pathogens *Xanthomonas oryzae* pv. oryzae and *Acidovorax oryzae*, consequently strengthening their antimicrobial characteristics and reducing infection on plant surfaces.[16,25] Kaziem et al. (2021) introduced an enzyme-responsive β-glucanase nanodelivery system for chlorothalonil modulation in the plant vascular system, resulting in enhanced bioactivity against rice blast.[70] Copper nanoparticles, applied at 50 mg/L, have demonstrated the ability to augment rice resistance against the fungal pathogen *Gibberella fujikuroi* by enhancing SA- and JA-mediated defenses.[71] Metal nanoparticles (Zn, Ag, Fe, Mn, and Cu) applied to wheat seeds prior to sowing induced defensive responses in seedlings infected with Pseudocercosporella herpotrichoides, as reported by Panyuta et al. (2016). In wheat, phytogenic silica nanoparticles triggered innate defense mechanisms and antioxidant systems in response to *Rhizoctonia solani*,[72] whereas silver nanoparticles triggered SA-mediated defense mechanisms in wheat to ward off *Puccinia striiformis* f. sp. tritici.[73] Plant pathogenic fungi including *Magnaporthe grisea* and *Bipolaris sorokiniana* have been found to be suppressed by a variety of silver ions and nanoparticles.[74] In watermelon, innate immune responses were triggered by manganese and copper nanoparticles against *Fusarium* wilt and bacterial fruit blotch, respectively.[58] According to Noman et al. (2021) watermelon's innate immune responses were triggered by manganese and copper nanoparticles against *Fusarium* wilt and bacterial fruit blotch, respectively.[58] Manganese nanoparticles modified the root surfaces of watermelon to prevent penetration of *Fusarium oxysporum* f. sp. niveum and changed the microbial communities in the soil.[58] According to Ma et al. (2020), silica and copper nanoparticles retained the integrity of the trichome in watermelon and soybeans, enhancing defenses against infections by *Acidovorax citrulli* and *Fusarium virguliforme*.[63] According to Noman et al. (2021), Bio-MnNPs at 100 μg/mL enhanced the Mn content in watermelon roots and shoots, promoting better growth and preventing *Fusarium* wilt by inhibiting Fon infection.[58] According to Kang et al. (2021), the application of 1500 mg/L of chemogenic silica nanoparticles improved watermelon's resistance against *Fusarium oxysporum* f. sp. niveum, hence mitigating the severity of the disease by increasing the concentration of silicon in plant tissues.[62] Additionally, *Fusarium* wilt in tomato plants was inhibited by chemogenic sulfur nanoparticles at a concentration of 30 mg/L by increasing plant biomass and sulfur construction. According to Abdallah et al. (2020), *Xanthomonas oryzae* pv. oryzae was significantly inhibited by phytogenic zinc and chitosan nanoparticles stabilized with tomato extract at a concentration of 16 μg/mL.[75] This resulted in the inhibition of pathogen growth, biofilm formation, and swarming motility. At a concentration of 16 μg/mL, magnesium oxide nanoparticles hindered spore germination and impaired cell division in *Phytophthora nicotianae* and *Tielaviopsis basicola*, hence hindering the growth of fungal infections. Additionally, Choudhary et al. (2017) tested Cu-chitosan nanomaterials for promoting plant growth and enhancing systemic resistance against *Curvularia* leaf spot disease in maize.[76] The incorporation of nanoparticles into crop disease management strategies marks a significant advancement in modern agriculture, holding promise for bolstering food security, advocating sustainable practices, and mitigating the environmental footprint associated with conventional methods.[77–79] Nano-pesticides have emerged as innovative alternatives to address challenges linked with traditional pesticides, offering effective control against plant pathogens while addressing concerns regarding toxicity, resistance, efficacy, and environmental impact.[80] Patil et al. (2016) documented the reduced catalytic activity of trypsin, an insect protease, upon exposure to latex-fabricated bioactive gold nanoparticles, showcasing their potential for insect control.[81] Zein-based nanocarriers, derived from maize storage proteins, have demonstrated high biocompatibility, efficacy, and low toxicity, serving as plant pest repellents. Moreover, Ag Idea's formulation of 2,4-dichlorophenoxyacetic acid as nanopesticides has been utilized for pest management.[82] Silver, aluminum oxide, zinc oxide, and TiO_2 are among the nanoparticles that have been utilized to protect rice against weevil infestation and grasseries disease in silkworms, which are caused by Baculovirus BmNPV and *Sitophilus oryzae*, respectively.[83] The stored food pests *Sitophilus oryzae* and *Rhyzopertha dominica* have been shown to be susceptible to the insecticidal abilities of nanostructured alumina, leading to their death in treated wheat after prolonged exposure.[84] Insects that cause damage to crops are the target of pesticides known as oil-in-water nanoemulsions.[85] Ag nanoparticles produced from *Turbinaria turbinata* and *Ulva lactuca* have shown nematicidal effectiveness against infestations of *Meloidogyne javanica* and *Meloidogyne incognita*, causing a reduction in egg production, growth inhibition, hatching rate reduction, and larval mortality.[86] Insects that feed on nectar, such as aphids, are particularly vulnerable to the systemic effects of pesticides that penetrate deeper into the plant thanks to formulations containing silica nanospheres. Furthermore, *Callosobruchus maculatus* has been shown to be fatally affected by silicon oxide nanoparticles at concentrations of 1–2 g/kg. This is due to the particles' ability to obstruct spiracles and tracheas or deteriorate the cuticle's protective wax layer, which finally results in the death of the insect.[87] To more precisely transfer herbicides into weed plants' roots and cause physiological disruption and plant death, nanoparticles encasing the herbicides have also been used.[88] Chaw et al. (2012) reported the development and testing of various types of nanoherbicides, such as glyphosate isopropylamine nano-emulsion, sodium triphosphate paraquat, clomazone/alginate/chitosan nanomaterials, and chitosan nano, to assess their impacts on plants.[89]

4.5 Potential of nanotechnology for abiotic stress management

Recent research suggests that plant reactions to various abiotic stresses are affected by factors such as the severity of the stress, the duration of exposure, and the developmental stage of the plant.[17,90,91] These factors have different effects on plant performance depending on the phenological stage of the plant.[4,7,8,92] Additionally, research has demonstrated the potential of nanotechnology in agriculture by investigating the use of nanoparticles (NPs) at suitable concentrations to lessen the adverse effects of abiotic factors.[3,37,38,93] According to Hu and Xianyu (2021) and Usman et al. (2020), nanomaterials reduce wasteful chemical usage and environmental impact by providing accuracy in the delivery of nutrients and insecticides complemented to plant demands.[3,94] In addition, improvements in genetic engineering devices have made it easier to distribute nanocarriers for genetic crop modification in a targeted manner, improving farming methods.[5,95,96] According to Mahakham et al. (2017), metallic-NP seed priming, like Ag-NP priming, has demonstrated better outcomes in raising seedling vigor and propagation rates, highlighting the potential of nanotechnology in promoting seed germination and early plant development.[97] As a common abiotic threat, drought impairs a number of physiological processes in plants, including photosynthetic frequency, stomatal conductance, and transpiration rate. As a result, water potential and leaf relative water content (RWC) are reduced.[5,6,98–100] Drought stress also impairs pollen viability, cell division, and can ultimately result in plant necrosis.[101] About half of annual yield losses are caused by drought stress, which is a serious threat to crop productivity, forecasts suggest that this problem will worsen in many areas.[5,6,102–105] The way that plants respond to drought stress is complex and depends on a number of variables, such as the length and intensity of the stress, the genetic composition of the plant, and the stage of development.[103,106] Utilizing carbon nanotubes (CNTs) and zinc oxide nanoparticles (ZnO-NPs) to reduce drought stress in rice and maize, respectively, has produced encouraging results in recent studies that have improved growth, yield, and physiological parameters.[107,108] Furthermore, it has been demonstrated that foliar application of ZnO-NPs and other nanomaterials improves a variety of physiological and biochemical parameters, hence increasing drought tolerance in eggplant and wheat.[109,110] According to Mustafa et al. (2021) and Taran et al. (2017), antioxidative activities, photosynthetic pigments, and grain yield have all improved in drought-stressed wheat when soil amendments containing TiO_2-NPs and colloidal solutions of Cu-Zn-NPs were applied.[111,112] In addition, research have shown that manganese and zinc nanoparticles (NPs) and nanochelates can effectively mitigate the impacts of drought in soybean and sorghum, respectively, improving growth, yield, and nutrient uptake.[113] These results highlight how nanotechnology might help crops become more drought-tolerant, opening up exciting new possibilities for environmentally friendly farming methods. Salinity stress, resulting from the accumulation of salts in soil, poses a significant challenge to crop productivity globally, with studies indicating a 25% decline in yield due to this stress.[114–116] Efforts to address this issue have led scientists to explore the potential of nanomaterials in mitigating the adverse effects of salinity stress on plants. For instance, $FeSO_4$ nanoparticles have been shown to enhance various growth-related traits in plants, including increased leaf area, shoot organic matter, and chlorophyll synthesis, while decreasing sodium concentration in leaves.[117] Similarly, the introduction of CeO_2 nanoparticles has been found to increase photosynthetic rates and productivity in plants under elevated temperature and saline conditions, attributed to their antioxidant properties. ZnO-NPs have demonstrated beneficial effects on wheat plants under salinity stress, improving growth-related traits and photosynthetic pigments.[118] Additionally, priming with S-NPs enhanced the antioxidant status and nitrogen metabolism in wheat subjected to salinity stress, while foliar application of Fe_3O_4-NPs mitigated the adverse effects of salinity stress by boosting various growth parameters and antioxidant defense systems in wheat.[119,120] Furthermore, priming maize seeds with TiO_2-NPs and rapeseed seeds with ZnO-NPs has been demonstrated to improve germination and seedling vigor under salinity stress.[121,122] Moreover, nano-priming with CeO_2-NPs and ZnO-NPs has been found to enhance seedling growth and upregulate gene expression associated with stress responses in rapeseed under salinity stress.[123,124] These findings highlight the potential of nanomaterials in ameliorating the detrimental effects of salinity stress on crop plants, offering promising strategies for enhancing agricultural productivity in saline environments. Under salt stress conditions, various nanomaterials have shown promising effects in enhancing the stress tolerance of plants. BioSe-NPs have demonstrated effectiveness in enhancing the phenotypic characteristics of rapeseed seedlings by activating antioxidant enzymes, reducing levels of reactive oxygen species (ROS) and malondialdehyde (MDA), and promoting lateral root development.[122] Multiwalled carbon nanotubes have been shown to promote growth in broccoli seedlings by enhancing aquaporin transduction and water uptake under salt stress conditions.[5,125] Similarly, in grape seeds, multiwalled carbon nanotubes improved salinity tolerance by increasing germination rates and seedling development while decreasing MDA content.[126] Graphene oxide-glycine betaine functionalized nanoparticles and chitosan nanoparticles have also exhibited the ability to alleviate the adverse effects of salinity stress in sweet basil and Catharanthus roseus, respectively, by enhancing photosynthetic components, maintaining membrane integrity, and bolstering antioxidative defense systems.[127,128] Additionally, priming cotton seeds with poly (acrylic acid)-coated cerium oxide nanoparticles improved root characteristics and reduced ROS production under salt stress.[129] Foliar application of ZnO nanoparticles enhanced chlorophyll content, root and shoot lengths, biomass, and antioxidative activities in tomato plants subjected to salinity stress.[130] Moreover, SiO_2 nanoparticles have been found to enhance seed germination, increase organic matter, and promote chlorophyll pigment accumulation in tomato and squash plants under NaCl stress.[131] Furthermore, nanomaterials have been investigated for their potential in detoxifying hazardous components in the environment, such as Cd stress in rice plants.[132] Overall, the application of nanomaterials shows promise in developing stress-tolerant plants, which are crucial for ensuring global food security and enhancing the resilience and efficiency of modern agriculture.[5,133]

4.6 Nano-mediated genome editing and nanocarriers

Nanotechnology stands as a promising frontier in revolutionizing agriculture, particularly in crop improvement and pest management. Through NP-mediated genetic engineering techniques, crops can be endowed with desirable traits, enhancing their resilience to changing climates and bolstering overall productivity.[134] Engineered nanomaterials (ENMs) offer tailored structural and physicochemical characteristics, facilitating precise delivery and controlled release of active ingredients within plants. This capability opens avenues for enhancing genome editing technologies, which traditionally faced challenges in efficient intracellular delivery. Revolutionary tools renowned for their high specificity and efficiency in genome editing include zinc-finger nucleases (ZFNs), transcription activator-like effector nucleases (TALENs), and clustered regularly interspaced short palindromic repeats (CRISPR).[5] Among these methods, CRISPR/Cas has given particularly encouraging results due to its ease of use and multiplexing range.[135,136] ENMs, including various nanocarriers, polymers, liposomes, and inorganic nanoparticles, show promise in delivering genetic materials and stabilizing agents like dsRNA, thereby enhancing their efficacy in pest control and crop protection. The convergence of advanced genetic engineering techniques with nanotechnology presents unprecedented opportunities for improving crop yield and resistance. Nanomaterials play a crucial role in facilitating targeted delivery of exogenous biomolecules for genome editing, thereby contributing to precise and efficient crop improvement strategies. In the realm of plant genetic engineering, nanotechnology has emerged as a promising avenue for delivering genetic materials, such as RNA, DNA, or their complexes with nanoparticles.[137]

Noteworthy is the utilization of carbon nanocarriers for effectively delivering CRISPR reagents like sgRNA into the plant genome, facilitating efficient gene knockdown.[138] Additionally, both the Cas-9 nanosystem carrier and lipid nanoparticles (LNPs) have demonstrated the capability to transport Cas9/sgRNA complexes and Cas9 ribonucleoprotein to tissues for genome editing, exhibiting high stability and editing efficiency.[139,140] Other methods utilize nanoparticles such as poly (lactic-co-glycolic acid) to encapsulate CRISPR/Cas9 plasmids for targeted delivery into tissues.[141] Moreover, nanocarriers offer pH-sensitive controlled release capabilities, exemplified by FLU-loaded glycine methyl ester-conjugated polysuccinimide nanoparticles, which mitigate disease intensity in banana plants by targeting FLU delivery to roots and rhizomes. In pest management, transgenic crops engineered to produce dsRNA exhibit resistance against various pests and pathogens. However, concerns over regulatory and acceptance issues with genetically modified crops have led to exploration of alternative methods such as topical application of dsRNA. Nanomaterials are instrumental in stabilizing and delivering exogenous dsRNA, ensuring its stability and successful release at the target site, thereby enhancing pathogen protection. Additionally, nanocarriers reduce the cost of dsRNA treatment and offer biodegradable options like chitosan, addressing concerns regarding genetic engineering and pesticide application accuracy. The application of nano-assisted genetic manipulation in agriculture represents a promising approach to sustainable crop improvement, although the long-term environmental impacts of nanocarriers and their cargo require careful consideration. Overall, the integration of nanotechnology with agriculture holds immense potential to revolutionize crop protection and enhancement, offering solutions to address global food security challenges.

4.7 Challenges

Nanotechnology holds immense promise for transforming agriculture, but it also presents challenges and concerns that need to be addressed for its effective and safe implementation. One major barrier is the lack of comprehensive understanding regarding the mechanisms underlying plant-engineered nanomaterial (ENM) interactions, particularly how ENMs directly interact with plants through soil, water, and the atmosphere. Additionally, the pathways for delivering nanomaterials into plant vascular systems and organelles are limited, hindering their efficacy. Moreover, the potential hazardous effects of nanomaterials on living organisms, including their entry into the food chain, remain only partially understood. Chemically synthesized nanomaterials often pose environmental risks due to their slow degradation, and the extent of toxicity depends on factors such as dose and exposure duration. For instance, Disfani et al. (2017) observed that while applying 15 mg/kg of nano Fe/SiO_2 to barley and maize seedlings increased shoot length, higher concentrations led to a marked reduction in growth, highlighting the concentration-dependent nature of crop responses to nanomaterial application.[48] To mitigate these toxic effects, current research efforts are primarily focused on understanding plant responses to nanomaterials and their interactions with environmental factors such as climatic conditions. By addressing these challenges and conducting further research, the potential benefits of nanotechnology in agriculture can be realized while minimizing associated risks.[15,40,48]

4.8 Future outlook and conclusion

Nanotechnology holds immense promise for revolutionizing agriculture by offering innovative solutions to enhance crop resilience in challenging environments. With its diverse applications in plant breeding, disease control, fertilizer application, and precision agriculture, nanotechnology offers unprecedented opportunities to improve the quality and quantity of agricultural products while addressing key challenges facing global food security. One of the most significant contributions of nanotechnology to agriculture lies in its ability to enable targeted delivery of agrochemicals, such as fertilizers and pesticides. Nano-based approaches facilitate the efficient and proper utilization of these inputs, minimizing wastage and environmental impact while maximizing their effectiveness in promoting plant growth and protecting crops from pests and diseases. By encapsulating nutrients and active ingredients within nanostructures, nanotechnology allows for controlled release mechanisms, ensuring that plants receive the required nutrients and protection over an extended period, even in harsh environmental conditions. Furthermore, nanomaterial-integrated biosensors represent a groundbreaking innovation in crop protection and management. These biosensors enable the real-time detection of microbial pathogens responsible for causing diseases in crops, allowing for timely intervention and preventive measures. The simplicity, rapidity, cost-effectiveness, sensitivity, specificity, and portability of nanobiosensors make them invaluable tools for farmers in monitoring and managing crop health, thereby minimizing yield losses and ensuring food security. Despite the significant advancements made in nanotechnology for agriculture, there are still several challenges that need to be addressed to fully realize its potential. Future efforts should focus on the development of highly innovative nanomaterials-based devices that are miniaturized, economical, safe, quick, efficient, and user-friendly. These devices should be tailored to meet the specific needs of farmers and agricultural practitioners, providing practical solutions to enhance crop productivity and resilience. Moreover, interdisciplinary collaborations between scientists, engineers, agronomists, and farmers are essential to translate laboratory-scale innovations into on-field applications. While many promising demonstrations of nanomaterials in agriculture have been conducted at the laboratory scale, there is a critical need for more field-scale trials and on-site demonstrations. Farmers stand to benefit greatly from the adoption of nanotechnology-enabled solutions, but widespread acceptance and adoption require evidence of their efficacy and feasibility under real-world conditions. Therefore, concerted efforts should be made to bridge the gap between research and implementation, facilitating the transfer of technology from the laboratory to the farm. Nanotechnology offers transformative opportunities to address the challenges facing modern agriculture and improve crop resilience in challenging environments. By harnessing the power of nanomaterials for targeted delivery of agrochemicals, development of nanobiosensors, and creation of innovative devices, we can create sustainable and resilient agricultural systems that ensure food security for future generations. However, to realize the full potential of nanotechnology in agriculture, collaborative efforts, investment in research and development, and on-site demonstrations are essential to facilitate technology adoption and integration into farming practices (Fig. 4.1).

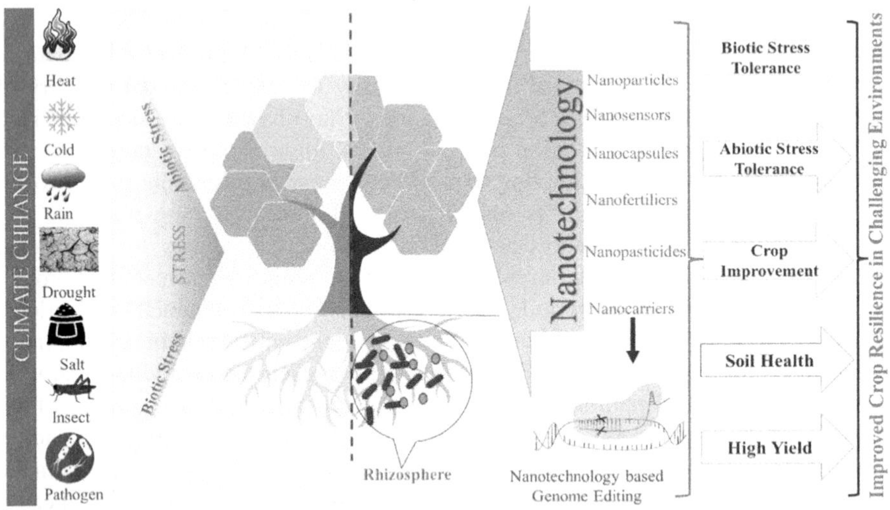

FIGURE 4.1

Diagrammatic representation of potential applications of Engineered Nanomaterials (ENMs) in sustainable agriculture for enhancing stress resilience in crop plants. Nanotechnology offers a promising avenue for enhancing stress resilience in crop plants within sustainable agriculture practices. Nanofertilizers can deliver nutrients more efficiently, promoting healthier plant growth while reducing environmental impact. Nanosensors enable precise monitoring of soil and plant conditions, allowing for timely interventions. Nanoparticles, including nanocapsules, hold potential for targeted delivery of agrochemicals like nanopesticides, minimizing wastage and ecological harm. Furthermore, nanoparticles can facilitate genome editing, fostering the development of stress-tolerant crop varieties suited for climate-smart agriculture. By integrating these technologies, we can optimize resource use, mitigate environmental pressures, and bolster global food security in a rapidly changing climate.

References

1. Rodell, M.; Velicogna, I.; Famiglietti, J. S. Satellite-based estimates of groundwater depletion in India. *Nature* **2009**, *460*, 999–1002. Available from: https://doi.org/10.1038/nature08238.
2. Yadav, N.; Garg, V. K.; Chhillar, A. K.; Rana, J. S. Recent advances in nanotechnology for the improvement of conventional agricultural systems: a review. *Plant. Nano Biol.* **2023**, *4*, 100032. Available from: https://doi.org/10.1016/j.plana.2023.100032.
3. Usman, M.; Farooq, M.; Wakeel, A.; Nawaz, A.; Cheema, S. A.; Rehman, H. ur, et al. Nanotechnology in agriculture: current status, challenges and future opportunities. *Sci. Total. Environ.* **2020**, *721*, 137778. Available from: https://doi.org/10.1016/j.scitotenv.2020.137778.
4. Ding, Y.; Yang, S. Surviving and thriving: how plants perceive and respond to temperature stress. *Dev. Cell* **2022**, *57*, 947–958. Available from: https://doi.org/10.1016/j.devcel.2022.03.010.
5. Raza, A.; Charagh, S.; Salehi, H.; Abbas, S.; Saeed, F.; Poinern, G. E. J., et al. Nano-enabled stress-smart agriculture: can nanotechnology deliver drought and salinity-smart crops? *J. Sustain. Agriculture Environ.* **2023**, *2*, 189–214. Available from: https://doi.org/10.1002/sae2.12061.
6. Mubarik, M. S.; Khan, S. H.; Sajjad, M.; Raza, A.; Hafeez, M. B.; Yasmeen, T., et al. A manipulative interplay between positive and negative regulators of phytohormones: a way forward for improving drought tolerance in plants. *Physiol. Plant.* **2021**, *172*, 1269–1290. Available from: https://doi.org/10.1111/ppl.13325.
7. Rivero, R. M.; Mittler, R.; Blumwald, E.; Zandalinas, S. I. Developing climate-resilient crops: improving plant tolerance to stress combination. *Plant. J.* **2022**, *109*, 373–389. Available from: https://doi.org/10.1111/tpj.15483.
8. Zandalinas, S. I.; Fritschi, F. B.; Mittler, R. Global warming, climate change, and environmental pollution: recipe for a multifactorial stress combination disaster. *Trends Plant. Sci.* **2021**, *26*, 588–599. Available from: https://doi.org/10.1016/j.tplants.2021.02.011.
9. Ijaz, M.; Khan, F.; Ahmed, T.; Noman, M.; Zulfiqar, F.; Rizwan, M., et al. Nanobiotechnology to advance stress resilience in plants: current opportunities and challenges. *Mater. Today Bio* **2023**, *22*, 100759. Available from: https://doi.org/10.1016/j.mtbio.2023.100759.
10. Farooq, M. A.; Hannan, F.; Islam, F.; Ayyaz, A.; Zhang, N.; Chen, W., et al. The potential of nanomaterials for sustainable modern agriculture: present findings and future perspectives. *Env. Sci. Nano* **2022**, *9*, 1926–1951. Available from: https://doi.org/10.1039/D1EN01124C.
11. Kumar, A.; Sharma, P. K.; Singh, S.; Verma, J. P. Impact of engineered nanoparticles on microbial communities. *Soil. Health Plants* **2021**, 201–215. Available from: https://doi.org/10.1007/978-3-030-66956-0_14.
12. Wang, Y.; Yao, Z.; Zhan, Y.; Zheng, X.; Zhou, M.; Yan, G., et al. Potential benefits of liming to acid soils on climate change mitigation and food security. *Glob. Chang. Biol.* **2021**, *27*, 2807–2821. Available from: https://doi.org/10.1111/gcb.15607.
13. Upadhyay, P.; Navvuru, S.; Yadav, P. K.; Lalotra, S.; Singh, A.; Rajput, V. D., et al. *Nano-omics-based abiotic and biotic stresses management. Harnessing NanoOmics and Nanozymes for Sustainable Agriculture;* IGI Global, 2024, pp. 348–371. Available from: https://doi.org/10.4018/979-8-3693-1890-4.ch017.

14. Mall, R. K.; Gupta, A.; Sonkar, G. *Effect of climate change on agricultural crops. Current Developments in Biotechnology and Bioengineering;* Elsevier, 2017, pp. 23–46. Available from: https://doi.org/10.1016/B978-0-444-63661-4.00002-5.
15. Husnain Gondal, A.; Tayyiba, L. Prospects of using nanotechnology in agricultural growth, environment and industrial food products. *Rev. Agric. Sci.* **2022**, *10*, 68–81. Available from: https://doi.org/10.7831/ras.10.0_68.
16. Ahmed, T.; Noman, M.; Manzoor, N.; Shahid, M.; Abdullah, M.; Ali, L., et al. Nanoparticle-based amelioration of drought stress and cadmium toxicity in rice via triggering the stress responsive genetic mechanisms and nutrient acquisition. *Ecotoxicol. Env. Saf.* **2021**, *209*, 111829. Available from: https://doi.org/10.1016/j.ecoenv.2020.111829.
17. Singh, A.; Agrawal, S.; Rajput, V. D.; Ghazaryan, K.; Yesayan, A.; Minkina, T., et al. Nanoparticles in revolutionizing crop production and agriculture to address salinity stress challenges for a sustainable future. *Discov. Appl. Sci.* **2024**, *6*, 317. Available from: https://doi.org/10.1007/s42452-024-06009-7.
18. Shang, Y.; Hasan, M. K.; Ahammed, G. J.; Li, M.; Yin, H.; Zhou, J. Applications of nanotechnology in plant growth and crop protection: a review. *Molecules* **2019**, *24*, 2558. Available from: https://doi.org/10.3390/molecules24142558.
19. Pandey, G. Challenges and future prospects of agri-nanotechnology for sustainable agriculture in India. *Env. Technol. Innov.* **2018**, *11*, 299–307. Available from: https://doi.org/10.1016/j.eti.2018.06.012.
20. Fathi, A.; Ghadirnezhad Shiade, S. R.; Parmoon, G.; Yaghoubian, Y.; Pirdashti, H.; Rajput, V. D., et al. *Bioremediation of heavy metals contaminated soils using nanotechnology. Bio-organic Amendments for Heavy Metal Remediation: Water, Soil and Plant Approaches and Technologies;* Elsevier, 2024, pp. 611–628. Available from: https://doi.org/10.1016/B978-0-443-21610-7.00004-5.
21. Khandelwal, A.; Joshi, R.; Mukherjee, P.; Singh, S. D.; Shrivastava, M. *Use of bio-based nanoparticles in agriculture. Nanotechnology for Agriculture;* Springer Singapore: Singapore, 2019, pp. 89–100. Available from: https://doi.org/10.1007/978-981-32-9370-0_6.
22. Vermeulen, S. J.; Aggarwal, P. K.; Ainslie, A.; Angelone, C.; Campbell, B. M.; Challinor, A. J., et al. Options for support to agriculture and food security under climate change. *Env. Sci. Policy* **2012**, *15*, 136–144. Available from: https://doi.org/10.1016/j.envsci.2011.09.003.
23. Giraldo, J. P.; Wu, H.; Newkirk, G. M.; Kruss, S. Nanobiotechnology approaches for engineering smart plant sensors. *Nat. Nanotechnol.* **2019**, *14*, 541–553. Available from: https://doi.org/10.1038/s41565-019-0470-6.
24. de Oliveira, J. L.; Campos, E. V. R.; Gonçalves da Silva, C. M.; Pasquoto, T.; Lima, R.; Fraceto, L. F. Solid lipid nanoparticles co-loaded with simazine and atrazine: preparation, characterization, and evaluation of herbicidal activity. *J. Agric. Food Chem.* **2015**, *63*, 422–432. Available from: https://doi.org/10.1021/jf5059045.
25. Noman.; Ahmed, M.; Shahid, T.; Nazir, M.; Azizullah, M. M.; Li, D., et al. Salicylic acid-doped iron nano-biostimulants potentiate defense responses and suppress Fusarium wilt in watermelon. *J. Adv. Res.* **2023**. Available from: https://doi.org/10.1016/j.jare.2023.06.011.
26. Khan, M. N.; Mobin, M.; Abbas, Z. K.; AlMutairi, K. A.; Siddiqui, Z. H. Role of nanomaterials in plants under challenging environments. *Plant. Physiol. Biochem.* **2017**, *110*, 194–209. Available from: https://doi.org/10.1016/j.plaphy.2016.05.038.

27. Vithanage, M.; Zhang, X.; Gunarathne, V.; Zhu, Y.; Herath, L.; Peiris, K., et al. Plant nanobionics: fortifying food security via engineered plant productivity. *Env. Res.* **2023**, *229*, 115934. Available from: https://doi.org/10.1016/j.envres.2023.115934.
28. Silva, G. A. Introduction to nanotechnology and its applications to medicine. *Surg. Neurol.* **2004**, *61*, 216−220. Available from: https://doi.org/10.1016/j.surneu.2003.09.036.
29. Scott, N. R. Nanotechnology and animal health. *Rev. Scientifique et. Tech. de. l'OIE* **2005**, *24*, 425−432. Available from: https://doi.org/10.20506/rst.24.1.1579.
30. Seeman, N. C. DNA in a material world. *Nature* **2003**, *421*, 427−431. Available from: https://doi.org/10.1038/nature01406.
31. Dziergowska, K.; Michalak, I. *The role of nanoparticles in sustainable agriculture. Smart Agrochemicals for Sustainable Agriculture;* Elsevier, 2022, pp. 225−278. Available from: https://doi.org/10.1016/B978-0-12-817036-6.00007-8.
32. Singh, D.; Gurjar, B. R. Nanotechnology for agricultural applications: facts, issues, knowledge gaps, and challenges in environmental risk assessment. *J. Env. Manage* **2022**, *322*.116033. Available from: https://doi.org/10.1016/j.jenvman.2022.116033.
33. Al-Antary, T.; Kahlel, A.; Ghidan, A.; Asoufi, H. Effects of nanotechnology liquid fertilizers on fruit set and pods of broad bean (*Vicia faba* L.). *Fresenius Env. Bull.* **2020**, *29*, 4794−4798.
34. Mittal, D.; Kaur, G.; Singh, P.; Yadav, K.; Ali, S. A. Nanoparticle-based sustainable agriculture and food science: recent advances and future outlook. *Front. Nanotechnol.* **2020**, *2*. Available from: https://doi.org/10.3389/fnano.2020.579954.
35. Gilbertson, L. M.; Pourzahedi, L.; Laughton, S.; Gao, X.; Zimmerman, J. B.; Theis, T. L., et al. Guiding the design space for nanotechnology to advance sustainable crop production. *Nat. Nanotechnol.* **2020**, *15*, 801−810. Available from: https://doi.org/10.1038/s41565-020-0706-5.
36. Pal, A.; Kaur, P.; Dwivedi, N.; Rookes, J.; Bohidar, H. B.; Yang, W., et al. Clay-nanocomposite based smart delivery systems: a promising tool for sustainable farming. *ACS Agric. Sci. & Technol.* **2023**, *3*, 3−16. Available from: https://doi.org/10.1021/acsagscitech.2c00140.
37. Zahedi, S. M.; Abdelrahman, M.; Hosseini, M. S.; Hoveizeh, N. F.; Tran, L.-S. P. Alleviation of the effect of salinity on growth and yield of strawberry by foliar spray of selenium-nanoparticles. *Environ. Pollut.* **2019**, *253*, 246−258. Available from: https://doi.org/10.1016/j.envpol.2019.04.078.
38. Zulfiqar, F.; Ashraf, M. Nanoparticles potentially mediate salt stress tolerance in plants. *Plant. Physiol. Biochem.* **2021**, *160*, 257−268. Available from: https://doi.org/10.1016/j.plaphy.2021.01.028.
39. Iavicoli, I.; Leso, V.; Beezhold, D. H.; Shvedova, A. A. Nanotechnology in agriculture: opportunities, toxicological implications, and occupational risks. *Toxicol. Appl. Pharmacol.* **2017**, *329*, 96−111. Available from: https://doi.org/10.1016/j.taap.2017.05.025.
40. Lowry, G. V.; Avellan, A.; Gilbertson, L. M. Opportunities and challenges for nanotechnology in the agri-tech revolution. *Nat. Nanotechnol.* **2019**, *14*, 517−522. Available from: https://doi.org/10.1038/s41565-019-0461-7.
41. Rehman, A.; Qunyi, T.; Usman, M.; Manzoor, M. F.; Zhao, L.; Feng, J., et al. *Nano-enabled agrochemicals for sustainable agriculture. Nano-enabled Agrochemicals in Agriculture;* Elsevier, 2022, pp. 291−306. Available from: https://doi.org/10.1016/B978-0-323-91009-5.00020-3.

42. Ditta, A.; Arshad, M. Applications and perspectives of using nanomaterials for sustainable plant nutrition. *Nanotechnol. Rev.* **2016**, *5*. Available from: https://doi.org/10.1515/ntrev-2015-0060.
43. Perea Vélez, Y. S.; Carrillo-González, R.; González-Chávez Ma, del C. A. Interaction of metal nanoparticles–plants–microorganisms in agriculture and soil remediation. *J. Nanopart. Res.* **2021**, *23*, 206. Available from: https://doi.org/10.1007/s11051-021-05269-3.
44. Ali, S.; Perveen, S.; Ali, M.; Jiao, T.; Sharma, A. S.; Hassan, H., et al. Bioinspired morphology-controlled silver nanoparticles for antimicrobial application. *Mater. Sci. Engineering: C.* **2020**, *108*, 110421. Available from: https://doi.org/10.1016/j.msec.2019.110421.
45. Kumaraswamy, R. V.; Kumari, S.; Choudhary, R. C.; Sharma, S. S.; Pal, A.; Raliya, R., et al. Salicylic acid functionalized chitosan nanoparticle: a sustainable biostimulant for plant. *Int. J. Biol. Macromol.* **2019**, *123*, 59–69. Available from: https://doi.org/10.1016/j.ijbiomac.2018.10.202.
46. Shojaei, T. R.; Salleh, M. A. M.; Tabatabaei, M.; Mobli, H.; Aghbashlo, M.; Rashid, S. A., et al. *Applications of nanotechnology and carbon nanoparticles in agriculture. Synthesis, Technology and Applications of Carbon Nanomaterials;* Elsevier, 2019, pp. 247–277. Available from: https://doi.org/10.1016/B978-0-12-815757-2.00011-5.
47. Prerna, D. I.; Govindaraju, K.; Tamilselvan, S.; Kannan, M.; Vasantharaja, R.; Chaturvedi, S., et al. Influence of nanoscale micro-nutrient α-Fe2O3 on seed germination, seedling growth, translocation, physiological effects and yield of rice (*Oryza sativa*) and maize (*Zea mays*). *Plant. Physiol. Biochem.* **2021**, *162*, 564–580. Available from: https://doi.org/10.1016/j.plaphy.2021.03.023.
48. Najafi Disfani, M.; Mikhak, A.; Kassaee, M. Z.; Maghari, A. Effects of nano Fe/SiO_2 fertilizers on germination and growth of barley and maize. *Arch. Agron. Soil. Sci.* **2017**, *63*, 817–826. Available from: https://doi.org/10.1080/03650340.2016.1239016.
49. Ashfaq, M.; Verma, N.; Khan, S. Carbon nanofibers as a micronutrient carrier in plants: efficient translocation and controlled release of Cu nanoparticles. *Env. Sci. Nano* **2017**, *4*, 138–148. Available from: https://doi.org/10.1039/C6EN00385K.
50. Siddiqui, M. H.; Al-Whaibi, M. H. Role of nano-SiO2 in germination of tomato (*Lycopersicum esculentum* seeds Mill.). *Saudi J. Biol. Sci.* **2014**, *21*, 13–17. Available from: https://doi.org/10.1016/j.sjbs.2013.04.005.
51. Kumar, R.; Ashfaq, M.; Verma, N. Synthesis of novel PVA–starch formulation-supported Cu–Zn nanoparticle carrying carbon nanofibers as a nanofertilizer: controlled release of micronutrients. *J. Mater. Sci.* **2018**, *53*, 7150–7164. Available from: https://doi.org/10.1007/s10853-018-2107-9.
52. Liu, R.; Lal, R. Synthetic apatite nanoparticles as a phosphorus fertilizer for soybean (*Glycine max*). *Sci. Rep.* **2014**, *4*, 5686. Available from: https://doi.org/10.1038/srep05686.
53. Wu, H.; Nißler, R.; Morris, V.; Herrmann, N.; Hu, P.; Jeon, S.-J., et al. Monitoring Plant health with near-infrared fluorescent H_2 O_2 nanosensors. *Nano Lett.* **2020**, *20*, 2432–2442. Available from: https://doi.org/10.1021/acs.nanolett.9b05159.
54. Ye, Y.; Cota-Ruiz, K.; Hernández-Viezcas, J. A.; Valdés, C.; Medina-Velo, I. A.; Turley, R. S., et al. Manganese nanoparticles control salinity-modulated molecular responses in *Capsicum annuum* L. through priming: a sustainable approach for agriculture. *ACS Sustain. Chem. Eng.* **2020**, *8*, 1427–1436. Available from: https://doi.org/10.1021/acssuschemeng.9b05615.

55. Singh, A.; Tomar, B.; Harutyunyan, A.; Rajput, P.; Minkina, T.; Mandzhieva, S., et al. *Utilizing nanotechnology in agriculture: a balancing approach between environmental health and risks. Nanotechnology Applications and Innovations for Improved Soil Health;* IGI Global, 2024, pp. 380−399. Available from: https://doi.org/10.4018/979-8-3693-1471-5.ch019.
56. Rajput, P.; Singh, A.; Agrawal, S.; Ghazaryan, K.; Rajput, V. D.; Movsesyan, H., et al. Effects of environmental metal and metalloid pollutants on plants and human health: exploring nano-remediation approach. *Stress. Biol.* **2024**, *4*, 27. Available from: https://doi.org/10.1007/s44154-024-00156-y.
57. Singh, A.; Sengar, R. S.; Shahi, U. P.; Rajput, V. D.; Minkina, T.; Ghazaryan, K. A. Prominent effects of zinc oxide nanoparticles on roots of rice (*oryza sativa* l.) grown under salinity stress. *Stresses* **2022**, *3*, 33−46. Available from: https://doi.org/10.3390/stresses3010004.
58. Noman, M.; Ahmed, T.; Shahid, M.; Niazi, M. B. K.; Qasim, M.; Kouadri, F., et al. Biogenic copper nanoparticles produced by using the Klebsiella pneumoniae strain NST2 curtailed salt stress effects in maize by modulating the cellular oxidative repair mechanisms. *Ecotoxicol. Env. Saf.* **2021**, *217*, 112264. Available from: https://doi.org/10.1016/j.ecoenv.2021.112264.
59. Peyraud, R.; Dubiella, U.; Barbacci, A.; Genin, S.; Raffaele, S.; Roby, D. Advances on plant−pathogen interactions from molecular toward systems biology perspectives. *Plant. J.* **2017**, *90*, 720−737. Available from: https://doi.org/10.1111/tpj.13429.
60. Dutta, P.; Kumari, A.; Mahanta, M.; Biswas, K. K.; Dudkiewicz, A.; Thakuria, D., et al. Advances in nanotechnology as a potential alternative for plant viral disease management. *Front. Microbiol.* **2022**, *13*. Available from: https://doi.org/10.3389/fmicb.2022.935193.
61. Noman, E. A.; Al-Gheethi, A.; Al-Sahari, M.; Saphira Radin Mohamed, R. M.; Crane, R.; Aziz, N. A. A., et al. Challenges and opportunities in the application of bioinspired engineered nanomaterials for the recovery of metal ions from mining industry wastewater. *Chemosphere* **2022**, *308*, 136165. Available from: https://doi.org/10.1016/j.chemosphere.2022.136165.
62. Kang, H.; Elmer, W.; Shen, Y.; Zuverza-Mena, N.; Ma, C.; Botella, P., et al. Silica Nanoparticle dissolution rate controls the suppression of *Fusarium Wilt* of Watermelon (*Citrullus lanatus*). *Env. Sci. Technol.* **2021**, *55*, 13513−13522. Available from: https://doi.org/10.1021/acs.est.0c07126.
63. Ma, C.; Borgatta, J.; Hudson, B. G.; Tamijani, A. A.; De La Torre-Roche, R.; Zuverza-Mena, N., et al. Advanced material modulation of nutritional and phytohormone status alleviates damage from soybean sudden death syndrome. *Nat. Nanotechnol.* **2020**, *15*, 1033−1042. Available from: https://doi.org/10.1038/s41565-020-00776-1.
64. Kumar, A.; Sharma, P. K.; Singh, S.; Verma, J. P. Impact of engineered nanoparticles on microbial communities. *Soil Health and Plants* **2021**, 201−215. Available from: https://doi.org/10.1007/978-3-030-66956-0_14.
65. Kim, J.-M.; To, T. K.; Matsui, A.; Tanoi, K.; Kobayashi, N. I.; Matsuda, F., et al. Acetate-mediated novel survival strategy against drought in plants. *Nat. Plants* **2017**, *3*, 17097. Available from: https://doi.org/10.1038/nplants.2017.97.
66. Kanto, T.; Miyoshi, A.; Ogawa, T.; Maekawa, K.; Aino, M. Suppressive effect of liquid potassium silicate on powdery mildew of strawberry in soil. *J. Gen. Plant. Pathol.* **2006**, *72*, 137−142. Available from: https://doi.org/10.1007/s10327-005-0270-8.

67. Singh, A.; Sharma, A.; Singh, O.; Rajput, V. D.; Movsesyan, H.; Minkina, T., et al. In-depth exploration of nanoparticles for enhanced nutrient use efficiency and abiotic stresses management: present insights and future horizons. *Plant. Stress.* **2024**, *14*, 100576. Available from: https://doi.org/10.1016/j.stress.2024.100576.
68. Singh, A.; Rajput, V. D.; Agrawal, S.; Ghazaryan, K.; Minkina, T.; Al Tawaha, A. R. M., et al. Nanoparticles mediated salt stress resilience: a holistic exploration of physiological, biochemical, and nano-omics approaches. *Rev. Env. Contam. Toxicol.* **2024**, *262*, 19. Available from: https://doi.org/10.1007/s44169-024-00070-4.
69. Savi, G. D.; Piacentini, K. C.; de Souza, S. R.; Costa, M. E. B.; Santos, C. M. R.; Scussel, V. M. Efficacy of zinc compounds in controlling Fusarium head blight and deoxynivalenol formation in wheat (*Triticum aestivum* L.). *Int. J. Food Microbiol.* **2015**, *205*, 98–104. Available from: https://doi.org/10.1016/j.ijfoodmicro.2015.04.001.
70. Kaziem, A. E.; Yang, L.; Lin, Y.; Kazem, A. E.; Xu, H.; Zhang, Z. Pathogenic Invasion-responsive carrier based on mesoporous silica/β-glucan nanoparticles for smart delivery of fungicides. *ACS Sustain. Chem. Eng.* **2021**, *9*, 9126–9138. Available from: https://doi.org/10.1021/acssuschemeng.1c02962.
71. Shang, H.; Ma, C.; Li, C.; White, J. C.; Polubesova, T.; Chefetz, B., et al. Copper sulfide nanoparticles suppress *Gibberella fujikuroi* infection in rice (*Oryza sativa* L.) by multiple mechanisms: contact-mortality, nutritional modulation and phytohormone regulation. *Env. Sci. Nano* **2020**, *7*, 2632–2643. Available from: https://doi.org/10.1039/D0EN00535E.
72. Abdelrhim, A.; Mazrou, Y.; Nehela, Y.; Atallah, O.; El-Ashmony, R.; Dawood, M. Silicon dioxide nanoparticles induce innate immune responses and activate antioxidant machinery in wheat against *Rhizoctonia solani*. *Plants* **2021**, *10*, 2758. Available from: https://doi.org/10.3390/plants10122758.
73. Bano, A.; Ummat-ul-Habib. Interactive effects of Ag-nanoparticles, salicylic acid, and plant growth promoting rhizobacteria on the physiology of wheat infected with yellow rust. *J. Plant. Pathol.* **2020**, *102*, 1215–1225. Available from: https://doi.org/10.1007/s42161-020-00626-y.
74. Jo, Y.-K.; Kim, B. H.; Jung, G. Antifungal activity of silver ions and nanoparticles on phytopathogenic fungi. *Plant. Dis.* **2009**, *93*, 1037–1043. Available from: https://doi.org/10.1094/PDIS-93-10-1037.
75. Abdallah, Y.; Liu, M.; Ogunyemi, S. O.; Ahmed, T.; Fouad, H.; Abdelazez, A., et al. Bioinspired green synthesis of chitosan and zinc oxide nanoparticles with strong antibacterial activity against rice pathogen *Xanthomonas oryzae* pv. oryzae. *Molecules* **2020**, *25*, 4795. Available from: https://doi.org/10.3390/molecules25204795.
76. Choudhary, R. C.; Kumaraswamy, R. V.; Kumari, S.; Sharma, S. S.; Pal, A.; Raliya, R., et al. Cu-chitosan nanoparticle boost defense responses and plant growth in maize (L.). *Sci. Rep.* **2017**, *7*, 9754. Available from: https://doi.org/10.1038/s41598-017-08571-0.
77. Elmer, W.; Ma, C.; White, J. Nanoparticles for plant disease management. *Curr. Opin. Env. Sci. Health* **2018**, *6*, 66–70. Available from: https://doi.org/10.1016/j.coesh.2018.08.002.
78. Munir, N.; Gulzar, W.; Abideen, Z.; Hancock, J. T.; El-Keblawy, A.; Radicetti, E. Nanotechnology improves disease resistance in plants for food security: applications and challenges. *Biocatal. Agric. Biotechnol.* **2023**, *51*, 102781. Available from: https://doi.org/10.1016/j.bcab.2023.102781.

79. Wang, D.; Saleh, N. B.; Byro, A.; Zepp, R.; Sahle-Demessie, E.; Luxton, T. P., et al. Nano-enabled pesticides for sustainable agriculture and global food security. *Nat. Nanotechnol.* **2022**, *17*, 347–360. Available from: https://doi.org/10.1038/s41565-022-01082-8.
80. Barik, T. K.; Sahu, B.; Swain, V. Nanosilica—from medicine to pest control. *Parasitol. Res.* **2008**, *103*, 253–258. Available from: https://doi.org/10.1007/s00436-008-0975-7.
81. Patil, C. D.; Borase, H. P.; Suryawanshi, R. K.; Patil, S. V. Trypsin inactivation by latex fabricated gold nanoparticles: a new strategy towards insect control. *Enzyme Microb. Technol.* **2016**, *92*, 18–25. Available from: https://doi.org/10.1016/j.enzmictec.2016.06.005.
82. Hofmann, T.; Lowry, G. V.; Ghoshal, S.; Tufenkji, N.; Brambilla, D.; Dutcher, J. R., et al. Technology readiness and overcoming barriers to sustainably implement nanotechnology-enabled plant agriculture. *Nat. Food* **2020**, *1*, 416–425. Available from: https://doi.org/10.1038/s43016-020-0110-1.
83. Goswami, A.; Roy, I.; Sengupta, S.; Debnath, N. Novel applications of solid and liquid formulations of nanoparticles against insect pests and pathogens. *Thin Solid. Films* **2010**, *519*, 1252–1257. Available from: https://doi.org/10.1016/j.tsf.2010.08.079.
84. Stadler, T.; Buteler, M.; Weaver, D. K. Novel use of nanostructured alumina as an insecticide. *Pest. Manag. Sci.* **2010**, *66*, 577–579. Available from: https://doi.org/10.1002/ps.1915.
85. Wang, L.; Li, X.; Zhang, G.; Dong, J.; Eastoe, J. Oil-in-water nanoemulsions for pesticide formulations. *J. Colloid Interface Sci.* **2007**, *314*, 230–235. Available from: https://doi.org/10.1016/j.jcis.2007.04.079.
86. Hamed, S. M.; Hagag, E. S.; El-Raouf, N. A. Green production of silver nanoparticles, evaluation of their nematicidal activity against Meloidogyne javanica and their impact on growth of faba bean. *Beni Suef Univ. J. Basic. Appl. Sci.* **2019**, *8*, 9. Available from: https://doi.org/10.1186/s43088-019-0010-3.
87. Rastogi, A.; Tripathi, D. K.; Yadav, S.; Chauhan, D. K.; Živčák, M.; Ghorbanpour, M., et al. Application of silicon nanoparticles in agriculture. *3 Biotech.* **2019**, *9*, 90. Available from: https://doi.org/10.1007/s13205-019-1626-7.
88. Duhan, J. S.; Kumar, R.; Kumar, N.; Kaur, P.; Nehra, K.; Duhan, S. Nanotechnology: the new perspective in precision agriculture. *Biotechnol. Rep.* **2017**, *15*, 11–23. Available from: https://doi.org/10.1016/j.btre.2017.03.002.
89. Chaw Jiang, L.; Basri, M.; Omar, D.; Abdul Rahman, M. B.; Salleh, A. B.; Raja Abdul Rahman, R. N. Z., et al. Green nano-emulsion intervention for water-soluble glyphosate isopropylamine (IPA) formulations in controlling Eleusine indica (*E. indica*). *Pestic. Biochem. Physiol.* **2012**, *102*, 19–29. Available from: https://doi.org/10.1016/j.pestbp.2011.10.004.
90. Ghazaryan, K.; Verma, A.; Rawat, S.; Warshini, A.; Rajput, P.; Minkina, T., et al. *Green remediation harnessing plant-based strategies for removal of emerging soil contaminants. Nanotechnology Applications and Innovations for Improved Soil Health;* IGI Global, 2024, pp. 57–71. Available from: https://doi.org/10.4018/979-8-3693-1471-5.ch004.
91. Singh, A.; Rajput, V. D.; Lalotra, S.; Agrawal, S.; Ghazaryan, K.; Singh, J., et al. Zinc oxide nanoparticles influence on plant tolerance to salinity stress: insights into physiological, biochemical, and molecular responses. *Env. Geochem. Health* **2024**, *46*, 148. Available from: https://doi.org/10.1007/s10653-024-01921-8.

92. Singh, A.; Rajput, V. D.; Varshney, A.; Ghazaryan, K.; Minkina, T. Small Tech, big impact: agri-nanotechnology journey to optimize crop protection and production for sustainable agriculture. *Plant. Stress.* **2023**, *10*, 100253. Available from: https://doi.org/10.1016/j.stress.2023.100253.
93. Abdelsalam, I. M.; Ghosh, S.; AlKafaas, S. S.; Bedair, H.; Malloum, A.; ElKafas, S. S., et al. Nanotechnology as a tool for abiotic stress mitigation in horticultural crops. *Biologia (Bratisl.)* **2022**, *78*, 163–178. Available from: https://doi.org/10.1007/s11756-022-01251-z.
94. Hu, J.; Xianyu, Y. When nano meets plants: a review on the interplay between nanoparticles and plants. *Nano Today* **2021**, *38*, 101143. Available from: https://doi.org/10.1016/j.nantod.2021.101143.
95. Demirer, G. S.; Silva, T. N.; Jackson, C. T.; Thomas, J. B.; W. Ehrhardt, D.; Rhee, S. Y., et al. Nanotechnology to advance CRISPR–Cas genetic engineering of plants. *Nat. Nanotechnol.* **2021**, *16*, 243–250. Available from: https://doi.org/10.1038/s41565-021-00854-y.
96. Yan, Y.; Ni, M.; Wang, F.; Yu, Y.; Gong, X.; Huang, Y., et al. Metal–organic framework-based biosensor for detecting hydrogen peroxide in plants through color-to-thermal signal conversion. *ACS Nano* **2022**, *16*, 15175–15187. Available from: https://doi.org/10.1021/acsnano.2c06481.
97. Mahakham, W.; Sarmah, A. K.; Maensiri, S.; Theerakulpisut, P. Nanopriming technology for enhancing germination and starch metabolism of aged rice seeds using phytosynthesized silver nanoparticles. *Sci. Rep.* **2017**, *7*, 8263. Available from: https://doi.org/10.1038/s41598-017-08669-5.
98. Hammond, W. M.; Williams, A. P.; Abatzoglou, J. T.; Adams, H. D.; Klein, T.; López, R., et al. Global field observations of tree die-off reveal hotter-drought fingerprint for Earth's forests. *Nat. Commun.* **2022**, *13*, 1761. Available from: https://doi.org/10.1038/s41467-022-29289-2.
99. Lamaoui, M.; Jemo, M.; Datla, R.; Bekkaoui, F. Heat and Drought stresses in crops and approaches for their mitigation. *Front. Chem.* **2018**, *6*. Available from: https://doi.org/10.3389/fchem.2018.00026.
100. Yang, Z.; Qin, F. The battle of crops against drought: genetic dissection and improvement. *J. Integr. Plant. Biol.* **2023**, *65*, 496–525. Available from: https://doi.org/10.1111/jipb.13451.
101. Fahad, S.; Bajwa, A. A.; Nazir, U.; Anjum, S. A.; Farooq, A.; Zohaib, A., et al. Crop Production under drought and heat stress: plant responses and management options. *Front. Plant. Sci.* **2017**, *8*. Available from: https://doi.org/10.3389/fpls.2017.01147.
102. Rasheed, A.; Zhao, L.; Raza, A.; Mahmood, A.; Xing, H.; Lv, X., et al. Role of molecular breeding tools in enhancing the breeding of drought-resilient cotton genotypes: an updated review. *Water (Basel)* **2023**, *15*, 1377. Available from: https://doi.org/10.3390/w15071377.
103. Varshney, R. K.; Barmukh, R.; Roorkiwal, M.; Qi, Y.; Kholova, J.; Tuberosa, R., et al. Breeding custom-designed crops for improved drought adaptation. *Adv. Genet.* **2021**, *2*. Available from: https://doi.org/10.1002/ggn2.202100017.
104. Wang, Y.; Chen, G.; Zeng, F.; Han, Z.; Qiu, C.; Zeng, M., et al. Molecular evidence for adaptive evolution of drought tolerance in wild cereals. *N. Phytologist* **2023**, *237*, 497–514. Available from: https://doi.org/10.1111/nph.18560.

105. Zhao, X.; Yuan, X.; Xing, Y.; Dao, J.; Zhao, D.; Li, Y., et al. A meta-analysis on morphological, physiological and biochemical responses of plants with PGPR inoculation under drought stress. *Plant. Cell Env.* **2023**, *46*, 199–214. Available from: https://doi.org/10.1111/pce.14466.
106. Salvi, P.; Manna, M.; Kaur, H.; Thakur, T.; Gandass, N.; Bhatt, D., et al. Phytohormone signaling and crosstalk in regulating drought stress response in plants. *Plant. Cell Rep.* **2021**, *40*, 1305–1329. Available from: https://doi.org/10.1007/s00299-021-02683-8.
107. Waqas Mazhar, M.; Ishtiaq, M.; Hussain, I.; Parveen, A.; Hayat Bhatti, K.; Azeem, M., et al. Seed nano-priming with zinc oxide nanoparticles in rice mitigates drought and enhances agronomic profile. *PLoS One* **2022**, *17*, e0264967. Available from: https://doi.org/10.1371/journal.pone.0264967.
108. Ghani, M. I.; Saleem, S.; Rather, S. A.; Rehmani, M. S.; Alamri, S.; Rajput, V. D., et al. Foliar application of zinc oxide nanoparticles: an effective strategy to mitigate drought stress in cucumber seedling by modulating antioxidant defense system and osmolytes accumulation. *Chemosphere* **2022**, *289*, 133202. Available from: https://doi.org/10.1016/j.chemosphere.2021.133202.
109. Semida, W. M.; Abdelkhalik, A.; Mohamed, G. F.; Abd El-Mageed, T. A.; Abd El-Mageed, S. A.; Rady, M. M., et al. Foliar application of zinc oxide nanoparticles promotes drought stress tolerance in eggplant (*Solanum melongena* L.). *Plants* **2021**, *10*, 421. Available from: https://doi.org/10.3390/plants10020421.
110. Akhtar, N.; Ilyas, N.; Mashwani, Z.-R.; Hayat, R.; Yasmin, H.; Noureldeen, A., et al. Synergistic effects of plant growth promoting rhizobacteria and silicon dioxide nano-particles for amelioration of drought stress in wheat. *Plant. Physiol. Biochem.* **2021**, *166*, 160–176. Available from: https://doi.org/10.1016/j.plaphy.2021.05.039.
111. Mustafa, H.; Ilyas, N.; Akhtar, N.; Raja, N. I.; Zainab, T.; Shah, T., et al. Biosynthesis and characterization of titanium dioxide nanoparticles and its effects along with calcium phosphate on physicochemical attributes of wheat under drought stress. *Ecotoxicol. Env. Saf.* **2021**, *223*, 112519. Available from: https://doi.org/10.1016/j.ecoenv.2021.112519.
112. Taran, N.; Storozhenko, V.; Svietlova, N.; Batsmanova, L.; Shvartau, V.; Kovalenko, M. Effect of zinc and copper nanoparticles on drought resistance of wheat seedlings. *Nanoscale Res. Lett.* **2017**, *12*, 60. Available from: https://doi.org/10.1186/s11671-017-1839-9.
113. Dimkpa, C. O.; Singh, U.; Bindraban, P. S.; Elmer, W. H.; Gardea-Torresdey, J. L.; White, J. C. Zinc oxide nanoparticles alleviate drought-induced alterations in sorghum performance, nutrient acquisition, and grain fortification. *Sci. Total. Environ.* **2019**, *688*, 926–934. Available from: https://doi.org/10.1016/j.scitotenv.2019.06.392.
114. Alam, H.; Khattak, J. Z. K.; Ksiksi, T. S.; Saleem, M. H.; Fahad, S.; Sohail, H., et al. Negative impact of long-term exposure of salinity and drought stress on native *Tetraena mandavillei* L. *Physiol. Plant.* **2021**, *172*, 1336–1351. Available from: https://doi.org/10.1111/ppl.13273.
115. Gupta, S.; Schillaci, M.; Walker, R.; Smith, P. M. C.; Watt, M.; Roessner, U. Alleviation of salinity stress in plants by endophytic plant-fungal symbiosis: current knowledge, perspectives and future directions. *Plant. Soil.* **2021**, *461*, 219–244. Available from: https://doi.org/10.1007/s11104-020-04618-w.

116. Melino, V.; Tester, M. Salt-tolerant crops: time to deliver. *Annu. Rev. Plant. Biol.* **2023**, *74*, 671–696. Available from: https://doi.org/10.1146/annurev-arplant-061422-104322.
117. Johnson, R.; Puthur, J. T. Seed priming as a cost effective technique for developing plants with cross tolerance to salinity stress. *Plant. Physiol. Biochem.* **2021**, *162*, 247–257. Available from: https://doi.org/10.1016/j.plaphy.2021.02.034.
118. Adil, M.; Bashir, S.; Bashir, S.; Aslam, Z.; Ahmad, N.; Younas, T., et al. Zinc oxide nanoparticles improved chlorophyll contents, physical parameters, and wheat yield under salt stress. *Front. Plant. Sci.* **2022**, *13*. Available from: https://doi.org/10.3389/fpls.2022.932861.
119. Saad-Allah, K. M.; Ragab, G. A. Sulfur nanoparticles mediated improvement of salt tolerance in wheat relates to decreasing oxidative stress and regulating metabolic activity. *Physiol. Mol. Biol. Plants* **2020**, *26*, 2209–2223. Available from: https://doi.org/10.1007/s12298-020-00899-8.
120. El-Saber, M. M.; Mahdi, A. A.; Hassan, A. H.; Farroh, K. Y.; Osman, A. Effects of magnetite nanoparticles on physiological processes to alleviate salinity induced oxidative damage in wheat. *J. Sci. Food Agric.* **2021**, *101*, 5550–5562. Available from: https://doi.org/10.1002/jsfa.11206.
121. Shah, T.; Latif, S.; Saeed, F.; Ali, I.; Ullah, S.; Abdullah Alsahli, A., et al. Seed priming with titanium dioxide nanoparticles enhances seed vigor, leaf water status, and antioxidant enzyme activities in maize (*Zea mays* L.) under salinity stress. *J. King Saud. Univ. Sci.* **2021**, *33*, 101207. Available from: https://doi.org/10.1016/j.jksus.2020.10.004.
122. El-Badri, A. M.; Batool, M.; Wang, C.; Hashem, A. M.; Tabl, K. M.; Nishawy, E., et al. Selenium and zinc oxide nanoparticles modulate the molecular and morpho-physiological processes during seed germination of Brassica napus under salt stress. *Ecotoxicol. Env. Saf.* **2021**, *225*, 112695. Available from: https://doi.org/10.1016/j.ecoenv.2021.112695.
123. Khan, M. N.; Li, Y.; Fu, C.; Hu, J.; Chen, L.; Yan, J., et al. CeO_2 Nanoparticles seed priming increases salicylic acid level and ROS scavenging ability to improve rapeseed salt tolerance. *Glob. Chall.* **2022**, *6*. Available from: https://doi.org/10.1002/gch2.202200025.
124. Hezaveh, T. A.; Pourakbar, L.; Rahmani, F.; Alipour, H. Interactive effects of salinity and ZnO nanoparticles on physiological and molecular parameters of rapeseed (*Brassica napus* L.). *Commun. Soil. Sci. Plant. Anal.* **2019**, *50*, 698–715. Available from: https://doi.org/10.1080/00103624.2019.1589481.
125. Martínez-Ballesta, M. C.; Zapata, L.; Chalbi, N.; Carvajal, M. Multiwalled carbon nanotubes enter broccoli cells enhancing growth and water uptake of plants exposed to salinity. *J. Nanobiotechnology* **2016**, *14*, 42. Available from: https://doi.org/10.1186/s12951-016-0199-4.
126. Li, Y.; Liu, M.; Yang, X.; Zhang, Y.; Hui, H.; Zhang, D., et al. Multi-walled carbon nanotubes enhanced the antioxidative system and alleviated salt stress in grape seedlings. *Sci. Hortic.* **2022**, *293*, 110698. Available from: https://doi.org/10.1016/j.scienta.2021.110698.
127. Ganjavi, A. S.; Oraei, M.; Gohari, G.; Akbari, A.; Faramarzi, A. Glycine betaine functionalized graphene oxide as a new engineering nanoparticle lessens salt stress impacts in sweet basil (*Ocimum basilicum* L.). *Plant. Physiol. Biochem.* **2021**, *162*, 14–26. Available from: https://doi.org/10.1016/j.plaphy.2021.02.028.

128. Hassan, F. A. S.; Ali, E.; Gaber, A.; Fetouh, M. I.; Mazrou, R. Chitosan nanoparticles effectively combat salinity stress by enhancing antioxidant activity and alkaloid biosynthesis in *Catharanthus roseus* (L.) G. Don. *Plant. Physiol. Biochem.* **2021**, *162*, 291–300. Available from: https://doi.org/10.1016/j.plaphy.2021.03.004.
129. An, J.; Hu, P.; Li, F.; Wu, H.; Shen, Y.; White, J. C., et al. Emerging investigator series: molecular mechanisms of plant salinity stress tolerance improvement by seed priming with cerium oxide nanoparticles. *Env. Sci. Nano* **2020**, *7*, 2214–2228. Available from: https://doi.org/10.1039/D0EN00387E.
130. Faizan, M.; Bhat, J. A.; Chen, C.; Alyemeni, M. N.; Wijaya, L.; Ahmad, P., et al. Zinc oxide nanoparticles (ZnO-NPs) induce salt tolerance by improving the antioxidant system and photosynthetic machinery in tomato. *Plant. Physiol. Biochem.* **2021**, *161*, 122–130. Available from: https://doi.org/10.1016/j.plaphy.2021.02.002.
131. Siddiqui, M. H.; Al-Whaibi, M. H.; Faisal, M.; Al Sahli, A. A. Nano-silicon dioxide mitigates the adverse effects of salt stress on *Cucurbita pepo* L. *Env. Toxicol. Chem.* **2014**, *33*, 2429–2437. Available from: https://doi.org/10.1002/etc.2697.
132. Wang, S.; Wang, F.; Gao, S. Foliar application with nano-silicon alleviates Cd toxicity in rice seedlings. *Environ. Sci. Pollut. Res.* **2015**, *22*, 2837–2845. Available from: https://doi.org/10.1007/s11356-014-3525-0.
133. Liu, M.; Pan, T.; Allakhverdiev, S. I.; Yu, M.; Shabala, S. Crop halophytism: an environmentally sustainable solution for global food security. *Trends Plant. Sci.* **2020**, *25*, 630–634. Available from: https://doi.org/10.1016/j.tplants.2020.04.008.
134. Cunningham, F. J.; Goh, N. S.; Demirer, G. S.; Matos, J. L.; Landry, M. P. Nanoparticle-mediated delivery towards advancing plant genetic engineering. *Trends Biotechnol.* **2018**, *36*, 882–897. Available from: https://doi.org/10.1016/j.tibtech.2018.03.009.
135. Kausch, A. P.; Nelson-Vasilchik, K.; Hague, J.; Mookkan, M.; Quemada, H.; Dellaporta, S., et al. Edit at will: genotype independent plant transformation in the era of advanced genomics and genome editing. *Plant. Sci.* **2019**, *281*, 186–205. Available from: https://doi.org/10.1016/j.plantsci.2019.01.006.
136. Saeed, F.; Hashmi, M. H.; Hossain, M. J.; Ali, M. A.; Bakhsh, A. *Transgenic technologies for efficient insect pest management in crop plants. Transgenic Technology Based Value Addition in Plant Biotechnology;* Elsevier, 2020, pp. 123–156. Available from: https://doi.org/10.1016/B978-0-12-818632-9.00006-X.
137. Mujtaba, M.; Wang, D.; Carvalho, L. B.; Oliveira, J. L.; Espirito Santo Pereira, A. do; Sharif, R., et al. Nanocarrier-mediated delivery of miRNA, RNAi, and CRISPR-Cas for plant protection: current trends and future directions. *ACS Agric. Sci. & Technol.* **2021**, *1*, 417–435. Available from: https://doi.org/10.1021/acsagscitech.1c00146.
138. Demirer, G. S.; Zhang, H.; Goh, N. S.; Pinals, R. L.; Chang, R.; Landry, M. P. Carbon nanocarriers deliver siRNA to intact plant cells for efficient gene knockdown. *Sci. Adv.* **2020**, *6*. Available from: https://doi.org/10.1126/sciadv.aaz0495.
139. Cheng, Q.; Wei, T.; Farbiak, L.; Johnson, L. T.; Dilliard, S. A.; Siegwart, D. J. Selective organ targeting (SORT) nanoparticles for tissue-specific mRNA delivery and CRISPR–Cas gene editing. *Nat. Nanotechnol.* **2020**, *15*, 313–320. Available from: https://doi.org/10.1038/s41565-020-0669-6.

140. Qiu, M.; Glass, Z.; Chen, J.; Haas, M.; Jin, X.; Zhao, X., et al. Lipid nanoparticle-mediated codelivery of Cas9 mRNA and single-guide RNA achieves liver-specific in vivo genome editing of *Angptl3*. *Proc. Natl Acad. Sci.* **2021**, *118*. Available from: https://doi.org/10.1073/pnas.2020401118.
141. Jo, A.; Ringel-Scaia, V. M.; McDaniel, D. K.; Thomas, C. A.; Zhang, R.; Riffle, J. S., et al. Fabrication and characterization of PLGA nanoparticles encapsulating large CRISPR–Cas9 plasmid. *J. Nanobiotechnology* **2020**, *18*, 16. Available from: https://doi.org/10.1186/s12951-019-0564-1.

CHAPTER

5 Nanomaterial interactions with plant systems: uptake, transport, and accumulation

Shobhna[1], Mani Salaria[1], Tarun Sharma[2], Diptanu Banik[2], Chandra Mohan Mehta[2], Ramesh K. Sadawarti[1] and Marina Burachevskaya[3]

[1]*Department of Horticulture, School of Agriculture, Lovely Professional University, Phagwara, Punjab, India*

[2]*Department of Agronomy, School of Agriculture, Lovely Professional University, Phagwara, Punjab, India*

[3]*Laboratory of Soil Chemistry and Ecology, Tula State Lev Tolstoy Pedagogical University (Tolstoy University), Tula, Russia*

5.1 Introduction

The science of nanotechnology is characterized as engineering of tiny objects between one and 100 nm (a nanometer is one billionth of a meter) (USDA). Because nanotechnology enhances plat breeding, helps with pesticide delivery systems, lowers input costs and lessens the environmental effect of contemporary agriculture, its application in agriculture might increase the yield and value of products.[1] Restrained sources as well as the swiftly growing human populace, which by 2050 is projected to reach 9.6 billion demands the evolution of a dynamic agriculture to decrease the global poverty and hunger. Chemical fertilizers are put in force to provide nutrients to plant for profitable increase and productivity. Currently, without the huge use of fertilizers the increasing demand of meals cannot be fulfilled.[2] In agriculture the utilization of nanotechnology contains liberation of few substances under control such as pesticides and fertilizers. Nanoparticles, characterized by minimum one dimension of 100 nm or less, occupy the intermediate realm between particular molecules or atoms and the bigger bulk substance. This positioning has the potential to significantly alter the physicochemical effects of the material, raising concerns about the possible induction of adverse biological effects in organisms.[3] However, the outcome of nanoparticles on uptake and use efficiency, growth, and metabolic functions in plant life ranges among genotypes. Through endocytosis, aquaporins and iconic channels nanoparticles filled with nutrients reach the cells of root by various internationalized pathways for protein transportation. Nanoparticles coated with carbon can transfer inside the entire plant to distinctive organs and tissues in various species of plant. On the contrary, plasmodesmata allow for the transfer of nanoparticles from one cell to the next inside the roots.[4]

Biogenic Nanoparticles. DOI: https://doi.org/10.1016/B978-0-443-33819-9.00009-3

Because of their size, these particles start to display characteristics that indicate large opposite numbers. Quantum mechanical interactions become more important than traditional mechanical forces because of their size, even though physical and chemical houses are superior due of their extraordinarily high surface-to-frame ratio. Upon reaching a particular concentration within the plant, nanoparticles pose a significant threat to crop yield and productivity. They exert their impact by altering morphological, physiological, anatomical, biochemical, and genetic system of crop. After being absorbed and moving through the plant system, the nanoparticles gather in distinct areas of the aerial plants, which raises the possibility that they will cycle across the ecosystem via multiple trophic levels. The excessive presence of nanoparticles inside the system of plant adversely effects the crop quality, which lead to reduction in germination rates, respiration rate, photosynthesis, total biomass and eventually yield of the crop. However, plants employ inherent defense mechanism, which includes both biochemical and nonbiochemical defense process to mitigate the infectiveness induced by stress conditions, including exposure to nanoparticles.[5] This chapter's primary objective is to explain how different NPs accumulate, translocate, and are absorbed inside plant systems, as well as how this affects plant life (Fig. 5.1).

5.2 Uptake and mechanism of different NPs within plant system

Nanoparticles, upon exposure to plants, undergo a complex process of penetration into the root epidermis, where they traverse the cellular wall and membrane. The

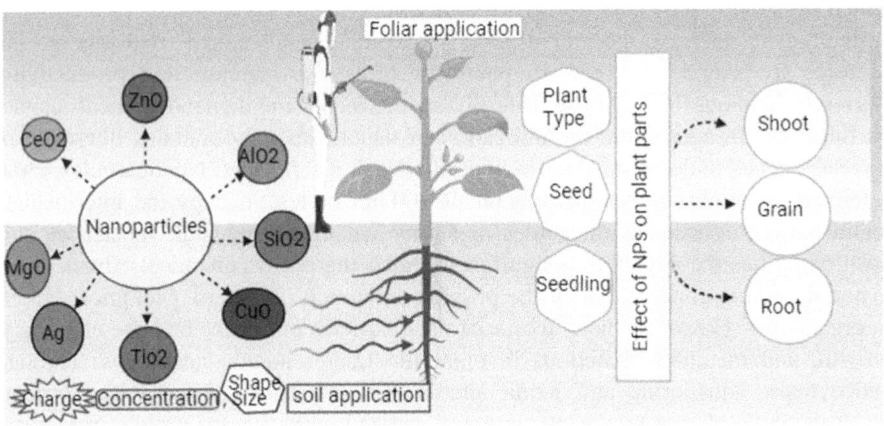

FIGURE 5.1

Overview of properties of NPs and their applications on plant.

Made by author.

movement of nanoparticles through these membranes is size-specific, occurring through openings in the cellular membrane. Prior to reaching the stele (central part of the root), nanoparticles must passively integrate via the apoplastic pathway within the endodermis. The xylem, serving as a pivotal transport system within plants, serves a vital function in the dispersion and translocation of nanoparticles. The wall of cell, characterized by a porous network of polysaccharide fiber matrix, facilitates the uptake of water molecules and other solutes, making it a key pathway for nanoparticle movement. This intricate journey underscores the significance of size-specific interactions and the involvement of various cellular and tissue structures in nanoparticle transportation within the plant.[6]

5.3 Foliar exposure

Plants can be subjected to nanoparticles (NPs) via two distinct modes: foliar exposure and root exposure.

5.3.1 Foliar exposure

The schematic diagram in Fig. 5.2 summarizes existing awareness of the foliar absorption and movement pathways of nanoparticles during exposure to foliage. The main natural barrier is the cuticle avoiding NPs from penetrating substrate tissue. The cuticular pathway encompasses two ways for uptake of solute: the nonpolar pathway, allowing diffusion and permeation of lipophilic solutes, and the hydrophilic route, enabling the entry of polar solutes through aqueous pores. These pores' effective sizes vary from 0.6 to 4.8 nm. As a result, nanoparticles 4.8 nm or less can directly breach the cuticular layer via the cuticle route. Despite this, numerous investigations have compiled the foliar intake and gathering of NPs more than 5 nm, and the specific mechanism for their uptake remains unclear. Therefore, it is imperative to thoroughly examine if nanoparticles (NPs), particularly those larger than 5 nm, may enter leaves through the cuticular route. While the structure of plant leaves varies across diverse species, the leaf surface typically consists of several features, as illustrated in Fig. 5.2. These features include stomata, hydrothode pores, and trichome hairs each serving particular role outlined beneath. Additionally, most of the leaf's exterior is enveloped by a cuticle, a lipophilic layer of wax with a thickness ranging from 100 to 250 nm. Notably, the surface of necrotized and trichomes cells may have a weaker or nonexistent cuticle.[7] Upon entry into the cuticle, the journey of nanoparticles goes through various functional groups present in a leaf. Several factors which could influence the NP translocation in leaves are:

1. **Functional groups:** The diverse functional groups in internal leaf cells may interact with the surfaces of NPs, potentially influencing their movement within the leaf.

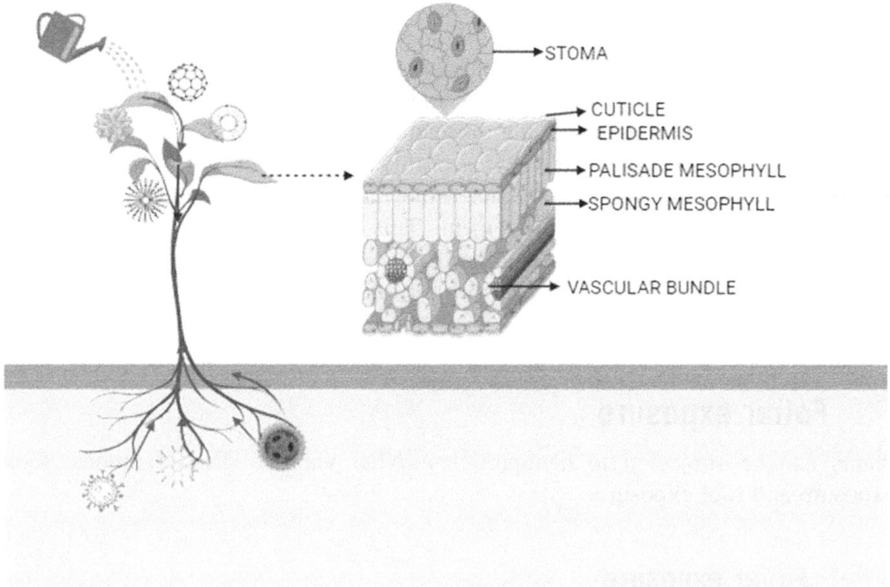

FIGURE 5.2

Diagrammatic representing of NP's foliar uptake.

Made by author.

2. **Hydrogen bonding**: The ability of NP surfaces to form hydrogen bonds with cell wall characteristics could affect their movement.
3. **Size exclusion limits:** Different features within internal leaf cells, such as apoplastic transport and endocytosis, may have size exclusion limits. The size and shape of nanoparticles could influence the route of cellular translocation (apoplastic versus symplastic).
4. **Apoplastic versus symplastic translocation:** The proposed path for NP movement (apoplastic vs symplastic) could be impacted by the shape and/or the size of the NPs.
5. **Critical property size:** The size of NPs is highlighted as a crucial characteristics that could influence their uptake and movement in the leaf vascular system.

Additionally, it's crucial to recognize that alternations to the cuticles of the leaves occur with damages from pest and diseases, life stages and weather potentially allowing the bypass cuticle tissue acts as the primary defense against NPs in certain instances.[8] Stoma has a vital function in regulating exchange of gas and water between the plant and its environment. These stomatal pores have dimensions in the tens of microns, such as spinach leaves have 5×13 μm, watermelon and tomato leaves have 5×7 μm, cucumber leaves have 21×13 μm, and lettuce leaves have 16 × 6 μm. The size of stomata allows for the penetration of nanoparticles (NPs) when open. However, their responsiveness, whether open or closed, is affected by several elements such as carbon dioxide concentrations, moisture levels, temperature, and light intensity. It is noteworthy that different plant species have different numbers and locations of stomata. Many species of plants contain just stomata on the bottom (abaxial) side of their leaves, while others have stomata on each side of their leaves. In instances where stomata are found on both surfaces however eudicots usually contain 1.4 times more stomata on the adaxial leaf surface compared to the adaxial surface. On the other hand, monocots have a comparable amount of stomata on both sides.[9] Direct application of agrochemicals to plant leaves is a practical and cost effective method of managing agriculture.[10]

5.3.2 Root uptake

Adsorption occurs on the root surface as the first contact between nanoparticles and plant roots. Due to the chemical compounds that are released, like organic acids or mucus, by root hairs, the root surface becomes negatively charged. As a result, nanoparticles carrying positive charges are easily and frequently generated in the root absorb onto surface of the root.[11] Mineral nutrient particulates may go via the xylem and phloem, respectively, to get mobilized and remobilized. The facilitation of nutrient internalization involves both the plasma membrane (apoplastic) and plasmodesma (symplastic) transport mechanisms, in conjunction with water. The roots draw water up from the soil, allowing it to be moved radially into xylem tracheary elements. The regulation of the speed of water transport is subsequently carried out by xylem structures, which play crucial roles as determinants. Diverse uptake kinetics for nanoparticles (NPs) can be validated by different xylem structures.[12] The extracellular gaps and the continuous cell walls of neighboring cells make up the apoplast architecturally. Symplastic transport refers to the flow of water from one cell to another through the plasma membrane and cytoplasm.[13] Nevertheless, nanoparticles (NPs) must transcend the Casparian strip barrier and travel along symplastic pathways in order to access the central vascular cylinder (Fig. 5.1). This happens when it binds to a protein carrier on the endodermal cell membrane through endocytosis, pore formation, and subsequent transport. NPs are absorbed inside the cytoplasm after moving through plasmodesmata from one cell to another. The NPs that are unable to internalize gather together on the Casparian strip. Meanwhile, NPs are transported to the shoots after passing via the xylem and then back to the root system through the phloem.[14] The sites of the apoplasmic bypass have been identified inside the lateral roots as well as in the regions where they originate. Deep within the main root is where lateral roots begin to grow. The tissues above experience cell separation, which leaves spaces amongst cells in the epidermis at the locations where lateral roots arise, preparing the way for the formation of medial root primordial.[15] Tiny holes in the cell wall of the root can obstruct the passage of big particles.[16] When the root's exodermis is absent, tiny particles can penetrate the root's xylem, or middle column. The development of lateral roots is essential in determining the structural makeup of the root system. The extent of root branching directly influences a plant's ability to efficiently absorb water, acquire nutrients, and establish firm anchorage. Therefore, a thorough comprehension of the mechanics governing lateral root development holds crucial significance in agronomy. The cellular and molecular foundations of the genesis of lateral roots have been thoroughly studied, with *Arabidopsis thaliana* (Arabidopsis) serving as a prominent model plant. Current developments in research have led to the identification of numerous Arabidopsis genes that govern various aspects of lateral root initiation, patterning, and emergence (Fig. 5.3).

One noteworthy discovery from these studies is the critical role of the plant's hormone auxin as a common indicator that unites various diverse but linked phases of development.[17] Numerous investigations have demonstrated the absorption of NPs by roots of the plants and their penetration into cells primarily via the hydrophilic pathway. However, because of the tiny size of the pore, the ability of nanoparticles to enter cells via this channel is severely limited.[18]Endocytosis is a critical process for the uptake of nanoparticles (NPs) in plant cells. The inward folding of the plasma membrane of plant cells facilitates the way that NPs enter cells. Research has indicated that protoplasts of plant can ingest particles smaller than 1 μm by the process of endocytosis. Consequently, nanoparticles absorbed

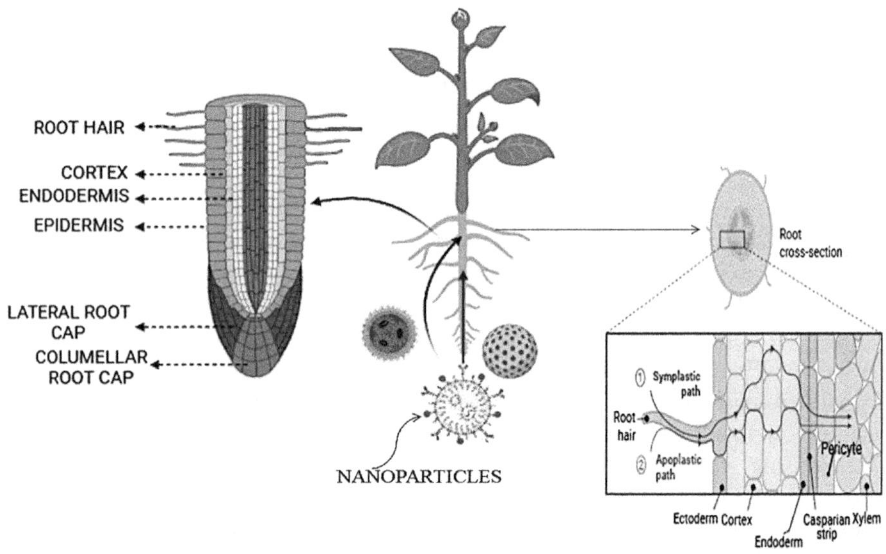

FIGURE 5.3

Diagrammatic illustrating of root uptake of nanoparticles.

Made by author.

through endocytosis exhibit no particle size selectivity.[19] Liu et al.[20] suggested that catharanthusroseus root cells absorb carbon-based nanoparticles and carbon nanotubes through endocytosis. Moreover, a method by which nanoparticles combine with proteins that travel on the epidermis's outer layer allows plants to absorb them.[21] Several key factors influence the uptake of nanoparticles through roots. Plant roots' ability to absorb nanoparticles is significantly influenced by several aspects, including the size, chemical content, and electricity on the particle surface.[22]

5.3.2.1 Size and chemic composition's effect

Several potential mechanisms exist for the entry of AuNPs into roots of plant. One potential scenario is their passage via pores present in the plant cell wall. Research suggests that the pores' limiting width in the cell walls of alive plants falls between 3.5 and 5 nm. Considering the hydrodynamic diameters of 6–10 nm and core sizes of around 2 nm for the AuNPs in this study, it is conceivable that they might pass via these holes and into root cells. Another scenario involves NPs enhancing the cell wall permeability in plants, creating "holes" that promote absorption into root cells, as observed in previous studies. It is known that NPs cause 15 to 40-nm diameter nanoscale membrane pores in cell plasma membranes. Furthermore, the apical meristematic tissue at the root tip, which is situated under the root cap, represents yet another potential entry point for NPs into plant roots.[23] The absorption of nanoparticles based on cerium oxide is contingent on energy and is facilitated through lipid raft endocytosis mediated by clathrin. These nanoparticles (CCNPs) exhibit broad distribution across various cell compartments, suggesting that their recent effectiveness as a novel antioxidant may be attributed, at least in part, to this extensive distribution.[24] According to several authors, following previous nanoparticle treatments, the average particle exclusion width of root cell wall pores dropped from 6.6 to 3.0 nm, suggesting that TiO_2 nanoparticles may not be taken by plants. Additionally, exposing plants to irrigation water containing well-characterized nanoparticle expulsions did not show detectable translocation. However, regarding *Arabidopsis thaliana* seedlings, it was noted that they took up TiO_2 NPs (< 5 nm) mixed with Alizarin red S nanoconjugate, distributing them within tissues and cells. The ability of *Arabidopsis thaliana* roots to secrete a mucilage that formed a pectin hydrogel capsule suggested that roots might either promote or prevent the entrance of TiO_2 NPs. Furthermore, wheat (*Triticum aestivum*) and the roots of cucumber (*Cucumis sativum*) were found to absorb 30–120 nm Ag_2S NPs, which were subsequently moved over to the shoots.[25]

5.3.3 Surface charge of nanoparticles

Chitosan derivative nanoparticles labelled with fluorescence, possessing both positive and negative charges, exhibited cell-line-dependent uptake. The endocytic process was found to be energy-dependent and involved actin-mediated mechanisms.[26] A positive cationic surface charge in most nanoparticles is associated with increased uptake by cells and heightened cytotoxicity in nonphagocytic cells. Cationic nanoparticles tend to induce greater disruption of the plasma membrane, while anionic nanoparticles are more likely to trigger apoptosis. In phagocytic cells, anionic nanoparticles are better internalized and exhibit higher cytotoxicity. Serum appears to increase phagocytic cell absorption of nanoparticles while decreasing it in nonphagocytic cells. When designing medicinal nanoparticles, these differences among phagocytic and nonphagocytic cells needs to be taken into account.[27] Positively charged nanoparticles coated with folate have demonstrated cellular uptake levels that are 20 times higher compared to nanoparticles coated with PEG. Gold nanostructures have found significant uses in therapeutics, identification of different diseases and delivery of genes and drugs. This is explained by their good biocompatibility, appropriate stability against oxidation, and surface plasmon resonance (SPR) characteristics.[28] A number of physicochemical variables, including as size, aggregated shape, and state, can impact the cell's absorption of nanoparticles as well as to surface charge.[29]

5.4 Nanoparticles types

Different many kinds of nanoparticles, such as metal-based nanoparticles, fullerol nanoparticles, fullerene nanoparticles, carbon-based nanoparticles, are discussed below.

5.4.1 Metal-based nanoparticles

Nanoparticles possess unique physiochemical properties that can cause harm to organs, tissues, cells, subcellular structures, and proteins. These features include their small dimensions, large surface area compared to volume ratio, chemical makeup, electronic, crystallinity characteristics, surface structure reactivity, groups of functions, both organic and inorganic coverings, the ability to dissolve, shape, and aggregation acts. The focus in metal nanoparticles has grown because of its extensive application in military, industrial, and medical domains. As the size of the particle reduces, certain metal-based nanoparticles exhibit heightened toxicity even when in bulk form, the substance is relatively inert, for example, Au, Cu, and Ag.[30]

5.4.1.1 Silica-based nanoparticles

Exposure of silver nanoparticles to plants (AgNPs) outcomes in a mixture of negative and positive effects. The effectiveness of AgNPs is concentration-dependent, as excessive amounts can impede the uptake of essential nutrients. Controlled application of AgNPs to plants may contribute to enhanced growth and development. Interestingly, AgNPs have been found to reduce the mound of lettuce, cabbage, *Cucurbita pepo*, and *Allium cepa*, among other culinary plants. Nevertheless, they also demonstrate a positive impact by inducing seed germination in various instances.[31] In plants, the transport of silver nanoparticles (AgNPs) occurs through long-distance transfer via vascular tissue and short-distance transport via intercellular spaces. NPs enter roots through the walls of cells and plasma membranes after being exposed to plants. This sets off a sequence of occurrences that lead them to enter the xylem, the vascular tissues found in plants, and then proceed in the direction of the stele. An essential pathway for the movement and dispersion of NPs is the xylem. AgNPs can be absorbed and transferred to leaves via the xylem. AgNPs are taken up by the roots of *Arabidopsis thaliana* and transported to the shoots. AgNPs are taken up by the root cap, columella, border cells, the first cells of the root meristem and epidermis among other root regions, where they progressively accumulate.[32] Another theory proposed that mesoporous silica nanoparticles (MSNs) might be a unique delivery strategy for introducing different sized biomolecules into plant systems through their translocation and increased internalization.[33] AgNPs can more easily penetrate cell wall openings and penetrate the plasma membrane due to their larger surface area and smaller size. The usual size of the cell wall pores that let a single nanoparticle to pass through is between 5 and 20 nm. Nevertheless, the interaction with the presence of nanoparticles causes the cell wall to produce new, bigger openings, thereby enhancing the internalization efficiency of the cell.[34] AgNPs were transferred to various types of tissues, suggesting the utilization of both stomatal and cuticular pathways.[35] Both nontransgenic and Bt-transgenic cotton's root biomass, cotton height, and shoot biomass were significantly impacted by the presence of SiO_2 nanoparticles. SiO_2 NPs significantly influenced nutrients found in Bt-transgenic cotton, including Cu, Mg, and Na in the shoots and roots, more so than in nontransgenic cotton. Furthermore, it was observed that SiO_2 nanoparticles enhanced the movement of magnesium in Bt-transgenic cotton xylem sap and iron in both nontransgenic and Bt-transgenic cotton xylem sap.[35] Silica nanoparticles (SiO_2 NPs), within the concentration range of 10–1000 mg/L, demonstrated build up and gathering on each plant's root surface. When the concentration of NP is greater (500 and 1000 mg/L), the accumulation became noticeably more prominent, while at 10 and 100 mg/L of $NSiO_2$, only few particles were seen to interact with the roots.[36]

5.4.1.2 Zinc-based nanoparticles

ZnO NPs are taken up by roots through soil solution when administered to the plant; first, they do so as Zn^{2+} ions or as compounds including chelates of organic acids. Then, through xylem, the NPs are transported to the above-ground section of the plant.[37] The root of the plant absorbs ZnO NPs and distributes them uniformly throughout the plant. Whereas the effectiveness of all engineering NPs may vary among different crops.[38] Examining how ZnO NPs are absorbed and accumulate (8 nm) glycine max, found in soybeans, at 500–400 mg/L were studied. A notably higher absorption of Zn NPs at 500 mg/L compared to 1000 mg/L concentration was exhibited by soybean seedlings (Martha et al., 2010). Despite the delivery methods Zn was shown to build up in rice plant tissues observed as shoot > root> leaves. Not only foliar but roots which was exposed to ZnO showed effects on physiological and biochemical parameters.[39] In contrast to standard zinc fertilizer ($ZnSO_4$), the impact that of ZnO NPs have on rice absorption, translocation, and growth was the subject of an inquiry conducted by[40] Rameshraddy et al. in 2017. Increase in seed vigour index was shown to be optimal when seed priming of ZnO at 1000 ppm was done. Significant growth was seen in potted rice plants treated with ZnO as opposed to $ZnSO_4$, especially in cases where foliar treatment was applied after seed priming. Plants exposed to nano ZnO showed improvements in height, biomass, tiller number, and chlorophyll content.[41] The goal of the study was to understand *Phaseolus vulgaris* (L.)'s physiological effects, biotransformation, and absorption of zinc. Nearby edge X-ray absorption spectroscopy revealed that roots absorbed ZnO at a rate of 40 nm as compared to 300 nm. Zn was found in the leaves as a mixture of $Zn_3(PO_4)^2$ and Zn-histidine complex, according to analysis. Conversely, Zn transfer through roots to shoots has been demonstrated by X-ray fluorescence spectroscopy to show a decreasing gradient of concentration and velocity, indicating simultaneous radial Zn and axial xylem transport. The lower stem tissue segment functioned as a buffer, keeping zinc from reaching the leaves, at concentrations less than 100 mg Zn/L. This formula was not observed at concentration of 1000 mg Zn/L $ZnSO_4$ (aq). Elevated expression levels of tonoplast-localized transporters were found in transcripts encoding metal carriers. This suggests that the mechanism leading to Zn accumulation in lower tissues may be related to improved Zn compartmentalization in vacuoles. The treatments adversely impacted transpiration, water conductance, and photosynthetic rate.[42]

5.4.1.3 Titanium dioxide-based nanoparticles

Globally TiO_2 nanoparticles are extensively produced in array of applications. Due to its promising applications in photocatalysis, solar cells and environmental pollution treatment nanostructure TiO_2 has gained significant interest in energy and environmental fields. Additionally, TiO_2 nanoparticles offer the advantages of being easily accessible, cost effective and having low toxicity.[43] There are three main theories that explain the advantages and disadvantages of giving plants nano-sized titanium dioxide ($nTiO_2$). The first mechanism explains how $nTiO_2$ modulates reactive oxygen species (ROS) signaling by acting as both an antioxidant and a prooxidant. The selective absorption of $nTiO_2$ has been convincingly demonstrated within the chloroplast of photosynthetic organisms, such as plants and algae. Since the chloroplast is a key site for ROS production in plants, alternations in $nTiO_2$-induced ROS levels may be linked to changes in chloroplast function, although the precise mechanism remains unclear. The second mechanism is centered on the increase of plant nitrogen metabolism mediated by $nTiO_2$, since $nTiO_2$ makes it easier for atmospheric nitrogen to be converted to nitrate when exposed to UV or sunshine. Furthermore, nitrate reductase, an essential enzyme for the absorption of nitrogen, functions better when supplemented with $nTiO_2$. The third mechanism relates to how $nTiO_2$'s size, shape, and surface properties affect how accessible it is to plants. However, higher concentrations of $nTiO_2$ may induce cytotoxicity and genotoxicity in plants. Hence, due to its various effects, accurate optimization of doses and application methods for $nTiO_2$ are essential before recommending its widespread use on crop plants.[44]

5.4.1.4 Copper-based nanoparticles

Cu NPs demonstrate a strong ability to hinder the growth of numerous pathogens, including *Salmonella typhi, Klebsiella pneumonia, Escherichia coli, Shigella dysenteriae, Bacillus subtilis*, and *Staphylococcus aureus*. Consequently, they are employed as an antibacterial additive in food packaging applications. The antifungal properties exhibited using copper nanoparticles against a range of fungus that cause plant disease positions them as promising candidates for managing plant diseases.[45] It takes involvement in hormone signaling, cell wall metabolism, mitochondrial respiration, photosynthetic electron transport, oxidative stress response and hormone signaling and structural protein control. Furthermore, it functions as a cofactor for a variety of enzymatic activities carried out by laccase, superoxide dismutase, amino oxidase, polyphenol oxidase, and plastocyanin. Copper is essential for the process of oxidative phosphorylation, which is the mechanism involved in signal trafficking and iron mobilization inside cells. Plants that lack copper exhibit curling of the leaves, bending of the petioles downward, mild chlorosis, and a chronic loss of turgor in the young leaves.[46] CuNPs showed a notable reduction in the levels of chlorophyll a and b, along with the elevated levels of H_2O_2 and malondialdehyde (MDA). Additionally, there was an increase in electrolyte leakage leading to the damage of plasma membrane of cucumber roots.[47] The transformation of copper ion to copper nanoparticles at the time of exposure to plants leaf extract can be followed by color change and UV-vis spectroscopy. Maximum observance was observed at 560 nm and often rise in intensity in relation to response time.[48] The application of CuO nanoparticles topically resulted in their accumulation in leaves of plants and their transportation from one part to another. The application of CuO nanoparticles hinder the growth of onion root tips, with a notable inhibition observed when roots were provided with 80 mg/L CuO nanoparticles causing a complete termination of growth after 72 hours of exposure. Both nano- and micro-CuO had a similar effect on shoot length reduction in cilantro (*Coriandrum sativum*) cultivated in soil with a high organic matter content. Comparable outcomes were observed in hydroponoically grown alfalfa (*Medicago sativa*). CuO nanoparticles have also been seen to shorten both the main and secondary roots of Indian mustard (*Brassica juncea* L.), inhibit growth, and reduce the total amount of carotenoid and chlorophyll. CuO NP buildup altered the ultrastructure of leaves, especially in the photosynthetic apparatus, resulting in a reduction in plastoglobules, starch content, stomatal aperture, and the amount of thylakoids per granum.[49]

5.4.1.5 Cerium-based nanoparticles

Applying CeO nanoparticles to the roots facilitates their entry into plant tissues by traversing the membrane of root epidermal cells. However, the size exclusion phenomenon restricts the plant cells' absorption of CeO_2 nanoparticles due to the limited opening of membrane pores. CeO_2 nanoparticles follow the apoplastic pathway via the cortex, epidermis and endodermis. However, they come across the lipophilic hydrocarbon-based casparian strip, which blocks their progress into the vascular cylinder. To access vascular tissue, nanoparticles must overcome several barriers including the Casparian strip, cuticle, epidermis, cortex, an endodermis, and finally reach shoots via the xylem. Nanoparticles larger than 20 nm face impediments in passing through pores in the cell wall due to size exclusion. Nevertheless, intact nanoparticles larger than 20 nm have been observed within intercellular spaces.[50] Based on the results of[51] Franziska et al. (2014), CeO_2 NPs dissolve easily in Hoagland mediums but are not stable there. We consolidate the four primary variables that affected CeO_2-NPs' ability to dissolve: the primary particle's size, the suspension's pH, the degree of agglomeration or aggregation in suspension, and the presence of reducing agents or chemical binding partners like phosphate or Fe (II)SO_4.[52] Sanghamitra et al. (2016) concluded that the natural organic matter present in the soil affects several soil features, including the actions of nanoparticles and how they affect the plant. This was clearly demonstrated that improvement of organic matter in the soil improves the mobility of the nano-CeO_2 within the plants. As a result, the nano-CeO_2 has easier access to the plants metabolic machinery, changing physiological processes related to the generation of food and energy as well as the health of the plant in the leaves. However, because of the opposite charge, the nano-CeO_2 aggregated to the soil grain surface in natural sandy loam soil with the low organic matter, mostly limiting it to the biomass below the earth.

5.4.1.6 Iron-based nanoparticles

Among the existing metal based nanoparticles, the most amazing new materials are iron nanoparticles because of their unique physiochemical properties, high magnetism, high catalytic activity, low toxicity, and ability to absorb microwave radiation. Hematite (α-Fe_2O_3), maghemite (γ-Fe_2O_3), and magnetite (Fe_3O_4) are examples of iron oxide nanoparticles. INPs fall into three primary categories: iron oxide hydroxide (FeOOH), zero-valent iron (ZVI), and iron oxide hydroxide (FeOOH).[53] It turned out that applying to the leaves worked better than exposing the roots. As particle size decreased for nonessential components, metal concentrations in shoots increased. However, for critical elements, metal concentrations varied with decreasing size. Shen et al.[54] studied that additionally, as MNPs'negative zeta potential grew, their uptake surged (particularly for Cu and Zn applying on the foliage proved to be more effective than exposing roots. The amount of metal in shoots increased with decreasing size of particles for nonessential components, but varied with decreasing particle size for key elements.

Furthermore, MNPs' uptake increased as their negative zeta potential increased (especially for Cu and Zn). According to[55] Justyna et al. (2023), the presence of distinct Fe-rich deposits in pollen grains—which are seen even in plants that have received an excess of emblematic Fe—strongly suggests that the tiny particles were absorbed by the plants and then transferred inside where they gathered.

5.4.1.7 Nickel-based nanoparticles

As the NP dosages were increased, the elements Cu, Ni, Zn, Y, Nd and Fe—which collectively make up NPs—became increasingly more prevalent in the leaf and root tissues.[56] The findings by[57] Huseyin et al. (2019) demonstrated that nanoparticles of $NiFe_2O_4$ are moved to the leaves and are absorbed by roots, changing mineral uptake as well as some plant growth metrics (such as RWC, soluble protein level, and chlorophyll content) without having any negative phytotoxic effects. Higher doses (1000 mg/L) on the other hand dramatically reduced biomass and plant growth. Therefore, lesser dosages (up to 500 mg/L) may be recommended for use in pythonanotechnological uses, such as the transportation of nanomaterials, fertilizers, as well as agricultural chemicals. XAS experiments revealed that plants treated with uncoated NPs had Ni NPs in their roots and shoots and a Ni (II)-organic acid type complex in their leaves. Plants exposed to coated NPs either prior to or following synthesis, however, only displayed Ni NPs in the roots and a combination of Ni (II) and organic acid in the petals and shoots.[58] Research has been conducted on the benefits of employing Ni NPs as a low-cost catalyst in Suzuki-Miyaura reactions to convert biomass to fuels, in addition to Ni nanoparticles benefits because of their numerous oxidation states. The benefits of employing nickel nanoparticles as a catalyst have been emphasized, but the drawbacks have not been listed.[59]

5.4.1.8 Aluminum-based nanoparticles

The most common metal on earth, aluminum, is used in many different industries. Since plants are the intermediary among human health and the environment, aluminum oxide nanoparticles are produced and applied to them in order to fully exploit the benefits of aluminum.[60] Over a wide temperature range, aluminum oxide nanoparticles, are stable crystalline particles. With oxygen atoms grouped hexagonally close to one another and aluminum ions filling octahedral gaps in two-thirds of the lattice, their structure has similarities to that of the crystals. Al_2O_3 nano-powders are generated utilizing a number of standard chemical and physical techniques, such as sol-gel, sputtering, mechanical milling, and hydrothermal. Even yet, the procedures used to produce alumina nanoparticles are not very environmentally friendly.[61] Gram-positive bacteria were more effectively targeted by Al_2O_3 NPs at 1000 mL as compared to Gram-negative bacteria. When Al_2O_3 NPs were coupled with imipenem and gentamicin, their synergistic/antibacterial activity demonstrated potential antibacterial activity against the examined species, which demonstrated stronger antibacterial action than Al_2O_3 NPs alone.[62]

5.4.1.9 Carbon-based nanoparticles

Nanomaterials based on carbon (CNMs) are naturally occurring substances found in the environment. Most often, they are inadvertently produced from coal and forest fire burning substances such as natural gas and/or methane. Nevertheless, it is challenging to fully extract and use natural CNMs on a big scale.[63] Many metal and metal oxide nanoparticle forms, such as copper oxide (CuO), silica (Si), magnesium oxide (MgO), gold (Au), titanium dioxide (TiO_2), silver (Ag), oxide of zinc (ZnO), magnesium oxide, calcium oxide (CaO), silver oxide (Ag_2O), and silver oxide (Ag), have been reported to exhibit antimicrobial activity. It is also commonly known that carbon nanoparticles possess potent antibacterial properties. Early studies found that single-walled carbon nanotubes (SWCNTs), fullerenes, and grapheme oxide (GO) nanoparticles all shown potent antimicrobial properties. These new allotropic types of carbon have been discovered in the last 20 years and have been used in many scientific domain.[64]

5.4.1.10 Fullerene nanoparticles

Ithas been discovered that fullerene, C60, and carbon nanotubes (CNTs) enhance agricultural fruit output, biomass, and water retention capacity.[65] This is an amazing application of nanotechnology in the agriculture industry. Applying fullerene to bitter melon seeds enhanced the production and concentration of phytomedicinal and other significant secondary metabolites, such as lycopene, cucurbitacin-B, insulin and charantin.[65] The foliar-applied fullerenol has to pass through the epidermis, or outer leaf surface, in order to enter cells. Taking into account the structural characteristics, the leaf epidermis has a waxy cuticle, which is an impermeable layer. Additionally, through turgor expansion, nanomaterials can pass through the 100–1300 stomata per mm^2 of epidermis that have size of pore 10–30 μm in length and 3–12 μm in breadth.[66] The polar pole hypothesis[67] states that the cuticle varied between 2 and 2.4 nm, while maximum size for ionic solutes to reach the stomata surface through diffusion surpasses 20 nm. Thus, we may presume that fullerenols (just like other nanomaterials) with sizes ranging upto few nanometers might penetrate the cuticle, whereas stomata could do so with ease by diffusion.After passing through the stomata or cuticle, fullerenol molecules must navigate the cell wall, where reported pore sizes range from 3.5 to 20 nm. In this regard, studies have shown that nanoparticles (NPs) can successfully overcome the barriers presented by the cuticle and cell wall. Once the cell wall is crossed, fullerenols typically enter the cell via endocytosis, a process that does not disrupt the lipid bilayer membrane. After entering the cells, fullerenols can be localized in various sub-cellular compartments, including the mitochondria, chloroplast thylakoid membranes, vacuoles, or possibly even the nucleus.[68] By several processes, the use of different surfactants (such as The Tweens 20 and 40) may improve fullerenol interaction with the surface of leaf via the hydrophobic waxy cuticle. The structure of solutions of amino acid fullerenes and sodium selenate, as well as C60 or C70 fullerenols and zinc sulfate, was compared to mineral salts of the indicated elements and control. These solutions were selected based on their increased efficiency in enriching plant products when different charges of molecules or functional groups of fullerene substitutes were used with selenium and zinc, respectively. Cucumber, tomato, and Chinese cabbage were among these plant items. In addition, the numerical characteristics of expansion, output, and/or quality of the produced goods improved and increased as appropriate. Further research and a thorough analysis of the mechanisms behind the effects of various trace element compounds and fullerene derivatives on plants should improve the management of these effects.[69]

Presently, the hydroxylated form of fullerenes, such as C60, C70, and so on, and fulllerols, are employed in many different fields, such as photovoltaics, medicine, and cosmetics. 1−4 tons to 1500 tons of these carbon-based nanomaterials are now manufactured annually. The intense photochemical reactivity of fullerenes, including their derivatives, is one of their most prominent features. Due to fullerenes' inherent hydrophobicity (C60 water solubility of $<10-9$ mg/L), several attempts have been undertaken to make them stable or soluble in water.[70] The majority of fullerenol nanoparticles in wheat were detected in the roots; very small levels were also found in the leaves and stems. Wheat plants may benefit from fullerenol nanoparticles as they enhanced the chlorophyll contents during the growth phases and encouraged the formation of wheat roots at the proper concentrations. Consequently, fullerenol nanoparticles in wheat roots may spread to the plant's stems and leaves, changing crop output and productivity and perhaps making their way up the food chain with unpredicted effects.[71] Under drought stress, foliar application of fullerol at a dosage of 100 mg/L raised aboveground DW more than other doses above the corresponding control. Various doses of fullerol treatments significantly increased leaf RWC relative to drought control (1 to 100 mg/L).[72]

5.4.1.11 Single-walled carbon nanotubes

Unbroken carbon nanotubes offer a multitude of beneficial possibilities. However, because of their special qualities, there may be a risk of toxicity to plants along with people and animals.[73] Through oxidative stress, SWCNTs negatively impact protoplasts and leaves, causing a certain level of planned cell death.[74]

5.4.1.12 Multiwalled carbon nanotubes

MWCNTs at doses of 10 and 50 mg/L might be beneficial in lessening the impact of Pb or Cd stress on the canola and sunflower seedlings' stunted shoot and root growth. MWCNTs 10 mg/L are more successful than MWCNTs 50 mg/L at hiding the adverse impacts of heavy metals on chlorophyll concentration. MWCNTs, frequently at a focus of 10 mg/L, enhance the build-up of cadmium and lead in cannabis, sunflower and canola seedlings. MWCNTs are also beneficial in preventing Pbbuildup in cannabis and canola seedlings. Based on the findings, the result of MWCNTs on heavy metal absorption and seedling development parameters vary on the kind of plant, MWCNT concentration, and heavy metal.[75]

5.5 Top of form

5.5.1 Translocation of NPs within plant system

One very practical method for absorbing extracellular molecules, such as NPs, is endocytosis. The transmembrane route known as endocytosis is created when the membrane of the cell folds. NPs can be endocytosed in two ways: clathrin dependently and independently. Numerous elements, including surface characteristics, charge, size, and others, are crucial to the NPs' uptake process.[76] Different complexes between nanoparticles and membrane transporters, root exudates, and plant growth are formed. It travels through the plant system either apoplastically or simplistically after entering cells. Plant cells are connected by plasmodesmata, which makes it simple for materials to move between the cells. Hydrophilic particles flow via polar channels through the cuticle areas, whereas lipophilic molecules cluster on anticlinal cell walls and the waxy layer of the cuticle before translocating with pectin (middle lamellae) into the apoplast. Lysed necrotic cells use the mucus entrapment mechanism to gather nanoparticles. Only a small percentage of the nanoparticles enter the system by translocation mechanisms; the majority either concentrate close to the root area. The pace at which nanoparticles translocate via xylem is primarily determined by several variables, including charge-charge interactions between nanoparticles and plants, root pressure, variables including temperature and humidity.[77] There are now around 20 different types of NPs detection techniques available. The first step is often to characterize NPs to comprehend their distribution, stability, and structure. Among the characterization techniques employed were a transmission electron microscopy (TEM), a Raman spectroscopy (RS), a technique called X-ray diffraction (XRD), and the scattering of dynamic light (DLS). The dimension and crystal makeup of CeO_2 NPs in dry powder samples are examined by Yang et al. using XRD, and the shape of CeO_2 NPs distributed in deionized water is examined using TEM. The molecular structure and particle size distribution of PS-NPs were measured using laserconfocal RS and DLS.[78] NPs are deposited in plant roots and then moved toward shoots and other sections of the plant. Furthermore, it has been shown that CuO nanoparticles are transportedfrom the roots to the shoots when rice was subjected to the CuONPs. Similar to this, Ag NPs were translocated into plant tissues such as shoots, which upset ionic homeostasis, severely damaged cells through oxidative stress, and stunted plant development. Furthermore, the antioxidative defense system, which includes APX, SOD and CAT; glutathione (GSH); phenolics; glutathione reductase (GR); guaiacol peroxidase (GPOX); ascorbate (AA); and others, is adversely regulated by the dispersion of metallic nanoparticles.[79] There are several factors that may influence the interaction on NPs within plant system.

5.5.2 Plant cell wall pore size

Compared to the cellular membrane, the cell wall is a little more flexible, robust, and long-lasting structure. The largest portion of the cellular wall is composed of lignin, hemicellulose, and cellulose.[80] Pectins and proteins form a cross linked network that makes up plant cell walls (Shoomer et al., 2003). The cell wall's size stops nanomaterials from moving any further.[81] Consequently, it is essential to understand the dimensions of pores of the cell wall.[82] The stover cell walls of maize (*Z. mays* L.) have pores that are between 10 and 100 nm in size, albeit this depends on the location.[83] The cell wall pores of wheat (*Triticum aestivum* L.) have a radius of 1.5–3 nm.[84]

5.5.3 Nanoparticles and cell wall interaction

According to research using transmission electron microscopy, ZnO NPs at a concentration of 25 mg/L bonded to Zn^{2+} in the root cell wall of Brassica (*Brassica napus*).[85] Ag + generated by Ag NPs can attach to hydroxy cellulose structures in the presence of ROS, breaking hydrogen bonds to encourage modifications in cell wall structure.[86] It is unclear how important cell wall potential is for plants to absorb NPs and whether relation among NPs and plant cell walls are related to certain cell types. Therefore, elements that also affect NP mobility inside the plant must be taken into account. These elements include the electrical gradient across the cellular structure, nanoparticles dimensions, engagements with components of the cellular structure, and nanoparticles density.[87]

5.5.4 Nanoparticle and cell membrane interaction

NPs must penetrate through the cell wall and then cross the membrane barrier.[88] Normally 5–10 nm thick, the plasma membrane is a border membrane that shields the internal components of the cell. Lipid exchange envelopes, transporters, and channels in the cell membrane are a few of the mechanisms by which cells can be absorbed.[89] The membrane potential, like plant cell walls, is essential for regulating the passage of NPs across the membrane. At about −120 mV, the potential of plasma membrane is higher than the cell wall potential. NPs with a high zeta potential can enter protoplasts. Nonetheless, it has been discovered that the electrical characteristics of Au NPs do not impede their passage over the plasma membrane. This could be attributed to the utilization of 5 nm AuNPs, which fall within the size range suitable for passing through the channels on the plasma membrane. By navigating organelle membranes, NPs can also enter organelles other than protoplasts. Though it may be one of the most fascinating, the interchange of NPs with membranes has not gotten much attentiveness.[90]

5.6 Effect of nanoparticles on plant system

A variety of seed conditioning techniques exist to promote seed germination, including immersion of the seed growth regulator therapy (hormo-priming), saline solutions (osmo-priming), water (hydro-priming), and temperature fluctuations (matrix-priming), dissolved organic matter treatment (bio-priming), and most recently, nanomaterial treatment (nano-priming). For agricultural and forestry purposes, several nanoparticles have been employed to reduce seed dormancy, stimulate germination, also increases germination vigour.[91] The development of seedlings was impacted by AgNPs at all doses, although for all parameters examined, this effect did not reach statistical significant at 10 mg/L concentration. The lack of preventive impact of AgNPs at 10 mg/L may be the consequence of some unidentified AgNP actions, which were sufficient to offset the AgNP-induced growth inhibitory effects. The elevated concentrations of AgNPs (100 and 1000 mg/L) notably inhibited both shoot and root development, particularly affecting dry weight. The suppression of shoot growth was more pronounced compared to root growth. In comparison to the no AgNP control, the dry weight of shoot decreased by 35%–78% with 100–1000 mg/L AgNPs, while the dry weight of roots decreased by 23%–26%[92] (Fig. 5.4, Table 5.1).

5.7 Imporatnce of sustainable goals in agriculture

The application of NPs improves nutrient and water intake, preserves the water relationship, and safeguards the membranes, all of which contribute to a noticeable boost in plant development under DS. Plants are better able to withstand defoliation (DS) when NPs safeguard the photosynthetic mechanisms and boost photosynthesis efficiency, osmolyte, hormone, and phenolic accumulation, antioxidant responses, and gene regulation.[109] With the help of cutting-edge technologies like nanotechnology, agriculture may experience sustainable expansion. The resources that were available were not completely utilized by conventional farming practices. Nanotechnology can be efficiently put in application to enhance crop quality along with yield while getting around these restrictions. Plant growth, soil and microbial stabilization, targeted chemical use, and, most importantly, significantly enhanced waste management are all outcomes of using nanotechnology to agriculture. The nanoparticles' increased surface area and small size are primarily responsible for their increased activity.[110] Global agriculture has differed formerly because of green nanotechnology (NT); nano-fertilizers are yet utilized in specific concentrations, in compliance with crop mineral needs while minimizing adverse effects. Therefore, the benefit of NT treatment on plant development, productivity, as well as restraint reveals the current perspective on NT implementation in agriculture, especially when plants are cultivated under stress. Hence,

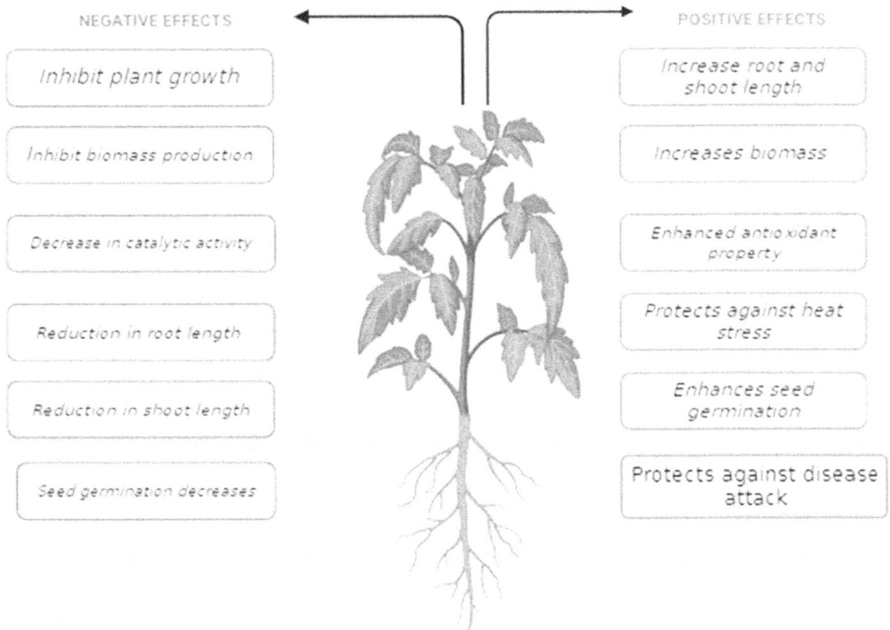

FIGURE 5.4

Illustration of negative and positive effects of NPs on plant

Made by author.

NT uses in agriculture may be successful provided that natural activities are impacted in a way that is more consistent with science for a reliable application.

5.8 Conclusion and gap analysis

Despite a lot of focus on NP absorption in plants over the past 5 years, there is still a dearth of reliable data about NP entry routes inside plants. Nonetheless, the recent findings on absorption and excretion barriers, routes and mechanism discussed in the preceding segments reveal significant advancements in understanding the qualitative aspects of NP uptake in plants. While the subsequent gap analysis was not exhaustive, it highlighted pressing scientific challenges that needed attention alongside typical concerns regarding NP absorption and translocation in plants.

Table 5.1 Effects of different nanoparticles with different concentration of dosages.

Crop	Scientific name	Nanoparticle	Dosage	Positive effects	Negative effects	References
Tomato	*Solanumlycopersicum* L.	Sio_2	1000 mg/L	Increased antioxidant enzyme activity,enhanced accumulation of essential nutrient	Reduced aerial biomass dry biomass, negative impact on nutrient uptake	93
		ZnO	400 and 800 mg/ dm^3	Enhanced antioxidant enzyme activity and transcriptions of respective genes	Inhibited Plant growth and biomass production	94
		TiO_2	5000 mg/L	Promoted root growth	Reduced root elongation at every concentration, No effects of phytotoxicity	95
Brinjal	*Solanummelongena*	ZnO	50 and 100 mg/L	Positive effects on yield	Phytotoxic effects	96
		NiO, CuO, ZnO	100–1000 mg/L	Increased ROS generation	Decrease chlorophyll content, suppression of seedling growth	97
Chilli	*Capsicum annum L.*	ZnO	0.75 and 0.25 g	Highest germination percentage	Reduction in root and shoot length	98
Broccoli	*Brassica oleracea* var. *Italicais*	ZnO	75 and 150 mg/L	Increase in sulforaphane content	Lower sulforphane content in case of 150 mg/L	99
Wheat	*Triticumaestivum*	Cu	0.2–1.0 ppm	No seed germination inhibited, increase in root weight	Seed germination decreases	100
		SiO_2	200 mg /L	Increase in antioxidant enzyme response	Decrease in root, shoot, and weight	101
		ZnO	50 and 100 mg/L	Increase in root, shoot and seedling length	Length and biomass reduced	102
		TiO_2,ZnO and chitosan	50 ppm, 300 ppm and chitosan	Protects against heat stress, enhances shoot and root length at 50 ppm	Reduction in growth parameters at 300 ppm	103

Maize	*Zea mays L.*	Fe	100 ppm	Improves photosynthesis	Lower content of chlorophyll, biomass, calcium, and proteins in second generation	104
		TiO_2	30–50 mg/L	+ve effects on macro and micro nutrients	Reduction in plant root, shoot, and chlorophyll content	105
		Pbs	5–50 mg/L (Pbs), 1.5 mg/L (Pb^{2+}ions)	Enhances phytotoxic effects	Inhibit biomass of shoot and root	106
Pearl millet	*Pennisetumglaucum L.*	Ag	20–80 ppm	Improves seed germination, seedling growth	Increased ROS production and toxicity	107
Common bean	*Phaseolus vulgaris*	ZnO	20 and 30 ppm	Enhances seed yield	Genotoxic effects on genome	108

Acknowledgment

The authors thank the Ministry of Education of Russia for their support (Agreement No. 073-00033-24-01 dated 02/09/2024).

References

1. Abdulraheem, M. I.; AbiodunYusuff., M. Effects of nanoparticles on improvement in quality and shelf life of fruits and vegetables. *J. Plant. Biol. Crop. Res.* **2021**, 2637−7721.
2. Zulfiqar, M.; Navarro, M. Nanofertilizer use for sustainable agriculture: Advantages and limitations. *Plant Sci* **2019**, *289*, 110270. Available from: https://doi.org/10.1016/j.plantsci.2019.110270.
3. Daohui, L.; Xing., B. Root uptake and phytotoxicity of ZnO nanoparticles. *Environ. Sci. Technol.* **2008**, *42*, 5580−5585.
4. Erika, M.-V.; Trejo-Téllez, L. I.; Gómez-Merino, F. C.; Sandoval-Villa, M.; Sánchez-García, P.; Aguilar-Méndez, M. Á. Nanophosphorus fertilizer stimulates growth and photosynthetic activity and improves P status in rice. *J. Nanomaterials* **2019**. Available from: https://doi.org/10.1155/2019/5368027.
5. Koushik, B.; Pragati, P.; Aniruddha, M.; Joshi, D. C.; Wani., S. H. Methods of using nanomaterials to plant systems and their delivery to plants (mode of entry, uptake, translocation, accumulation, biotransformation and barriers). *Adv. Phytonanotechnology* **2019**. Available from: https://doi.org/10.1016/B978-0-12-815322-2.00005-5.
6. Durgesh, K. T.; Gaur, S.; Singh, S.; Singh, S.; Pandey, R.; Singh, V. P., et al. An overview on manufactured nanoparticles in plants: uptake, translocation, accumulation and phytotoxicity. *Plant. Physiol. Biochem.* **2016**. Available from: https://doi.org/10.1016/j.plaphy.2016.07.030.
7. Astrid, A.; Yun, J.; Morais, B. P.; Clement, E. T.; Rodrigues, S. M.; Lowry, G. V. Critical Review: role of inorganic nanoparticle properties on their foliar uptake and in planta translocation. *Environ. Sci. Technol.* **2021**. Available from: https://pubs.acs.org/action/showCitFormats?doi = 10.1021/acs.est.1c00178&ref = pdf.
8. Jitao, L.; Christie, P.; Zhang, S. Uptake, translocation, and transformation of metal-based nanoparticles in plants: recent advances and methodological challenges. *R. Soc. Chem.* **2019**, *6*, 41−59. Available from: https://doi.org/10.1039/C8EN00645H.
9. Yiming, S.; Ashworth, V.; Kim, C.; Adeleye, A. S.; Rolshausen, P.; Roper, C., et al. Delivery, uptake, fate, and transport of engineered nanoparticles in plants: a critical review and data analysis. *R. Soc. Chem.* **2019**. Available from: https://doi.org/10.1039/c9en00461k.
10. Yu, D. The Value of Certification to Consumers: Evidence from the US. *Org. Food Mark* **2022**. Available from: https://doi.org/10.2139/ssrn.4299523.
11. Wang, X.; Xie, H.; Wang, P., et al. Nanoparticles in plants: uptake, transport and physiological activity in leaf and root. *Materials* **2023**, *16* (8), 3097.
12. Shukla, P. K.; Misra, P.; Kole, C. *Uptake, translocation, accumulation, transformation, and generational transmission of nanoparticles in plants.* In *Plant Nanotechnology;* Kole, C., Kumar, D., Khodakovskaya, M., Eds.; Springer: Cham, 2016; pp 183−218. Available from: https://doi.org/10.1007/978-3-319-42154-4_8

13. Hischemöller, A.; Nordmann, J.; Ptacek, P., et al. In-vivo imaging of the uptake of upconversion nanoparticles by plant roots. *J. Biomed. Nanotechnol* **2009**, 278–284. Available from: https://doi.org/10.1166/jbn.2009.1032.
14. Ali, S.; Mehmood, A.; Khan, N. Uptake, translocation, and consequences of nanomaterials on plant growth and stress adaptation. *J. Nanomater* **2021**, 6677616. Available from: https://doi.org/10.1155/2021/6677616.
15. Li, L.; Luo, Y.; Li, R., et al. Effective uptake of submicrometre plastics by crop plants via a crack-entry mode. *Nat Sustain* **2020**, 929–937. Available from: https://doi.org/10.1038/s41893-020-0567-9.
16. Pérez-de-Luque, A. Interaction of nanomaterials with plants: what do we need for real applications in agriculture? *Front Environ Sci* **2017**, *5*. Available from: https://doi.org/10.3389/fenvs.2017.00012.
17. Péret, B.; De Rybel, B.; Casimiro, I., et al. Arabidopsis lateral root development: an emerging story. *Trends Plant Sci* **2009**, 399–408. Available from: https://doi.org/10.1016/j.tplants.2009.05.002.
18. Behzadi, S.; Serpooshan, V.; Tao, W., et al. Cellular uptake of nanoparticles: journey inside the cell. *Chem Soc Rev* **2017**, *46* (14), 4218–4244.
19. Ma, X.; Yan, J. Plant uptake and accumulation of engineered metallic nanoparticles from lab to field conditions. *Curr Opin Environ Sci Health* **2018**, 16–20. Available from: https://doi.org/10.1016/j.coesh.2018.07.008.
20. Liu, W., Zeb, A., Lian, J., Wu, J., Xiong, H., Tang, J., et al. (2020). Interactions of Metal-based and Metal-oxide-based Nanoparticles (MBNPs and MONPs) with Crop Plants: A Critical Review of Research Progress and Prospects.
21. Wang, X.; Xie, H.; Wang, P., et al. Nanoparticles in plants: uptake, transport and physiological activity in leaf and root. *Materials* **2023**, *16* (8), 3097. Available from: https://doi.org/10.3390/ma16083097.
22. Rani, S.; Kumari, N.; Sharma, V. Uptake, translocation, transformation and physiological effects of nanoparticles in plants. *Arch Agron Soil Sci* **2023**, 1579–1599. Available from: https://doi.org/10.1080/03650340.2022.2103549.
23. Zhu, Z. J.; Wang, H.; Yan, B., et al. Effect of surface charge on the uptake and distribution of gold nanoparticles in four plant species. *Environ Sci Technol* **2012**, *46* (22), 12391–12398. Available from: https://doi.org/10.1021/es301977w.
24. Sanjay, S. S.; Pandey, A. C.; Singh, M.; Prasad, S. M. Effects of functionalized ZnO nanoparticles on the phytohormones: growth and development of *Solanummelongena* L. (*Brinjal*) plant. *World J. Pharm. Res.* **2015**, *4* (05).
25. Liu, W.; Zeb, A.; Lian, J., et al. Interactions of metal-based nanoparticles (MBNPs) and metal-oxide nanoparticles (MONPs) with crop plants: a critical review of research progress and prospects. *Environ Rev* **2020**, *28* (3), 294–310. Available from: https://doi.org/10.1139/er-2019-0085.
26. He, C.; Hu, Y.; Yin, L., et al. Effects of particle size and surface charge on cellular uptake and biodistribution of polymeric nanoparticles. *Biomaterials* **2010**, *31* (13), 3657–3666. Available from: https://doi.org/10.1016/j.biomaterials.2010.01.065.
27. Fröhlich, E. The role of surface charge in cellular uptake and cytotoxicity of medical nanoparticles. *Int J Nanomed* **2012**, 5577–5591. Available from: https://doi.org/10.2147/IJN.S36111.

28. Salatin, S.; MalekiDizaj, S.; YariKhosroushahi, A. Effect of the surface modification, size, and shape on cellular uptake of nanoparticles. *Cell Biol Int* **2015**, 881–890. Available from: https://doi.org/10.1002/cbin.10459.
29. Jeon, S.; Clavadetscher, J.; Lee, D. K., et al. Surface charge-dependent cellular uptake of polystyrene nanoparticles. *Nanomaterials* **2018**, *8* (12), 1028. Available from: https://doi.org/10.3390/nano8121028.
30. Schrand, A. M.; Rahman, M. F.; Hussain, S. M., et al. Metal-based nanoparticles and their toxicity assessment. *WIREs Nanomed Nanobiotechnol* **2010**, *2*, 544–568. Available from: https://doi.org/10.1002/wnan.103.
31. Siddiqi, K. S.; Husen, A. Plant response to silver nanoparticles: a critical review. *Crit Rev Biotechnol* **2021**, *42* (7), 973–990. Available from: https://doi.org/10.1080/07388551.2021.1975091.
32. Yan, A.; Chen, Z. Impacts of silver nanoparticles on plants: a focus on the phytotoxicity and underlying mechanism. *Int J Mol Sci* **2019**, *20* (5), 1003. Available from: https://doi.org/10.3390/ijms20051003.
33. Sun, D.; Hussain, H. I.; Yi, Z.; Siegele, R.; Cresswell, T.; Kong, L., et al. Uptake and cellular distribution, in four plant species, of fluorescently labelled mesoporous silica nanoparticles. *Plant. Cell Rep.* **2014**, *33*, 1389–1402. Available from: https://doi.org/10.1007/s00299-014-1624-5.
34. Tripathi, D. K.; Tripathi, A.; Shweta., et al. Uptake, accumulation and toxicity of silver nanoparticle in autotrophic plants, and heterotrophic microbes: a concentric review. *Front Microbiol* **2007**, *8*. Available from: https://doi.org/10.3389/fmicb.2017.00007.
35. Le, V. N.; Rui, Y.; Gui, X.; Li, X.; Liu, S.; Han, Y. Uptake, transport, distribution and bio-effects of SiO2 nanoparticles in Bt-transgenic cotton. *J. Nanobiotechnology* **2014**, *12*.
36. Zarafshar, M.; Akbarinia, M.; Askari, H.; Hosseini, S. M.; Rahaie, M.; Struve, D. Toxicity assessment of SiO2 nanoparticles to pear seedlings. *Int. J. Nanosci. Nanotechnol.* **2015**, *11* (1), 13–22.
37. Sturikova, H.; Krystofova, O.; Huska, D., et al. Zinc, zinc nanoparticles and plants. *J Hazard Mater* **2018**, 101–110. Available from: https://doi.org/10.1016/j.jhazmat.2018.01.040.
38. Priester, J. H.; Ge, Y.; Mielke, R. E.; Holden, P. A. Soybean susceptibility to manufactured nanomaterials with evidence for food quality and soil fertility interruption. *Proc. Natl Acad. Sci.* **2012**, *109* (37), E2451–E2456. Available from: https://doi.org/10.1073/pnas.1205431109.
39. Afzal, S.; Aftab, T.; Singh, N. K. Impact of zinc oxide and iron oxide nanoparticles on uptake, translocation, and physiological effects in Oryzasativa L. *J Plant Growth Regul* **2022**, *41*, 1445–1461. Available from: https://doi.org/10.1007/s00344-021-10388-1.
40. Rameshraddy.; Pavithra, G. J.; Rajashekar Reddy, B. H. Zinc oxide nano particles increases Zn uptake, translocation in rice with positive effect on growth, yield and moisture stress tolerance. *Indian J Plant Physiol* **2017**, *22*, 287–294. Available from: https://doi.org/10.1007/s40502-017-0303-2.
41. Rameshraddy, K.; Pavithra, G. J.; Rajashekar Reddy, B. H.; Salimath, M.; Geetha, K. N.; Shankar, A. G. Zinc oxide nanoparticles increase Zn uptake, translocation in rice with positive effect on growth, yield and moisture stress tolerance. *Indian. J. Plant. Physiol.* **2017**, *22* (3), 287–294. Available from: https://doi.org/10.1007/s40502-017-0303-2.

42. da Cruz, T. N.; Savassa, S. M.; Montanha, G. S., et al. Pereira de Carvalho HW. A new glance on root-to-shoot in vivo zinc transport and time-dependent physiological effects of ZnSO4 and ZnO nanoparticles on plants. *Sci Rep* **2019**, *9*, 10416. Available from: https://doi.org/10.1038/s41598-019-46796-3.
43. Sungur, Ş. Titanium dioxide nanoparticles. In *Handbook of nanomaterials and nanocomposites for energy and environmental applications;* Springer: Cham, 2021; pp 713–730.
44. Abdel Latef, A. A.; Srivastava, A. K.; El-sadek, M. S., et al. Titanium dioxide nanoparticles improve growth and enhance tolerance of broad bean plants under saline soil conditions. *Land Degrad. Dev* **2018**, 1065–1073. Available from: https://doi.org/10.1002/ldr.2780.
45. Kasana, R. C.; Panwar, N. R.; Kaul, R. K., et al. Biosynthesis and effects of copper nanoparticles on plants. *Environ. Chem. Lett* **2017**, 233–240. Available from: https://doi.org/10.1007/s10311-017-0615-5.
46. Shobha, G.; Moses, V.; Ananda, S. Biological synthesis of copper nanoparticles and its impact. *Int. j. pharm. sci. Invent.* **2014**, *3* (8), 6–28.
47. Mosa, K. A.; El-Naggar, M.; Ramamoorthy, K., et al. Copper nanoparticles induced genotoxicty, oxidative stress, and changes in superoxide dismutase (SOD) gene expression in cucumber (Cucumissativus) plants. *Front. Plant Sci* **2018**. Available from: https://doi.org/10.3389/fpls.2018.00872.
48. Lee, H.-J., Lee, G., Jang, N.R., Yun, J.H., Song, J.Y., & Kim, B.S. (2011). Biological synthesis of copper nanoparticles using plant extract. ISBN 978-1-4398-7142-3. 1.
49. Rajput, V. D.; Minkina, T.; Suskova, S.; Mandzhieva, S.; Tsitsuashvili, V.; Chapligin, V., et al. Effects of copper nanoparticles (CuO NPs) on crop plants: a mini review. *BioNanoScience* **2018**, *8* (1), 36–42. Available from: https://doi.org/10.1007/s12668-017-0466-3.
50. Prakash, V.; Peralta-Videa, J.; Tripathi, D. K.; Ma, X.; Sharma, S. Recent insights into the impact, fate, and transport of cerium oxide nanoparticles in the plant-soil continuum. *Ecotoxicol. Environ. Saf.* **2021**, *221*. 112403. Available from: https://doi.org/10.1016/j.ecoenv.2021.112403.
51. Schwabe, F.; Schulin, R.; Rupper, P., et al. Dissolution and transformation of cerium oxide nanoparticles in plant growth media. *J. Nanopart. Res* **2014**, *16*, 1–11. Available from: https://doi.org/10.1007/s11051-014-2668-8.
52. Majumdar, S.; Peralta-Videa, J. R.; Trujillo-Reyes, J., et al. Soil organic matter influences cerium translocation and physiological processes in kidney bean plants exposed to cerium oxide nanoparticles. *Sci. Total Environ* **2016**, *569–570*, 201–211. Available from: https://doi.org/10.1016/j.scitotenv.2016.06.087.
53. Ebrahiminezhad, A.; Zare-Hoseinabadi, A.; Sarmah, A. K., et al. Plant-mediated synthesis and applications of iron nanoparticles. *Sci. Total Environ* **2018**, *60*, 154–168. Available from: https://doi.org/10.1016/j.bcab.2023.102950.
54. Shen, C.-X.; Zhang, Q.-F.; Li, J.; Bi, F.-C.; Yao, N. Induction of programmed cell death in Arabidopsis and rice by single-wall carbon nanotubes. *Am. J. Botany* **2010**, *97* (10), 1602–1609. Available from: https://doi.org/10.3732/ajb.1000073.
55. Wojcieszek, J.; Chay, S.; Jiménez-Lamana, J., et al. Study of the stability, uptake and transformations of zero valent iron nanoparticles in a model plant by means of an optimised single particle ICP-MS/MS method. *Nanomaterials* **2023**, *13*, 1736. Available from: https://doi.org/10.3390/nano13111736.

56. Tombuloglu, H.; Ercan, I.; Alshammari, T., et al. Incorporation of micro-nutrients (nickel, copper, zinc, and iron) into plant body through nanoparticles. *J. Soil Sci. Plant Nutr* **2020**, *20*, 1872–1881. Available from: https://doi.org/10.1007/s42729-020-00258-2.
57. Tombuloglu, H.; Slimani, Y.; Tombuloglu, G., et al. Tracking of NiFe2O4 nanoparticles in barley (Hordeumvulgare L.) and their impact on plant growth, biomass, pigmentation, catalase activity, and mineral uptake. *Environ. Nanotechnol. Monit. Manag* **2019**, *11*, 100223. Available from: https://doi.org/10.1016/j.enmm.2019.100223.
58. Parsons, J. G.; Lopez, M. L.; Gonzalez, C. M., et al. Toxicity and biotransformation of uncoated and coated nickel hydroxide nanoparticles on mesquite plants. *Environ. Toxicol. Chem.* **2010**, *29* (5), 1146–1154. Available from: https://doi.org/10.1002/etc.146.
59. Jaji, N.-D.; Lee, H. L.; Hussin, M. H.; Akil, H. M.; Zakaria, M. R.; Othman, M. B. H. Advanced nickel nanoparticles technology: from synthesis to applications. *Nanotechnol. Rev.* **2020**, *9*, 1456–1480. Available from: https://doi.org/10.1515/ntrev-2020-0109.
60. Qidwai, Y. N.; Vishvakarma, R.; Farooqui, A., et al. *Aluminum oxide nanoparticles: plant response, interaction, phytotoxicity, and defense mechanism. Nanomaterials and Nanocomposites Exposures to Plants. Smart Nanomaterials Technology;* Springer Nature: Singapore, 2023 285–300. Available from: *https://doi.org/10.1007/978-981-99-2419-6_14.*
61. Nagarajan, P.; Subramaniyan, V.; Elavarasan, V., et al. Biofabricated aluminium oxide nanoparticles derived from Citrus aurantium L.: antimicrobial, anti-proliferation, and photocatalytic efficiencies. *Sustainability* **2023**, *15* (2), 1743. Available from: https://doi.org/10.3390/su15021743.
62. Sabah, T.; Jawad, K. H.; Al-Attar, N. Synthesis and biomedical activity of aluminium oxide nanoparticles by laser ablation technique. *Res. J. Pharm. Technol* **2023**, *16* (3), 1267–1273. Available from: https://doi.org/10.52711/0974-360X.2023.00209.
63. Wu, Q.; Fan, C.; Wang, H., et al. Biphasic impacts of graphite-derived engineering carbon-based nanomaterials on plant performance: Effectiveness vs. nanotoxicity. *Adv. Agrochem* **2023**, *2* (2), 113–126. Available from: https://doi.org/10.1016/j.aac.2023.01.001.
64. MalekiDizai, S.; Mennati, A.; Jafari, S., et al. Antimicrobial activity of carbon-based nanoparticles. *Adv Pharm Bull* **2015**, *5* (1), 19–23. Available from: http://doi.org/10.5681/apb.2015.003.
65. Jha, S.; Yadav, A. *Assessment of carbon and fullerene nanomaterials for sustainable crop plants growth and production.* Engineered Nanomaterials for Sustainable Agricultural Production, Soil Improvement and Stress Management, *Plant Biology, sustainability and climate change*, Academic Press, 2023145–160. *Available from: https://doi.org/10.1016/B978-0-323-91933-3.00015-5.*
66. Chichiriccò, G.; Poma, A. Penetration and toxicity of nanomaterials in higher plants. *Nanomaterials* **2015**, *5* (2), 851–873. Available from: https://doi.org/10.3390/nano5020851.
67. Schreiber, L. Polar Paths of diffusion across plant cuticles: new evidence for an old hypothesis. *Ann. Botany* **2005**, *95* (7), 1069–1073. Available from: https://doi.org/10.1093/aob/mci122.
68. Shafq, F.; Iqbal, M.; Raza, S. H., et al. Fullerenol [60] nano cages for protection of crops against oxidative stress: A critical review. *J. Plant Growth Regul* **2023**, *42*, 1267–1290. Available from: https://doi.org/10.1007/s00344-022-10629-x.

69. Panova, G. G.; Semenov, K. N.; Zhuravleva, A. S., et al. Obtaining vegetable production enriched with minor micronutrients using fullerene derivatives. *Horticulturae* **2023**, *9* (7), 828. Available from: https://doi.org/10.3390/horticulturae9070828.
70. Navarro, D. A.; Kookana, R. S.; McLaughlin, M. J., et al. Fullerol as a potential pathway for mineralization of fullerene nanoparticles in biosolid-amended soils. *Environ. Sci. Technol. Lett* **2016**, *3* (1), 7–12. Available from: https://doi.org/10.1021/acs.estlett.5b00292.
71. Wang, C.; Zhang, H.; Ruan, L., et al. Bioaccumulation of ^{13}C-fullerenolnanomaterials in wheat. *Environ. Sci. Nano* **2016**, *3* (4), 799–805. Available from: https://doi.org/10.1039/C5EN00276A.
72. Xiong, J. L.; Li, J.; Wang, H. C., et al. Fullerol improves seed germination, biomass accumulation, photosynthesis and antioxidant system in Brassica napus L. under water stress. *Plant Physiol. Biochem* **2018**, *129*, 130–140. Available from: https://doi.org/10.1016/j.plaphy.2018.05.026.
73. Cañas, J. E.; Long, M.; Nations, S., et al. Effects of functionalized and nonfunctionalized single-walled carbon nanotubes on root elongation of select crop species. *Environ. Toxicol. Chem* **2008**, *27* (9), 1922–1931. Available from: https://doi.org/10.1897/08-117.1.
74. Shen, C. X.; Zhang, Q. F.; Li, J., et al. Induction of programmed cell death in Arabidopsis and rice by single-wall carbon nanotubes. *Am. J. Bot* **2010**, *97* (10), 1062–1609. Available from: https://doi.org/10.3732/ajb.1000073.
75. Oloumi, H.; Mousavi, E. A.; Nejad, R. M. Multi-wall carbon nanotubes effects on plant seedlings growth and cadmium/lead uptake in vitro. *Russian J. Plant. Physiol.* **2018**, *65* (2), 260–268. Available from: https://doi.org/10.1134/S102144371802019X.
76. Azim, Z.; Singh, N. B.; Singh, A., et al. A review summarizing uptake, translocation and accumulation of nanoparticles within the plants: current status and future prospectus. *J. Plant Biochem. Biotechnol* **2023**, *32*, 211–224. Available from: https://doi.org/10.1007/s13562-022-00800-6.
77. Dev, A.; Srivastava, A. K.; Karmakar, S. Nanomaterial toxicity for plants. *Environ. Chem. Lett* **2018**, *16*, 85–100. Available from: https://doi.org/10.1007/s10311-017-0667-6.
78. Gao, M.; Chang, J.; Wang, Z., et al. Advances in transport and toxicity of nanoparticles in plants. *J. Nanobiotechnol* **2023**, *21*, 75. Available from: https://doi.org/10.1186/s12951-023-01830-5.
79. Basit, F.; He, X.; Zhu, X., et al. Uptake, accumulation, toxicity, and interaction of metallic-based nanoparticles with plants: current challenges and future perspectives. *Environ. Geochem. Health* **2023**, *45*, 4165–4179. Available from: https://doi.org/10.1007/s10653-023-01561-4.
80. Zeng, Y.; Himmel, M. E.; Ding, S. Y. Visualizing chemical functionality in plant cell walls. *Biotechnol. Biofuels* **2017**, *10*, 263. Available from: https://doi.org/10.1186/s13068-017-0953-3.
81. Wu, H.; Li, Z. Nano-enabled agriculture: how nanoparticles cross barriers in plants? *Plant. Commun.* **2022**, (6), 100346. Available from: https://doi.org/10.1016/j.xplc.2022.100346.
82. Lew, T. T. S.; Wong, M. H.; Kwak, S. Y.; Sinclair, R.; Koman, V. B.; Strano, M. S. Rational design principles for the transport and subcellular distribution of nanomaterials into plant protoplasts. *Small* **2018**, *14* (29).1802086. Available from: https://doi.org/10.1002/smll.201802086.

83. Chundawat, S. P. S.; Donohoe, B. S.; Sousa, L. D.; Dale, B. E. Multiscale visualization and characterization of lignocellulosic plant cell wall deconstruction during thermochemical pretreatment. *Energy & Environ. Sci.* **2011**, *4* (3), 973–984. Available from: https://doi.org/10.1039/c0ee00574f.
84. Chesson, A.; Gardner, P. T.; Wood, T. J. Cell wall porosity and available surface area of wheat straw and wheat grain fractions. *J. Sci. Food Agriculture* **1997**, *75* (3), 289–295. Available from: https://doi.org/10.1002/(ISSN)1097-0010.
85. Molnár, Á.; Rónavári, A.; Bélteky, P., et al. ZnO nanoparticles induce cell wall remodeling and modify ROS/RNS signalling in roots of Brassica seedlings. *Ecotoxicol. Environ. Saf* **2020**, *206*, 111158. Available from: https://doi.org/10.1016/j.ecoenv.2020.111158.
86. de PaivaPinheiro, S. K.; Miguel, T. B.; de Medeiros Chaves, M., et al. Silver nanoparticles (AgNPs) internalization and passage through the Lactucasativa (Asteraceae) outer cell wall. *Funct. Plant Biol* **2021**, *48* (11), 1113–1123. Available from: https://doi.org/10.1071/FP21161.
87. Zhao, W.; Zhou, P.; Lou, B., et al. Translocation and transformation of engineered nanomaterials in plant cells and their effect on metabolism. *Biocell* **2023**, *47* (3), 493–502. Available from: https://doi.org/10.32604/biocell.2023.025740.
88. Wu, H.; Li, Z. Nano-enabled agriculture: How do nanoparticles cross barriers in plants? *Plant Commun* **2022**, *3* (6), 2590–3462. Available from: https://doi.org/10.1016/j.xplc.2022.100346.
89. Lv, J. T.; Zhang, S. Z.; Luo, L.; Zhang, J.; Yang, K.; Christie, P. Accumulation, speciation and uptake pathway of ZnO nanoparticles in maize. *Environ. Science-Nano* **2015**, *2*, 68–77. Available from: https://doi.org/10.1039/C4EN00064A.
90. Zhao, W.; Zhou, P.; Lou, B., et al. Translocation and transformation of engineered nanomaterials in plant cells and their effect on metabolism. *Biocell* **2023**, *47* (3), 493–502. Available from: https://doi.org/10.32604/biocell.2023.025740.
91. Santás-Miguel, V.; Arias-Estévez, M.; Rodríguez-Seijo, A., et al. Use of metal nanoparticles in agriculture. A review on the effects on plant germination.*Environmental. Pollution* **2023**, *334*, 122222. Available from: https://doi.org/10.1016/j.envpol.2023.122222.
92. Thuesombat, P.; Hannongbua, S.; Akasit, S., et al. Effect of silver nanoparticles on rice (Oryzasativa L. cv. KDML 105) seed germination and seedling growth. *Ecotoxicol. Environ. Saf* **2014**, *104*, 302–309. Available from: https://doi.org/10.1016/j.ecoenv.2014.03.022.
93. González-Moscoso, M.; Martínez-Villegas, N.; Cadenas-Pliego, G., et al. Effect of silicon nanoparticles on tomato plants exposed to two forms of inorganic arsenic. *Agronomy* **2022**, *12* (10), 2366. Available from: https://doi.org/10.3390/agronomy12102366.
94. Wang X.P., Li Q.Q., Pei Z.M.,Wang S.C. 2018. Available from: https://doi.org/10.1007/s10535-018-0813-4, https://link.springer.com/article/.
95. Song, U.; Jun, H.; Waldman, B., et al. Functional analyses of nanoparticle toxicity: a comparative study of the effects of TiO2 and Ag on tomatoes (*Lycopersicon esculentum*). *Ecotoxicology and environmental safety* **2013**, *93*, 60–67. Available from: https://doi.org/10.1016/j.ecoenv.2013.03.033.
96. Sanjay, S. S.; Pandey, A. C.; Singh, M., et al. Effects of functionalized ZnO nanoparticles on the phytohormones: growth and development of Solanummelongena L. (Brinjal) plant. *World J. Pharm. Res* **2015**, *21* (5), 1990–2009.

97. Baskar, V.; Nayeem, S.; Kuppuraj, S. P.; Muthu, T.; Ramalingam, S. k Assessment of the effects of metal oxide nanoparticles on the growth, physiology and metabolic responses in in vitro grown eggplant (*Solanummelongena*). *3 Biotech.* **2018**, *8* (8), 362. Available from: https://doi.org/10.1007/s13205-018-1386-9.
98. Afrayeem, S. M.; Chaurasia, A. K. Effect of zinc oxide nanoparticles on seed germination and seed vigour in chilli (*Capsicum annuum* L.). *J. Pharmacogn. Phytochem* **2017**, *6* (5), 1564–1566. Available from: https://www.phytojournal.com/archives/2017/vol6issue5/PartW/6-5-100-866.pdf.
99. Shakir, W. M.; Muhamed, M.; AL-Mahdawe, I.; &Hammadi, M. Effect of ZnO nanoparticles on the content of sulforaphane in broccoli plant. *NeuroQuantology* **2022**, *20* (5), 361–367. Available from: https://doi.org/10.14704/nq.2022.20.5.NQ22182.
100. Hafeez, A.; Razzaq, A.; Mahmood, T., et al. Potential of copper nanoparticles to increase growth and yield of wheat. *J NanosciAdv Technol* **2015**, *1* (1), 6–11. Available from: https://doi.org/10.24218/jnat.2015.02.
101. Karimi, J.; Mohsenzadeh, S. Effects of silicon oxide nanoparticles on growth and physiology of wheat seedlings. *Russian J. Plant. Physiol.* **2016**, *63* (1), 119–123. Available from: https://doi.org/10.1134/S1021443716010106.
102. Srivastav, A.; Ganjewala, D.; Singhal, R. K., et al. Effect of ZnO nanoparticles on growth and biochemical responses of wheat and maize. *Plants* **2021**, *10* (12), 2556. Available from: https://doi.org/10.3390/plants10122556.
103. Rawat, P. S.; Kumar, R.; Ram, P., et al. Effect of nanoparticles on wheat seed germination and seedling growth. *Int. J. Agric. Biosyst. Eng* **2018**, *12* (1), 13–16.
104. Jalali, M.; Ghanati, F.; ModaressSanavi, A. M.; Khosgoftarmanesh, A. H. Physiological effects of repeated foliar application of magnetite nanoparticles on maize plants. *J. Agro Crop. Sci.* **2017**, 1–10. Available from: https://doi.org/10.1111/jac.12208.
105. Daghan, H. Effects of TiO2 nanoparticles on maize (*Zea Mays L.*) growth, chlorophyll content, and nutrient uptake. *Appl. Ecol. Environ. Res.* **2018**, *16*, 6873–6883. Available from: https://doi.org/10.15666/aeer/1605_68736883.
106. Ullah, H.; Li, X.; Peng, L., et al. In vivo phytotoxicity, uptake, and translocation of PbS nanoparticles in maize (Zea mays L.) plants. *Sci. Total Environ* **2020**, *737*, 139558. Available from: https://doi.org/10.1016/j.scitotenv.2020.139558.
107. Salama, D. M.; Osman, S. A.; Abd El-Aziz, M. E.; AbdElwahed, M. S. A.; Shaaban, E. A. Effect of zinc oxide nanoparticles on the growth, genomic DNA, production, and the quality of common dry bean (*Phaseolus vulgaris*). *Biocatalysis Agric. Biotechnol.* **2019**, *18*, 10108. Available from: https://doi.org/10.1016/j.bcab.2019.101083.
108. Salama, D. M.; Osman, S. A.; Abd El-Aziz, M. E., et al. Effect of zinc oxide nanoparticles on the growth, genomic DNA, production and the quality of common dry bean (Phaseolus vulgaris). *Biocatal. Agric. Biotechnol* **2019**, *18*, 101083. Available from: https://doi.org/10.1016/j.bcab.2019.101083.
109. Rasheed, A.; Li, H.; Tahir, M. M., et al. The role of nanoparticles in plant biochemical, physiological, and molecular responses under drought stress: A review. *Front. Plant Sci* **2022**, *13*, 976179. Available from: https://doi.org/10.3389/fpls.2022.976179.
110. Sahoo, A.; Sethi, J.; Satapathy, K. B., et al. Nanotechnology for precision and sustainable agriculture: recent advances, challenges and future implications. *Nanotechnol. Environ. Eng* **2023**, *8*, 775–787. Available from: https://doi.org/10.1007/s41204-022-00277-7.

CHAPTER

Nanoparticle-induced stress in crops: impact on growth, yield, and human nutrition

6

Anita Jaswal and Chandra Mohan Mehta

Department of Agronomy, School of Agriculture, Lovely Professional University, Phagwara, Punjab, India

6.1 Introduction

The increasing global population, along with the depletion of natural resources like arable land, water, and soil, highlights the necessity for agricultural growth to be economically viable, environmentally responsible, and efficient. This modification will be crucial for attaining several objectives this year.

Economic expansion has a substantial impact on agricultural nutrient balances. The enhancement of soil fertility holds significant importance in developing nations. The progression of agriculture is a crucial undertaking in order to eliminate poverty and hunger from the present circumstances. Therefore, it is imperative that we implement a resolute measure to foster agricultural progress. The majority of individuals on this planet reside below the poverty threshold, predominantly in rural regions where agricultural progress has shown to be ineffectual. Presently, there is a significant emphasis on examining the interplay between agriculture, poverty, and the nutritional aspects of food acquisition. Therefore, it is imperative for new technologies to give priority to enhancing agricultural production. The concepts of food and nutritional security have emerged as essential elements within contemporary understanding. These factors ought to be explicitly delineated within objectives that are geared toward specific goals. Sustainable agriculture provides an apparent possibility to alleviate poverty and hunger. Agriculture is improving; thus, it is necessary to focus on environmental performance and involve food chain ecosystems in production of food. The sustainable development of agriculture relies heavily on novel methods such as nanotechnology. We are naturally curious about the nature of this significant technology. Since 1959, Feynman's lecture on "Plenty of Room at the Bottom" has initiated the development of nanotechnology processes. Professor Norio Tanaguchi introduced the term "nanotechnology" in 1974. Nanotechnology has advanced due to the development of more sophisticated tools that can accurately consider or isolate nanomaterials.

Biogenic Nanoparticles. DOI: https://doi.org/10.1016/B978-0-443-33819-9.00010-X

1 nm is equivalent to one billionth of a meter. Why does this nanoparticle change its characteristics? The phenomenon is attributed to alterations in the atomic structure, leading to the emergence of magnetic characteristics. The lower size of nanomaterial is anticipated to have a bigger surface area and show increased activity. The altered property of nanomaterials is associated with increased reactivity in various sectors, particularly in biological processes. This advanced technique exhibits distinctive electrical, plasmonic, and visual characteristics that are influenced by quantum imprisonment effects. Currently, there is a significant need for efficient, dependable, and economically viable approaches to identify, monitor, and diagnose genetic host molecules in the agricultural industry. Chemically synthesized nanomaterials are now viewed as harmful. As an alternative, nanomaterials can be synthesized from plant systems, known as green nanotechnology. The method employed is both secure and efficient in terms of energy which leads to waste reduction and a drop in greenhouse gas emissions. Utilizing renewable materials in manufacturing these items is advantageous, resulting in little environmental impact. Presently, there is an increasing inclination toward the rapid implementation of green nanotechnology in order to harness its potential. It remains uncertain how green nanotechnology will achieve environmental sustainability in the future. Agriculture relies heavily on agrochemicals like pesticides and fertilizers for sustainable output and efficiency. Agrochemicals provide potential risks, including water contamination and residues on food products, which can adversely affect human well-being and the environment. Effective management and control of inputs can help minimize these risks. The application of sophisticated intelligent nanotools in advanced farming has the potential to alter agricultural methods, perhaps reducing the ecological consequences of contemporary agriculture while enhancing the value and production of crops. Nanotechnology is essential for the advancement of biosensors, using its diverse advantages in this domain. The distinctive characteristics of nanomaterials have the potential to significantly augment the sensitivity and efficacy of biosensors. Furthermore, the utilization of these materials has facilitated the advancement of novel signal transduction techniques in the field of biosensors. The utilization of nanomaterials has facilitated the diminution in dimensions of diverse sensors, resulting in the creation of compact devices such as nanosensors and other nanosystems.

6.2 Definition and characteristics of nanoparticles

The term "nanoparticle" originates from the Greek word "nanos," which signifies a diminutive sized entity, and the Latin word "particulum," which denotes a particle. Within the realm of scientific terminology, the term "nano" is employed to precisely indicate a magnitude of 10^{-9} in the metric system. The nanometer is a unit of measurement that is equal to one billionth of a meter. It can be used to indicate volume, weight, or time. Their synthesis can be classified into two primary branches: the bottom-up and the top-down method. In the first method, nanoparticles are created by assembling atoms and molecules, while in the second method, nanoparticles are produced by disintegrating bulk material. The bottom-up technique is more efficient because it allows for rapid and large-scale synthesis with a more uniform structure and reduced defect rate. These various types of nanoparticles will be explored in the following Table 6.1.

Table 6.1 Types of nanoparticles and their characteristics.

S. no.	Type of nanoparticle	Characteristics	References
1.	Carbon Nanotubes (CNTs)	Carbon-based nanoparticles can be broadly classified into two types: fullerenes CNTs. Fullerenes are nanomaterials composed of hollow cage-like structures, which are different kinds of carbon allotropes.	1
2.	Polymeric nanoparticles	Organic nanoparticles, primarily in the form of nanospheres or nanocapsules, offer full drug protection and are both biocompatible and biodegradable.	2
3.	Silver nanoparticles	Electrical, antibacterial, and optical characteristics are influenced by the form and size.	3
4.	Lipid-based NPs	The lipid nanoparticle is commonly characterized by its spherical shape, with a diameter that spans from 10 to 1000 nm. The structure possesses a robust central region comprised of lipids and an outside layer that encompasses soluble lipophilic substances.	4
5.	Iron oxide nanoparticles	Highly reactive with oxidizing chemicals, increased surface area, surface-to-volume ratio, and simple separation method	5
6.	Metal NPs	These NPs, which consist of metal precursors, display unique optoelectrical properties as a result of their Surface Plasmon Resonance (LSPR) characteristics. Copper, silver, and gold nanoparticles, which are alkali and noble metals, exhibit a wide absorption range within the visible portion of the solar electromagnetic energy spectrum.	6
7.	Ceramics NPs	The formation of inorganic nonmetallic solids by the process of heating followed by subsequent cooling.	7
8.	Semiconductor NPs	These materials possess features that lie between those of metals and nonmetals. They have wide bandgaps, which leads to significant alterations in their properties when the bandgap is modified.	8

6.3 The positive implications of NPs on crop yield

Nanomaterials are currently utilized in various fields. Nanomaterials used in agriculture improve soil health, increase crop productivity, and reduce water requirements for farmers. Pesticides and fertilizers are the main considerations for crop cultivation. Therefore, the advancements of nanopesticides and nanofertilizers are elaborated upon.

6.3.1 Enhancement of crop quality and quantity

6.3.1.1 Nanofertilizers

The rapid population growth has resulted in a considerable increase in the production of conventional fertilizers to improve agricultural output and secure crops. Consequently, this phenomenon leads to a decrease in soil fertility and a deterioration in food quality. The utilization of synthetic fertilizers poses problems to the delicate environment and has the potential to adversely affect human health if not completely harnessed. The United Nations' Food and Agricultural Organization has projected a global surge in the demand for fertilizers in the future years. Therefore, the implementation of innovative solutions, such as nanotechnology, is of utmost importance in the context of sustainable agriculture. Nanofertilizers are a class of environmentally sustainable solutions that have been developed with the aim of enhancing the efficacy of fertilizer application and mitigating nutrient loss, particularly in relation to phosphorus and nitrogen. Implementing a controlled and timed release of nutrients at a designated location minimizes the likelihood of water bodies and the environment becoming polluted. Nanofertilizers are substances that, even at lower dosages, give plants nutrients or increase the efficiency of fertilizers. Nano-encapsulating fertilizers can improve nutrient absorption, reduce nutrient wastage, enhance crop quality and yield, and minimize environmental harm. Nanofertilizers, when administered on leaves, have demonstrated a capacity to alleviate plant stress. According to Liu and Lal,[9] NPs have the potential to function as efficient fertilizers for field crops. Nanoparticles including micronutrients enhance nutrient uptake and fertilizer efficacy in comparison to conventional ions and salts. The effective enhancement of soybean plants' pliability to drought stress has been achieved through the development of a combination of three micronutrient nanoparticles, namely, ZnO, CuO, and B_2O_3.[10] Nanoparticulate fertilizers include additional nanoparticles such CNTs, TiO_2, and SiO_2, which promote plant development. The utilization of (TiO_2) and (SiO_2) in soybeans resulted in an augmentation of nitrogen fixation, stimulation of growth, and improvement of seed germination.

6.3.1.2 Nanopesticides

Enhancing plant productivity and maintaining their health is crucial due to over partial of the global inhabitants depending on plants and their derivatives for sustenance. Pathogens, nutrient deficiencies, and soil/air pollution are biotic and abiotic variables that can equally impact plant health by causing harm, ultimately reducing production. Plant disease is often caused by a wide variety of pathogens or pests. Hence, it is crucial to prevent them from further inducing plant withering. Insects play a vital role in agriculture by damaging crops, infesting stored food, and spreading plant diseases, ultimately leading to a decline in food quality.[11] Pesticides are utilized to increase agricultural output and efficiency by safeguarding plants from harmful variables. The utilization of pesticides has been found to be detrimental and toxic to the environment, presenting hazards to the health of both humans and animals due to its perilous characteristics. This pertains to various crucial issues such as the excessive and unselective application of pesticides in high concentrations, which negatively impact the environment by increasing bioaccumulation, leading to soil infertility, and disturbing its microbiota.[12] Several factors need to be considered to enhance agronomic output and efficiency while minimizing environmental impact. Synthesizing efficient and safe pesticides can be challenging and costly, but nanotechnology offers a new and enhanced option. An exemplary nanopesticide is nanostructured alumina (NSA), a nanoengineered material designed to act as an insecticide. It is produced by oxidizing metal that has a consistent electric charge. Nanotechnology has been advanced to enable the NSA to attack insects by utilizing its system. Insects exhibit electric charges generated by triboelectrification. Moreover, NSA utilizes insect tribocharging in a two-step mechanical process. NSA particles with a negative charge first connect with the insect's positive charge through strong electrical bonding. The contact dehydrates the insect by absorbing moisture with the intense sorptive characteristic of the NSA particle, causing the detachment of the insect's chitinous cuticle layer and ultimately leading to the insect's death due to dehydration. The efficacy of NSA powder or particle as an insecticide is attributed to its ability to adhere to the cuticle of insects, thereby perturbing their water equilibrium and inducing an increase in humidity levels. The potent insecticidal effectiveness of NSA is attributed to its diminutive size, inherent electric charge, absorbent properties, and expansive surface area.[13] TiO_2 and ZnO nanoparticles, which are photochemically active, operate as antibacterial agents when tomato plants suffer from spot disease. According to Zhao et al.,[14] the illumination of TiO_2 and ZnO nanoparticles, which exhibit photochemical activity, results in the generation of excited electrons. These excited electrons, when exposed to oxygen, facilitate the direct transfer of electrons, leading to the formation of superoxide radicals. Titanium dioxide nanoparticles are designed to possess antibacterial capabilities through photochemical processes, making them effective as nanoinsecticides in agriculture. It exhibits significant photocatalytic activity. Conventional pesticides are being packaged with NPs or polymers to improve efficiency, increase production, reduce excess runoff, and protect the environment by releasing active ingredients slowly over time.

6.3.1.3 Nanoherbicides

Weeds are seen as a major hazard to crops since they utilized nutrients that could be used by plants in larger quantities. Conventional methods, such as manual weed removal, for eliminating weeds are labor-intensive and time-consuming. Many herbicides available in the market can eliminate weeds in a field, but they can also harm crops and reduce soil fertility. Nanoherbicides offer a more environmentally friendly option for controlling weeds efficiently, without causing any harmful residues in the soil. Continued and regular use of one pesticide can lead to weeds developing immunity to it. Poly (epsiloncaprolactone) Poly(caprolactone) (PCL) nanocapsules have been employed as a vehicle for the traditional herbicide atrazine. The postemergence herbicidal effects of the nanocapsules containing atrazine were shown to be more potent compared to a conventional atrazine formulation.[15]

6.3.1.4 Nanofungicides

Fungal diseases are prevalent and cause over 70% of harm to important crops. These diseases impact the world by significantly reducing crop produce and affecting the economy. Traditional fungicides reduce these losses but can harm nonspecific organisms and disrupt biodiversity. Hence, we must explore alternative methods to improve the accuracy of fungal disease control. An effective technique against fungal diseases is the creation of NPs. Due to their antibacterial capabilities, Ag NPs are extensively utilized for disinfection reasons.[16] Both the ionic and nanoparticulate versions showed efficacy against fungus and notably decreased illness severity after a 3-hour application.[17]

6.3.1.5 Nanoemulsions

Nanoemulsions are oil-in-water emulsions where pesticides are spread as nanosized droplets in water. Producing nanoemulsions enhances the solubility of pesticidal active chemicals. Research has demonstrated that nanoemulsions of neem oil with smaller droplets are more effectively absorbed, resulting in a lower LC50.[18] Emamectin-benzoate nanoemulsion was shown to be more efficient against the rice borer than coarse emulsions. Some studies did not yield noteworthy outcomes when testing pesticide colloids. In the case of cotton bollworm, it was shown that there was no significant difference in LC50 values between a nanoemulsion of chlorpyrifos and a commercially available coarse emulsion. Another crucial factor is the gradual and regulated release of substances from the colloidal formulations, leading to extended and continuous exposure.[19]

6.4 Types of stress

Several factors impact agricultural productivity, such as the climate. Agriculture is essential for human well-being, and numerous groups and individuals are worried about the change in climate impact on agriculture. It is impacting important agricultural crops due to rising yearly temperatures, shifting rainfall patterns, floods, and decreasing water supplies. Climate change induces several biotic and abiotic pressures on plants, adversely impacting their growth and leading to diminished output. To mitigate the adverse effects of climate change, various measures have been implemented, including the use of resistant genotypes, the application of diverse plant growth regulators, and the employment of organic fertilizers.

6.4.1 Drought stress

Drought is a prolonged nonliving stress that impacts the growth and progress of crops, causing almost 70% of the possible decrease in worldwide production. Global warming trends are significantly affecting soil moisture levels and environmental conditions, leading to more severe droughts. Water shortage in plants causes a rise in solute concentration in the cytosol and extracellular matrix due to reduced water potential and turgor in plant cells, resulting in growth inhibition and reproductive failure. Drought-induced water stress in plants impacts stomatal functioning, reducing gaseous exchange, which in turn decreases transpiration rate and carbon assimilation as indicated in Fig. 6.1. Plants can decrease resource use and control growth in reaction to unfavorable environmental circumstances. In certain dynamic situations, it is necessary to improve plants' tolerance to drought. To improve water usage efficiency in plants under drought stress from

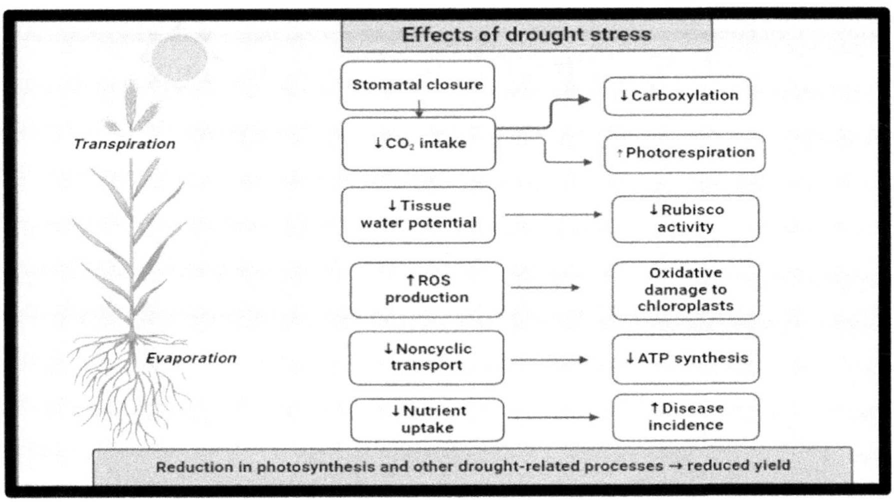

FIGURE 6.1

Illustrates the impact of drought stress on plants, including their growth, phenology, physiology, reproduction, and yield.

deteriorating root and leaf health, genetic modification can be utilized by integrating genes that encode specific proteins and signals through crosstalk, thereby activating diverse genes according to distinct regulatory processes. Ahmed et al.,[20] enhancing drought resistance in plants and adopting cost-effective agricultural practices are crucial to meet future food demands.

6.4.2 Temperature stress

The mean temperature of the Earth has exhibited an upward trend since the onset of the 21st century. Global warming can have detrimental effects on the environment due to the rise in Earth's temperature. Crops produce lower yields in locations with temperatures beyond the optimal range. Severe climate change can lead to elevated temperatures and drought, resulting in significant harm to agriculture and endangering tree populations as indicated in Fig. 6.2.

6.4.3 Salinity

Research communities must focus on addressing soil salinity to attain sustainable crop production. Roughly 20% of cultivated land worldwide is at risk of salt accumulation, and this percentage is increasing. It refers to soil that exhibits a significant abundance of water-soluble salts, surpassing a concentration of 4 decisiemens per meter. The phenomenon of urbanization and the increasing worldwide demand for food have led to a shift in agricultural practices toward drier or marginal lands. Consequently, this shift has resulted in a reduction of

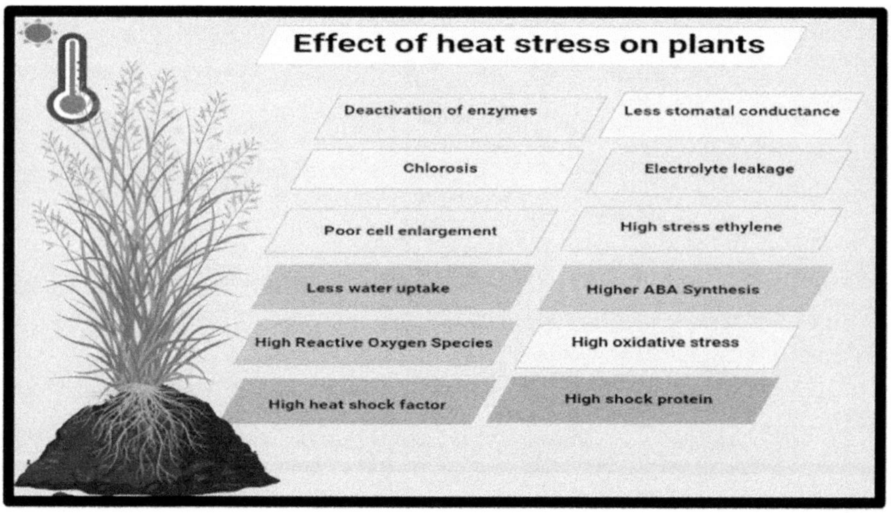

FIGURE 6.2

Effect of heat stress on plants physiology and metabolic activities.

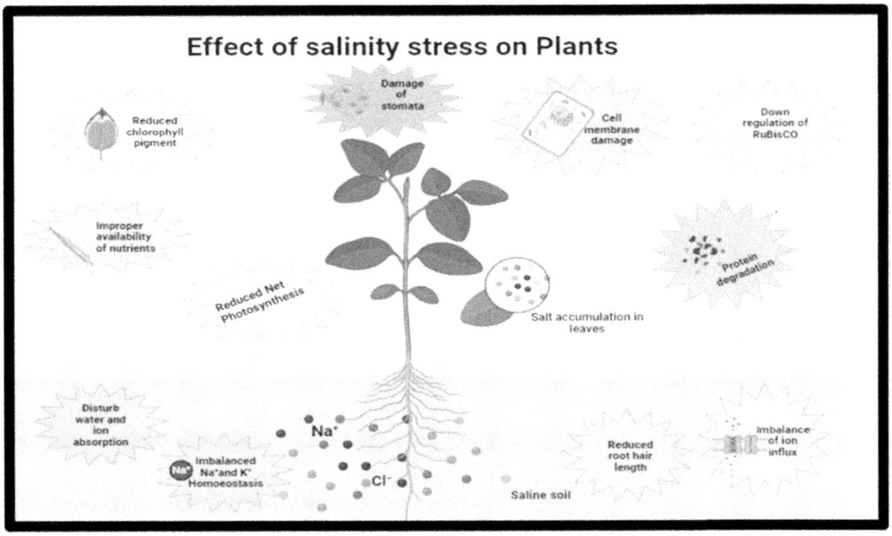

FIGURE 6.3

Effect of salt stress on the plant morphology and physiology.

accessible land and water resources required for the purpose of augmenting food production.[21] Salinity stress has a significant influence on plants, mostly resulting from diminished soil osmotic potential, imbalances in nutrient levels, and increased ionic toxicity, as depicted in Fig. 6.3.

6.4.4 Heavy metals (HMs)

Over the past few decades, the process of rapid industrialization has significantly increased the pressure on the Earth's ecosystem as a result of the excessive emission of greenhouse gas emissions. There is a rising worry about deteriorating global environmental circumstances due to an escalation in droughts and water scarcity. Industrial activities and urban growth are causing heavy metal pollution, posing a hazard to both the ecology and human health. Climate conditions face continuous risks, and the obstacles posed by a growing population make achieving food security in the 21st-century challenging.[22] The agricultural industry continually faces the problem of HMs buildup in soil due to the increasing industrial development, which includes mining and tannery activities. The utilization of excess fertilizers and pesticides in agricultural practices has resulted in detrimental environmental and health consequences through the emission of volatile chemicals and heavy metals into the atmosphere, hydrosphere, and lithosphere. Heavy metals exert a detrimental influence on the inherent composition of soil. According to Zhou et al.,[23] the agrobiological systems of plants are primarily adversely affected by Cr, Cd, Ni, Hg, Pb, and Cu. Due to their oxidative states,

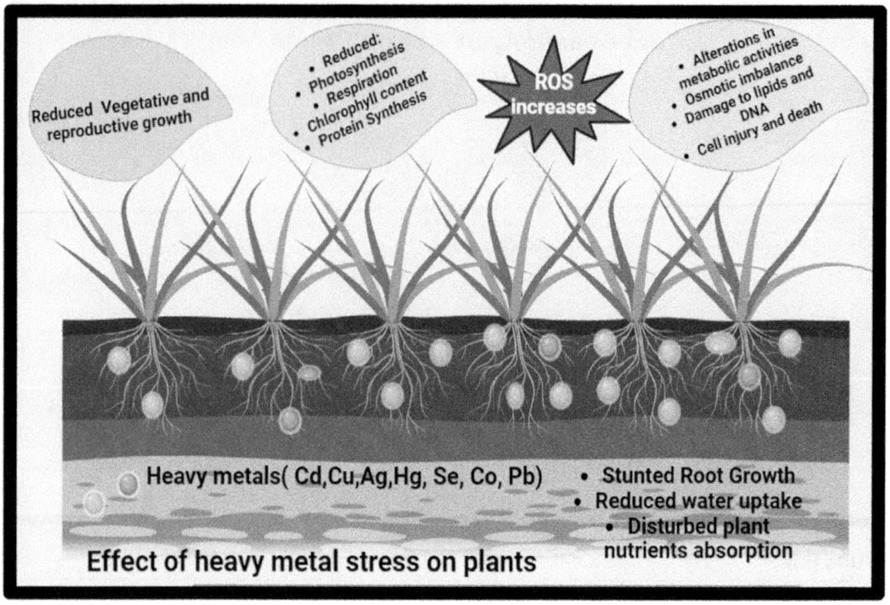

FIGURE 6.4

Effect of heavy metal stress on plant growth, nutrient uptake and other biochemical traits.

heavy metals possess a high degree of reactivity, hence inducing molecular and cellular modifications. This can result in changes in the plant's physiology by deactivating enzymes, denaturing proteins, substituting essential metals, and damaging membranes[24] as represented in Fig. 6.4.

6.5 Mechanism of NPs interaction

NPs are recognized for their distinct form, adjustable pore size, and heightened responsiveness due to increased surface area. Nanoparticles are widely regarded as a promising and influential approach for managing crop productivity and mitigating current and future limitations in agricultural production through the augmentation of plants' resilience mechanisms under adverse environmental circumstances. The uptake of nanoparticles by plants is influenced by various factors, including the characteristics of the nanoparticles, the physiological mechanisms of the plants, and the interactions between nanomaterials and the surrounding environment. The properties of nanoparticles will greatly influence their behavior and, thus, their absorption by plants. Size is a significant limitation for nanoparticles to enter plant tissues, with reports indicating that plants typically allow particles up to 40–50 nm to move and gather inside cells. The type of nanoparticle, its chemical content, and form are factors that influence uptake. Modifying and covering the surface of nanomaterials can significantly impact their characteristics, affecting how plants absorb and accumulate them. Plant species exhibit variability in their physiology, leading to changes in the uptake of nanoparticles. Nevertheless, the manner in which nanomaterials are applied plays a crucial role in influencing the plant's ability to absorb them.[25] Nanoparticles can interact with environmental components, potentially altering their properties and characteristics for plant assimilation.

6.6 Overview of NP absorption by plants

6.6.1 Nanoparticle absorption within plant roots

Nanoparticles initially adhere to the surface of plant roots by adsorption. The chemical compounds released by root hairs i.e. mucus or organic acids, resulting in the presence of negative charges on the surface of the root. This enhances the likelihood of NPs with positive charges to accumulate on the surface of the root and be readily absorbed. Lateral root growth can establish a new surface for nanoparticles to adhere to, allowing them to penetrate the root structure. The structure and role of the plant root epidermis closely resemble those of the plant leaf surface.[26] Root hair and the plant's surface epidermis at the tips of primary and secondary roots are still underdeveloped. Nanoparticles come into direct touch with the root epidermis and penetrate it when exposed to this region. The root cell wall's epidermal cells are semi-permeable as represented in Fig. 6.5. The primary cell wall has tiny openings that can block the movement of big particles. NPs pass in the root's core column or xylem if the root does not an exodermis.[27] Ion channels, endocytosis, protein–protein interactions, and physical damage are some of the methods in which these nanoparticles can be taken up by plant cells. Scientists found that plants mostly used the hydrophilic pathway for nanoparticle uptake and cellular uptake. However, nanoparticles cannot enter cells through this pathway due to the small pore size. Nanoparticle absorption by plant cells also relies on endocytosis. Plant cell plasma membrane invagination facilitates the entry of NPs into cells. Research has demonstrated that plant protoplasts can uptake particles smaller than 1 micron through endocytosis, indicating that endocytosis does not exhibit selectivity based on particle size.[28]

6.6.2 Nanoparticle absorption in plant leaves: mechanisms for foliar uptake

As part of agricultural treatments, nanoparticles are sprayed onto leaves, where they are absorbed by plants through the cuticle and stomata. The waxy cuticle found on the surface of leaves is primarily made up of pectin, cutin,

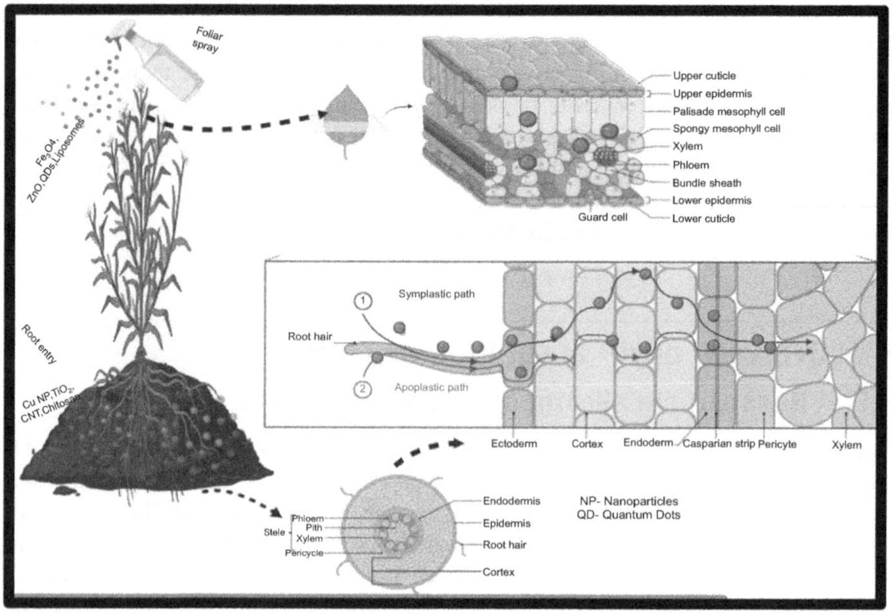

FIGURE 6.5

Illustrates a schematic depiction of the process of nanoparticle absorption and transportation throughout various plant components, specifically the roots and leaves.

and wax. It is the principal natural barrier that prevents nanoparticles from penetrating the leaves of growing plants and also keeps the leaves from drying out. Hydrophilic and lipophilic channels, distinct from one another, are present on the surface of the waxy stratum corneum.[29] Lipophilic channels on the cuticle surface facilitate the absorption of lipophilic nanoparticles by leaves by diffusion and penetration.

6.7 Variables influencing the assimilation and intake of nutrients

6.7.1 Impact of size and chemical composition and its impacts on interactions

Extensive study has been done on the influence of size on the absorption of metal-based nanoparticles. Research findings reveal that metal-based NPs with a size less than 50 nm have the ability to pass through the stomatal channel in plant leaves. As particle size increased, leaves' ability to absorb nanoparticles diminished. Several investigations have recognized the absorption of nanoparticles through leaves. ZnO nanoparticles were observed to primarily enter wheat leaves through the epidermis via the stomata channel and thereafter gather in chloroplasts using confocal microscopy. They also studied how stomatal aperture modulation affects the uptake of ZnO nanoparticles. Venzhik et al.[30] utilized Au nanoparticles of different size (3, 10, 50 nm) that were coated to interact with leaves of wheat. The study discovered that wheat leaves could absorb coated gold nanoparticles of various sizes. In a separate study, researchers verified the absorption of MgO in watermelon leaves using transmission electron microscopy (TEM). Scientists have found that plant leaves may absorb nanoparticles made of silica, polymer, and natural substances, in addition to nanoparticles made of metals. Their study findings showed that the nanoparticle type significantly influenced the crucial size that plant leaves could absorb. Researchers recently discovered that chitosan-based silicon nanoparticles, averaging 166 nm in size, were taken up by rice leaves and spread throughout the leaf cells. Nanoparticles interacting with the cell wall create huge pores that enhance the entry of nanoparticles.[31] Additionally, research has shown the buildup of larger nanoparticles in the cell wall and then in the cytoplasm. However, there is still a difference between the reported pore size of the cell wall (which is influenced by various factors) and the accumulation of larger nanoparticles. Additionally, this method avoids the challenges that come when examining the absorption and measurement of nanoparticles within plants, mainly caused by the plant cell wall's internal fluorescence and the plants' resistance to specific staining dyes.[32] Once nanoparticles enter the plants, it is crucial to identify the pathway they take for transportation to determine potential accumulation sites. The movement of plant cells throughout the plant is facilitated by two primary channels, namely the apoplast and the symplast. The symplastic system facilitates mobility across the cytoplasm of neighboring cells, whereas the apoplastic pathway permits movement via the extracellular regions, such as the cell walls of neighboring cells and xylem arteries. Cell-to-cell migration occurs through plasmodesmata, which act as cytoplasmic bridges enabling particle transfer between adjacent cells. The particles are around 20–50 nm in diameter and are encircled by 3-nanometer particles.[33] Due to size constraints, metal-based nanoparticles larger than 100 nm are commonly thought to have difficulty being taken up by plants by their roots.[34]

6.7.2 Impact of shape and surface charge

The dimensions, morphology, and surface charge of nanoparticles affect their penetration into plant mesophyll tissue. The various interfacial properties of nanoparticles of varying shapes influence their surface area and contact angle with plant surfaces, thereby affecting their absorption. Plant leaves can absorb nanoparticles with positive or negative charges. Nanoparticles exhibit a significant degree of autonomy, mostly contingent upon their absorption, uptake, and translocation within plants.[35] The energy and surface charge of nanomaterials enhance their interaction with plants, resulting in alterations to surface receptors, transporters, and specific membrane proteins. Nanoparticles with varying surface charges exhibit distinct aggregation capabilities and surface characteristics relative to their original states. The leaves contain a biological membrane known as the cell wall, characterized by hydrophobic and hydrophilic components, along with an uneven distribution of fixed negative charges. The negatively charged cell wall can facilitate the absorption of positively charged nanoparticles in the tissues. The negatively charged plant cell walls promote ion exchange, perhaps increasing the uptake of cationic nanoparticles compared to anionic ones.[36]

6.8 Translocation and interaction within plant system

6.8.1 Efficient translocation of NP through plant systems

Nanoparticles in the soil experience many biological and geological changes that influence their bioavailability and toxicity. Following their interaction with plant roots, the nanoparticles migrate to the aboveground regions. Numerous nanoparticles were investigated by researchers, who proposed that the process of plant uptake occurs through root absorption and subsequent dissemination within plant tissues. A number of mechanisms, including osmotic pressure, capillary pressures, and direct penetration through the root epidermal cells, allow for the passage of tiny nanoparticles, usually measuring 3 to 5 nm, into plant roots.[37] The epidermal cells of the root cell wall exhibit semi permeability, characterized by the presence of minuscule apertures that impede the transit of large nanoparticles. The nanoparticles generated novel apertures in the epidermal cell wall, facilitating their penetration. After traversing cell walls, nanoparticles (NPs) exhibit migration through extracellular gaps until they ultimately reach the central vascular cylinder, facilitating unidirectional upward movement inside the xylem. However, in order to access the core vascular cylinder, nanoparticles (NPs) need to navigate through the Casparian strip barrier in a symplastic fashion. According to Shukla et al.,[38] the process described involves the binding of carrier proteins to the endodermal cell membrane by endocytosis, pore formation, and transit. Nanoparticles translocate into cells by traversing the plasmodesmata and undergoing cytoplasmic absorption. In the Casparian strip, nonpermeable nanoparticles buildup, while those that have pierced the xylem are transported to the shoots and then back to the roots via the phloem. Nuclei, cytoplasm of cortical cells, and the cell wall of the epidermis are all potential sites of nanoparticle absorption by plants. Soil aggregates' root surfaces can be affected by nonparticulate debris, which can affect nutrient absorption. Nanoparticles can be directly absorbed by seeds when they pass through the parenchymal intercellular gaps, penetrate the seed coat, and then diffuse throughout the cotyledon. When nanoparticles are applied to leaves, they are able to pass through the cuticles and stomata. The cuticle functions as the primary protective layer of the leaf, effectively restricting the penetration of nanoparticles to dimensions below 5 nm. Nanoparticles with dimensions above

10 nm are able to penetrate the plant through stomata and thereafter undergo transportation into the plant via apoplastic and symplastic pathways, ultimately reaching the vascular system.[39] Intercellular transport of larger nanoparticles (50–200 nm) occurs. The phloem sieve tubes transport the internalized nanoparticles in the sugar flow. Nanoparticles can move in both directions and gather in roots, stems, fruits, grains, and young leaves through phloem transport due to these organs being effective storage sites for the plant's fluids. The apoplastic pathway is recognized as a route that lacks selectivity and offers minimal resistance.

6.8.2 Impact on plant pathways

Due to their beneficial capabilities as nano-carriers and nanopesticides, the use of nanoparticles in agriculture has grown significantly. These advantages encompass their compact dimensions, user-friendly application, extended storage capacity, and improved efficacy of agricultural chemicals. Nanoparticles can exert both valuable and detrimental effects on plants, influencing their physical composition, development, and agricultural productivity. The influence of nanoparticles on plants is contingent upon various aspects, including the specific plant species, the methods employed for application, the dosage and concentration of nanoparticles.

6.8.3 Potential impact of nanoparticles on photosynthesis

One of the most important ways plants on The Earth changes light energy into chemical energy via the process of photosynthesis. Photosynthesis is essential for the survival of all organisms, serving as either their major energy source or as an alternative. The process of converting light energy into chemical energy is facilitated by chloroplasts, which utilize chlorophyll, H_2O, and CO_2 to synthesize sugar molecules. Multiple studies indicate that applying metal nanoparticles to plant leaves significantly enhances chlorophyll content.[40] This allows plants to produce more developments for light absorption, leading to increased light energy absorption and improved photosynthesis. Research has shown that TiO_2 NPs enhance the photosynthetic efficiency of spinach by improving light absorption and speeding up light energy transfer and conversion. Additionally, the delayed photosynthetic activity of chloroplasts will be prevented by TiO_2 NPs, therefore inhibiting chloroplast aging. Zinc is a necessary element for plants, and zinc oxide nanoparticles have enhanced the chlorophyll levels in peanuts to 1000 mg/kg. Zinc oxide nanoparticles at specific concentrations enhance the quality and quantity of plant features and exhibit increased PSII fluorescence kinetics. Moreover, the single-walled carbon nanotubes (SWCNTs) enhanced photosynthesis by threefold associated to the control group.[41]

6.9 Adaptation to stressful environmental factors and stressful conditions caused by living organisms induced by living organisms

Nanoparticles alleviate drought stress by triggering and controlling the activity of genes associated with drought response and tolerance. Nanoparticles boost the photosynthetic process in plants experiencing drought by enhancing root growth, increasing aquaporin expression, modifying water metabolism within cells, accumulating compatible solutes, and maintaining ionic balance. These mechanisms help decrease osmotic stress caused by water scarcity.[42] Nanoparticles are mostly transported from soil to the plant body through lateral roots. Nanoparticles move via xylem tissues and distribute throughout the entire plant. The translocation relies upon the size of the NPs. Nanoparticles have been reported to have a good effect on reducing abiotic stress. Various nanoparticles have been utilized in field applications, like nanoagrochemicals, to enhance agricultural productivity. Nonproteinogenic amino acids have a crucial role in plants while experiencing stress, aiding in their ability to withstand abiotic challenges, particularly temperature stress. Applying nanoparticles in various quantities boosted plant growth and hydration to mitigate the impact of heat stress.[43] Plants show antioxidative effects at low concentrations of nanoparticles, but experience oxidative stress at high quantities. The heat shock protein plays a crucial role in the stabilization of other proteins under adverse conditions and subsidizes to the progress of heat stress resistance. According to Iqbal et al.,[44] the utilization of multiwall CNTs has been found to augment the gene expression of heat shock proteins such as HSP90, as well as facilitate the manufacturing of heat shock proteins. The presence of CeO_2 nanoparticles in maize results in an excessive production of H_2O_2 and an increase in the expression of HSP70. Nanoparticles alleviate the impacts of heat stress by controlling the opening of stomata. The rising contamination of soil, water, and air has led to a growing interest in using nanoparticles for remediation due to their minimal environmental impact. Nanoparticle remediation is more efficient and environmentally benign compared to bioremediation and chemical remediation. It does not produce hazardous by-products and is not as time-consuming or dependent on microbes as bioremediation, nor does it rely on the kinetic pace of reactions like chemical remediation. Nanotechnology is becoming increasingly popular in different sectors because of its sustainable advantages and adaptability. Nanotechnology is increasingly being utilized in farming through the use of nanofertilizers and nanoinsecticides.[45] Nanoparticles (NPs) have gained significant attention in recent years as a valuable tool for enhancing growth and survival in conditions of salt stress. The administration of NPs impacts many developmental stages in both beneficial and detrimental ways. Nanoparticles have been found to have significant impacts on the physical, functional, and chemical characteristics of plant species. Nanoparticles have been shown to influence how plants respond to salinity as indicated in Fig. 6.6.[46] Aside from the size, shape, and quantities of nanoparticles used, these impacts may alter depending on environmental conditions or plant species. The impact of

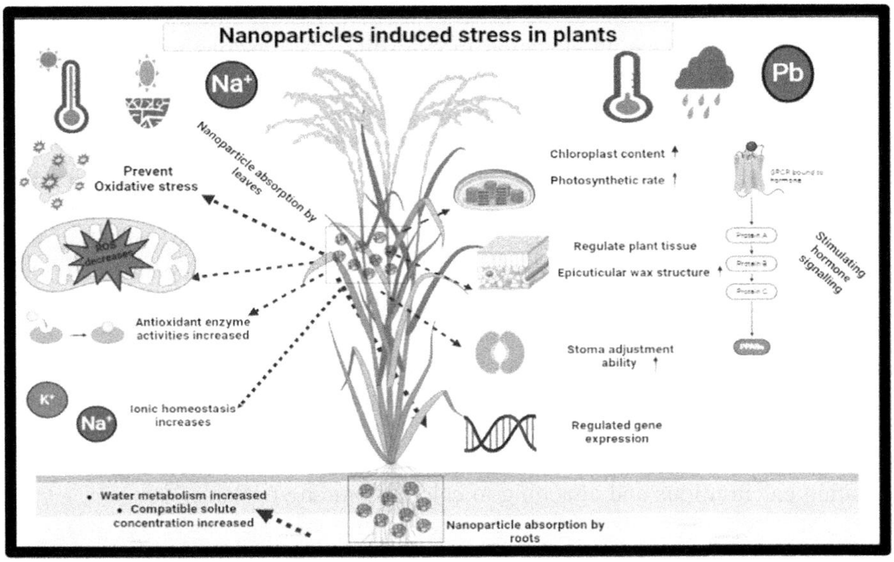

FIGURE 6.6

Representing how the inducing of nanoparticles helps to reduce the stress in plant in various aspects.

nanoparticles on different metabolic processes can be influenced by factors such as their concentration, size, application method, uptake by plants, characteristics, responsiveness, and movement into various tissues. Plants uptake nanoparticles through roots and leaves via many pathways.[47]

6.10 Mechanism of nanoparticles in alleviating environmental stresses at the molecular level

Plants experience a range of environmental challenges during entire lifetime that prompt them to adapt their defensive systems against these stresses. This adaptation involves modifying genetic, biochemical, and physiological processes at different stages. To cope with stresses, plants employ molecular pathways that include substantial modifications in gene expressions. Numerous research have provided evidence indicating that the effect of nanoparticles on the progress and expansion of plants is dependent on the concentration of these particles. The signaling network in plants triggers the defense system, which then initiates the molecular mechanism to respond to specific stress conditions. According to Almutairi,[48] stress signals elevate the concentration of calcium ions in the cytosol via activating Ca^+ channels. The detection of this rise is facilitated by calcium ion-binding proteins, which subsequently induce alterations in gene expression and enable plants to adapt to stress. The phenomenon of nitric oxide (NO) has been observed to induce an elevation in cytosolic calcium ions within plant cells under various biotic and abiotic stress conditions, resulting in the generation of nitric oxide through the interaction with calcium ions. The application of Ag nanoparticle therapy on the roots of *O. sativa* has been observed to play a role in the regulation of various biological processes. The interaction between Ag nanoparticles and calcium ion channels and Ca1/Na1 ion pumps through calcium ion receptors has been observed to have an impact on cellular metabolism.[49] Furthermore, the functional regulation of Ca1/calmodulin-dependent protein kinase II was facilitated by the existence of C60 nanocrystals. In A. thaliana, the introduction of cadmium sulfide quantum dots resulted in an upregulation of the expression levels of calcium-binding protein CML45 and calcium-dependent protein kinase 23. The role of calcium-binding proteins in regulating stress responses has been proven, and their upregulation has improved plant resistance to several abiotic stressors. Nanoparticles induce the activation of stress-responsive genes by imitating calcium ions and attaching to calcium-binding proteins.[46]

6.11 Toxicity of nanoparticles to plants

Despite its numerous advantages, toxicity of nanoparticles in plants must be acknowledged. Nanoparticles can harm plants by inducing oxidative stress and causing physical damage, like stomatal closure caused by nanoparticle aggregation. High quantities of nanoparticles sprayed on plant leaves can accumulate on the leaf surface, leading to stomatal obstruction that impairs gas exchange and photosynthesis in plants. Copper oxide nanoparticles are harmful to *Hordeum sativum*, foremost reduction in the plants' photosynthetic rate and hindering the development of roots and shoots.[50] Other research has also investigated the effect of polymer NPs on plant seedling growth, in addition to their impacts on seed germination. The onion root length fell by 41.5% after exposure to 50 nm polystyrene nanoparticles for 72 hours. The root length of Lemna minor decreased due to the presence of polyethylene nanoparticles. The inhibition of *Arabidopsis thaliana* seedlings becomes more pronounced with larger concentrations of nanoparticles. Enhanced comprehension of how nanoparticles impact plants can aid in evaluating their toxicity.[51]

6.12 Altered structures, form and functions of plants: changes in plant morphology due to NPs interactions

Abiotic stresses, such as salinity, drought, severe temperatures, and heavy metals, exert significant influence on multiple aspects of plant morphology. Remédios et al.[52] conducted a study wherein they noticed that the utilization of nanoparticles at modest quantities resulted in enhancements in the length of the root, shoot diameter, and seed germination rate. Silver nanoparticles obtained from plants enhance the growth of above-ground structures at high temperatures. Titanium nanoparticles (TiNPs) documented to improve the growth of *Dracocephalum moldavica* plants under conditions of salinity stress. Iron nanoparticles (FeNPs) have confirmed the ability to enhance plant development and total biomass yield in *Brassica napus* when exposed to drought conditions.[53]

6.12.1 Anatomical alterations caused by nanoparticles

Plants experience different anatomical alterations in reaction to stresses. Structural responses vary based on the specific abiotic stressor. During drought conditions, stomata structure changes by developing huge substomatal chambers. Heat stress leads to a reduction in size of cell, closing of stomata, and a restricted transcription rate. Furthermore, there is a noticeable rise in the density of stomata and trichomatous cells, as well as the presence of bigger xylem arteries in both the roots and shoots. Heat stress has caused the death of mesophyll cells and increased permeability of the plasma membrane in *Vitis vinifera L. CV. Jingx.* When subjected to elevated temperatures, *Zygophyllum qatarense* demonstrates a reduction in transpirational water loss through the manifestation of bimodal stomatal activity. Nanoparticles assist plants in surviving stress circumstances by contributing to the plant's anatomical adaptations. TiO_2 nanoparticles effectively controlled opening of stomata in *Zea mays* plants exposed to heat stress, hence decreasing its impact. Treating strawberry plants with SiO_2 nanoparticles helped counteract the harmful effects of salinity by enhancing the epicuticular wax layer. Zinc oxide nanoparticles were effective in minimizing harm to the outer layer and blood vessels in *Sorghum bicolor* when exposed to high salt levels. A new study has been published on the impact of CeO_2 NPs on *Brassica napus L.* under salt stress. The former was found to be able to reduce root apoplastic barriers.

6.12.2 Functional modifications in response to NPs exposure

Stressed plants display several physiological reactions including alterations in photosynthetic processes, transpiration, absorption of minerals and water, lipid degradation, and seed sprouting. Plants experience reduced seed germination, pollen fertility, seed production, and chlorophyll content during cold stress, leading to a negative impact on photosynthesis.[54] Heat stress can reduce the speed of nitrogen absorption in plants. Cadmium accumulation in soil affects the uptake of macronutrients and micronutrients by plants because they are unable to absorb water when under cadmium stress. Drought stress hinders photosynthesis by impacting the thylakoid membranes. Excessive soil salinity hinders plant germination. Heavy metals inhibit seed germination, photosynthesis, and other plant physiological activities.

6.12.3 Significant biochemical alterations to withstand abiotic stress

Abiotic stress can result in protein and enzyme denaturation, leading to enzyme activation, decreased protein synthesis, heightened fluidity of membrane lipids, and compromised membrane integrity. The carbohydrate levels vary in response to different biotic stress factors. The carbohydrate concentration reduces with salt stress and increases with decreasing temperature stress.[55] This physiological response ultimately leads to a reduction in cellular damage. NPs are reported to cause several fluctuations in the structure, function, and chemical processes of plants in order to improve their resistance to drought. Applying Ag NPs reduced protein misfolding caused by flooding stress, enhancing the stress tolerance of soybean seedlings.[56] Applying Zn NPs to leaves increased the concentrations of proline, glycine betaine, free amino acids, and carbohydrates during drought circumstances. Proline concentration is elevated under different abiotic stressors. Proline is believed to safeguard enzymes and cellular structures in high amounts by acting as an osmoticum. Accumulating osmolytes helps cells maintain their redox potential during stressful conditions and serves as a signal for antioxidant response.[57]

6.13 Potential threats of inadvertent NPs release

Extensive investigation has yielded few advancements in grasping the risk and health dangers associated with NM. Efforts have been concentrated on creating unified regulatory frameworks for nanosafety. Plant-based nanosafety experiments has primarily concentrated on deliberate or targeted exposure of food crops to certain nanomaterials such as nanopesticides, nanofertilizers, and nanoherbicides. Additionally, it has also considered indirect or accidental exposure through phenomena like spray drifting, leaching, and runoff of nanoparticles. It is crucial to thoroughly examine the physiochemical characteristics of nanomaterials before utilizing them in any application, as these properties can significantly influence their potential health and environmental risks, thereby restricting their full beneficial potential. Size and concentration are the primary qualities of nanoparticles that are significant in toxicology. Increased surface area due to decreased size can lead to more absorption and contact with the plant system, thereby raising the likelihood of harmful consequences. This suggests that smaller NPs can be absorbed more quickly by plants and may cause more harm even at lower concentrations. The bioaccumulation of graphene oxide (GO) in plants is reduced due to the impeded translocation from roots to shoots at bigger sizes. Concentration-dependent effects play a crucial role in determining the beneficial or harmful effects of nanoparticles. The primary cause of nanoparticles' toxicity, particularly metal nanoparticles, is the generation of high levels of reactive oxygen species (ROS) through the release of ions in the biological system, leading to harmful effects.[58] It disrupts the electron transport pathway in mitochondria and chloroplasts. Various stress variables impact carbon fixation, leading to photoinhibition and increased generation of superoxide anion radicals and H_2O. The excess ROS produced interacts with biological components, leading to protein changes, lipid peroxidation, and DNA damage. It also has beneficial consequences by aiding in reducing stress-related problems in plants, highlighting the significance of balance between ROS generation and scavenging. The elevated levels of reactive oxygen species (ROS) significantly affect the function of antioxidative enzymes in plants exposed to nitrogen and phosphorus stress.[59] An excessive concentration of nanoparticles can significantly impact photosynthesis, perhaps leading to inhibited plant growth or plant mortality. Based on the mentioned research, certain elements must be taken into account before effectively utilizing NPs in the agricultural sector. Studying the physicochemical features of the soil system is crucial due to its significant impact on agricultural production through plant-soil interaction. The soil is seen as the primary repository for nanoparticles, and hence, their interaction with soil components could significantly influence the destiny of nanoparticles. pH, organic content, and cation exchange capacity have been found to correlate with an elevated risk of toxicity.[60] Zinc oxide nanoparticles by themselves lead to a notable decrease in plant biomass. Nanoparticles (NPs) affect the soil microbial community structure directly in a dose-dependent way. Nitrogen-fixing taxa are decreasing while the community taxa involved in decomposing organic contaminants and biopolymers are growing.[61]

6.14 Potential threats to human health

It is conceivable that nanomaterials (NMs) may be introduced into the atmosphere by agronanotechnologies, subsequently traversing the food chain and entering the human exposure pathway. Current obstacles in applying nanochemicals in agriculture stem from the possible movement of nanoparticles through the food chain, necessitating prompt safety evaluations. Humans can be exposed to nanoagrochemicals through different pathways, including the increased need for food crops due to more population and higher call for animal-based products, wastage in affluent industries, and various agricultural activities related to food crop production. Plants play a crucial role in transferring nanoparticles to various trophic levels throughout the food web because they are a major food source for lower trophic-level species, animals, and humans.[62] Understanding the one health idea is crucial, particularly in the context of agro nanotechnology. The excessive use and improper application of chemicals as growth stimulants for food crops greatly enhance the movement of NMs in the food chain. An intelligent testing strategy framework is needed to evaluate the risk of agronanotechnologies due to nanotoxicity and associated human health concerns. Conduct an initial evaluation of existing knowledge and areas where information is lacking. Classify and combine data to create a technique for organizing nanomaterials according to their characteristics and evaluate their effects on living organisms. The purpose of this endeavor is to develop sophisticated risk-assessment methodologies and guarantee safety in light of the potential hazards linked to this particular technology.[63]

6.15 Risk posed to plant systems and ecosystem

Nanoparticles are employed in the agricultural sector to mitigate environmental strain; nonetheless, concerns have arisen over their accumulation and their introduction into the food web. The utilization of certain synthetic nanoscale materials in agriculture has the potential to provide hazards to human well-being and the ecosystem, despite the generally accepted nontoxic nature of nanoscale components.[64] Plants show a crucial part in ecological systems since they are responsible for the transportation of environmental nanoparticles (ENPs), which have the potential to stick to roots of plant and pose physicochemical hazards. Attaching ENPs to plant roots can induce the formation of large holes in cell walls, facilitating the absorption of large ENPs through them. There are further apprehensions pertaining to the potential chemical damage inflicted upon plants and the enduring ramifications of nanoparticles on the ecological food web. According to Rajput et al.,[65] the introduction of nanoparticles into the food chain has the potential to present risks to both people and animals.

6.16 Controlled applications and mitigation strategies

Agricultural land is a significant receiver of the NPs emitted into the environment. Soil is a diverse substance that allows farmers to cultivate a range of crops due to its varied nature. The current impact of nanoparticles on soil is insufficient, and comprehending their influence necessitates understanding the transportation, dynamics, and alteration of nanoparticles within the soil. When nanoparticles are introduced into soil, they are either physically trapped or adsorbed chemically by soil particles. The nanoparticles are typically 1 to 100 nm in size. Engineered nanoparticles that are persistent and have a slow degradation rate in the environment present a notable hazard.[66] Preserving the ecology and microbial population of the soil is a significant problem due to the unavoidable exposure to NP. NPs effect on soil hinge on on the nanoparticle concentration, soil type, and the biogeochemical processes in the soil. Pesticides and fertilizers have generated a range of environmental and health problems due to their excessive and unselective application in the past. Nanomaterials are anticipated to have a significant rise in usage in agriculture in the next decade because of so many applications. Despite the growth in nanotechnology, safety measures have not seen proportional enhancement. Okeke et al.[67] reported instances of reckless utilization and disposal of nanomaterials that harm the environment. The nanoparticles utilized in creating nanopesticides or nanofertilizers have a high likelihood of causing negative impacts on the ecosystem and animal life. Metallic nanoparticles present an increased hazard as they release metal atoms/ions during decomposition. Soil is crucial for agriculture, and using nano-fertilizers or nanopesticides repeatedly may result in the buildup of nanoparticles in the soil. This can subsequently impact seed germination, plant development, and biomass production. NPs face increased risks related to both the environment and health. Below are some crucial future aspects being mentioned.

- To evaluate the risk linked to nanoparticles, globally recognized and precise methodologies for detecting and characterizing them must be developed.
- It is necessary to establish acceptable thresholds for different nanoparticles in different environments.
- Greater emphasis should be placed on conducting research and allocating resources toward exploring the underlying mechanism by which nanoparticles interact with the food chain.
- Other factor to consider is the amount of fertilizers, which must be determined by taking into account its destiny, delivery, persistence in diverse media, toxicity, and impact on nontarget organisms.
- The large volume to mass ratio of nanoparticles enables them to readily interact with biomolecules due to their unique characteristics. Thus, it is essential to comprehend toxico dynamics and toxicokinetic investigations in relation to agroecosystems. Research in agronanotechnology has to focus on topics including the behavior of nanoparticles, their spread, movement, and availability in biological systems.

6.17 Conclusion

This chapter specifically examines the diverse roles of nanoparticles in the area of farming and livestock. Nanotechnology in contemporary farming enhances the global economy by providing assistance and progress. Nanopesticides (NPs) have emerged as a crucial tool in tackling various concerns, including population expansion, plant and animal illnesses, and climate change. Nanoparticles (NPs) have been found to enhance crop productivity while minimizing environmental impact, in contrast to conventional pesticides and fertilizers. The introduction of NPs has significantly enhanced effectiveness and agronomic efficiency in comparison to conventional resources. Various nanoinsecticides and nanofertilizers are currently being developed, with some already available for purchase. The investigation of the interplay between nanoparticles and plants is crucial for understanding the physiological and biochemical reactions that plants undergo. New methods are being created to identify and measure nanoparticles as they move through plants, providing insights into their changes and safety within an intricate scheme. Agriculture is the sole source of human food production, utilizing both transitional and final inputs with established technology. Therefore, acquiring contemporary agricultural knowledge is essential. Although there are certain gains in the agricultural process, emerging countries still face a significant shortage of essential food goods. Although there is abundant information on individual nanomaterials, the toxicity of numerous NPs remains indeterminate. Consequently, the utilization of these constituents is restricted by the insufficient understanding of risk assessments and their impact on human health. Nanomaterials have the capacity to enhance plants' ability to withstand environmental stress. Nanoparticles exhibit a wide range of effects, such as boosting water absorption in seeds, metabolizing starch reserves, promoting photosynthesis, changing phytohormone levels, regulating oxidative stress, and influencing nutrient uptake. Most study has focused on comprehending a single sort of stressful condition. Future study should prioritize investigating stress conditions that closely resemble those found in real-life situations. Many researchers have shown the good effect of NPs on plant health, although the specific molecular processes responsible for the enhanced plant tolerance are not yet fully understood. Additional research is required to investigate the impact of NPs on the antioxidant system of plant cells and how this can enhance plant resilience to different stresses. This comprehension could assist in developing advanced smart nanoparticles to decrease stress and assure long-lasting agricultural productivity. The practical use of many newly developed substances is now restricted by variations in ambient circumstances, soil types, target plants, and especially the physicochemical characteristics of the new metallic/nonmetallic chemicals. It is crucial to do future research to assess the toxicologic effects on certain microorganisms, plants, and animals in order to effectively apply nanotechnology in practical environments.

6.18 Future perspective

Sustainable agriculture ought to be perceived as an ecological framework, wherein abiotic and biotic species reside in a state of equilibrium, ensuring a stable food chain and its corresponding energy equilibriums. The achievement of optimal agricultural output is facilitated through the application of novel technology, modernization efforts, enhanced utilization of nano-chemicals, specialization strategies, and adherence to government rules. Establishing modern technologies in the food business is essential to address the dilemma. Nanotechnology is an advanced technological field that possesses distinctive characteristics that have the potential to significantly transform multiple facets of the global food supply chain. The forthcoming study in the realm of agricultural nanotechnology or nanofoods may necessitate heightened attention toward particular regions. Novel transportation systems for the transportation of exact diet/feed components, nutrients in plant, and other chemicals with a focus on environmental and safety considerations. These systems also have potential applications in the pharmaceutical field. Nanotechnology-based biosensors has played a vital role in controlling insects and enhancing the quality of agricultural food items.

Consumers can receive actual information around condition of exact diet products with keen packaging of food equipped through nanosensors. Novel analytical techniques must be developed to identify, confirm, and evaluate the impacts of individual nanomaterials/nanofoods on entire ecosystems. Conducting life-cycle study on nanomaterials and nanofoods is necessary. Enhancing a comprehensive database and fostering international cooperation for the management of this information require attention. To thoroughly analyze the impact of this technology on human life, economy, and science, it is imperative to establish new communication channels and engage in debates involving many stakeholders. This technology possesses the capacity to provide novel and economically efficient avenues for worldwide human nourishment in the forthcoming years. Nanotechnology significantly impacts various sectors of the food industry. While it shows potential uses, it is important to note that the majority of our understanding of it comes from laboratory tests. It is difficult to grasp the practical application without comprehending its impact on environmental toxicity. Therefore, there is a need to conduct extensive trials on identified information gaps. Here, we suggest six crucial fundamental characteristics.

1. The primary goal of the exposed administration consciousness initiative is to teach the general public about food nanotechnology and its various uses. This will be achieved by creating a comprehensive database and providing supporting evidence. The program aims to benefit both the public and the food industry.
2. An in-depth analysis of future studies is necessary to assess the risks of using NP-based products in agriculture and develop a more practical strategy.
3. Conduct a concentration-dependent inquiry to regulate the minimal nontoxic amount dose of the nanoparticle (NP) in the soil.
4. In order to gain an inclusive considerate of the bioaccumulation capacity of nanoparticles in farm applications, its influence on nanotoxicity, the mechanisms of trophic chain transfer within ecological systems, and the identification of appropriate safety protocols.
5. Comprehensive knowledge of how soil physicochemical properties interact with nanoparticles to minimize harm to the soil microbiota. Additional research is warranted to explore the potential of soil rotation as a means to modify the destiny, movement, and accessibility of nanoparticles, with the aim of mitigating their subsequent toxicity and guaranteeing their dependable and advantageous utilization in the field of agriculture.
6. Finally, we recommend replacing biosynthesized nanoparticles with synthetic ones and conducting thorough investigation. Utilizing the environmentally sustainable protocol may result in minimal or negligible toxicity. Therefore, future research should prioritize investigating their functional efficacy. It is imperative to conduct pilot research in natural settings, such as cultivating plants in soil, in order to effectively demonstrate the environmental ramifications of nanoparticles (NPs). Furthermore, some defense precautions should be taken into account.

a. It is essential to align the nanosafety legislation frameworks with safety-by-design synthesis methodologies to prevent excessive use at dangerous levels.

b. Physical–chemical characteristics of a new nanoparticle-based pesticide should be thoroughly examined in extended tests before being used in practical applications due to human health concerns and ecological risks that could hinder its beneficial capabilities. Research on nanosafety with an environmental focus should primarily investigate the unintentional release of nanoparticles during the production, application, and disposal of nanoformulated agricultural products.

Acknowledgment

The research was not supported financially by anyone. The credit goes to all the authors.

References

1. Astefanei, A.; Núñez, O.; Galceran, M. T. Characterisation and determination of fullerenes: a critical review. *Analytica Chim. acta* **2015**, *882*, 1–21. Available from: https://doi.org/10.1016/j.aca.2015.03.025.
2. Crucho, C. I.; Barros, M. T. Polymeric nanoparticles: a study on the preparation variables and characterization methods. *Mater. Sci. Engineering: C.* **2017**, *80*, 771–784. Available from: https://doi.org/10.1016/j.msec.2017.06.004.
3. Ravindran, A.; Chandran, P.; Khan, S. S. Biofunctionalized silver nanoparticles: advances and prospects. *Colloids Surf. B: Biointerfaces* **2013**, *105*, 342–352. Available from: https://doi.org/10.1016/j.colsurfb.2012.07.036.
4. Gujrati, V. B.; Jon, S. Bioengineered bacterial outer membrane vesicles: what is their potential in cancer therapy? *Nanomedicine* **2014**, *9* (7), 933–935. Available from: https://doi.org/10.2217/nnm.14.56.
5. Ali, A.; Zafar, H.; Zia, M.; ul Haq, I.; Phull, A. R.; Ali, J. S.; Hussain, A. Synthesis, characterization, applications, and challenges of iron oxide nanoparticles. *Nanotechnology, Sci. Appl.* **2016**, 49–67. Available from: https://doi.org/10.2147/NSA.S99986.
6. Dreaden, E. C.; Alkilany, A. M.; Huang, X.; Murphy, C. J.; El-Sayed, M. A. The golden age: gold nanoparticles for biomedicine. *Chem. Soc. Rev.* **2012**, *41* (7), 2740–2779. Available from: https://doi.org/10.1039/C1CS15237H.
7. Tsuchiya, S.; Abe, S.; Era, Y.; Nakagawa, Y.; Yamagata, S.; Nakamura, M.; Iida, J. Assessment of size and morphological effects of ceramics nanoparticles on their biocompatibility. *Nanosci. Nanotechnol. Lett.* **2016**, *8* (4), 329–334. Available from: https://doi.org/10.1166/nnl.2016.2139.
8. Terna, E. E.; Elemike, J. I. The future of semiconductors nanoparticles: Synthesis, properties and applications. *Mat. Sci. Eng. A. 2021*, B n.d., 272, 115363. Available from: https://doi.org/10.1016/j.mseb.2021.115363.

9. Liu, R.; Lal, R. Potentials of engineered nanoparticles as fertilizers for increasing agronomic productions. *Sci. Total. Environ.* **2015**, *514*, 131–139. Available from: https://doi.org/10.1016/j.scitotenv.2015.01.104.
10. Dimkpa, C. O.; Bindraban, P. S. Nanofertilizers: new products for the industry? *J. Agric. Food Chem.* **2017**, *66* (26), 6462–6473. Available from: https://doi.org/10.1021/acs.jafc.7b02150.
11. Chaud, M.; Souto, E. B.; Zielinska, A.; Severino, P.; Batain, F.; Oliveira-Junior, J.; Alves, T. Nanopesticides in agriculture: benefits and challenge in agricultural productivity, toxicological risks to human health and environment. *Toxics* **2021**, *9* (6), 131. Available from: https://doi.org/10.3390/toxics9060131.
12. Ding, Y.; Wang, Q.; Zhu, G.; Zhang, P.; Rui, Y. Application and perspectives of nanopesticides in agriculture. *J. Nanopart. Res.* **2023**, *25* (8), 159. Available from: https://doi.org/10.1007/s11051-023-05811-5.
13. Stadler, T.; Buteler, M.; Valdez, S. R.; Gitto, J. G. Particulate nanoinsecticides: a new concept in insect pest management. *Insecticides: Agric. Toxicol.* **2018**, 83.
14. Zhao, W.; Liu, Y.; Zhang, P.; Zhou, P.; Wu, Z.; Lou, B.; Tan, Z. Engineered Zn-based nano-pesticides as an opportunity for treatment of phytopathogens in agriculture. *NanoImpact* **2022**, *28*.100420. Available from: https://doi.org/10.1016/j.impact.2022.100420.
15. Sangeetha, J.; Hospet, R.; Thangadurai, D.; Adetunji, C. O.; Islam, S.; Pujari, N.; Al-Tawaha, A. R. M. S. Nanopesticides, nanoherbicides, and nanofertilizers: the greener aspects of agrochemical synthesis using nanotools and nanoprocesses toward sustainable agriculture. *Handb. Nanomaterials Nanocomposites Energy Environ. Appl.* **2021**, 1663–1677.
16. Periakaruppan, R.; Palanimuthu, V.; Abed, S. A.; Danaraj, J. New perception about the use of nanofungicides in sustainable agriculture practices. *Arch. Microbiology* **2023**, *205* (1), 4. Available from: https://doi.org/10.1007/s00203-022-03324-8.
17. Malandrakis, A. A.; Kavroulakis, N.; Chrysikopoulos, C. V. Use of copper, silver and zinc nanoparticles against foliar and soil-borne plant pathogens. *Sci. Total. Environ.* **2019**, *670*, 292–299. Available from: https://doi.org/10.1016/j.scitotenv.2019.03.210.
18. Díaz-Blancas, V.; Medina, D. I.; Padilla-Ortega, E.; Bortolini-Zavala, R.; Olvera-Romero, M.; Luna-Bárcenas, G. Nanoemulsion formulations of fungicide tebuconazole for agricultural applications. *Molecules* **2016**, *21* (10), 1271. Available from: https://doi.org/10.3390/molecules21101271.
19. Ravichandran, M.; Samiappan, S. C.; Rangaraj, S.; Murugan, K.; Al-Dhabi, N. A.; Karuppiah, P. *Nanoemulsion formulations with plant growth promoting rhizobacteria (PGPR) for sustainable agriculture. Bio-based nanoemulsions for agri-food applications;* Elsevier, 2022207–223*https://doi.org/10.1016/B978-0-323-89846-1.00017-6.*
20. Ahmed, T.; Noman, M.; Manzoor, N.; Shahid, M.; Abdullah, M.; Ali, L.; Li, B. Nanoparticle-based amelioration of drought stress and cadmium toxicity in rice via triggering the stress responsive genetic mechanisms and nutrient acquisition. *Ecotoxicol. Environ. Saf.* **2021**, *209*.111829. Available from: https://doi.org/10.1016/j.ecoenv.2020.111829.
21. Negrão, S.; Schmöckel, S. M.; Tester, M. J. A. O. B. Evaluating physiological responses of plants to salinity stress. *Ann. botany* **2017**, *119* (1), 1–11. Available from: https://doi.org/10.1093/aob/mcw191.

22. Ghori, N. H.; Ghori, T.; Hayat, M. Q.; Imadi, S. R.; Gul, A.; Altay, V.; Ozturk, M. Heavy metal stress and responses in plants. *Int. J. Environ. Sci. Technol.* **2019**, *16*, 1807–1828. Available from: https://doi.org/10.1007/s13762-019-02215-8.
23. Zhou, P.; Adeel, M.; Shakoor, N.; Guo, M.; Hao, Y.; Azeem, I.; Rui, Y. Application of nanoparticles alleviates heavy metals stress and promotes plant growth: an overview. *Nanomaterials* **2020**, *11* (1), 26. Available from: https://doi.org/10.3390/nano11010026.
24. Venkatachalam, P.; Jayaraj, M.; Manikandan, R.; Geetha, N.; Rene, E. R.; Sharma, N. C.; Sahi, S. V. Zinc oxide nanoparticles (ZnONPs) alleviate heavy metal-induced toxicity in *Leucaena leucocephala* seedlings: a physiochemical analysis. *Plant. Physiol. Biochem.* **2017**, *110*, 59–69. Available from: https://doi.org/10.1016/j.plaphy.2016.08.022.
25. Yang, J.; Cao, W.; Rui, Y. Interactions between nanoparticles and plants: phytotoxicity and defense mechanisms. *J. plant. Interact.* **2017**, *12* (1), 158–169. Available from: https://doi.org/10.1080/17429145.2017.1310944.
26. Cifuentes, Z.; Custardoy, L.; de la Fuente, J. M.; Marquina, C.; Ibarra, M. R.; Rubiales, D.; Pérez-de-Luque, A. Absorption and translocation to the aerial part of magnetic carbon-coated nanoparticles through the root of different crop plants. *J. Nanobiotechnology* **2010**, *8*, 1–8.
27. Wang, X.; Xie, H.; Wang, P.; Yin, H. Nanoparticles in plants: uptake, transport and physiological activity in leaf and root. *Materials* **2023**, *16* (8), 3097. Available from: https://doi.org/10.3390/ma16083097.
28. Hu, T.; Li, H.; Li, J.; Zhao, G.; Wu, W.; Liu, L.; Guo, Y. Absorption and biotransformation of selenium nanoparticles by wheat seedlings (*Triticum aestivum* L.). *Front. Plant. Sci.* **2018**, *9*, 597. Available from: https://doi.org/10.3389/fpls.2018.00597.
29. Hong, J.; Wang, C.; Wagner, D. C.; Gardea-Torresdey, J. L.; He, F.; Rico, C. M. Foliar application of nanoparticles: mechanisms of absorption, transfer, and multiple impacts. *Environ. Science: Nano* **2021**, *8* (5), 1196–1210. Available from: https://doi.org/10.1039/D0EN01129K.
30. Venzhik, Y. V.; Deryabin, A. N.; Popov, V. N.; Dykman, L. A.; Titov, A. F.; Moshkov, I. E. *Influence of gold nanoparticles on the tolerance of wheat to low temperature. Doklady Biochemistry and Biophysics*, 502. Pleiades Publishing: Moscow, 20225–9*https://doi.org/10.1134/S1607672922010100*No. 1.
31. Sun, D.; Hussain, H. I.; Yi, Z.; Rookes, J. E.; Kong, L.; Cahill, D. M. Mesoporous silica nanoparticles enhance seedling growth and photosynthesis in wheat and lupin. *Chemosphere* **2016**, *152*, 81–91. Available from: https://doi.org/10.1016/j.chemosphere.2016.02.096.
32. Torney, F.; Trewyn, B. G.; Lin, V. S. Y.; Wang, K. Mesoporous silica nanoparticles deliver DNA and chemicals into plants. *Nat. Nanotechnol.* **2007**, *2* (5), 295–300. Available from: https://doi.org/10.1038/nnano.2007.108.
33. Dietz, K. J.; Herth, S. Plant nanotoxicology. *Trends plant. Sci.* **2011**, *16* (11), 582–589. Available from: https://doi.org/10.1016/j.tplants.2011.08.003.
34. Leopold, L. F.; Coman, C.; Clapa, D.; Oprea, I.; Toma, A.; Iancu, . D.; Coman, V. The effect of 100–200 nm ZnO and TiO2 nanoparticles on the in vitro-grown soybean plants. *Colloids Surf. B: Biointerfaces* **2022**, *216*.112536. Available from: https://doi.org/10.1016/j.colsurfb.2022.112536.

35. Ma, X.; Quah, B. Effects of surface charge on the fate and phytotoxicity of gold nanoparticles to *Phaseolus vulgaris*. *J. Food Chem. Nanotechnol.* **2016**, *2* (1), 57–65. Available from: https://doi.org/10.17756/jfcn.2016-011?.
36. Milewska-Hendel, A.; Zubko, M.; Stróż, D.; Kurczyńska, E. U. Effect of nanoparticles surface charge on the *Arabidopsis thaliana* (L.) roots development and their movement into the root cells and protoplasts. *Int. J. Mol. Sci.* **2019**, *20* (7), 1650. Available from: https://doi.org/10.3390/ijms20071650.
37. Lv, J.; Christie, P.; Zhang, S. Uptake, translocation, and transformation of metal-based nanoparticles in plants: recent advances and methodological challenges. *Environ. Science: Nano* **2019**, *6* (1), 41–59. Available from: https://doi.org/10.1039/C8EN00645H.
38. Shukla, P. K.; Misra, P.; Kole, C. Uptake, translocation, accumulation, transformation, and generational transmission of nanoparticles in plants. *Plant. Nanotechnology: Princ. Pract.* **2016**, 183–218. Available from: https://doi.org/10.1007/978-3-319-42154-4_8.
39. Mendez, O. E.; Astete, C. E.; Cueto, R.; Eitzer, B.; Hanna, E. A.; Salinas, F.; Sabliov, C. M. Lignin nanoparticles as delivery systems to facilitate translocation of methoxyfenozide in soybean (*Glycine max*). *J. Agriculture Food Res.* **2022**, *7*.100259. Available from: https://doi.org/10.1016/j.jafr.2021.100259.
40. Tripathi, D. K.; Singh, S.; Singh, S.; Dubey, N. K.; Chauhan, D. K. Impact of nanoparticles on photosynthesis: challenges and opportunities. *Mater. Focus.* **2016**, *5* (5), 405–411. Available from: https://doi.org/10.1166/mat.2016.1327.
41. Cao, Z.; Stowers, C.; Rossi, L.; Zhang, W.; Lombardini, L.; Ma, X. Physiological effects of cerium oxide nanoparticles on the photosynthesis and water use efficiency of soybean (*Glycine max* (L.) Merr.). *Environ. Science: Nano* **2017**, *4* (5), 1086–1094. Available from: https://doi.org/10.1039/C7EN00015D.
42. Sreelakshmi, B.; Induja, S.; Adarsh, P. P.; Rahul, H. L.; Arya, S. M.; Aswana, S.; Vishnudasan, D. Drought stress amelioration in plants using green synthesised iron oxide nanoparticles. *Mater. Today: Proc.* **2021**, *41*, 723–727. Available from: https://doi.org/10.1016/j.matpr.2020.05.801.
43. Thakur, S.; Asthir, B.; Kaur, G.; Kalia, A.; Sharma, A. Zinc oxide and titanium dioxide nanoparticles influence heat stress tolerance mediated by antioxidant defense system in wheat. *Cereal Res. Commun.* **2021**, 1–12. Available from: https://doi.org/10.1007/s42976-021-00190-w.
44. Iqbal, M.; Raja, N. I.; Mashwani, Z. U. R.; Hussain, M.; Ejaz, M.; Yasmeen, F. Effect of silver nanoparticles on growth of wheat under heat stress. *Iran. J. Sci. Technology, Trans. A: Sci.* **2019**, *43*, 387–395. Available from: https://doi.org/10.1007/s40995-017-0417-4.
45. Konate, A.; He, X.; Zhang, Z.; Ma, Y.; Zhang, P.; Alugongo, G. M.; Rui, Y. Magnetic (Fe3O4) nanoparticles reduce heavy metals uptake and mitigate their toxicity in wheat seedling. *Sustainability* **2017**, *9* (5), 790. Available from: https://doi.org/10.3390/su9050790.
46. Etesami, H.; Fatemi, H.; Rizwan, M. Interactions of nanoparticles and salinity stress at physiological, biochemical and molecular levels in plants: a review. *Ecotoxicol. Environ. Saf.* **2021**, *225*.112769. Available from: https://doi.org/10.1016/j.ecoenv.2021.112769.
47. Nasrallah, A. K.; Kheder, A. A.; Kord, M. A.; Fouad, A. S.; El-Mogy, M. M.; Atia, M. A. Mitigation of salinity stress effects on Broad bean productivity using calcium

phosphate nanoparticles application. *Horticulturae* **2022**, *8* (1), 75. Available from: https://doi.org/10.3390/horticulturae8010075.

48. Almutairi, Z. M. Plant molecular defense mechanisms promoted by nanoparticles against environmental stress. *Int. J. Agric. Biol.* **2019**, *21* (2), 259–270. Available from: https://doi.org/10.17957/IJAB/15.0890.
49. Bhatt, D.; Bhatt, M. D.; Nath, M.; Dudhat, R.; Sharma, M.; Bisht, D. S. Application of nanoparticles in overcoming different environmental stresses. *Protective Chem. agents amelioration plant. abiotic stress: Biochemical Mol. Perspect.* **2020**, 635–654. Available from: https://doi.org/10.1002/9781119552154.ch32.
50. Mustafa, G.; Komatsu, S. Toxicity of heavy metals and metal-containing nanoparticles on plants. *Biochimica et. Biophysica Acta (BBA)-Proteins Proteom.* **2016**, *1864* (8), 932–944. Available from: https://doi.org/10.1016/j.bbapap.2016.02.020.
51. Rizwan, M.; Ali, S.; ur Rehman, M. Z.; Riaz, M.; Adrees, M.; Hussain, A.; Rinklebe, J. Effects of nanoparticles on trace element uptake and toxicity in plants: a review. *Ecotoxicol. Environ. Saf.* **2021**, *221*.112437. Available from: https://doi.org/10.1016/j.ecoenv.2021.112437.
52. Remédios, C.; Rosário, F.; Bastos, V. Environmental nanoparticles interactions with plants: morphological, physiological, and genotoxic aspects. *J. Botany* **2012**. Available from: https://doi.org/10.1155/2012/751686.
53. Aqeel, U.; Aftab, T.; Khan, M. M. A.; Naeem, M.; Khan, M. N. A comprehensive review of impacts of diverse nanoparticles on growth, development and physiological adjustments in plants under changing environment. *Chemosphere* **2022**, *291*, 132672.
54. Hossain, Z.; Mustafa, G.; Komatsu, S. Plant responses to nanoparticle stress. *Int. J. Mol. Sci.* **2015**, *16* (11), 26644–26653. Available from: https://doi.org/10.3390/ijms161125980.
55. Demirel, U.; Morris, W. L.; Ducreux, L. J.; Yavuz, C.; Asim, A.; Tindas, I.; Hancock, R. D. Physiological, biochemical, and transcriptional responses to single and combined abiotic stress in stress-tolerant and stress-sensitive potato genotypes. *Front. Plant. Sci.* **2020**, 169. Available from: https://doi.org/10.3389/fpls.2020.00169.
56. Arora, S.; Jha, P. N. Impact of plant-associated microbial communities on host plants under abiotic stresses. *Microb. Interventions Agriculture Environment: Volume 2: Rhizosphere, Microbiome Agro-ecology* **2019**, 303–340. Available from: https://doi.org/10.1007/978-981-13-8383-0_10.
57. Khan, M.; Khan, M. S. A.; Borah, K. K.; Goswami, Y.; Hakeem, K. R.; Chakrabartty, I. The potential exposure and hazards of metal-based nanoparticles on plants and environment, with special emphasis on ZnO NPs, TiO2 NPs, and AgNPs: a review. *Environ. Adv.* **2021**, *6*.100128. Available from: https://doi.org/10.1016/j.envadv.2021.100128.
58. Dimkpa, C. O.; McLean, J. E.; Britt, D. W.; Anderson, A. J. Bioactivity and biomodification of Ag, ZnO, and CuO nanoparticles with relevance to plant performance in agriculture. *Ind. Biotechnol.* **2012**, *8* (6), 344–357. Available from: https://doi.org/10.1089/ind.2012.0028.
59. Yang, J.; Jiang, F.; Ma, C.; Rui, Y.; Rui, M.; Adeel, M.; Xing, B. Alteration of crop yield and quality of wheat upon exposure to silver nanoparticles in a life cycle study. *J. Agric. Food Chem.* **2018**, *66* (11), 2589–2597. Available from: https://doi.org/10.1021/acs.jafc.7b04904.
60. Wang, P.; Lombi, E.; Sun, S.; Scheckel, K. G.; Malysheva, A.; McKenna, B. A.; Kopittke, P. M. Characterizing the uptake, accumulation and toxicity of silver sulfide

nanoparticles in plants. *Environ. Science: Nano* **2017**, *4* (2), 448–460. Available from: https://doi.org/10.1039/C6EN00489J.

61. Liu, X.; Wang, F.; Shi, Z.; Tong, R.; Shi, X. Bioavailability of Zn in ZnO nanoparticle-spiked soil and the implications to maize plants. *J. Nanopart. Res.* **2015**, *17*, 1–11. Available from: https://doi.org/10.1007/s11051-015-2989-2.
62. Simkó, M.; Mattsson, M. O. Risks from accidental exposures to engineered nanoparticles and neurological health effects: a critical review. *Part. fibre Toxicol.* **2010**, *7*, 1–15. Available from: https://doi.org/10.1186/1743-8977-7-42.
63. Ahamed, M.; AlSalhi, M. S.; Siddiqui, M. K. J. Silver nanoparticle applications and human health. *Clinica Chim. Acta* **2010**, *411* (23-24), 1841–1848. Available from: https://doi.org/10.1016/j.cca.2010.08.016.
64. Servin, A. D.; White, J. C. Nanotechnology in agriculture: next steps for understanding engineered nanoparticle exposure and risk. *NanoImpact* **2016**, *1*, 9–12. Available from: https://doi.org/10.1016/j.impact.2015.12.002.
65. Rajput, V.; Minkina, T.; Sushkova, S.; Behal, A.; Maksimov, A.; Blicharska, E.; Barsova, N. ZnO and CuO nanoparticles: a threat to soil organisms, plants, and human health. *Environ. Geochem. Health* **2020**, *42*, 147–158. Available from: https://doi.org/10.1007/s10653-019-00317-3.
66. Sajid, M.; Ilyas, M.; Basheer, C.; Tariq, M.; Daud, M.; Baig, N.; Shehzad, F. Impact of nanoparticles on human and environment: review of toxicity factors, exposures, control strategies, and future prospects. *Environ. Sci. Pollut. Res.* **2015**, *22*, 4122–4143. Available from: https://doi.org/10.1007/s11356-014-3994-1.
67. Okeke, E. S.; Chukwudozie, K. I.; Nyaruaba, R.; Ita, R. E.; Oladipo, A.; Ejeromedoghene, O.; Okoye, C. O. Antibiotic resistance in aquaculture and aquatic organisms: a review of current nanotechnology applications for sustainable management. *Environ. Sci. Pollut. Res.* **2022**, *29* (46), 69241–69274. Available from: https://doi.org/10.1007/s11356-022-22319-y.

CHAPTER

7 Metal resilience: biogenic nanomaterials and interaction with heavy metal stress

Juhi Sharma[1], Kratika Pathak[1], Ayushi Bhatnagar[1], Ankita Mitra[2] and Nikita Sangwan[1]

[1]*School of Life and Basic Sciences, Jaipur National University, Jaipur, Rajasthan, India*
[2]*School of Medicine Stanford University, Stanford, CA, United States*

7.1 Introduction

The term "heavy metals (HMs)" encompasses metals and metalloids with an atomic density exceeding 5 g/cm^3.[1] Vital heavy metals, including molybdenum (Mo), manganese (Mn), cobalt (Co), and nickel (Ni), are essential for numerous biological processes and development.[2] These metals enter the human body through ingestion, inhalation, and water consumption, where they play significant roles in maintaining various biological functions.[3] Nonetheless, industrial activities frequently introduce harmful heavy metals into the environment, leading to soil and water pollution. Such contamination can detrimentally affect soil health, microbial activity, and crop yields, and it poses considerable health risks to humans through the intake of contaminated food.[4]

Heavy metals interfere with plant water regulation, membrane integrity, and result in the depletion of essential osmolytes due to increased production of malondialdehyde (MDA) and hydrogen peroxide. They also lead to elevated levels of reactive oxygen species (ROS), which can cause damage to proteins, lipids, and DNA.[5] Factors such as heightened human activities, geochemical weathering of rocks, and environmental events like volcanic eruptions, acid rain, and dust storms have substantially raised soil concentrations of heavy metals, including cadmium (Cd), copper (Cu), chromium (Cr), lead (Pb), mercury (Hg), aluminum (Al), nickel (Ni), and zinc (Zn). High concentrations of these metals can impede seed germination, harm antioxidant enzymes and cellular membranes, induce chromosomal damage, and potentially result in plant mortality.[6]

The increase in industrialization and human activities significantly threatens crop production by deteriorating soil fertility and productivity.[7] Recent progress in bioremediation has integrated genomics, transcriptomics, metabolomics, and proteomics to improve in-situ remediation strategies.[8] While metals like iron (Fe),

Biogenic Nanoparticles. DOI: https://doi.org/10.1016/B978-0-443-33819-9.00011-1

copper (Cu), cobalt (Co), manganese (Mn), selenium (Se), and nickel (Ni) are essential for various biological functions at low concentrations, excessive levels can be toxic.[9] For example, high concentrations of nickel, cadmium, and lead can cause phytotoxicity and reduce the effectiveness of halophytic plants in extracting heavy metals.[10] The toxicity of heavy metals in plants largely stems from disruptions in redox balance, leading to the excessive production of reactive oxygen species (ROS) such as superoxide anions (O_2^-), hydrogen peroxide (H_2O_2), and hydroxyl radicals (OH^-). This disturbance in redox homeostasis plays a central role in heavy metal toxicity.[11]

Research indicates that the accumulation of heavy metals in food crops can negatively impact human health.[12] When harmful heavy metals accumulate in plant tissues, they increase the production of reactive oxygen species (ROS), which disrupts the redox balance.[13] To counteract heavy metal toxicity, plants utilize several strategies (Fig. 7.1), including the scavenging of ROS through metallothioneins and phytochelatins, sequestering metals in cell walls and intracellular compartments, and reducing the uptake of contaminants from the soil.[14] Additionally, mercury can convert into the highly toxic compound methylmercury when it interacts with aquatic sediments.[15] Methylmercury then enters the human food chain primarily through contaminated fish, seafood, and wildlife, which absorb it from toxic microorganisms. Once inside the human body, methylmercury travels through the bloodstream and can cause various neurological disorders.[16]

Lead, recognized as a carcinogen, damages DNA repair processes, disrupts tumor-suppressing genes, and alters chromosomal structure and function through the production of reactive oxygen species (ROS). It also interferes with transcription by displacing zinc from regulatory proteins.[17] Both short-term and long-term exposure to lead can severely affect the immune system, increasing the risk of allergies, infections, autoimmune diseases, and cancer.[18]

Heavy metals (HMs) are transported from plant roots to their above-ground parts, where they accumulate and interfere with cellular metabolism. This disruption adversely affects plant growth and development.[19]

7.2 Nanotechnology and defined engineering

Nanotechnology involves the precise production and manipulation of materials at a scale between 1 nm and 100 nm. This enables the creation of nano-systems with advanced functionalities by altering or reassembling particulate matter.[20]

Nanoparticles, at the forefront of material science advancements, are only slightly larger than individual atoms. They display unique characteristics, such as increased stability and the ability to self-assemble, which can be customized for specific uses. Their high surface area relative to conventional materials makes them exceptionally versatile.[21]

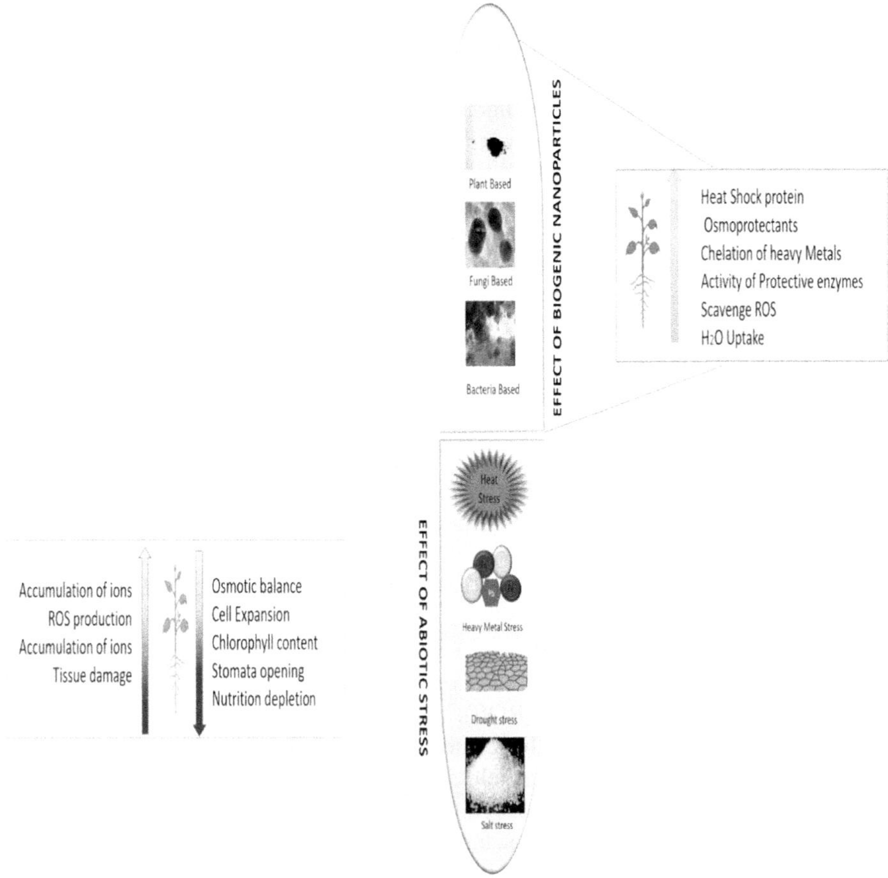

FIGURE 7.1

Effects of nanoparticles on plant growth.

Over the past twenty years, nanotechnology has progressed from academic research to widespread industrial use. It is projected to significantly impact the global economy, potentially contributing up to three trillion dollars by 2020.[22] Additionally, nanoparticles are increasingly used in diagnostic applications, including as contrast agents in medical imaging and for detecting specific biomolecules in biological samples.[23]

Historically, nanoparticles, particularly gold, have been employed for thousands of years in ancient China and Egypt for both therapeutic and decorative purposes. For instance, gold nanoparticles were used in art and glassmaking, as seen in artifacts such as the fifth-century AD Lycurgus Cup. The study of colloidal nanoparticles has been documented since the 17th century.[24]

In modern applications, silver nanoparticles (AgNP) are valued for their natural antibacterial properties. They are commonly integrated into food packaging materials, including biodegradable wraps for fruits, meats, and poultry, or mixed into polymer matrices to improve functionality.[23]

Nanotechnology is increasingly applied in food preservation, extending the shelf life of products like orange juice, strawberries, and liquid egg albumen.[25] In the cosmetics sector, nanomaterials offer various advantages. For example, zinc oxide and titanium dioxide nanoparticles are frequently used in sunscreens for their effective UV filtration while reducing potential health risks.[26]

The influence of nanotechnology in medicine is significant, especially in drug delivery systems. Advanced systems using both natural and synthetic nanoparticles have transformed disease treatment. Notably, researchers at Harvard University's Wyss Institute have created a "nano-robot" designed to specifically target cancer cells for delivering anticancer drugs.[27]

Nanotechnology in medicine aims to overcome challenges such as the poor bioavailability and solubility of conventional drugs, which can hinder their absorption and effectiveness. This field has also given rise to nano-toxicology, which investigates the potential toxic effects of nanoparticles.[28] Studies indicate that the same characteristics that make nanoparticles useful for therapeutic purposes can also lead to their toxicity.[29]

Nanoparticles have the potential to penetrate body tissues and cells, which can disrupt cellular functions and increase oxidative stress due to excessive production of reactive oxygen species (ROS).[30] This oxidative stress may cause DNA damage, interfere with cell signaling, and lead to cell death. Additionally, modifications to the surfaces of nanoparticles, designed to enhance their functionality, may unintentionally heighten their cytotoxicity by altering their interactions with biomolecules and disrupting cellular balance. For example, charged gold nanoparticles (AuNPs) have been shown to be more cytotoxic than neutral ones, leading to increased oxidative stress and DNA damage.[31]

Advancements in lipid-based drug delivery systems (DSSs) aim to enhance the targeted delivery of therapeutics, reducing harm to surrounding tissues that can occur with free drugs. Significant progress has been made in developing various DSSs for treating diseases such as cancer and neurodegenerative disorders.[32]

Drug delivery systems (DSSs) based on lipids come in various formulations and sizes, primarily classified as micelles and liposomes. Micelles are nano-sized vesicles formed through the self-assembly of lipid molecules in water, typically ranging from 5 to 50 nm in diameter.[33]

Liposomes, although generally having a short circulation time, have benefited from innovations such as PEGylation, which helps prolong their presence in the bloodstream.[34]

Recent interest has focused on polymer-based nanoparticle DSSs, which are constructed from repeating units of specific polymers. Commonly researched polymers include PEG, chitosan, poly(lactic-co-glycolic acid) (PLGA), and polylactic acid (PLA).[35] While PEG, PLGA, and PLA have been extensively studied, chitosan is gaining attention for its biocompatibility, low immunogenicity, and minimal toxicity.[36] For example, pH-sensitive polyamines can release drugs like doxorubicin in the acidic environment of tumor tissues, effectively targeting tumors without needing specific targeting molecules. Similarly, dextran nanoparticles with arylboronic esters, which degrade in the presence of reactive oxygen species (ROS), have been utilized to deliver ovalbumin to murine dendritic cells, enhancing antigen presentation.[37]

Peptides, both synthetic and naturally derived, also play a significant role in nanoparticle DSSs. These peptides often serve as ligands for cell surface receptors, utilizing specific sequences that bind to these receptors. Their ease of synthesis and low immunogenicity make peptides advantageous for drug delivery.[38] For example, somatostatins, which are peptides expressed in various cancer cells, have been conjugated with anticancer drugs such as doxorubicin, methotrexate, and paclitaxel to provide targeted therapy against cancer cells.[39]

7.3 Class of nanoparticles

Nanoparticles (NPs) are cataloged based on their morphology, size, and chemical characteristics (Fig. 7.2).[40] The main types of NPs are mentioned below.

7.3.1 Carbon-based nanoparticles

7.3.1.1 Fullerenes

Fullerenes are spherical carbon molecules with carbon atoms connected through sp^2 hybridization. They vary in size from 8.2 nm for single-layer forms to 4–36 nm for multilayered versions, consisting of 28 to 1500 carbon atoms.[41]

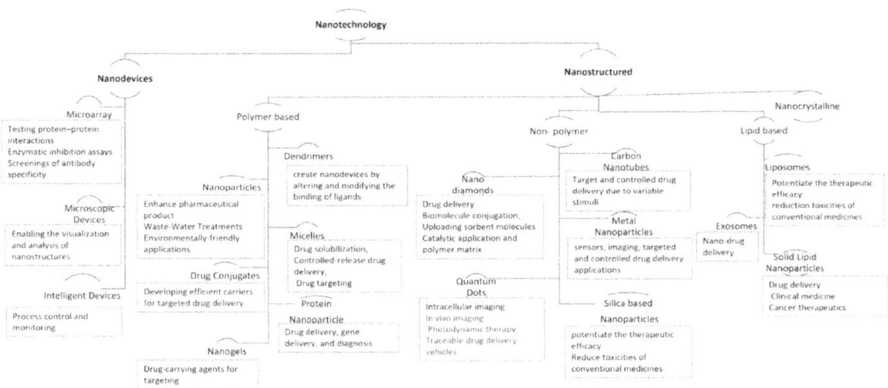

FIGURE 7.2

Types and categories of nanoparticles.

Their hollow, spherical structure provides excellent electrical conductivity, strength, and electron affinity, making them highly valuable commercially.[42]

7.3.1.2 Graphene

Graphene is a two-dimensional sheet of carbon atoms arranged in a hexagonal lattice. Each layer of graphene is about 1 nm thick and is renowned for its exceptional strength and electrical conductivity.[43]

7.3.1.3 Carbon nanotubes (CNTs)

CNTs are cylindrical structures created by rolling graphene sheets into tubes. They can be single-walled (SWNTs), double-walled (DWNTs), or multiwalled (MWNTs), with diameters ranging from 0.7 nm to 100 nm and lengths extending from micrometers to millimeters. CNTs are typically produced using methods such as vapor deposition or arc discharge.[44,45]

7.3.1.4 Carbon nanofibers

These are similar to CNTs but are wound into cone-shaped structures rather than cylindrical tubes. They are manufactured using techniques similar to those used for graphene and CNTs.[46]

7.3.1.5 Carbon black

Carbon black is an amorphous carbon form with spherical particles ranging from 20 to 70 nm in diameter. These particles often cluster together, forming larger aggregates that can reach up to 500 nm.[47]

7.3.1.6 Metal nanoparticles

These are nanoparticles made from various metals and are synthesized using either destructive or constructive processes. Almost any metal can be turned into nanoparticles.[48]

7.3.1.7 Organic nanoparticles

This category includes ferritin, micelles, dendrimers, and liposomes. Organic nanoparticles are typically nontoxic, biodegradable, and can form hollow spheres, making them ideal for drug delivery. They are also known as polymeric nanoparticles and are used extensively for targeted drug delivery.[49] The common forms include nanospheres and nanocapsules, where the matrix is solid, and the outer boundary can adsorb or encapsulate other molecules.[50]

7.3.1.8 Inorganic nanoparticles

These do not contain carbon. They are generally biocompatible, hydrophilic, and more stable than organic nanoparticles. Inorganic nanoparticles are mainly classified into metal and metal oxide nanoparticles.[51]

7.4 Classification by dimension

7.4.1 One-dimensional nanomaterials

This category includes materials such as nanotubes, nanobelts, and nanowires, with dimensions ranging from 1 nm to 100 nm.[52,53] These materials are pivotal in various research and application areas, including electronics, storage systems, and light-emitting diodes (LEDs).[54]

7.4.2 Two-dimensional nanomaterials

These materials possess a planar structure and extend beyond the nanoscale in one direction. They are essential components in a range of nanodevices.[55]

7.4.3 Three-dimensional nanomaterials

The properties and performance of three-dimensional nanomaterials are significantly influenced by their shape, size, and structure. These characteristics are crucial for determining their applications and overall effectiveness in different fields.

These classifications highlight the diverse applications and characteristics of nanoparticles across disciplines such as chemistry, pharmaceuticals, electronics, and engineering.

7.5 Biogenic nanoparticles

Microorganisms have the capability to produce a range of distinctive nanostructures, which has led to growing interest in utilizing these organisms for nanoparticle synthesis. Bacteria, fungi, and algae can generate inorganic nanoparticles through biological processes.[56] By carefully managing these biological synthesis methods, researchers can create nanoparticles with specific shapes and compositions. Nonetheless, challenges remain in achieving precise control over particle geometry and scaling up production. Microbes interact differently with metal precursors, resulting in a variety of nanoparticle forms.[57] For example, bacteria and fungi can produce nanoparticles either intracellularly or extracellularly, each method involving different pathways. Intracellular synthesis entails transporting metal ions through the cell wall, where they react with cellular enzymes to form nanoparticles. An example of this is the synthesis of gold nanoparticles using *Lactobacillus kimchicus*.[58]

7.5.1 Bacteria

Bacteria are widely available, easy to cultivate, and adaptable to different environments, making them suitable for nanoparticle synthesis. By adjusting bacterial culture conditions such as temperature and oxygen levels, nanoparticles of various sizes can be produced to meet specific needs.[59] For instance, gold nanoparticles (15–35 nm) produced extracellularly by the marine bacterium *Paracoccus hauseri* have been found to be nontoxic to human cells and effective in inhibiting cancer cell growth.[60] Similarly, silver nanoparticles (AgNPs) synthesized with *Bacillus subtilis* have shown notable antibacterial properties. Research by Mathivanan et al. demonstrated that crude AgNPs were more effective against pathogenic bacteria than calcined AgNPs. The process involves the reduction of silver ions through enzymatic action, with nanoparticle formation indicated by color changes.[61]

7.5.1.1 Biochemical mechanisms

The synthesis of nanoparticles by microorganisms involves various biochemical processes, such as changes in solubility and toxicity, bioadsorption, bioaccumulation, metal precipitation, and efflux systems.[62] For instance, some bacteria, such as *Gluconacetobacter*, are capable of producing cellulose nanocomposites,[63] while magnetotactic bacteria generate magnetosomes,[64,65] which are utilized in applications like cancer therapy and molecular imaging.[66] Despite these advancements, challenges persist in purifying nanoparticles and precisely controlling their geometry, largely due to an incomplete understanding of the underlying biochemical mechanisms.[67]

7.5.2 Fungi

Fungi utilize enzymes present in their cytoplasm and cell walls to transform metal ions into nanoparticles.[68] Using fungi for nanoparticle synthesis offers several benefits, including scalability, cost efficiency, and environmental sustainability, as they can cover large surface areas.[69] For example, Joshi et al. successfully synthesized gold nanoparticles using the fungus *Cladosporium cladosporioides*, which was isolated from the seaweed *Sargassum wightii*.[70] Similarly, *Fusarium oxysporum* has been used in various studies to produce nanoparticles, including silver nanoparticles (10–15 nm) with significant antibacterial activity against *Escherichia coli* and *Pseudomonas aeruginosa*.[71] Other nanoparticles produced by *Fusarium oxysporum* include copper nanoparticles.[72]

Fungi are particularly known for their extracellular synthesis methods, which involve secreting molecules that assist in nanoparticle formation. This approach is relatively straightforward and cost-effective. For instance, Manjunath and Joshi showed that *Cladosporium cladosporioides* can produce silver nanoparticles ranging from 30 to 60 nm.[73] Additionally, fungi such as *Aspergillus* are capable of producing magnetite, a common iron oxide with magnetic properties.[74]

7.5.3 Algae

Microalgae are valued for their ability to convert metal ions into nanoparticles through an eco-friendly process. This transformation is facilitated by active compounds in their cell walls, such as alginate and laminarin, which contain reactive groups.[75,76] The general method for algae-assisted nanoparticle synthesis involves preparing algae extracts in solution at elevated temperatures, mixing with reagents, and stirring the combined solutions for a set period. The initiation of the reaction is often indicated by a color change, signaling the formation and growth of nanoparticles into various shapes.[77]

Seaweeds like *Sargassum wightii* and *Fucus vesiculosus* have been employed for nanoparticle synthesis,[78] while *Tetraselmis kochinensis* has demonstrated the ability to produce intracellular gold nanoparticles.[79] Additionally, microalgae such as diatoms (*Nitzschia atomus* and *Diatom gallica*) have been utilized to produce gold nanoparticles and gold-silica bionanocomposites.[80]

Nanoparticles exhibit a wide range of applications across various fields due to their distinctive nanoscale properties. Here are some notable uses:

7.6 Applications

7.6.1 Medical applications

Advances in nanoparticle research have significantly broadened their applications in the biomedical field, including diagnostics, drug delivery, biosensing, and bioimaging.[81] Nanoparticles are increasingly recognized for their ability to deliver targeted drug doses, improve therapeutic effectiveness, and reduce adverse side effects.[82] For example, magnetic nanoparticles such as magnetite (Fe_2O_3) are utilized in various medical applications.[83] Semiconductor and metal nanoparticles, with their unique light scattering and absorption characteristics, hold considerable promise for cancer detection and treatment. Gold nanoparticles, for instance, can convert electromagnetic energy into heat, selectively targeting cancerous tissues. Research has investigated the use of gold-silica nanoparticles in treating prostate cancer.[83]

7.6.2 Materials and mechanical applications

Nanomaterials possess properties that differ markedly from their bulk forms, making them useful for a range of functional applications. Engineering materials at the nanoscale allows for customization of their electrical, physical, optical, and imaging properties to meet specific requirements.[84] In the textile industry, nanoparticles are used to enhance fabric properties, offering improvements such as increased ballistic protection, reduced wrinkling, and antimicrobial effects. Innovative "smart fabrics" integrate nanoscale sensors and electronic components, enabling functions like health monitoring, solar energy capture, and energy harvesting through movement.[85] Additionally, nanoparticles contribute to the development of mechanically stronger devices with enhanced Young's modulus, stress, and strain capabilities. Coatings made from alumina, titania, and carbon-based nanoparticles have been shown to improve hardness and wear resistance.[40]

7.6.3 Environmental applications

The growing use of nanoparticles in industrial and consumer products raises concerns about their potential environmental impact, particularly with regard to leakage.[86] However, nanoparticles offer significant advantages in environmental remediation. For instance, copper nanoparticles integrated into paper filters have shown effective bacteria reduction in water with minimal copper leaching (1 ppm).[87] They are also used to tackle oil spills; finely engineered carbon particles can stabilize oil-in-water emulsions.[88] Functionalized carbon black nanoparticles are nontoxic to brine shrimp and excel in adsorbing benzene. Furthermore, magnetic nanoparticles coated with hydrophobic biocomponents from *Anthemis pseudocotula* have been found effective in collecting heavy crude oil.[89]

7.6.4 Applications in energy technology

Nanoparticles are advancing several energy technologies, including water splitting through electrolysis, thermal processes, and piezoelectrics.[90] They enhance fuel production processes by improving catalysis, leading to reduced fuel consumption and increased combustion efficiency in vehicles and power plants. Nanoparticles are also pivotal in energy storage technologies.[91]

7.6.5 Applications in the petroleum industry

In the oil and gas sector, nanoparticles are becoming increasingly valuable. They are used to improve drilling fluids, which are essential for maintaining borehole stability and efficiency. For example, CNT-polymer nanocomposites have been studied to enhance borehole stability in high-temperature drilling environments.[92] In enhanced oil recovery, nanoparticles are utilized in various injection techniques—chemical, gas, and thermal—to boost oil recovery rates. Aluminum oxide nanoparticles, for instance, have been applied to reduce oil viscosity,[93] while iron oxide nanoparticles have been used to improve the performance of xanthan gum-based drilling fluids.[94]

1. **Medicine**
    a. **Drug delivery**: Nanoparticles can be engineered to deliver drugs precisely to targeted cells or tissues, enhancing therapeutic efficacy while minimizing side effects. For example, liposomes and polymeric

   nanoparticles are used to deliver chemotherapy drugs specifically to cancer cells.
   b. **Diagnostic imaging**: Nanoparticles such as quantum dots and gold nanoparticles are used as contrast agents in imaging techniques like MRI, CT scans, and optical imaging, providing clearer and more detailed images.
   c. **Theranostics**: Combining therapeutic and diagnostic functions, theranostic nanoparticles allow for simultaneous disease diagnosis and treatment, improving patient outcomes through personalized medicine.
   d. **Cancer treatment**: Nanoparticles can be used in targeted cancer therapies, such as photothermal therapy, where nanoparticles absorb light and convert it to heat, killing cancer cells while sparing surrounding healthy tissue.
2. **Environmental science**
   a. **Water purification**: Nanoparticles, including silver and titanium dioxide, are used to remove contaminants from water through processes like photocatalysis, adsorption, and disinfection.[93,95,96]
   b. **Soil remediation**: Nanoparticles can help in the cleanup of contaminated soils by breaking down pollutants or binding to heavy metals, thus reducing soil toxicity and improving agricultural productivity.[97–101]
   c. **Air filtration**: Nanoparticles are employed in air filters to capture and neutralize pollutants, including particulate matter and hazardous gases.[97,102]
3. **Electronics and computing**
   a. **Semiconductors**: Nanoparticles and nanowires are used in the fabrication of semiconductors for electronic devices, enabling smaller, faster, and more efficient components.
   b. **Sensors**: Nanomaterials are incorporated into sensors for detecting gases, chemicals, and biological agents with high sensitivity and accuracy, useful in environmental monitoring and healthcare diagnostics.
   c. **Data storage**: Nanotechnology enhances data storage capacities by enabling the creation of smaller and more efficient memory devices, such as hard drives and flash memory.
4. **Energy**
   a. **Solar cells**: Nanoparticles are used in photovoltaic cells to improve light absorption and energy conversion efficiency, leading to more efficient and cost-effective solar energy technologies.
   b. **Batteries**: Nanotechnology enhances battery performance by increasing the surface area of electrodes and improving conductivity, resulting in batteries with higher capacity, faster charging times, and longer lifespans.
   c. **Fuel cells**: Nanocatalysts are employed in fuel cells to accelerate chemical reactions and improve energy conversion efficiency, making fuel cells more viable for clean energy applications.

5. **Agriculture and food Industry**
    a. **Pesticides and fertilizers**: Nanoparticles can enhance the delivery and effectiveness of pesticides and fertilizers, reducing the amount needed and minimizing environmental impact.[97,101,102]
    b. **Food packaging**: Nanomaterials are used in food packaging to improve shelf life, protect against contamination, and provide antimicrobial properties.
    c. **Food quality monitoring**: Nanoparticles are incorporated into sensors and diagnostic tools to monitor food quality, detect contaminants, and ensure food safety.
6. **Textiles**
    a. **Smart fabrics**: Nanotechnology enables the development of textiles with enhanced properties such as water resistance, stain resistance, and UV protection.
    b. **Self-cleaning fabrics**: Nanoparticles like titanium dioxide are used to create fabrics that can self-clean by breaking down organic stains and contaminants when exposed to sunlight.
7. **Cosmetics**
    a. **Sunscreens**: Nanoparticles of zinc oxide and titanium dioxide are used in sunscreens to provide effective UV protection without leaving a white residue on the skin.
    b. **Antiaging products**: Nanoparticles are employed in anti-aging creams and lotions to improve the penetration of active ingredients and enhance their efficacy.
8. **Aerospace**
    a. **Lightweight materials**: Nanocomposites are used to create lightweight yet strong materials for aerospace applications, improving fuel efficiency and performance of aircraft and spacecraft.
    b. **Coatings**: Nanocoatings provide protective layers for aerospace components, enhancing resistance to wear, corrosion, and extreme temperatures.
9. **Pharmaceuticals**
    a. **Gene therapy**: Nanoparticles can be designed to deliver genetic material into cells for gene therapy applications, potentially treating genetic disorders and diseases at the molecular level.
10. **Biomedicine**
    a. **Biosensors**: Nanoparticles are used in biosensors to detect biological molecules with high sensitivity, aiding in disease diagnosis and monitoring.

Nanotechnology continues to advance, opening new possibilities for innovation across these diverse fields. As research progresses, the scope and impact of nanoparticles are likely to expand, driving technological advancements and improving various aspects of daily life.

7.7 Recommendation

To mitigate the adverse effects of heavy metal contamination and harness the potential of nanotechnology, a multifaceted approach should be adopted.

1. **Enhanced monitoring and regulation**: Governments and environmental agencies should implement stricter regulations and regular monitoring of industrial activities to prevent heavy metal pollution. This includes controlling emissions from industries and ensuring proper waste management practices.
2. **Bioremediation strategies**: Invest in research to improve bioremediation techniques using advanced genomics, transcriptomics, and proteomics. This will help in developing more efficient methods to clean contaminated soils and waters, thus protecting plant and human health.
3. **Sustainable agricultural practices**: Adopt sustainable farming practices that minimize heavy metal uptake by crops. Techniques such as phytoremediation, the use of soil amendments, and crop rotation can help reduce soil contamination and improve soil health.
4. **Nanotechnology in pollution control**: Explore and invest in the application of nanotechnology for environmental cleanup and pollution control. Nano-engineered materials can be used to remove heavy metals from contaminated sites, enhance soil fertility, and improve crop resilience.
5. **Research and development**: Support ongoing research into the dual role of nanoparticles in medicine and environmental science. Investigate the potential health risks associated with nanoparticles and develop strategies to mitigate any adverse effects.
6. **Public awareness and education**: Increase public awareness about the sources and risks of heavy metal contamination and the benefits and limitations of nanotechnology. Educational programs can empower individuals and communities to make informed decisions and advocate for environmental health.

7.8 Conclusion

Heavy metals present a significant challenge due to their potential toxicity and impact on both environmental and human health. While essential metals play crucial roles in biological processes, excessive levels can lead to severe disruptions in plant growth, soil fertility, and human health. The integration of nanotechnology into environmental management offers promising solutions for mitigating heavy metal pollution and enhancing soil and crop productivity. However, the benefits of nanotechnology must be carefully weighed against its potential risks. As this field continues to evolve, it is essential to balance innovation with caution, ensuring that technological advancements do not inadvertently introduce new hazards. By adopting a comprehensive approach that includes strict regulation, improved bioremediation, sustainable practices, and continued research, it is possible to address the challenges posed by heavy metal contamination and leverage nanotechnology for positive environmental and health outcomes.

References

1. Ghuge, S. A.; Nikalje, G. C.; Kadam, U. S.; Suprasanna, P.; Hong, J. C. Comprehensive mechanisms of heavy metal toxicity in plants, detoxification, and remediation. *J. Hazard. Mat.* **2023**, *450*, 131039.
2. Tiwari, S.; Lata, C. Heavy metal stress, signaling, and tolerance due to plant-associated microbes: an overview. *Front. Plant. Sci.* **2018**, *9*, 452.
3. Chasapis, C. T.; Loutsidou, A. C.; Spiliopoulou, C. A.; Stefanidou, M. E. Zinc and human health: an update. *Arch. Toxicol.* **2012**, *86*, 521–534.
4. Atuchin, V. V.; Asyakina, L. K.; Serazetdinova, Y. R.; Frolova, A. S.; Velichkovich, N. S.; Prosekov, A. Y. Microorganisms for bioremediation of soils contaminated with heavy metals. *Microorg.* **2023**, *11* (4), 864.
5. Hassan, S. E.; Hijri, M.; St-Arnaud, M. Effect of arbuscularmycorrhizal fungi on trace metal uptake by sunflower plants grown on cadmium contaminated soil. *N. Biotechnol.* **2013**, *30* (6), 780–787.
6. Feng, Z.; Ji, S.; Ping, J.; Cui, D. Recent advances in metabolomics for studying heavy metal stress in plants. *TrAC. Trends Anal. Chem.* **2021**, *143*, 116402.
7. Yaashikaa, P. R.; Kumar, P. S. Bioremediation of hazardous pollutants from agricultural soils: a sustainable approach for waste management towards urban sustainability. *Environ. Pollut.* **2022**, *312*, 120031.
8. Villegas-Plazas, M.; Sanabria, J.; Junca, H. A composite taxonomical and functional framework of microbiomes under acid mine drainage bioremediation systems. *J. Environment. Manag.* **2019**, *251*.109581.
9. Bharti, R.; Sharma, R. Effect of heavy metals: An overview. *Mater. Today: Proc.* **2022**, *51*, 880–885.
10. Amari, T.; Ghnaya, T.; Abdelly, C. Nickel, cadmium and lead phytotoxicity and potential of halophytic plants in heavy metal extraction. *SouthAfrican J. Bot.* **2017**, *111*, 99–110.
11. Pourrut, B.; Shahid, M.; Dumat, C.; Winterton, P.; Pinelli, E. ead uptake, toxicity, and detoxification in plants. *Rev. Environ. Contaminat Toxicol.* **2011**, *213*, 113–136.
12. Uzu, G.; Sauvain, J. J.; Baeza-Squiban, A.; Riediker, M.; Sánchez Sandoval Hohl, M.; Val, S.; Dumat, C. In vitro assessment of the pulmonary toxicity and gastric availability of lead-rich particles from a lead recycling plant. *Env. Sci. & Technol.* **2011**, *45* (18), 7888–7895.
13. AbdElgawad, H.; Zinta, G.; Hamed, B. A.; Selim, S.; Beemster, G.; Hozzein, W. N.; Abuelsoud, W. Maize roots and shoots show distinct profiles of oxidative stress and antioxidant defense under heavy metal toxicity. *Environ. Pollut.* **2020**, *258*, 113705.
14. Feng, D.; Wang, R.; Sun, X.; Liu, P.; Tang, J.; Zhang, C.; Liu, H. Heavy metal stress in plants: ways to alleviate with exogenous substances. *Sci. Total. Environ.* **2023**.165397.
15. Gworek, B.; Dmuchowski, W.; Baczewska-Dąbrowska, A. H. Mercury in the terrestrial environment: a review. *Environ. Sci. Europe* **2020**, *32* (1), 128.

16. Rice, K. M.; Walker, E. M., Jr; Wu, M.; Gillette, C.; Blough, E. R. Environmental mercury and its toxic effects. *J. Prevent. Med. Public. Health* **2014**, *47* (2), 74.
17. Silbergeld, E. K.; Waalkes, M.; Rice, J. M. Lead as a carcinogen: experimental evidence and mechanisms of action. *Am. J. Indust. Med.* **2000**, *38* (30), 316–323.
18. Hsiao, C. L.; Wu, K. H.; Wan, K. S. Effects of environmental lead exposure on T-helper cell-specific cytokines in children. *J. Immunotoxicol.* **2011**, *8* (4), 284–287.
19. Wang, D.; Hu, B.; Hu, C.; Zhu, F.; Liu, X.; Zhang, J.; Peng, Z. Clinical characteristics of 138 hospitalized patients with 2019 novel coronavirus–infected pneumonia in Wuhan, China. *jama* **2020**, *323* (11), 1061–1069.
20. Nasrollahzadeh, M.; Sajjadi, M.; Soufi, G. J.; Iravani, S.; Varma, R. S. Nanomaterials and nanotechnology-associated innovations against viral infections with a focus on coronaviruses. *Nanomaterials* **2020**, *10* (6), 1072.
21. Cheng, Y. J.; Wolkenhauer, M.; Bumbu, G. G.; Gutmann, J. S. A facile route to reassemble titania nanoparticles into ordered chain-like networks on substrate. *Macromolecul. rapid comm.* **2012**, *33* (3), 218–224.
22. Roco, M.C. Overview: affirmation of nanotechnology between 2000 and 2030. Nanotechnology commercialization: manufacturing processes and products. 2017, 1–23.
23. Carbone, M.; Donia, D. T.; Sabbatella, G.; Antiochia, R. Silver nanoparticles in polymeric matrices for fresh food packaging. *J. King Saud. Uni.-Sci.* **2016**, *28* (4), 273–279.
24. Jia, H. Y.; Liu, Y.; Zhang, X. J.; Han, L.; Du, L. B.; Tian, Q.; Xu, Y. C. Potential oxidative stress of gold nanoparticles by induced-NO releasing in serum. *J. Am. Chem. Soc.* **2009**, *131* (1), 40–41.
25. Bajpai, V. K.; Kamle, M.; Shukla, S.; Mahato, D. K.; Chandra, P.; Hwang, S. K.; Han, Y. K. Prospects of using nanotechnology for food preservation, safety, and security. *J. Food And. Drug. Anal.* **2018**, *26* (4), 1201–1214.
26. Morganti, P. Use and potential of nanotechnology in cosmetic dermatology. *Clinical, cosmetic investigate. dermatology* **2010**, 5–13.
27. Douglas, S. M.; Bachelet, I.; Church, G. M. A logic-gated nanorobot for targeted transport of molecular payloads. *Science* **2012**, *335* (6070), 831–834.
28. Khadka, P.; Ro, J.; Kim, H.; Kim, I.; Kim, J. T.; Kim, H.; Lee, J. Pharmaceutical particle technologies: an approach to improve drug solubility, dissolution and bioavailability. *Asian J. Pharma. Sci.* **2014**, *9* (6), 304–316.
29. Gatoo, M. A.; Naseem, S.; Arfat, M. Y.; Mahmood Dar, A.; Qasim, A. K.; Zubair, S. Physicochemical properties of nanomaterials: implication in associated toxic manifestations. *BioMed. Res. Int.* **2014**.(1).498420.
30. Fu, P. P.; Xia, Q.; Hwang, H. M.; Ray, P. C.; Yu, H. Mechanisms of nanotoxicity: generation of reactive oxygen species. *J. Food Drug. Anal.* **2014**, *22* (1), 64–75.
31. Schaeublin, N. M.; Braydich-Stolle, L. K.; Schrand, A. M.; Miller, J. M.; Hutchison, J.; Schlager, J. J.; Hussain, S. M. Surface charge of gold nanoparticles mediates mechanism of toxicity. *Nanoscale* **2011**, *3* (2), 410–420.
32. SilindirGunay, M.; YektaOzer, A.; Chalon, S. Drug delivery systems for imaging and therapy of Parkinson's disease. *Curr. Neuropharmacol.* **2016**, *14* (4), 376–391.
33. Wang, J.; Mongayt, D.; Torchilin, V. P. Polymeric micelles for delivery of poorly soluble drugs: preparation and anticancer activity in vitro of paclitaxel incorporated into mixed micelles based on poly (ethylene glycol)-lipid conjugate and positively charged lipids. *J. drug. Target.* **2015**, *13* (1), 73–80.

34. Milla, P.; Dosio, F.; Cattel, L. PEGylation of proteins and liposomes: a powerful and flexible strategy to improve the drug delivery. *Curr. Drug. Metab.* **2012**, *13* (1), 105–119.
35. Nicolas, J.; Mura, S.; Brambilla, D.; Mackiewicz, N.; Couvreur, P. Design, functionalization strategies and biomedical applications of targeted biodegradable/biocompatible polymer-based nanocarriers for drug delivery. *Chem. Soc. Rev.* **2013**, *42* (3), 1147–1235.
36. Cheung, R. C. F.; Ng, T. B.; Wong, J. H.; Chan, W. Y. Chitosan: an update on potential biomedical and pharmaceutical applications. *Mar. drugs* **2015**, *13* (8), 5156–5186.
37. Broaders, K. E.; Grandhe, S.; Fréchet, J. M. A biocompatible oxidation-triggered carrier polymer with potential in therapeutics. *J. Am. Chem. Soc.* **2011**, *133* (4), 756–758.
38. Kim, S. K.; Huang, L. Nanoparticle delivery of a peptide targeting EGFR signaling. *J. controlled Rel.* **2012**, *157* (2), 279–286.
39. Sun, L. C.; Coy, D. H. Somatostatin receptor-targeted anti-cancer therapy. *Curr. drug. delivery* **2011**, *8* (1), 2–10.
40. Khan, I.; Saeed, K.; Khan, I. Nanoparticles: properties, applications and toxicities. *Arab. J. Chemistry* **2019**, *12* (7), 908–931.
41. Patel, K. D.; Singh, R. K.; Kim, H. W. Carbon-based nanomaterials as an emerging platform for theranostics. *Mater. Horiz.* **2019**, *6* (3), 434–469.
42. Mallikarjunaiah, S.; Pattabhiramaiah, M.; Metikurki, B. Application of nanotechnology in the bioremediation of heavy metals and wastewater management. *Nanotechnol. Food, Agri. Environ.* **2020**, 297–321.
43. Bandyopadhyay, A.; Jana, D. A review on role of tetra-rings in graphene systems and their possible applications. *Rep. Prog. Phys.* **2020**, *83* (5), 056501.
44. Jana, S.; Bandyopadhyay, A.; Datta, S.; Bhattacharya, D.; Jana, D. Emerging properties of carbon based 2D material beyond graphene. *J. Physics: Condens. Matter* **2021**, *34* (5), 053001.
45. Mohamed, A. E. M. A.; Mohamed, M. A. *Carbon nanotubes: synthesis, characterization, and applications. Carbon Nanomaterials for Agri-Food and Environmental Applications;* Elsevier, 202021–32.
46. Foong, L. K.; Foroughi, M. M.; Mirhosseini, A. F.; Safaei, M.; Jahani, S.; Mostafavi, M.; Khatami, M. Applications of nano-materials in diverse dentistry regimes. *Rsc Adv.* **2020**, *10* (26), 15430–15460.
47. Mkhari, O.; Ntuli, T. D.; Coville, N. J.; Nxumalo, E. N.; Maubane-Nkadimeng, M. S. A comparison of fluorescent N-doped carbon dots supported on the surface of hollow and solid carbon spheres, and solid silica spheres. *Diam. Relat. Mater.* **2021**, *118*.108500.
48. Ijaz, I.; Gilani, E.; Nazir, A.; Bukhari, A. Detail review on chemical, physical and green synthesis, classification, characterizations and applications of nanoparticles. *Green. Chem. Lett. Rev.* **2020**, *13* (3), 223–245.
49. Mansha, M.; Khan, I.; Ullah, N.; Qurashi, A. Synthesis, characterization and visible-light-driven photoelectrochemical hydrogen evolution reaction of carbazole-containing conjugated polymers. *Int. J. Hydrog. Energy* **2017**, *42*, 10952–10961.
50. Rao, J. P.; Geckeler, K. E. Polymer nanoparticles: Preparation techniques and size-control parameters. *Prog. Polym. Sci.* **2011**, *36*, 887–913.

51. Liu, J.; Huang, Y.; Kumar, A.; Tan, A.; Jin, S.; Mozhi, A.; Liang, X. J. pH-sensitive nano-systems for drug delivery in cancer therapy. *Biotechnol. Adv.* **2014**, *32*, 693–710.
52. Hangarter, C. M.; Bangar, M.; Mulchandani, A.; Myung, N. V. Conducting polymer nanowires for chemiresistive and FET-based bio/chemical sensors. *J. Mater. Chem.* **2010**, *20*, 3131–3140.
53. Arroyo, C. R.; Debut, A.; Vaca, A.; Stael, C.; Guzman, K.; Cumbal, B. K. L. Reliable tools for quantifying the morphological properties at the nanoscale. *Biol. Med.* **2016**, *8*, 1 3.
54. Gopi, S.; Amalraj, A.; Thomas, S. Effective drug delivery system of biopolymers based on nanomaterials and hydrogels—a review. *Drug. Des.* **2016**, *5*, 129 2169-0138.
55. Iwamoto, T. Clinical application of drug delivery systems in cancer chemotherapy: review of the efficacy and side effects of approved drugs. *Biol. Pharm. Bull.* **2013**, *36*, 715–718.
56. Moitra, P.; Alafeef, M.; Dighe, K.; Frieman, M. B.; Pan, D. Selective naked-eye detection of SARS-CoV-2 mediated by N gene targeted antisense oligonucleotide capped plasmonic nanoparticles. *ACS Nano* **2020**, *14*, 7617–7627.
57. Fang, X.; Wang, Y.; Wang, Z.; Jiang, Z.; Dong, M. Microorganism assisted synthesized nanoparticles for catalytic applications. *Energies* **2019**, *12*, 190.
58. Mathivanan, K.; Selva, R.; Chandirika, J. U.; Govindarajan, R. K.; Srinivasan, R.; Annadurai, G.; Duc, P. A. Biologically synthesized silver nanoparticles against pathogenic bacteria: synthesis, calcination and characterization. *Biocatal. Agric. Biotechnol.* **2019**, *22*.101373.
59. Poreddy, R.; Engelbrekt, C.; Riisager, A. Copper oxide as efficient catalyst for oxidative dehydrogenation of alcohols with air. *Catal. Sci. Technol.* **2015**, *5*, 2467–2477.
60. Bruins, M. R.; Kapil, S.; Oehme, F. W. Microbial resistance to metals in the environment. *Ecotoxicol. Environ. Saf.* **2000**, *45*, 198–207.
61. Patil, M. P.; Kang, M. J.; Niyonizigiye, I.; Singh, A.; Kim, J. O.; Seo, Y. B.; Kim, G. D. Extracellular synthesis of gold nanoparticles using the marine bacterium Paracoccus haenudaensis BC74171T and evaluation of their antioxidant activity and antiproliferative effect on normal and cancer cell lines. *Colloids Surf. Biointerfaces* **2019**, *183*.110455.
62. Liu, K.; Catchmark, J. M. Enhanced mechanical properties of bacterial cellulose nanocomposites produced by co-culturing Gluconacetobacter hansenii and Escherichia coli under static conditions. *Carbohydr. Polym.* **2019**, *219*, 12–20.
63. Peigneux, A.; Valverde-Tercedor, C.; López-Moreno, R.; Pérez-González, T.; Fernández-Vivas, M. A.; Jiménez-López, C. Learning from magnetotactic bacteria: A review on the synthesis of biomimetic nanoparticles mediated by magnetosome-associated proteins. *J. Struct. Biol.* **2016**, *196*, 75–84.
64. Yan, L.; Da, H.; Zhang, S.; López, V. M.; Wang, W. Bacterial magnetosome and its potential application. *Microbiol. Res.* **2017**, *203*, 19–28.
65. Boucher, M.; Geffroy, F.; Prévéral, S.; Bellanger, L.; Selingue, E.; Adryanczyk-Perrier, G.; Mériaux, S. Genetically tailored magnetosomes used as MRI probe for molecular imaging of brain tumor. *Biomaterials* **2017**, *121*, 167–178.
66. Roda, A.; Cevenini, L.; Borg, S.; Michelini, E.; Calabretta, M. M.; Schüler, D. Bioengineered bioluminescent magnetotactic bacteria as a powerful tool for chip-based whole-cell biosensors. *Lab. Chip* **2013**, *13*, 4881–4889.

67. Mukherjee, P.; Senapati, S.; Mandal, D.; Ahmad, A.; Khan, M. I.; Kumar, R.; Sastry, M. Extracellular synthesis of gold nanoparticles by the fungus Fusarium oxysporum. *Chem. Bio Chem* **2002**, *3*, 461–463.
68. Chatterjee, S.; Mahanty, S.; Das, P.; Chaudhuri, P.; Das, S. Biofabrication of iron oxide nanoparticles using manglicolous fungus Aspergillus niger BSC-1 and removal of Cr(VI) from aqueous solution. *Chem. Eng. J.* **2020**, *385*.123790.
69. Syed, A.; Ahmad, A. Extracellular biosynthesis of platinum nanoparticles using the fungus *Fusarium oxysporum*. *Colloids Surf. B: Biointerfaces* **2012**, *97*, 27–31.
70. Bhargava, A.; Jain, N.; Khan, M. A.; Pareek, V.; Dilip, R. V.; Panwar, J. Utilizing metal tolerance potential of soil fungus for efficient synthesis of gold nanoparticles with superior catalytic activity for degradation of rhodamine B. *J. Environ. Manage.* **2016**, *183*, 22–32.
71. Pham, N. D.; Duong, M. M.; Le, M. V.; Hoang, H. A. Preparation and characterization of antifungal colloidal copper nanoparticles and their antifungal activity against Fusarium oxysporum and Phytophthora capsici. *Comptes Rendus Chimie* **2019**, *22*, 786–793.
72. Naimi-Shamel, N.; Pourali, P.; Dolatabadi, S. Green synthesis of gold nanoparticles using *Fusarium oxysporum* and antibacterial activity of its tetracycline conjugant. *J. Mycol. Med.* **2019**, *29*, 7–13.
73. Catal. B Environ. 2019, 261, 118224.
74. Mughal, B.; Zaidi, S. Z. J.; Zhang, X.; Hassan, S. U. Biogenic nanoparticles: Synthesis, characterisation and applications. *Appl. Sci.* **2021**, *11*, 2598.
75. Venkatesan, J.; Manivasagan, P.; Kim, S. K.; Kirthi, A. V.; Marimuthu, S.; Rahuman, A. A. Marine algae-mediated synthesis of gold nanoparticles using a novel *Ecklonia cava*. *Bioprocess. Biosyst. Eng.* **2014**, *37*, 1591–1597.
76. Sharma, A.; Sharma, S.; Sharma, K.; Chetri, S. P. K.; Vashishtha, A.; Singh, P.; Kumar, R.; Rathi, B.; Agrawal, V. Algae as crucial organisms in advancing nanotechnology: a systematic review. *J. Appl. Phycol.* **2016**, *28*, 1759–1774.
77. Kumaresan, M.; Anand, K. V.; Govindaraju, K.; Tamilselvan, S.; Kumar, V. G. Seaweed Sargassum wightii mediated preparation of zirconia (ZrO_2) nanoparticles and their antibacterial activity against Gram positive and Gram negative bacteria. *Microb. Pathog.* **2018**, *124*, 311–315.
78. Asmathunisha, N.; Kathiresan, K. A review on biosynthesis of nanoparticles by marine organisms. *Colloids Surf. B: Biointerfaces* **2013**, *103*, 283–287.
79. Ali, D. M.; Sasikala, M.; Gunasekaran, M.; Thajuddin, N. Biosynthesis and characterization of silver nanoparticles using marine cyanobacterium *Oscillatoria willei* NTDM01. *Dig. J. Nanomater. Biostruct.* **2011**, *6*, 385–390.
80. Jacob, J. M.; Ravindran, R.; Narayanan, M.; Samuel, S. M.; Pugazhendhi, A.; Kumar, G. Microalgae: a prospective low cost green alternative for nanoparticle synthesis. *Curr. Opin. Environ. Sci. Health* **2021**, *20*.100163.
81. Ramos, A. P.; Cruz, M. A.; Tovani, C. B.; Ciancaglini, P. Biomedical applications of nanotechnology. *Biophys. Rev.* **2017**, *9*, 79–89.
82. Ali, A.; Zafar, H.; Zia, M.; ulHaq, I.; Phull, A. R.; Ali, J. S.; Hussain, A. Synthesis, characterization, applications, and challenges of iron oxide nanoparticles. *Nanotechnology, Sci. Appl.* **2016**, *9*, 49–67.
83. Khlebtsov, N. G.; Dykman, L. A. Optical properties and biomedical applications of plasmonic nanoparticles. *J. Quant. Spectrosc. Radiat. Transf.* **2010**, *111*, 1–35.

84. Atwah, A. A.; Khan, M. A. Influence of microscopic features on the self-cleaning ability of textile fabrics. *Text. Res. J.* **2023**, *93*, 450–467.
85. Shateri-Khalilabad, M.; Yazdanshenas, M. E.; Etemadifar, A. Fabricating multifunctional silver nanoparticles-coated cotton fabric. *Arab. J. Chem.* **2017**, *10*, S2355–S2362.
86. Houtman, C. J. Emerging contaminants in surface waters and their relevance for the production of drinking water in Europe. *J. Integr. Environ. Sci.* **2010**, *7*, 271–295.
87. Dankovich, T. A.; Smith, J. A. Incorporation of copper nanoparticles into paper for point-of-use water purification. *Water Res.* **2014**, *63*, 245–251.
88. Rodd, A. L.; Creighton, M. A.; Vaslet, C. A.; Rangel-Mendez, J. R.; Hurt, R. H.; Kane, A. B. Effects of surface-engineered nanoparticle-based dispersants for marine oil spills on the model organism *Artemia franciscana*. *Environ. Sci. Technol.* **2014**, *48*, 6419–6427.
89. Abdullah, M. M.; Atta, A. M.; Allohedan, H. A.; Alkhathlan, H. Z.; Khan, M.; Ezzat, A. O. Green synthesis of hydrophobic magnetite nanoparticles coated with plant extract and their application as petroleum oil spill collectors. *Nanomaterials* **2018**, *8*, 855.
90. Li, D.; Baydoun, H.; Verani, C. N.; Brock, S. L. Efficient water oxidation using CoMnP nanoparticles. *J. Am. Chem. Soc.* **2016**, *138*, 4006–4009.
91. Sagadevan, S. A review on role of nanofluids for solar energy applications. *Am. J. Nano Res. Appl.* **2015**, *8*, 53–61.
92. Ahmad, H.M.; Kamal, M.S.; Al-Harthi, M.A.; Elkatatny, S.M.; Murtaza, M.M. Synthesis and experimental investigation of novel CNT-polymer nanocomposite to enhance borehole stability at high temperature drilling applications. SPE Kingdom of Saudi Arabia Annual Technical Symposium and Exhibition. 2018, April, pp. SPE-192352.
93. Abdullahi, M. B.; Rajaei, K.; Junin, R.; Bayat, A. E. Appraising the impact of metal-oxide nanoparticles on rheological properties of HPAM in different electrolyte solutions for enhanced oil recovery. *J. Petrol. Sci. Eng.* **2019**, *172*, 1057–1068.
94. Alvi, M.A.; Belayneh, M.; Saasen, A.; Aadnøy, B.S. The effect of micro-sized boron nitride BN and iron trioxide Fe_2O_3 nanoparticles on the properties of laboratory bentonite drilling fluid, SPE Norway Subsurface Conference. 2018, April, p. D011S011R002.
95. Singh, A.; Rajput, V. D.; Agrawal, S.; Ghazaryan, K.; Minkina, T.; Tawaha, A. R. M. A., et al. Nanoparticles mediated salt stress resilience: a holistic exploration of physiological, biochemical, and nano-omics approaches. *Rev. Environ. Contamination Toxicol.* **2024**, *262*. Available from: https://doi.org/10.1007/s44169-024-00070-4.
96. Margaryan, G., et al. Unveiling the salinity tolerance potential of Armenian Dandur (*Portulaca oleracea* L.) genotypes: enhancing sustainable agriculture and food security. *J. King Saud. Univ.—Sci.* **2024**, *36*.103332. Available from: https://doi.org/10.1016/j.jksus.2024.103332.
97. Singh, A.; Sharma, A.; Singh, O.; Rajput, V. D.; Movsesyan, H. S.; Minkina, T., et al. In-depth exploration of nanoparticles for enhanced nutrient use efficiency and abiotic stresses management: present insights and future horizons. *Plant. Stress.* **2024**.100576. Available from: https://doi.org/10.1016/j.stress.2024.100576.
98. Sharma, A., et al. Revolutionizing agricultural sustainability and food security and management to achieve SDGs goals via nanotechnology. *Adv. Environ. Eng. green. Technol. book. Ser.* **2024**, 276–288. Available from: https://doi.org/10.4018/979-8-3693-1471-5.ch013.

99. Ghazaryan, K.; Verma, A.; Rawat, S.; Rajput, P.; Minkina, T.; Mandzhieva, S., et al. Implementing sustainable nano-bioremediation for emerging pollutants. *Adv. Environ. Eng. green. Technol. book. Ser.* **2024**, 316–332. Available from: https://doi.org/10.4018/979-8-3693-1471-5.ch016.
100. Fathi, A.; Shiade, S. R. G.; Parmoon, G.; Yaghoubian, Y.; Pirdashti, H.; Rajput, V. D., et al. *Bioremediation of Heavy Metals Contaminated Soils Using Nanotechnology*; Elsevier eBooks, **2024**, 611–628. Available from: https://doi.org/10.1016/b978-0-443-21610-7.00004-5.
101. Verma, A.; Singh, A.; Rawat, S.; Rajput, P.; Minkina, T.; Mandzhieva, S., et al. Unveiling nanomaterial-induced toxicity. *Adv. Environ. Eng. green. Technol. book. Ser.* **2024**, 236–253. Available from: https://doi.org/10.4018/979-8-3693-1471-5.ch011.
102. Upadhyay, P., et al. Nano-omics-based abiotic and biotic stresses management. *Adv. Environ. Eng. green. Technol. book. Ser.* **2024**, 348–371. Available from: https://doi.org/10.4018/979-8-3693-1890-4.ch017.

CHAPTER

8 Mitigating drought stress: biogenic nanoparticles in agriculture and their interaction with the environmental components

Vandna Kushwaha[1], Shambhoo Prasad[1], Ashwini Kumar[1], Rahul Kumar Maurya[1], Sumit Jangra[2] and Yogesh Pratap[1]

[1]*Department of Molecular Biology and Biotechnology, Acharya Narendra Deva University of Agriculture and Technology, Kumarganj, Ayodhya, Uttar Pradesh, India*

[2]*Tropical Research and Education Center, Homestead, Florida*

8.1 Introduction

It is estimated that the global population will reach 9.6 billion By the end of 2050. To feed this growing population, food output must rise by 70%–100%.[1,2] According to models and predictions, changes in global precipitation patterns and climate change would cause deserts and xeric scrublands to overtake other terrestrial biomes as the largest one.[3] Due to its increased susceptibility to biotic and abiotic challenges, climate change is seen as a serious issue for agricultural crops.[4] A vital source of the raw materials required by the food and feed industry is agricultural produce. They supply food, fuel, textiles, feedstock, furniture, and other necessities for human populations.[5] Agricultural crops' productivity and output may be increased through breeding research. In addition, the farmer has access to a variety of chemical pesticides, fungicides, fertilizers, and other treatments to enhance crop productivity and nutritional quality.[6] The growing use of chemical pesticides and fertilizers raised considerable concerns about their possible implications on human health, and breeding efforts may not yield the intended results right away. To fulfill the future demand for agricultural products, it is imperative that alternative methods and technologies be explored. One of the cutting-edge strategies that might be employed to support agriculture is nanotechnology.[7]

Furthermore, drought and a lack of water have a detrimental effect on plant physiology and morphology, as well as the quantitative and qualitative production of crops.[8] During the growth state, using fertilizers and nutritional compounds is an important way to lessen the effect of drought on crops. Micronutrients reduce

Biogenic Nanoparticles. DOI: https://doi.org/10.1016/B978-0-443-33819-9.00012-3

crop drought stress.[9–11] Zinc is a microelement that plants need for several processes, such as the production of proteins, enzymes, and chlorophyll, as well as for increasing crop productivity.[12,13] Furthermore, studies have shown that adding zinc to plants in the form of zinc sulfate, zinc chelate, and zinc nanoparticles enhances their morphological, physiological, and biochemical traits and increases their resilience to environmental stress.[14–17] Nutrient absorption, growth, and improved resistance to abiotic stress are all facilitated by nanoparticles, which are defined as particles with a single dimension of at least 100 nm, as reported by Jordan et al.,[18] and Dimkpa et al.,[11] In addition, the application of nanoparticles as micronutrients and fertilizers enhances both the qualitative and quantitative output of crops.

Nanoparticles can control and restrict the release of nutrients, which can improve plant development and stress tolerance.[9,10] It has been demonstrated that NPs improve photosynthetic parameters, control stomatal conductance and transpiration rate, and improve germination parameters, growth rate, biomass, and yield. Additionally, they decrease membrane ion leakage and enhance the absorption of carbon dioxide in leaves.[19–21] Moreover, it has been shown that NPs control defensive mechanisms by increasing the activities of enzymatic and nonenzymatic antioxidants such as glutathione (GSH), ascorbate peroxidase (APX), superoxide dismutase (SOD), catalase (CAT), and peroxidase (POD).[22,23] Some studies show that NPs help prevent oxidative damage brought on by the production of ROS (H_2O_2 and O_2) and increase the levels of osmolytes and osmoprotectants, which include soluble sugars, amino acids, prolines, and glycine betaine.[24,25] The benefits of metal- and carbon-based nanoparticles (NPs) vary depending on a variety of factors, including the type of plant, how they are applied, their concentration, form, and surface properties.

The development of resilient crops with long-term production has been underlined for future prospectives. Both organic and inorganic nanoparticles having widespread use in a wide range of products and their commercial utilization, concerns have been raised about the toxicological and environmental impacts of NPs.[26] On the other hand, when they are used excessively, plants experience oxidative stress and physiological abnormalities that reduce their antioxidant capacities and gas exchange qualities.[27] In addition, NPs reduce the mitotic index and impede cell division and root growth.[28] NPs also cause indirect toxicity by changing the growing medium and the microbial activities in the soil. Besides, the increasing number of uses for nanoparticles in household and industrial items raises the possibility of environmental contamination.[29] Many research have reported on the positive impacts of NPs on growth under DS; however, the connection between NPs and intercellular processes in DS plants remains unclear.[30] Numerous physiological, pharmacological, and molecular mechanisms that NPs mediate to allow plants to acquire DS tolerance. To increase plant yield characteristics, we have summarized the most recent research on NP-mediated drought tolerance and mitigation mechanisms in this chapter.

NPs, or nanoparticles, are any particles with at least one dimension smaller than 100 nm.[31] Due to its extraordinary small size and special qualities, it has created new opportunities to increase agricultural output under a variety of environmental stressors. Fundamental characteristics for creating the nanoparticles (NPs) employed in agriculture are composition and size. Nanoparticles can come from manmade or natural sources. Natural or manmade sources can be found in nanoparticles. Engineered nanoparticles or ENPs, are a class of particles that include surface-modified clay and are classed as organic, inorganic, and hybrid materials. They are created synthetically or through modification of naturally existing nanoparticles.[32]

8.2 Types of nanoparticles

NPs fall into several categories based on their composition, traits, and potential applications.[33] Based on their composition, NPs are commonly divided into three groups which are as follows

8.2.1 Silver

Silver nanoparticles (AgNPs) have been the focus of extensive research for a range of applications and have several properties that make them effective in mitigating drought stress in plants:

- Antioxidant activity: AgNPs enhance the antioxidant activity in plants, reducing oxidative damage caused by drought-induced reactive oxygen species (ROS). This helps maintain cellular integrity and promotes plant growth under stress conditions.
- Water uptake and retention: AgNPs improve water uptake and retention in plants by influencing root morphology and enhancing the efficiency of water transport systems.
- Stomatal regulation: AgNPs can regulate the opening and closing of stomata, which helps conserve water during drought stress.
- Photosynthetic efficiency: AgNPs increase chlorophyll content and improve photosynthetic efficiency, which is often compromised under drought stress.
- Gene expression: AgNPs modulate the expression of drought-responsive genes, enhancing the plant's ability to cope with water scarcity.
- Phytohormone regulation: AgNPs influence the biosynthesis of phytohormones such as abscisic acid, which play crucial roles in drought tolerance.

Case Study *International Journal of Farming and Allied Sciences*.
Available online at http://www.ijfas.com
©2016 IJFAS Journal-2016-5-3/208–212/ 31 March, 2016 ISSN 2322–4134 ©2016 IJFAS.
The effect of silver nanoparticle on lentil Seed Germination under drought stress.
Seyed Saeid Hojjat.[34]
Result: It is showed that AgNPs has significant effect on germination percent, the length of the shoot and the root on the one percent probable level ($P > 0/01$). Silver nanoparticle caused an increase in germination parentage and germination rate (table 2) and the highest values achieved by applying 10 μg m/L of silver nanoparticle about 89.5% and 6.88%, respectively.

These properties collectively contribute to the enhanced drought tolerance observed in plants treated with silver nanoparticles.

8.2.2 Gold

Another metal that has been extensively studied is gold nanoparticles (AuNPs) and have several properties that make them effective in mitigating drought stress in plants:

- Size and surface area: AuNPs have a high surface area-to-volume ratio, which enhances their interaction with plant cells and tissues, improving water and nutrient uptake.
- Catalytic activity: AuNPs can act as catalysts, enhancing various biochemical reactions in plants, including those involved in stress responses.
- Antioxidant properties: AuNPs help in reducing oxidative stress by scavenging reactive oxygen species (ROS), which are typically elevated under drought conditions.
- Stomatal regulation: AuNPs can influence the opening and closing of stomata, helping plants to conserve water during drought stress.
- Photosynthetic efficiency: AuNPs can increase chlorophyll content and improve the efficiency of photosynthesis, which is often compromised under drought stress.
- Gene expression: AuNPs can modulate the expression of drought-responsive genes, enhancing the plant's ability to cope with water scarcity.
- Phytohormone regulation: AuNPs can influence the biosynthesis of phytohormones such as abscisic acid, which plays a crucial role in drought tolerance.

These properties collectively contribute to the enhanced drought tolerance observed in plants treated with gold nanoparticles.

8.2.3 Metal oxides

Studies have been conducted on a wide range of metal oxides, such as aluminum (Al_2O_3), magnesium oxide (MgO), zinc oxide (ZnO), copper oxide (CuO), iron oxide (Fe_3O_4), and many more. Because of its magnetic characteristics, iron oxide has emerged as a novel and intriguing nanoparticles. Applications for iron oxide (Fe_3O_4) nanoparticles include biological, catalytic, and environmental safety.

8.3 Biogenic synthesis of nanoparticles

A vast array of unique nanoparticles can be produced by microorganisms. The desire to employ these microbes to produce nanoparticles for various applications has increased among researchers as a result. Both bacteria and fungi are capable of producing inorganic chemicals through biologically driven and induced synthesis.[35] The biological synthesis of nanoparticles still has limits with regard to process scalability and particle geometry controllability, despite advances in the physicochemical synthesis of nanoparticles.[36] However, physiologically induced synthesis offers a broad variety of composition and has enabled scientists to synthesis inorganic nanoparticles using traditional metal precursors.[37] The usage of bacteria, algae, and fungi among other microbes to create nanoparticles is expanding. For example, bacteria and fungi can produce both externally and intracellularly, and the pathways for both processes differ amongst microbes. During intracellular synthesis, metal ions are carried through the cell wall, where the positively charged ions interact with the negatively charged wall. The enzymes in the cells transform these ions into metal NPs.[38]

8.3.1 Plant-based (green) synthesis of nanoparticles

Plant-based or green synthesis of nanoparticles (NPs) is the use of biological synthesis techniques as a safe and ecologically acceptable replacement for conventional physical and chemical manufacturing processes. In this rapidly expanding field, commonly referred to as "green technology" or "green nanotechnology," the production of nanoparticles (NPs) gives careful consideration to sustainable and ecologically friendly processes.[39] The primary objective of green nanotechnology is to produce nanoparticles (NPs) that mitigate environmental problems while minimizing detrimental impacts on the environment and public health. This approach is anticipated to produce industrial processes that are both environmentally and energy-efficient.[40] Three essential components drive the environmentally responsible synthesis of nanoparticles: the use of solvents that are safe for the environment, strong reducing agents, and non-hazardous stabilizing compounds[41] (Fig. 8.1).

8.3.2 Mechanism of uptake and transport of nanoparticles

Su et al. identified three primary stages in which NPs interact with plants Deposition, Access, and Translocation.[42] When NPs are sprayed on leaves, plant cells absorb them through stomata or cuticles on the leaf surface, a process known

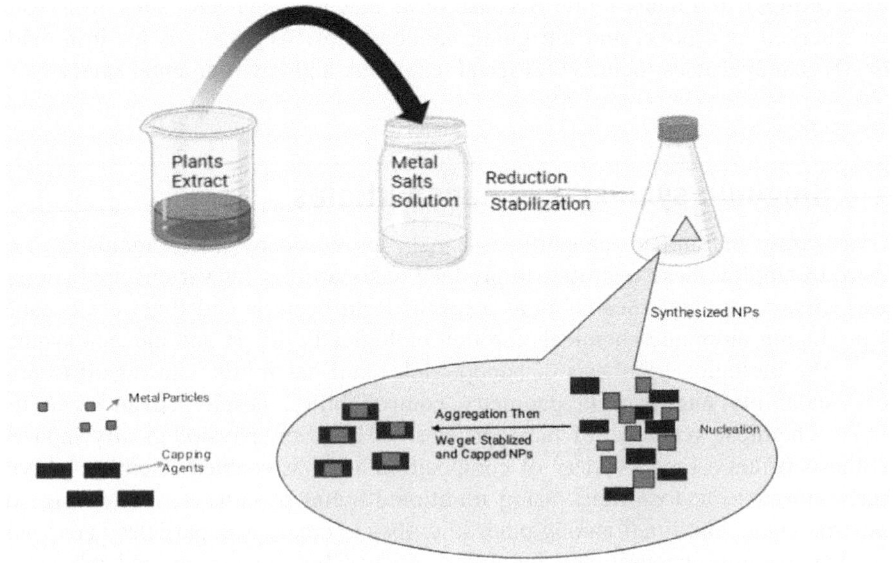

FIGURE 8.1

Plant-based synthesis of nanoparticles.

as endocytosis. Smaller NPs can enter through a cuticular channel, while larger NPs can be absorbed by stomatal pores, which have micrometer-sized pores.[43] The foliar pathway of NPs uptake is influenced by leaf shape in addition to pore size and stomatal density. Furthermore, NPs' capacity to pierce the cuticle and leaf mesophyll tissue is also influenced by their size, chemical coating, and hydrophobicity. PVP-coated gold (Au) NPs (3 nm) showed a higher cuticular penetration rate than citrate-coated Au NPs, according to Avellan et al.[43]

8.4 Interaction of NPs with phytohormone

Gibberellins (GA) and indole acetic acid (IAA), which improve plant development under drought stress, are produced in greater quantities when NPs are applied.[44,45] The effect of NPs on the production and concentration of certain plant hormones including cytokinin (CK), ethylene (ET), and ABA has been studied. It may help crops fend off drought stress by increasing endogenous hormones. Applying 40 ppm TiO_2-NPs resulted in a 31.32% decrease in ABA levels and an increase of 49.07% and 27.43% in IAA and GA, respectively.[29] Plants under drought stress that were treated with nanoparticles displayed increased proline, glycine betaine, total proteins, and soluble sugar levels. Proline, glycine betaine, and amino acid levels rose in drought-stressed cucumbers after applying zinc oxide nanoparticle therapy. Furthermore, ZnO NP treatment prevented the decline of soluble sugar. ZnO is said to boost proline biosynthesis by controlling gene expression in proline biosynthesis. Proline is an important osmolyte that aids in restoring osmotic balance.[46]

8.5 Effects of biogenic NPs at the biochemical and molecular level

NPs can significantly increase the expression of genes that respond to stress. These genes are crucial for how plants respond to stress and have the power to change a plant's physiology, metabolism, and growth to boost stress tolerance.[47] Biogenic NPs exposure modifies the expression of genes linked to stress and defense processes. For example, overexpression of enzymes including POD, CAT, APX, and SOD when biogenic TiO_2 NPs (leaf extract from *Moringa oleifera*) are treated to wheat.[48] Zinc oxide nanoparticles upregulate the expression of genes produce proteins involved in ion transport, ROS detoxification, signaling, and other processes.

Stress-responsive gene expression is influenced by the transcription factors' activity, which can be changed by nanoparticles. By controlling the activity of transcription factors involved in stress responses, seven crucial regulatory genes were found to be more abundant in plants treated with NP during drought circumstances, including GmWRKY27, GmMYB118, GmMYB174, GmNAC11, GmRD20A, GmERD1, and GmDREB2. Soybeans treated with Fe, Cu, Co, and ZnO NP showed increased ABA content in drought stress. The synthesis of ABA and signaling pathways linked to the ABA hormone are regulated by increased GmWRKY27 gene expression.

8.6 The promising role of nanoparticles in mitigating drought stress

Water is necessary for plants from the moment of seed germination until they reach the blooming or fruiting stage of their lives. Plants that don't get enough water either through transpiration or a damaged root system. A water deficiency can have adverse effects on the morphological, physiological, biochemical, and molecular processes of plants.[49,50] Abiotic factors like salt and drought have a substantial impact on plant development and productivity; nanotechnology may provide new ways to counteract these effects. Certain types of nanoparticles, such as hydrogel nanoparticles, can retain a lot of water and release it gradually to the plants, encouraging plant development in situations where there is a shortage of water.[51] NPs can improve a plant's resistance to drought by altering its physiology and biochemistry. For example, wheat plants respond to silver nanoparticles (AgNPs) with increased antioxidant activity, which reduces oxidative damage and encourages plant development.[52] ZnO NPs may improve plant water usage efficiency and help plants maintain their water balance during droughts by controlling stomatal conductance and transpiration rate.[53] TiO_2 NPs enhance photosynthetic efficiency under drought stress by increasing the chlorophyll content and increase plant growth and productivity.[54,55] Besides, TiO_2 NPs also have the same capacity as other types of nanoparticles to boost the activity of antioxidant enzymes.

8.6.1 Improvement of membrane stability

Nanoparticles stimulate the overproduction of ROS, because oxidation can lead to oxidative damage to cell membranes, which is one of the main effects of drought. AgNPs and TiO_2 NPs are two examples of nanoparticles that can increase the activity of antioxidant enzymes, reducing oxidative damage and preserving membrane stability.[56]

8.6.2 Nanoparticles improve phenolic compounds accumulation

Nanoparticles enhance the production of more phenolic compounds, which are secondary metabolites with important functions in drought stress responses because they are antioxidants, absorb UV light, and have signaling capabilities.[57] Under drought stress, NPs can increase the accumulation of these phenolic compounds, increasing the stress tolerance of plants. According to Rasheed et al.,[58] NPs can promote the manufacture of phenolic chemicals. for instance, recent researchers have reported that applying AgNPs and ZnONPs can increase the production of phenolic compounds in plants that are experiencing drought stress.

8.6.3 Enhancement of plant—water relationship

Sustainable agricultural practices are under threat from environmental issues and water scarcity, which makes the function of nanoparticles in improving plant-water relationships crucial.[59] The interactions between plants and water are based on how plants absorb, move, and use water processes that are essential to plant development, photosynthesis, and overall productivity (Fig. 8.2).

Innovative methods to increase plant resilience and water efficiency are provided by nanotechnology, which addresses a number of factors: Because of their tiny size and large surface area, NPs have the ability to change the physical characteristics of soil, increasing its ability to hold water.[60]

8.6.4 Nanoparticles improve growth, yield, and quality

NPs have shown great potential in improving the growth, yield, and quality of plants under drought and salinity stress conditions. Their unique properties and interactions with plants can lead to beneficial effects on plant physiology,

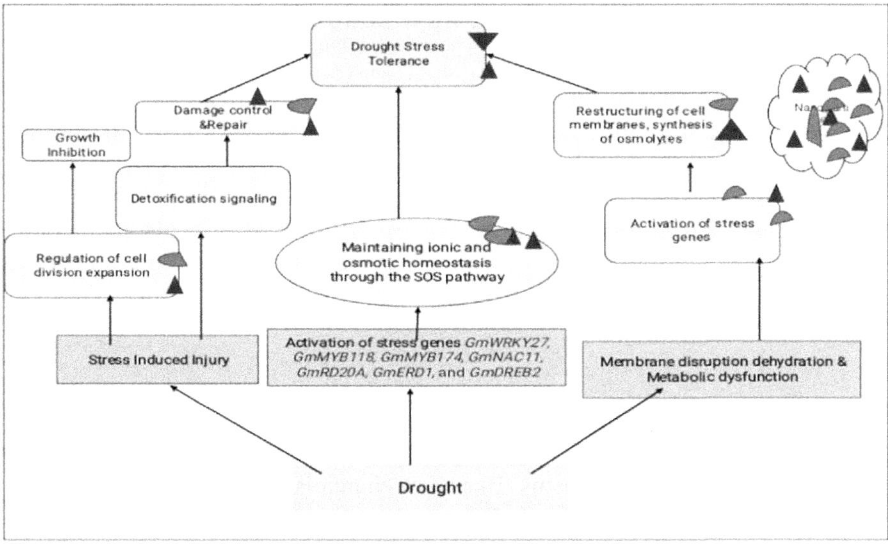

FIGURE 8.2

Role of nanoparticles during drought stress depicted in the flowchart.

metabolism, and overall performance. NPs can promote plant growth under stress challenges by improving nutrient uptake, regulating hormone levels, and mitigating stress-induced damage.[50]

8.7 Nanoparticles/nanofertilizers in agriculture

In general, nanoparticles are referred to as nanofertilizers and nanopesticides, respectively, because they can be utilized in agriculture as fertilizers or insecticides. The efficiency of nutrient consumption in agriculture may be increased by the use of nanoparticles as nanofertilizers.[61,62] Furthermore, applying nanoparticles as nanopesticides could shield crops from bacterial and fungal diseases.[63] However, the long-term impacts of nanoparticle use on microorganisms linked with plants are still unknown. Although microbial communities are important and sensitive markers of the environmental hazards associated with nanoparticles.[64,65]

In agriculture, crop development is difficult to be maintained without the application of fertilizers, herbicides, and insecticides for crop protection. Nanofertilizers support soil fertility while nano herbicides and pesticides protect plants from early deterioration, fight pests for longer periods of time, and increase solubility. Pesticides can be sprayed directly on grains and seeds to prevent plant diseases.[66]

8.7.1 Methods of nanoparticles application in agriculture

8.7.1.1 Foliar application

NPs can penetrate plant cells by foliar application via the cuticular (lipophilic) and stomatal (hydrophilic) pathways; bigger NPs (>5 nm) enter through stomata, while smaller NPs (<4.8 nm) enter directly through the cuticle. In order to prevent nutrient loss, foliar application of NPs is used to apply herbicides, fertilizers, and nutritional supplements directly to leaves. This method has advantages over soil-applied fertilizers, which have low utilization rates due to leaching and adsorption, which can cause environmental problems like eutrophication.[67,68]

8.7.1.2 Root application

In addition to entering by physical damage, NPs enter the plant and move through it through mechanisms like ion channels, endocytosis, or contact with cell membrane proteins. Size-independent NP absorption is made possible by endocytosis; the uptake of carbon NPs in Madagascar periwinkle is one example of this process. Moreover, NPs can attach to transport proteins on the exterior epidermis of the root, providing an additional pathway for absorption.[69]

8.7.1.3 Stem injection or stem feeding

By providing NPs with direct access to the plant's vascular network, stem injection, often referred to as stem feeding, has the advantage of efficiently distributing NPs throughout the plant. By removing obstacles from the outside, such as cuticles, this technique ensures instantaneous NP availability, reduces loss, and boosts distribution efficiency.[70] Stem injection bypasses the external barriers that usually impede NP uptake in plants, enabling higher NP concentration delivery to targeted targets within the plant. Once inside, NPs quickly diffuse throughout the plant's tissues through the xylem and phloem. An application of stem injection is in disease control, such as utilizing copper NPs to manage citrus canker in citrus trees, effectively resolving plant health issues.[71] There are some crops in which nanoparticles applied and functions has been described in Table 8.1.

It is possible that this elevated antioxidant activity protects plant cells from the oxidative damage caused by ROS.[78,79] Abiotic stresses like drought stress are countered by plants primarily through the phytohormone ABA. ABA assists plants under stress from drought in overcoming the stress of the drought and achieving tolerance by regulating root development, leaf elongation, and plant expansion.[4] Water-stressed sorghum seedlings with improved total phenolic compounds and noticeably increased flavonoid content were produced by Si NP-treated soil.

Table 8.1 Description about the response due to application of nanoparticles in various crops during drought stress.

Crops	Nanoparticles	NPs applied	Functions	References
1. Wheat (*Triticum aestivum* L.)	Titanium dioxide (TiO_2) nanoparticles	Spraying throughout the stages of stem elongation and flowering (0.02%)	Applied at stem elongation: Increased yield, starch and gluten content, number of ears, and seeds.	72
			Use to raise test weight, ear weight, harvest index, and gluten content during the flowering period.	
		Application period: 4 weeks; 0.1%	Improved growth kinetics and parameters related to leaf growth.	73
2. Sorghum (*Sorghum bicolor* L.)	Silicon	200 mL of silicon per liter of potassium silicate (K_2O_3Si) in each kilogram of soil	Increases the chlorophyll content, specific leaf weight, shoot dry weight, root dry weight, and leaf dry weight.	74
			Decrease in the shoot-to-root (S/R) ratio and the leaf water potential.	
	ZnO NPs, or zinc oxide nanoparticles	5 mg Zn/kg soil	Improved crop growth, total K acquisition, grain K, and grain N translocation in addition to increased grain output. Edible plant parts are biofortified with zinc to decrease the accumulation of total P.	13
3. Barley (*Hordeum vulgare* L.)	Nanoparticles of silicon dioxide (SiO_2) and titanium dioxide (TiO_2)	20 ppm each	Stimulating effects on the growth and yield components. Less obvious effects are seen with SiO_2 NP than with TiO_2 NP.	75

(*Continued*)

Table 8.1 Description about the response due to application of nanoparticles in various crops during drought stress. *Continued*

Crops	Nanoparticles	NPs applied	Functions	References
4. Lentil (*Lens culinaris* L.)	AgNPs, or silver nanoparticles	10 μg mL/L; apply for 7 days, pausing every other day	Rise in the radical's percentage of germination, germination rate, radical length, and dry weight.	34
5. Soybean (*Glycinemax* L.)	Iron nanoparticles (Fe NP)	50 mg/L; 30 g of seeds steeped for 30 minutes in a 10 mL suspension	An increase in the root length, specific root length index (SRL), shoot dry weight, dry weight, and relative water content of the plants (DTI). Low rate of biomass loss.	76
	Zinc oxide nanoparticles (ZnO)	1 g/L	Increase in the length, fresh and dry weight, rate of germination, and proportion of radicles that germinate. Reduction in the seeds' fresh and dried residual weight.	77

8.8 Interaction of nanoparticles with the environment component

There are numerous opportunities to use this new technology. In science, technology, and business today, nanotechnology is recognized as a significant and impactful technological advancement. Examples of how nanotechnology is being used to create systems and important applications in environmental challenges include the manufacture of nanomaterials, nanotubes, nanocomposites, nanofilters, and nanoparticles.[80] Using nanotechnology to create nanocomposites has resulted in the creation of extremely strong and light raw materials that can take the place of heavy metal components, drastically reducing the weight of machinery and vehicle parts, and ultimately lowering energy consumption and air pollution[81] (Fig. 8.3).

An efficient way to save energy and safeguard the environment in this regard is to utilize nanofilters. Microfilters, ultrafilters, and nanofilters are the three categories of filters based on the size of their pores. Since nanofiltration essentially

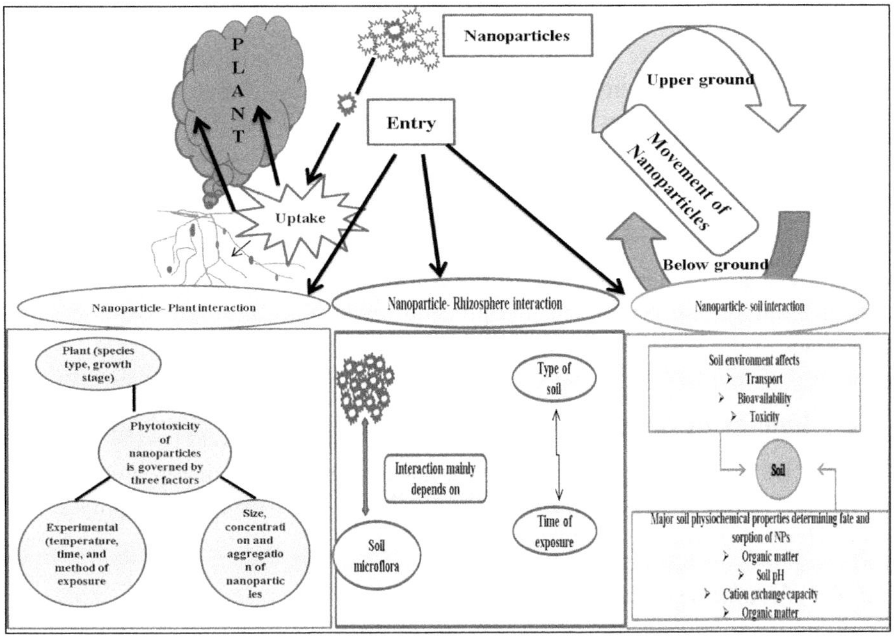

FIGURE 8.3

Nanoparticles interaction with different components of the environment.

involves lower pressure filtration than reverse osmosis, nanofilters are more reasonably priced. Furthermore, nanofilters can be utilized to eliminate pollutants from agricultural and drinking water by eliminating bacteria and viruses.[82]

8.8.1 Nanoparticle emission to the environment

These NPs of distinct properties may be dispersed into the environment in a variety of ways at various times. In order to manage the release of nanomaterials into the environment, it is essential to assess the entire life cycle of NP, from manufacture to disposal. The ecosystem may be impacted by this life cycle at various stages.[83]

8.8.1.1 Manufacture and employment

Nanoparticles can readily be released during manufacturing if the protocol is not closely followed. Indirect emissions may result from this inadvertent release. Unintentional emissions of NPs can come from a variety of causes, including fireworks, combustion, and nanoparticle spills. Moreover, manmade operations, also referred to as a cause of emissions of nanoparticles, into the environment. The emission characteristics of NPs depends on the different type of products such as solid, liquid, and spray products and the use of nanoparticles occurs at various phases.[84]

8.8.1.2 Disposal

Landfills are commonly used for the disposal of nanomaterials, which poses a danger of degradation and the release of nanoparticles (NPs) into the environment. Besides, the ash particles from incineration could leak into the surrounding area. As a result, incinerator facilities use specific filters to efficiently regulate and reduce ash emissions.[85]

8.8.1.3 Recycling process

Resistance and emissions happen as a result of the product-specific recycling process. It is important to perform this procedure while according to specific safety guidelines.[86]

8.8.2 Impact of nanoparticles on the outdoor

The toxicity of biogenic nanoparticles is an important area of research, as their impact on the environment and living organisms needs to be thoroughly understood. Here are some key points:

8.8.2.1 Mechanisms of toxicity

Biogenic nanoparticles can interact with biological systems at the cellular and molecular levels. They may generate reactive oxygen species (ROS), which can cause oxidative stress, damage cellular components, and lead to cell death. The size, shape, surface charge, and concentration of nanoparticles play crucial roles in determining their toxicity. Smaller nanoparticles have a larger surface area relative to their volume, which can increase their reactivity and potential to cause harm. The method of synthesis and the type of biological material used can also influence the toxicity. For example, nanoparticles synthesized using plant extracts might have different properties compared to those produced by bacteria.

8.8.2.2 Environmental impact

Once released into the environment, biogenic nanoparticles can accumulate in soil and water, potentially affecting microorganisms, plants, and animals. Their long-term effects on ecosystems are still being studied.

8.8.2.3 Human health concerns

Inhalation, ingestion, or skin contact with nanoparticles can pose health risks to humans. Studies have shown that nanoparticles can penetrate biological barriers and reach various organs, potentially causing inflammation, toxicity, and other adverse effects.

I. Positive impact
   - ✓ NPs are applied in many different domains, such as nanofiltration, and nanodelivery systems for agriculture, which improve crop nutrition and are employed in numerous agro-industries.[87–92]

- ✓ For seed germination and root elongation, NPS is also given a lot of attention. They are used in cosmetics, such as skin treatments, that fight skin aging by using proteins produced from stem cells.
- ✓ NPs are used in waste water treatment, a method that is used all around the world. To supply clean water for drinking and irrigation, they lessen pesticides, fertilizers, and heavy metals.[93]

II. Negative impact

- ✓ Depending on the situation, nanoparticles released into environment cause harm to living things. Skin and gastrointestinal system harm can result from living things absorbing and dispersing nanoparticles through their respiratory systems.[94]
- ✓ Both carbon nanotubes and carbon nanoparticles, such as fullerenes, have the ability to penetrate the body and cause toxicity. The primary target of carbon nanotubes is the mitochondria. It has been demonstrated that short-term exposure to carbon nanotubes in amphibians can cause genotoxicity, which damages DNA and causes oxidative stress.[95]
- ✓ AgNPs-containing consumer goods have the potential to contaminate aquatic habitats. Aquatic species like fish and algae may become poisonous when AgNP dissolves. Particle characteristics, such as concentration, pH, size, and shape, all affect the reactivity of the nanoparticle and have a significant impact on AgNP toxicity. According to Frenk et al., CuO and Fe_3O_4 nanoparticles showed differing degrees of toxicity in two distinct sand soils at different concentrations, with CuO exhibiting greater toxicity than Fe_3O_4.[96]
- ✓ For water filtration and the stabilization of erosion effects, burrowing is an essential technique. Because earthworms make up a sizable fraction of soil biomass, they are among the burrowing species in the soil that may be impacted by nanoparticles.[97]
- ✓ Plants are susceptible to cellular toxicity caused by zinc nanoparticles, which can result in decreased photosynthesis, growth retardation, and changed gene expression in all impacted plants. The main cause of the decline in photosynthesis is higher quantities of nanoparticles. Plant growth can also be inhibited by AgNP. Depending on the type of metal a particle contains, its characteristics vary. The environment may suffer from a shortage of skilled laborers and engineers. Emissions of nanoparticles happen at several stages of the life cycle, such as batch processing, experimentation, and decomposition.[98]
- ✓ It has been observed that superparamagnetic Fe_3O_4 NPs cause disturbances in the functions of the kidney, liver, and heart in mice, resulting in symptoms such as nausea and vomiting. Tissue injury can occur in fish that ingest carbon nanotubes at higher quantities than humans.[99]

In recent years, a large number of investigations and tests have been carried out.

8.9 Regulatory and safety measures

To mitigate risks, it is essential to establish guidelines for the safe production, use, and disposal of biogenic nanoparticles.[100–105] To ensure safer nanoparticles with minimal environmental and health impacts during the use of biogenic nanoparticles is subject to various regulations to ensure safety and efficacy. Here are some key points:

1. FDA regulations: In the United States, the FDA regulates nanotechnology products, including biogenic nanoparticles, under existing statutory authorities. The FDA evaluates the safety and effectiveness of these products based on their unique properties and intended use.[106]
2. Environmental and health safety: Regulations often focus on the potential environmental and health impacts of nanoparticles. This includes assessing toxicity, environmental persistence, and bioaccumulation.[102]
3. International guidelines: Different countries may have their own guidelines and regulations. For example, the European Union has specific regulations for nanomaterials under REACH (Registration, Evaluation, Authorisation, and Restriction of Chemicals).[107]
4. Consultation and compliance: Manufacturers are advised to consult with regulatory bodies early in the development process to ensure compliance with all relevant regulations and to address any scientific and regulatory issues.

References

1. Rodrigues, M. S.; Demokritou, P.; Dokoozlian, N.; Hendren, C. O.; Karn, B.; Mauter, S. M.; Lowry, G. V. Nanotechnology for sustainable food production: promising opportunities and scientific challenges. *Environ Science: Nano* **2017**, *4* (4), 767–781. Available from: https://doi.org/10.1039/c6en00573j.
2. Alabdallah, M. N.; Hasan, M. M.; Hammami, I.; Alghamdi, A. I.; Alshehri, D.; Alatawi, H. A. Green synthesized metal oxide nanoparticles mediate growth regulation and physiology of crop plants under drought stress. *Plants* **2021**, *10* (8), 1730. Available from: https://doi.org/10.3390/plants10081730.
3. Guo, X.; Chen, R.; Thomas, D. S.; Li, Q.; Xia, Z.; Pan, Z. Divergent processes and trends of desertification in Inner Mongolia and Mongolia. *Land Degrad & Dev* **2021**, *32* (13), 3684–3697. Available from: https://doi.org/10.1002/ldr.3825.
4. Sallam, A.; Alqudah, A. M.; Dawood, M. F.; Baenziger, P. S.; Börner, A. Drought stress tolerance in wheat and barley: advances in physiology, breeding and genetics research. *Int J Mol Sci* **2019**, *20* (13), 3137. Available from: https://doi.org/10.3390/ijms20133137.
5. Yunlong, C.; Smit, B. Sustainability in agriculture: a general review. *Agriculture, Ecosyst & Environ* **1994**, *49* (3), 299–307. Available from: https://doi.org/10.1016/0167-8809(94)90059-0.

6. Francis, D. V.; Sood, N.; Gokhale, T. Applications of metal nanoparticles in agriculture. *Prog Prosp Nanosci Today* **2020**,)157.
7. Mishra, S.; Singh, A.; Keswani, C.; Singh, H. B. Nanotechnology: exploring potential application in agriculture and its opportunities and constraints. *Biotech Today* **2014**, *4* (1), 9–14.
8. Bayat, H.; Moghadam, A. N. Drought effects on growth, water status, proline content and antioxidant system in three *Salvia nemorosa* L. cultivars. *Acta Physiologiae Plant* **2019**, *41* (9), 149.
9. Adrees, M.; Khan, Z. S.; Ali, S.; Hafeez, M.; Khalid, S.; Rehman, M. Z.; Rizwan, M. Simultaneous mitigation of cadmium and drought stress in wheat by soil application of nanoparticles. *Chemosphere* **2020**, *238*, 124681. Available from: https://doi.org/10.1016/j.chemosphere.2019.124681.
10. Bashir, M. A.; Silvestri, C.; Ahmad, T.; Hafiz, I. A.; Abbasi, N. A.; Manzoor, A.; Rugini, E. Osmotin: a cationic protein leads to improve biotic and abiotic stress tolerance in plants. *Plants* **2020**, *9* (8), 992. Available from: https://doi.org/10.3390/plants9080992.
11. Dimkpa, C. O.; Andrews, J.; Sanabria, J.; Bindraban, P. S.; Singh, U.; Elmer, W. H.; White, J. C. Interactive effects of drought, organic fertilizer, and zinc oxide nanoscale and bulk particles on wheat performance and grain nutrient accumulation. *Sci Total Environ* **2020**, *722*, 137808. Available from: https://doi.org/10.1016/j.scitotenv.2020.137808.
12. Singh, A.; Singh, N. Á.; Afzal, S.; Singh, T.; Hussain, I. Zinc oxide nanoparticles: a review of their biological synthesis, antimicrobial activity, uptake, translocation and biotransformation in plants. *Journal of materials science* **2018**, *53* (1), 185–201.
13. Dimkpa, C. O.; Singh, U.; Bindraban, P. S.; Elmer, W. H.; Gardea-Torresdey, J. L.; White, J. C. Zinc oxide nanoparticles alleviate drought-induced alterations in sorghum performance, nutrient acquisition, and grain fortification. *Sci Total Environ* **2019**, *688*, 926–934.
14. Mahdieh, M.; Sangi, M. R.; Bamdad, F.; Ghanem, A. Effect of seed and foliar application of nano-zinc oxide, zinc chelate, and zinc sulphate rates on yield and growth of pinto bean (*Phaseolus vulgaris*) cultivars. *J Plant Nutr* **2018**, *41* (18), 2401–2412.
15. Farooq, M.; Romdhane, L.; Al Sulti, M. K.; Rehman, A.; Al-Busaidi, W. M.; Lee, D. J. Morphological, physiological and biochemical aspects of osmopriming-induced drought tolerance in lentil. *J Agron Crop Sci* **2020**, *206* (2), 176–186.
16. Faizan, M.; Bhat, J. A.; Chen, C.; Alyemeni, M. N.; Wijaya, L.; Ahmad, P.; Yu, F. Zinc oxide nanoparticles (ZnO-NPs) induce salt tolerance by improving the antioxidant system and photosynthetic machinery in tomato. *Plant Physiol Biochem* **2021**, *161*, 122–130. Available from: https://doi.org/10.1016/j.plaphy.2021.02.002.
17. Tariverdizadeh, N.; Mohebodini, M.; Chamani, E.; Ebadi, A. Iron and zinc oxide nanoparticles: an efficient elicitor to enhance trigonelline alkaloid production in hairy roots of fenugreek. *Ind Crop Products* **2021**, *162*, 113240.
18. Jordan, J. T.; Singh, K. P.; Cañas-Carrell, J. E. Carbon-based nanomaterials elicit changes in physiology, gene expression, and epigenetics in exposed plants: a review. *Curr OpEnviron Sci & health* **2018**, *6*, 29–35.
19. Aghdam, M. T. B.; Mohammadi, H.; Ghorbanpour, M. Effects of nanoparticulate anatase titanium dioxide on physiological and biochemical performance of Linum usitatissimum (Linaceae) under well-watered and drought stress conditions. *Braz J botany* **2016**, *39*, 139–146.

20. Borišev, M.; Borišev, I.; Župunski, M.; Arsenov, D.; Pajević, S.; Ćurčić, Ž.; Djordjevic, A. Drought impact is alleviated in sugar beets (*Beta vulgaris* L.) by foliar application of fullerenol nanoparticles. *PLoS One* **2016**, *11* (11), e0166248. Available from: https://doi.org/10.1371/journal.pone.0166248.
21. Semida, W. M.; Abdelkhalik, A.; Mohamed, G. F.; Abd El-Mageed, T. A.; Abd El-Mageed, S. A.; Rady, M. M.; Ali, E. F. Foliar application of zinc oxide nanoparticles promotes drought stress tolerance in eggplant (*Solanum melongena* L.). *Plants* **2021**, *10* (2), 421. Available from: https://doi.org/10.3390/plants10020421.
22. Taran, N.; Storozhenko, V.; Svietlova, N.; Batsmanova, L.; Shvartau, V.; Kovalenko, M. Effect of zinc and copper nanoparticles on drought resistance of wheat seedlings. *Nanoscale Res Lett* **2017**, *12*, 1−6. Available from: https://doi.org/10.1186/s11671-017-1839-9.
23. Janaguiraman, M. D.; Nair, R.; Giraldo, J. P.; Prasad, P. V. V. Cerium oxide nanoparticles decrease drought-induced oxidative damage in sorghum leading to higher photosynthesis and grain yield. *ACS omega* **2018**, *3* (10), 14406−14416.
24. Mustafa, H.; Ilyas, N.; Akhtar, N.; Raja, N. I.; Zainab, T.; Shah, T.; Ahmad, P. Biosynthesis and characterization of titanium dioxide nanoparticles and its effects along with calcium phosphate on physicochemical attributes of wheat under drought stress. *Ecotoxicol Environ Saf* **2021**, *223*, 112519. Available from: https://doi.org/10.1016/j.ecoenv.2021.112519.
25. Van Nguyen, D.; Nguyen, H. M.; Le, N. T.; Nguyen, K. H.; Nguyen, H. T.; Le, H. M.; VanHa, C. Copper nanoparticle application enhances plant growth and grain yield in maize under drought stress conditions. *J Plant Growth Regul* **2022**, *41* (1), 364−375.
26. Seabra, A. B.; Durán, N. Nanotoxicology of metal oxide nanoparticles. *Metals* **2015**, *5* (2), 934−975. Available from: https://doi.org/10.3390/met5020934.
27. Wang, X. P.; Li, Q. Q.; Pei, Z. M.; Wang, S. C. Effects of zinc oxide nanoparticles on the growth, photosynthetic traits, and antioxidative enzymes in tomato plants. *Biologia Plant* **2018**, *62*, 801−808.
28. Kumari, M.; Khan, S. S.; Pakrashi, S.; Mukherjee, A.; Chandrasekaran, N. Cytogenetic and genotoxic effects of zinc oxide nanoparticles on root cells of Allium cepa. *J Hazard Mater* **2011**, *190* (1-3), 613−621. Available from: https://doi.org/10.1016/j.jhazmat.2011.03.095.
29. Houtman, C. J. Emerging contaminants in surface waters and their relevance for the production of drinking water in Europe. *J Integr Environ Sci* **2010**, *7* (4), 271−295. Available from: https://doi.org/10.1080/1943815X.2010.511648.
30. Ge, Y.; Priester, J. H.; Van De Werfhorst, L. C.; Walker, S. L.; Nisbet, R. M.; An, Y. J.; Holden, P. A. Soybean plants modify metal oxide nanoparticle effects on soil bacterial communities. *Environ Sci & Technol* **2014**, *48* (22), 13489−13496. Available from: https://doi.org/10.1021/es5031646.
31. Thomas, R.; Jasim, B.; Mathew, J.; Radhakrishnan, E. K. Extracellular synthesis of silver nanoparticles by endophytic *Bordetella* sp. isolated from Piper nigrum and its antibacterial activity analysis. *Nano Biomed Eng* **2012**, *4* (4), 183−187. Available from: https://doi.org/10.5101/nbe.v4i14.p183-187.
32. Saranya, S.; Aswani, R.; Remakanthan, A.; Radhakrishnan, E. K. In agriculture. *Nanotechnol agriculture: Adv Sustain agriculture* **2019**, 1−17.
33. Abdal Dayem, A.; Hossain, M. K.; Lee, S. B.; Kim, K.; Saha, S. K.; Yang, G. M.; Cho, S. G. The role of reactive oxygen species (ROS) in the biological activities of

metallic nanoparticles. *Int J Mol Sci* **2017**, *18* (1), 120. Available from: https://doi.org/10.3390/ijms18010120.

34. Hojjat, S. S.; Ganjali, A. The effect of silver nanoparticle on lentil seed germination under drought stress. *Int J Farm Allied Sci* **2016**, *5* (3), 208–212.
35. Moitra, P.; Alafeef, M.; Dighe, K.; Frieman, M. B.; Pan, D. Selective naked-eye detection of SARS-CoV-2 mediated by N gene targeted antisense oligonucleotide capped plasmonic nanoparticles. *ACS nano* **2020**, *14* (6), 7617–7627. Available from: https://doi.org/10.1021/acsnano.0c03822.
36. Fang, X.; Wang, Y.; Wang, Z.; Jiang, Z.; Dong, M. Microorganism assisted synthesized nanoparticles for catalytic applications. *Energies* **2019**, *12* (1), 190. Available from: https://doi.org/10.3390/en12010190.
37. Grasso, G.; Zane, D.; Dragone, R. Microbial nanotechnology: challenges and prospects for green biocatalytic synthesis of nanoscale materials for sensoristic and biomedical applications. *Nanomaterials* **2019**, *10* (1), 11. Available from: https://doi.org/10.3390/nano10010011.
38. Mathivanan, K.; Selva, R.; Chandirika, J. U.; Govindarajan, R. K.; Srinivasan, R.; Annadurai, G.; Duc, P. A. Biologically synthesized silver nanoparticles against pathogenic bacteria: synthesis, calcination and characterization. *Biocatalysis Agric Biotechnol* **2019**, 22.101373. Available from: https://doi.org/10.1016/j.bcab.2019.101373.
39. Patra, J. K.; Baek, K. H. Green nanobiotechnology: factors affecting synthesis and characterization techniques. *J Nanomaterials* **2014**, *2014* (1).417305. Available from: https://doi.org/10.1155/2014/417305.
40. Srivastava, S.; Bhargava, A.; Srivastava, S.; Bhargava, A. Green nanotechnology: an overview. *Green Nanoparticles: Future Nanobiotechnology* **2022**, 1–13.
41. Sarkar, S.; Ponce, N. T.; Banerjee, A.; Bandopadhyay, R.; Rajendran, S.; Lichtfouse, E. Green polymeric nanomaterials for the photocatalytic degradation of dyes: a review. *Environ Chem Lett* **2020**, *18* (5), 1569–1580.
42. Su, L. X.; Ma, X. L.; Zhao, K. K.; Shen, C. L.; Lou, Q.; Yin, D. M.; Shan, C. X. Carbon nanodots enhancing the stress resistance of peanut plants. *ACS Omega* **2018**, *3* (12), 17770–17777. Available from: https://doi.org/10.1021/acsomega.8b02604.
43. Avellan, A.; Yun, J.; Zhang, Y.; Spielman-Sun, E.; Unrine, J. M.; Thieme, J.; Lowry, G. V. Nanoparticle size and coating chemistry control foliar uptake pathways, translocation, and leaf-to-rhizosphere transport in wheat. *ACS nano* **2019**, *13* (5), 5291–5305. Available from: https://doi.org/10.1021/acsnano.8b09781.
44. Churilova, V. V.; Nazarova, A. A.; Polishchuk, S. D. Influence of biodrugs with nanoparticles of ferrum, cobalt and cuprum on growth, development, yield and phytohormone status of fodder and red beets. *Nano Hybrids Compos* **2017**, *13*, 149–155.
45. Li, M.; Zhang, P.; Adeel, M.; Guo, Z.; Chetwynd, A. J.; Ma, C.; Rui, Y. Physiological impacts of zero valent iron, Fe_3O_4 and Fe_2O_3 nanoparticles in rice plants and their potential as Fe fertilizers. *Environ Pollut* **2021**, *269*, 116134. Available from: https://doi.org/10.1016/j.envpol.2020.116134.
46. Ghani, M. I.; Saleem, S.; Rather, S. A.; Rehmani, M. S.; Alamri, S.; Rajput, V. D.; Liu, M. Foliar application of zinc oxide nanoparticles: an effective strategy to mitigate drought stress in cucumber seedling by modulating antioxidant defense system and osmolytes accumulation. *Chemosphere* **2022**, *289*, 133202. Available from: https://doi.org/10.1016/j.chemosphere.2021.133202.

47. Prajapati, P.; Gupta, P.; Kharwar, R. N.; Seth, C. S. Nitric oxide mediated regulation of ascorbate-glutathione pathway alleviates mitotic aberrations and DNA damage in *Allium cepa* L. under salinity stress. *Int J Phytoremediat* **2023**, *25* (4), 403–414. Available from: https://doi.org/10.1080/15226514.2022.2086215.
48. Satti, S. H.; Raja, N. I.; Ikram, M.; Oraby, H. F.; Mashwani, Z. U. R.; Mohamed, A. H.; Omar, A. A. Plant-based titanium dioxide nanoparticles trigger biochemical and proteome modifications in *Triticum aestivum* L. under biotic stress of *Puccinia striiformis*. *Molecules* **2022**, *27* (13), 4274. Available from: https://doi.org/10.3390/molecules27134274.
49. Seleiman, M. F.; Al-Suhaibani, N.; Ali, N.; Akmal, M.; Alotaibi, M.; Refay, Y.; Battaglia, M. L. Drought stress impacts on plants and different approaches to alleviate its adverse effects. *Plants* **2021**, *10* (2), 259. Available from: https://doi.org/10.3390/plants10020259.
50. Hayat, F.; Khanum, F.; Li, J.; Iqbal, S.; Khan, U.; Javed, H. U.; Chen, J. Nanoparticles and their potential role in plant adaptation to abiotic stress in horticultural crops: a review. *Sci Horticulturae* **2023**, *321*, 112285. Available from: https://doi.org/10.3390/plants10020259.
51. Nair, P. M. G.; Chung, I. M. Physiological and molecular level effects of silver nanoparticles exposure in rice (*Oryza sativa* L.) sedlings. *Chemosphere* **2014**, *112*, 105–113. Available from: https://doi.org/10.1016/j.chemosphere.2014.03.056.
52. El-Saadony, M. T.; Saad, A. M.; Soliman, S. M.; Salem, H. M.; Desoky, E. S. M.; Babalghith, A. O.; AbuQamar, S. F. Role of nanoparticles in enhancing crop tolerance to abiotic stress: a comprehensive review. *Front plant Sci* **2022**, *13*.946717. Available from: https://doi.org/10.3389/fpls.2022.946717.
53. Seleiman, M. F.; Al-Selwey, W. A.; Ibrahim, A. A.; Shady, M.; Alsadon, A. A. Foliar applications of ZnO and SiO2 nanoparticles mitigate water deficit and enhance potato yield and quality traits. *Agronomy* **2023**, *13* (2), 466. Available from: https://doi.org/10.3390/agronomy13020466.
54. Ramadan, T.; Sayed, S. A.; Abd-Elaal, A. K.; Amro, A. The combined effect of water deficit stress and TiOs nanoparticles on cell membrane and antioxidant enzymes in *Helianthus annuus* L. *Physiol Mol Biol Plants* **2022**, *28* (2), 391–409.
55. Thabet, S. G.; Sallam, A.; Moursi, Y. S.; Karam, M. A.; Alqudah, A. M.; Wu, H. Genetic factors controlling nTiO2 nanoparticles stress tolerance in barley (*Hordeum vulgare*) during seed germination and seedling development. *Funct Plant Biol* **2021**, *48* (12), 1288–1301. Available from: https://doi.org/10.1071/FP21129.
56. Al-Khayri, J.M.; The role of nanoparticles in response of plants to abiotic stress at physiological, biochemical, and molecular levels. 2023, 12 (2), 292.
57. Lin, D., et al. An overview of plant phenolic compounds and their importance in human nutrition and management of type 2 diabetes. *Molecules* **2016**, *21* (10), 1374.
58. Rasheed, A.; The role of nanoparticles in plant biochemical, physiological, and molecular responses under drought stress: a review. Front Plant Sci. 2022, 13, https://doi.org/10.3389/fpls.2022.976179.
59. Rehman, A.; Exploring the nano-wonders: unveiling the role of nanoparticles in enhancing salinity and drought tolerance in plants. Front Plant Sci. 2024, 14, https://doi.org/10.3389/fpls.2023.1324176.
60. Khan, I.; Saeed, K. Nanoparticles: Properties, applications and toxicities. *Arab J Chem* **2019**, *12* (7), 908–931. Available from: https://doi.org/10.1016/j.arabjc.2017.05.011.

61. Toksha, B.; Sonawale, V. A. M.; Vanarase, A.; Bornare, D.; Tonde, S.; Hazra, C.; Chatterjee, A. Nanofertilizers: A review on synthesis and impact of their use on crop environment. *Environ Technol & Innov* **2021**, *24*, 101986. Available from: https://doi.org/10.1016/j.eti.2021.101986.
62. Rabalao, T. M. B.; Ndaba.; Roopnarain, A.; Vatsha, B. Towards a circular economy: The influence of extraction methods on phytosynthesis of metallic nanoparticles and their impact on crop growth and protection. *JSFA Rep* **2022**, *2* (5), 208–221. Available from: https://doi.org/10.1002/jsf2.44.
63. Yadav, S. A.; Suvathika, G.; Alghuthaymi, M. A.; Abd-Elsalam, K. A. Fungal-derived nanoparticles for the control of plant pathogens and pests. *Fungal Cell factories Sustain nanomaterials Prod Agric Appl* **2023**, 755–784. Available from: https://doi.org/10.1016/B978-0-323-99922-9.00009-X.
64. Brookes, P. C. The use of microbial parameters in monitoring soil pollution by heavy metals. *Biol Fertil soils* **1995**, *19*, 269–279.
65. Holden, P. A.; Schimel, J. P.; Godwin, H. A. Five reasons to use bacteria when assessing manufactured nanomaterial environmental hazards and fates. *Curr OpBiotechnol* **2014**, *27*, 73–78. Available from: https://doi.org/10.1016/j.copbio.2013.11.008.
66. Prabhakar, P. K.; Lakhanpal, J. Recent advances in the nucleic acid-based diagnostic tool for coronavirus. *Mol Biol Rep* **2020**, *47*, 9033–9041.
67. Sun, H.; Du, W.; Peng, Q.; Lv, Z.; Mao, H.; Kopittke, P. M. Development of ZnO nanoparticles as an efficient Zn fertilizer: using synchrotron-based techniques and laser ablation to examine elemental distribution in wheat grain. *J Agric food Chem* **2020**, *68* (18), 5068–5075. Available from: https://doi.org/10.1021/acs.jafc.0c00084.
68. Tighe-Neira, R.; Carmora, E.; Recio, G.; Nunes-Nesi, A.; Reyes-Diaz, M.; Alberdi, M.; Inostroza-Blancheteau, C. Metallic nanoparticles influence the structure and function of the photosynthetic apparatus in plants. *Plant Physiol Biochem* **2018**, *130*, 408–417. Available from: https://doi.org/10.1016/j.plaphy.2018.07.024.
69. Grillo, R.; Mattos, D. B.; Antunes, D. R.; Forini, M. M.; Monikh, F. A.; Rojas, O. J. Foliage adhesion and interactions with particulate delivery systems for plant nanobionics and intelligent agriculture. *Nano Today* **2021**, *37*, 101078. Available from: https://doi.org/10.1016/j.nantod.2021.101078.
70. Su, Y.; Ashworth, V.; Kim, C.; Adeleye, S. A.; Rolshausen, P.; Roper, C.; Jassby, D. Delivery, uptake, fate, and transport of engineered nanoparticles in plants: a critical review and data analysis. *Environ Science: Nano* **2019**, *6* (8), 2311–2331. Available from: https://doi.org/10.1039/C9EN00461K.
71. Atiq, M.; Mazhar, H. M. R.; Rajput, N. A.; Ahmad, U.; Hameed, A.; Lodhi, A. Green synthesis of silver and copper nanoparticles from leaves of eucalyptus globulus and assessment of its antibacterial potential towards *Xanthomonas citri pv.citri* causing citrus canker. *Appl Ecol. Env Res* **2022**, *20* (3), 2205–2213.
72. Jaberzadeh, A.; Moaveni, P.; Moghadam, H. R. T.; Zahedi, H. Influence of bulk and nanoparticles titanium foliar application on some agronomic traits, seed gluten and starch contents of wheat subjected to water deficit stress. *Not botanicae horti agrobotanici cluj-napoca* **2013**, *41* (1), 201–207. Available from: https://doi.org/10.15835/nbha4119093.
73. Dawood, M. A.; Koshio, S.; Zaineldin, A. I.; Van Doan, H.; Moustafa, E. M.; Abdel-Daim, M. M.; Hassaan, M. S. Dietary supplementation of selenium nanoparticles modulated systemic and mucosal immune status and stress resistance of red sea bream (Pagrus major*). Fish Physiol Biochem* **2019**, *45*, 219–230.

74. Ahmad, A.; Hassim, M. F. N. Effects of silica nanoparticles on morpho-histological and antioxidant activities of rice seedlings under drought stress. *South Afr J Botany* **2024**, *168*, 497–508.
75. Ghorbanian, H.; Janmohammadi, M.; Ebadi-Segherloo, A.; Sabaghnia, N. Genotypic response of barley to exogenous application of nanoparticles under water stress condition. *Annales Universitatis Mariae Curie-Sklodowska, sectio C–Biologia* **2017**, *72* (2). Wydawnictwo Uniwersytetu Marii Curie-Skłodowskiej.
76. Linh, T. M.; Mai, N. C.; Hoe, P. T.; Lien, L. Q.; Ban, N. K.; Hien, L. T. T.; Van, N. T. Metal-based nanoparticles enhance drought tolerance in soybean. *J Nanomaterials* **2020**, *2020*, 1–13. Available from: https://doi.org/10.1155/2020/4056563.
77. Sedghi, M.; Sheikhnavaz Jahed, P.; Gholi-Tolouie, S. Zinc oxide nano particles alleviate drought stress effects on soybean antioxidant system during germination. *Iran J Plant Physiol* **2021**, *11* (4), 3769–3778. Available from: https://doi.org/10.30495/ijpp.2021.684985.
78. Cevik, S. TiO 2 nanoparticles alleviates the effects of drought stress in tomato seedlings. *Bragantia* **2023**, *82*, e20220203. Available from: https://doi.org/10.1590/1678-4499.20220203.
79. Samanta, S.; Seth, C. S.; Roychoudhury, A. The molecular paradigm of reactive oxygen species (ROS) and reactive nitrogen species (RNS) with different phytohormone signaling pathways during drought stress in plants. *Plant Physiol biochemistry: PPB* **2023**, *206*, 108259. Available from: https://doi.org/10.1016/j.plaphy.2023.108259 -108259.
80. Fang, Y.; Guo, Y. Copper-based non-precious metal heterogeneous catalysts for environmental remediation, Cuihua Xuebao/Chinese. *J Catal* **2018**, *39*, 566–582.
81. Dong, J.; Xu, W.; Liu, S.; Du, L.; Chen, Q.; Yang, T.; Gong, Y.; Li, M.; Tan, X.; Liu, Y. Recent advances in applications of nonradical oxidation in water treatment: mechanisms, catalysts and environmental effects. *J Clean Prod* **2021**, *321*. Available from: https://doi.org/10.1016/j.jclepro.2021.128781.
82. Mansas, C.; Atfane-Karfane, L.; Petit, E.; Mendret, J.; Brosillon, S.; Ayral, A. Functionalized ceramic nanofilter for wastewater treatment by coupling membrane separation and catalytic ozonation. *J Env Chem Eng* **2020**, *8*, 104043. Available from: https://doi.org/10.1016/j.jece.2020.104043.
83. Dong, H.; Li, L.; Wang, Y.; Ning, Q.; Wang, B.; Zeng, G. Aging of zero-valent iron-based nanoparticles in an aqueous environment and the consequent effects on their reactivity and toxicity. *Water Env Res* **2020**, *92*, 646–661. Available from: https://doi.org/10.1002/wer.1265.
84. Tada, S.; Fujiwara, K.; Yamamura, T.; Nishijima, M.; Uchida, S.; Kikuchi, R. Flame spray pyrolysis makes highly loaded Cu nanoparticles on ZrO2 for CO2-to-methanol hydrogenation. *Chem Eng J* **2020**, *381*, 122750. Available from: https://doi.org/10.1016/j.cej.2019.122750.
85. Nowack, B. Evaluation of environmental exposure models for engineered nanomaterials in a regulatory context. *Nano Impact* **2017**, *8*, 38–47. Available from: https://doi.org/10.1016/j.impact.2017.06.005.
86. Mitrano, D. M.; Motellier, S.; Clavaguera, S. B.; Nowack. Review of nanomaterial aging and transformations through the life cycle of nano-enhanced products. *Env Int* **2015**, *77*, 132–147. Available from: https://doi.org/10.1016/j.envint.2015.01.013.
87. Zhang, L.; Mazouzi, Y.; Salmain, M.; Liedberg, B.; Boujday, S. Antibody-gold nanoparticle bioconjugates for biosensors: synthesis, characterization and selected applications. *Biosens Bioelectron* **2020**.112370. Available from: https://doi.org/10.1016/j.bios.2020.112370.

88. Cox, A.; Venkatachalam, P.; Sahi, S.; Sharma, N. Reprint of Silver and titanium dioxide nanoparticle toxicity in plants: a review of current research. *Plant Physiol Biochem PPB* **2017**, *110*, 33–49. Available from: https://doi.org/10.1016/j.plaphy.2016.05.022.
89. Dasgupta, N.; Ranjan, S.; Mundekkad, D.; Ramalingam, C.; Shanker, R.; Kumar, A. Nanotechnology in agro-food: from field to plate. *Food Res Int* **2015**, *69*, 381–400. Available from: https://doi.org/10.1016/j.foodres.2015.01.005.
90. Awad, M. A.; Eisa, N. E.; Virk, P.; Hendi, A. A.; Ortashi, K. M. O. O.; Mahgoub, A. S. A.; Elobeid, M. A.; Eissa, F. Z. Green synthesis of gold nanoparticles: preparation, characterization, cytotoxicity, and anti-bacterial activities. *Mater Lett* **2019**, *256*, 126608. Available from: https://doi.org/10.1016/j.matlet.2019.126608Getrightsandcontent.
91. Takeuchi, M. T.; Kojima, M.; Luetzow, M. State of the art on the initiatives and activities relevant to risk assessment and risk management of nanotechnologies in the food and agriculture sectors. *Food Res Int* **2014**, *64*, 976–981. Available from: https://doi.org/10.1016/j.foodres.2014.03.022.
92. Marchiol, L.; Mattiello, A.; Pošćić, F.; Giordano, C.; Musetti, R. In vivo synthesis of nanomaterials in plants: location of silver nanoparticles and plant metabolism. *Nanoscale Res Lett* **2014**, *9*, 101.
93. López-Valdez, F.; FernándezLuqueño, F., Eds. *In Agricultural Nanobiotechnology: Modern Agriculture for a Sustainable Future;* Springer International Publishing: Cham, Switzerland, 2018; pp 3–16. ISBN 978-3-319- 96719-6.
94. Prabhakar, P. K., et al. Formulation and evaluation of polyherbal anti-acne combination by using in-vitro model. *Biointerface Res Appl Chem* **2020**, *10* (1), 4747–4751.
95. Mottier, A. F.; Mouchet.; Pinelli, É.; Gauthier, L.; Flahaut, E. Environmental impact of engineered carbon nanoparticles: from releases to effects on the aquatic biota. *Curr Opin Biotechnol* **2017**, *46*, 1–6. Available from: https://doi.org/10.1016/j.copbio.2016.11.024.
96. Frenk, S.; Ben-Moshe, T.; Dror, I.; Berkowitz, B.; Minz, D. Effect of metal oxide nanoparticles on microbial community structure and function in two different soil types. *PLoS ONE* **2013**, *8*, 84441. Available from: https://doi.org/10.1371/journal.pone.0084441.
97. Lapied, E.; Moudilou, E.; Exbrayat, J. M.; Oughton, D. H.; Joner, E. J. Silver nanoparticle exposure causes apoptotic response in the earthworm Lumbricus terrestris (Oligochaeta). *Nanomedicine* **2010**, *5* (6), 975–984. Available from: https://doi.org/10.2217/nnm.10.58.
98. K. Syamala Devi1, A. Alakanandana2 and V. Vijaya Lakshmi, Vol: I. Issue LVIV, January 2018 ISSN (Print): 2320-5504.
99. UshaRani, P.; Rajasekharreddy, P. Green synthesis of silver-protein (core-shell) nanoparticles using *Piper betle* L. leaf extract and its ecotoxicological studies on Daphnia Magna. *Colloids Surf A Physicochem Eng Asp* **2011**, *389*, 188–194. Available from: https://doi.org/10.1016/j.colsurfa.2011.08.028.
100. Singh, A.; Rajput, V. D.; Agrawal, S.; Ghazaryan, K.; Minkina, T.; Tawaha, A. R. M. A., et al. Nanoparticles mediated salt stress resilience: a holistic exploration of physiological, biochemical, and nano-omics approaches. *Rev Environ Contamination Toxicol* **2024**, *262*. Available from: https://doi.org/10.1007/s44169-024-00070-4.
101. Margaryan, G., et al. Unveiling the salinity tolerance potential of Armenian Dandur (*Portulaca oleracea* L.) genotypes: enhancing sustainable agriculture and food security. *J King Saud Univ - Sci* **2024**, *36*, 103332. Available from: https://doi.org/10.1016/j.jksus.2024.103332.

102. Singh, A.; Sharma, A.; Singh, O.; Rajput, V. D.; Movsesyan, H. S.; Minkina, T., et al. In-depth exploration of nanoparticles for enhanced nutrient use efficiency and abiotic stresses management: present insights and future horizons. *Plant Stress* **2024**, 100576. Available from: https://doi.org/10.1016/j.stress.2024.100576.
103. Sharma, A., et al. Revolutionizing agricultural sustainability and food security and management to achieve SDGs goals via nanotechnology. *Adv Environ Eng green Technol book Ser* **2024**, 276–288. Available from: https://doi.org/10.4018/979-8-3693-1471-5.ch013.
104. Upadhyay, P., et al. Nano-omics-based abiotic and biotic stresses management. *Adv Environ Eng green Technol book Ser* **2024**, 348–371. Available from: https://doi.org/10.4018/979-8-3693-1890-4.ch017.
105. Fathi, A.; Shiade, S. R. G.; Parmoon, G.; Yaghoubian, Y.; Pirdashti, H.; Rajput, V. D., et al. *Bioremediation of heavy metals contaminated soils using nanotechnology*. Elsevier eBooks, 2024611–628*https://doi.org/10.1016/b978-0-443-21610-7.00004-5*.
106. Verma, A.; Singh, A.; Rawat, S.; Rajput, P.; Minkina, T.; Mandzhieva, S., et al. Unveiling nanomaterial-induced toxicity. *Adv Environ Eng green Technol book Ser* **2024**, 236–253. Available from: https://doi.org/10.4018/979-8-3693-1471-5.ch011.
107. Ghazaryan, K.; Verma, A.; Rawat, S.; Rajput, P.; Minkina, T.; Mandzhieva, S., et al. Implementing sustainable nano-bioremediation for emerging pollutants. *Adv Environ Eng green Technol book Ser* **2024**, 316–332. Available from: https://doi.org/10.4018/979-8-3693-1471-5.ch016.

CHAPTER

9 Foliar and soil application dynamics: investigating the transport and mechanism of biogenic nanoparticles in the plant tissues

Shivangee Solanki[1], Shreni Agrawal[1], Indrani Bhattacharya[1], Abhishek Singh[2], Darbinyan Nare[2], Rupesh Kumar Singh[3], Francisco Roberto Quiroz-Figueroa[4], João Ricardo Sousa[3], Henrique Trindade[3], Mohamed S. Elshikh[5], Arunava Pradhan[6,7] and Karen Ghazaryan[2]

[1]*Department of Life Sciences, Parul Institute of Applied Sciences, Parul University, Waghodia, Gujarat, India*

[2]*Faculty of Biology, Yerevan State University, Yerevan, Armenia*

[3]*Centro de Investigação e Tecnologias Agroambientais e Biológicas (CITAB), UTAD, Vila Real, Portugal*

[4]*Laboratorio de Fitomejoramiento Molecular, Centro Interdisciplinario de Investigación para el Desarrollo Integral Regional-Unidad Sinaloa, Instituto Politécnico Nacional, Guasave, Sinaloa, Mexico*

[5]*Department of Botany and Microbiology, College of Science, King Saud University, Riyadh, Saudi Arabia*

[6]*Department of Biology, Centre of Molecular and Environmental Biology (CBMA), Aquatic Research Network (ARNET), Associate Laboratory, University of Minho, Campus of Gualtar, Braga, Portugal*

[7]*Institute for Science and Innovation for Bio-Sustainability (IB-S), University of Minho, Campus of Gualtar, Braga, Portugal*

9.1 Introduction

Nanotechnology, an interdisciplinary field poised at the convergence of physics, chemistry, biology, and engineering, represents a revolutionary leap in scientific innovation with profound implications across diverse sectors. At its core, nanotechnology involves the manipulation of matter at the nanometer scale—typically between 1 and 100 nm—to create structures, devices, and systems with unique properties and functions.[1] The marvels of nanotechnology stem from the ability to engineer materials with unprecedented precision, enabling unprecedented control over their optical, electronic, and mechanical properties. This precision facilitates breakthroughs in fields ranging from medicine, where targeted drug delivery and

Biogenic Nanoparticles. DOI: https://doi.org/10.1016/B978-0-443-33819-9.00013-5

advanced imaging techniques are transforming patient care, to materials science, where the development of nanocomposites and nano-coatings is driving advances in durability and functionality. Additionally, nanotechnology holds promise for addressing critical challenges in energy and environmental sustainability, such as enhancing the efficiency of solar cells and enabling advanced water purification systems. As we continue to explore and harness the potential of nanotechnology, its wonders underscore a transformative shift in our capacity to innovate, solve complex problems, and enhance the quality of life through scientific and technological progress.

But obviously, nanotechnology has pioneered a paradigm shift in the agrarian sector as well. Owning to an array of biological, chemical, electrical, magnetic, optical and physical attributes, nanomaterials have garnered considerable intrigue.[2] In the realm of plant science, nanotechnology offers transformative possibilities, including the development of advanced nanofertilizers that optimize nutrient delivery with minimal environmental impact, and nanopesticides that enhance targeted pest control while reducing the need for broad-spectrum chemical applications. Furthermore, nanoscale sensors and imaging techniques are providing unprecedented insights into plant physiology and stress responses, facilitating real-time monitoring and management of crops. The integration of nanotechnology into agricultural practices promises to enhance crop yield, sustainability, and resilience against abiotic and biotic stresses, ultimately contributing to global food security. As research progresses, the wonders of nanotechnology in plants and agriculture are poised to address critical challenges, driving innovations that blend precision with environmental stewardship to meet the needs of a growing global population.

The promising prospects and pragmatic utilities of NPs (nanoparticles) has been adequately discussed in the scientific literature database, as it has been in previous chapters as well. This scoping work essentially elucidates biomigration and mechanistic action of biogenic nanoparticles in plant tissues, as the efficacy and biocompatibility of nanoparticles is largely attributed to the parameters.

9.1.1 Rudiments of biogenic nanoparticles

The realization via experimental evidences that microorganisms are propitious candidates for fabrication of distinct nanostructures has driven a favorable shift in the preference of biogenic nanoparticles over those synthesized via chemical and physical methods. In light of same, nanoparticles fabricated through biological process pose minimal threat to the environment are nontoxic and least-interfering with their application, economically feasible, provide freedom of customization by means of genetically engineered microbes and conform to principles of sustainability.[3] Furthermore, it is remarkably simple to exercise control over process regulatory parameters, particle geometries and crystallinity in case of biological synthesis.[4] Nevertheless, beneath the glorious perks of biogenic nanoparticles lie major shortcomings including excessive time consumption, cumbersome preparatory phase of microbe cultivation, sluggish biosynthesis rate and inability to produce monodispersed nanoparticles.[3,5] New discoveries concerning biological synthesis of nanoparticles have expanded our comprehension of underlying mechanisms and revealed novel microbial abilities that have opened fresh avenues for nanotechnological applications.[5,6]

Bacteria, algae, and fungi are widely utilized to synthesize NPs and are popularly recognized as microbial bioreactors for production of NPs. Both, intracellular and extracellular synthesis of NPs are mediated by distinct metabolic pathways. During intracellular synthesis, precursor metal ions are transported across the cell wall by virtue of electrostatic interactions between unlike charges of the cell wall and metal ions. Following which, metal ions are subject to the process of nucleation catalyzed by cellular enzymes. Nanoclusters, thus formed, are finally transported out of the cell.[7,8] Similarly, numerous experimental studies were undertaken to elucidate the mechanism of extracellular synthesis of NPs. Precursor metal ions aggregate on the cell surface and are acted on by reducing ions and enzymes to generate NPs.[8] Aforementioned mechanisms are largely observed in bacteria and fungi. NP synthesis in algal systems essentially involves algal extract (prepared at high temperature) and reagents, followed by incubation accompanied with consistent stirring for a set amount of time.[9] Technical factors such as extract preparation and physiological variables (pH, temperature) predominantly regulate the yield of nanoparticles; hence, these control factors ought to be extensively researched to improve process output and scalability.

9.2 Biological entry points for nanoparticles: uptake mechanics

In the intricate web of ecological interactions, the uptake of biogenic nanoparticles through plant organs stands as a captivating phenomenon. Passive diffusion, driven by concentration gradients, facilitates the entry of nanoparticles into roots, exploiting intercellular spaces and water channels such as aquaporins. Active transport mechanisms, including endocytosis, vesicular transport, and protein-mediated translocation, orchestrate precise nanoparticle uptake and distribution within plant tissues. The interplay of ion channels, transporters, and receptor-mediated pathways intricately regulates the uptake and translocation of biogenic nanoparticles, ensuring the integration of these nanomaterials into vital cellular processes, as illustrated in Fig. 9.1.

9.2.1 Leaves: pathways and factors affecting NP absorption

Leaf surfaces are primary interfaces for absorption of exogenously applied (foliar) nanoparticles. The outermost cuticle layer of the leaves is enriched with wax, cutin, and pectin. The waxy stratum corneum is impermeable to external molecules for the most part,[10,11] except for hydrophilic and lipophilic channels littered across its surface. Provided the channels range between 0.6 nm and 4.8 nm in

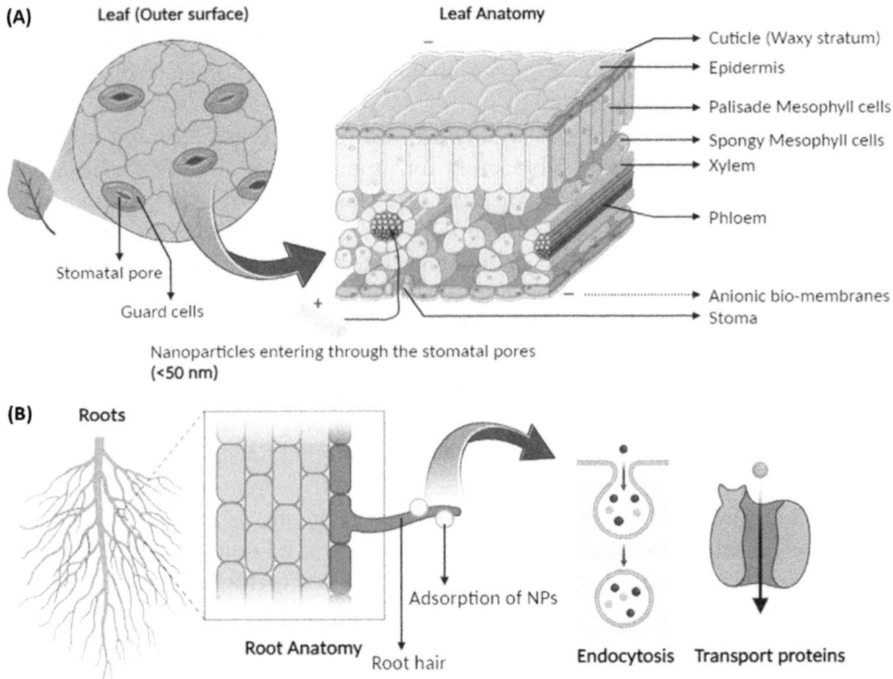

FIGURE 9.1

Mechanisms underlying of uptake of nanoparticles through (A) leaves (foliar application) and (B) roots (soil treatment).

diameter,[12] hydrophobic and/or lipophilic nanoparticles of size less than 4.8 nm can diffuse through their corresponding channels.[13] However, size restricted posed by these pore channels limits the absorption of NPs. Besides the epidermal pathway, foliar NPs are further internalized via the stomatal pathway. The stomatal apparatus is essentially responsible for water regulation and gaseous exchange. It must be noted that construction and configurational geometry of these pores vary across different plant species. Hence, despite stomata serving as potential entry points for NPs, the actual size exclusion limit (SEL) of the stomatal aperture for diffusion of NPs remains unidentified.[13]

While understanding mechanisms underlying uptake of NPs, it is imperative to elucidate the factors governing the rate and efficiency of uptake. An intricate interplay of properties of plant species, NPs (surface charge, size, chemical composition, surface modifications), and the environment fosters optimum uptake of NPs via leaves.[14] In accordance to research findings, metal-derived NPs of size 50 nm or less can easily penetrate the leaf epidermis via the stomatal pathway.[13] Experimental data implies that an inversely proportional relationship exists between NP size and absorption frequency of leaves. Likewise, stomatal diameter is directly linked to the accumulation of NPs within chloroplasts and cytoplasm. A study wherein absorption of Zn-NPs was observed to reduce by 33.2% in chloroplasts and by 8.3% in cytoplasm upon reduction in stomatal diameter of wheat leaf cells, justified the aforementioned claim.[15] Several studies have been undertaken to highlight the significance of NP size in context of their uptake and translocation throughout the plant vascular system, conclusions of which have been summarized as follows. Wheat leaves (*Triticum aestivum*) readily absorb Ag-NPs of a wide range of sizes: 3 nm, 10 nm and 50 nm;[16] however, MgO NPs of size ranging between 27 and 35 nm were absorbed by watermelon (*Citrullus lanatus*). Leaves of *Arabidopsis thaliana* are permeable to SiO_2 NPs of size 54 nm[17] while those of cucumber are permeable to MSNs (mesoporous silica nanoparticles) of size 200–300 nm. Furthermore, 93.6 nm wide polystyrene nanoplastic particles possess the ability to penetrate plant phloem.[18] Correspondingly, such trends have been observed for NP shape and surface charge. Interfacial properties of NPs such as surface area and contact area are largely governed by the shape of NPs.[19,20] While keeping the geometric dimensions constant, rod-shaped NPs were readily absorbed in contrast to spherical NPs.[21] In context of surface charge, both, anionic and cationic NPs are absorbed by plant tissues however, provided the anionic nature of biomembranes, positively-charged NPs are more strongly absorbed than their counter-ions.[20] Owning to a greater number of stomatal pores, entry of NPs is more efficient in dicotyledonous plants than in monocotyledonous species.[22] Additionally, leaves with large extracellular spaces and high stomatal density offer a conducive environment for entry of NPs.[23,24] Apart from the properties, NP absorption is further indirectly influenced by abiotic variables such as temperature, humidity, light, wind velocity, water, and more.[25] Table 9.1 enumerates important properties of NPs and target species, and their effect on uptake of NPs via leaves.

9.2.2 Roots: uptake and factors affecting NP absorption

In case of foliar application of NPs, leaves function as fundamental receptors and site of entry for NPs. However, in case of soil treatment, accessible NPs are predominantly accumulated within the roots. Initially, root hairs facilitate adsorption of NPs onto the surface of roots, which are subsequently internalized via transport proteins (transmembrane channels), the hydrophilic pathway, endocytosis, or physical damage. High efficacy of NP uptake in roots is credited to the anionic nature of root surface, large adsorption interface due to presence of lateral roots,[28] semipermeability of root epidermal cell walls, partially developed epidermis on surface of root hairs and root tips[23] and absence of root exodermis.[29] It must be noted that pores of the root cell wall restrict the entry of large-diameter molecules. However, some NPs possess the ability to induce reconstruction of wider pores to accommodate large NPs.[30] Infiltration of small-diameter NPs is facilitated by the hydrophilic pathway.[14] By virtue of no particle size selectivity, NPs unable to penetrate plant tissues via the hydrophilic pathway are absorbed via endocytosis (invagination of plasma membrane).[31]

Table 9.1 Factors affecting uptake of nanoparticles in leaves and roots.

Sr. no.	Plant organ	Property	Magnitude	Efficiency of NP absorption	Refs.
1.	Leaves	Size of NPs	Increases	Decreases	15
		Positive charge on surface of NPs	Increases	Increases/favored	20
		Negative charge on surface of NPs	Increases	Decreases/comparatively less favored	20
		Number of stomata	Increases	Increases	22
		Stomatal density	Increases	Increases	23,24
		Diameter of stomatal aperture	Increases	Increases	15
2.	Roots	Positive charge on surface of NPs	Increases	Accumulation at epidermis increases, absorption into root tissue decreases	26
		No charge on surface of NPs	–	Accumulation is not affected, absorption increases	27
		Negative charge on surface of NPs	Increases	Accumulation decreases, absorption increases	27

As witnessed during foliar uptake of NPs, a myriad of factors (particularly, size, chemical composition, and surface charge) affects the uptake of NPs via roots. After several imperative evaluations, it was established that metal-based nanoparticles of size less than 10 nm are more efficiently absorbed by those exceeding 100 nm.[13] Interestingly, silicon-based and natural polymer-based nanoparticles are anomalies as they can be absorbed despite being larger than 100 nm.[32] Furthermore, positively charged NPs abundantly accumulate at the epidermis due to its anionic nature and electrostatic attraction, but are unable to enter the root tissues.[26] On the contrary, neutral, and anionic NPs were promptly absorbed, in conformance to findings of O'Reilly and Napier et al.[27] The properties affecting uptake of NPs in roots is tabulated in Table 9.1.

9.3 Biological migration of nanoparticles: transport mechanics

At the core of plant transport are two primary systems: the xylem and the phloem, which serve distinct but complementary roles in the movement of essential substances throughout the plant. The xylem primarily facilitates the unidirectional ascent of water and dissolved minerals from the root system to the aerial parts of the plant. This movement is driven by a combination of root pressure, capillary action, and the transpiration-driven pull from the leaf surfaces, which creates a negative pressure gradient that assists in water transport. In contrast, the phloem is involved in the bidirectional transport of photosynthates, such as sugars, from source tissues (primarily leaves) to sink tissues (such as roots and developing fruits), utilizing pressure flow mechanisms wherein the osmotic pressure gradients generated by active transport of solutes into the phloem sieve tubes drive the bulk flow of nutrient-rich sap. These transport dynamics is regulated by complex cellular and molecular mechanisms, including ion channels, transport proteins, and signal transduction pathways that ensure efficient nutrient uptake, distribution, and utilization. In the xylem, the efficiency of water transport is influenced by vessel structure, hydraulic conductivity, and the regulation of transpiration. In the phloem, the distribution and movement of nutrients are regulated by source-sink relationships, phloem loading and unloading processes, and the integrity of the sieve element-companion cell complex.

The efficiency of nanoparticle delivery and function in plants is intricately linked to the plant's transport mechanisms. Their effectiveness is profoundly influenced by the plant's internal transport processes, specifically those of the xylem and phloem. At the heart of nanoparticle transport lies the intricate interplay between nanomaterials and biological entities. A wide array of studies has detected presence of NPs in leaves, roots, stem, fruits and subcellular compartments post application, suggesting that nanomaterials exhibit the property of biomigration/translocation. Hence, the investigation of pathways and routes adopted by NPs to navigate through the complex physiology of a plant became a subject of great interest and prospect. Additionally, precision of NP-mediated deliveries as well as NPs themselves is largely governed by the efficacy of nanomaterials to migrate from sites of entry (leaves and roots) to the target location without jeopardizing their functional activity (Fig. 9.2).

9.3.1 Transport routes in leaves and stem

Upon internalization and accumulation in the mesophyll cells, if NPs undergo long-distance translocation in plant tissues through extracellular spaces (such as but not limited to cell wall, longitudinal channels between cell walls, xylem, intermediate lamella), it is referred as extracellular transit, and the pathway is identified as extracellular or the plastid pathway.[33] However, if NPs transvers plant tissues by means of intracellular channels and plasmodesmata (typically, 2–20 nm wide), the transit route is known as the intracellular/protoplast pathway.[34] The route is further classified as symplast pathway (characterized by plasmodesmata and cytosol) and apoplast pathway (through cell wall and hindered by impermeable casparian strips). Experimental evidences indicate that NPs smaller than 50 nm are likely to be transported via the plastid pathway and NPs sizing between 50 and 200 nm through the apoplast pathway.[35] Subsequent leaf-to-root migration of NPs (size of about

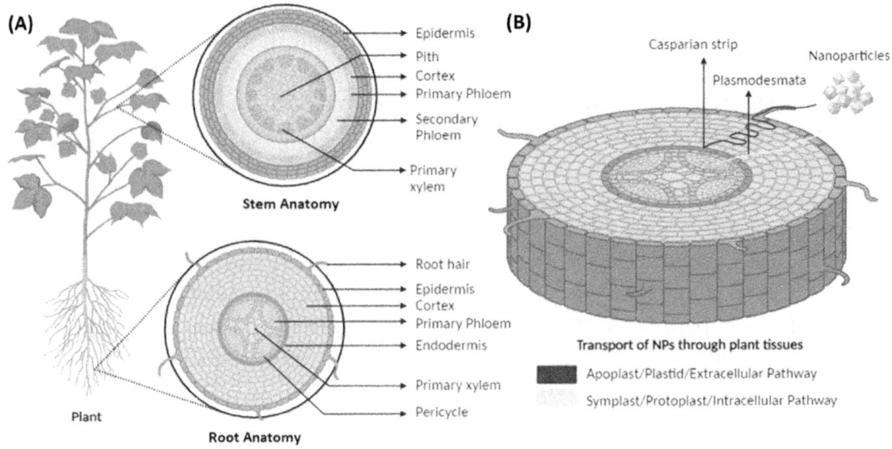

FIGURE 9.2

(A) Plant shoot and root anatomy, depicting all physical barriers crossed by nanoparticles during intrinsic transport. (B) Symplastidic and apoplastic routed adopted by nanoparticles to navigate through plant tissues.

0.405 μm) is ascribed to the vascular column (phloem), by virtue of large diameter of phloem sieve tubes.[36] A study confirmed that NPs, after leaving the leaves, accumulate in the stem. Only then do they make their descend to the roots. Subsequently, NPs travel to other young leaves of the plant.[37] Despite positive change being a hindrance in proficient absorption of NPs, in context of transportation, cationic nature of NPs facilitates easier migration through the extracellular pathway, given electrostatic attraction to the negatively charged cell wall.[38] However, an experimental study employing amino-modified positively charged PS-NH_2 NPs and carboxyl-modified negatively charged PS-COOH NPs indicated that the migration of PS-COOH NPs from leaves to roots via the vascular bundles was favored over that of PS-NH_2 NPs.[19]

9.3.2 Transport routes is roots

Bibliographic and statistical analyzes reveal that the current scientific database holds more literature concerning transport of NPs via roots than leaves. The physiology of the root is divided in four distinct concentric layer: the innermost pericycle along with vascular tissues (xylem and phloem), cortex (consisting of the endothelium and exodermis), and the outermost epidermis.[39] The ascend of NPs from roots to the aerial parts of the plant is facilitated by xylem vessels.[40] The two predominant pathways responsible for NP transit, beginning from the roots, are the exoplast and eutectic pathway. As previously discussed, the exoplast pathway is restricted by an impermeable barrier known as the casparian strip.

However, it is poorly developed in the lateral root junctions thereby allowing NPs to penetrate into the vascular system.[41] The roots present a smaller number of barriers to intrinsic translocation of NPs when compare to leaves. Mucus and certain root secretions serve as the primary, negatively charged barriers to NPs, followed by the casparian strip in the apoplast pathway and biomembranes of the cell and organelles.

9.3.3 Biological action of nanoparticles: functional mechanics

The utilization of nanoparticles in agricultural practices, particularly their application as nanofertilizers, nanopesticides, and nanosensors, has garnered considerable attention due to their potential to enhance crop productivity, mitigate environmental stresses, complement plant growth and facilitate nutrient uptake in plants. However, the intricate cross-talk between nanoparticles and plant systems remains a subject of intense investigation. Nanomaterials can interact with the plant system chemically, mechanically, or genetically, as demonstrated in Fig. 9.3. Once internalized, their mechanistic action is governed by their physicochemical properties such as size, shape, surface charge, and composition. In roots, nanoparticles can interact with cell membranes and cellular structures, potentially altering ion fluxes and cellular signaling pathways. In the xylem and phloem, nanoparticles may affect nutrient transport and distribution, either enhancing or inhibiting the flow of essential substances. Within plant cells, nanoparticles can influence biochemical pathways by interacting with organelles such as chloroplasts, mitochondria, and the endoplasmic reticulum, thereby impacting processes like photosynthesis, respiration, and

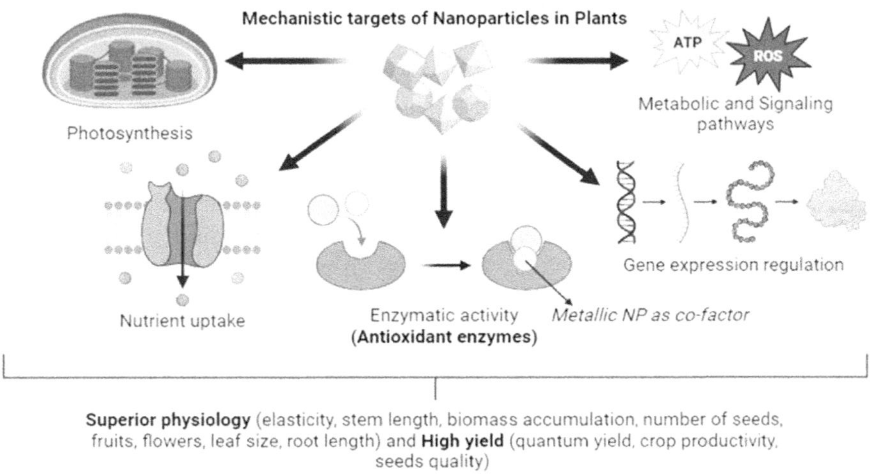

FIGURE 9.3

Mechanisms through which nanoparticles exert their advantageous and favorable effects on plant physiology, metabolism, and productivity.

stress responses. Furthermore, nanoparticles can act as carriers for targeted delivery of bioactive molecules, including growth regulators or genetic materials, enabling precise modulation of plant growth and development. Understanding these mechanistic interactions is crucial for optimizing the application of nanoparticles in agriculture, ensuring their efficacy while minimizing potential disruptions to plant health and development. The following section elucidates mechanistic targets of NPs at the cellular as well as molecular level in order to exert advantageous and favorable effects on plants.

9.3.3.1 Photosynthetic apparatus

Numerous experimental evidences have suggested that photosynthesis is one of the primary targets of NPs. NPs, such as ZnO-NPs, have been indicated to actively participate in the synthesis of photosynthetic pigments such as chlorophyll, as reported by several studies.[42] TiO_2-NPs can enhance the energy conversion efficacy of the light harvesting system thereby increasing light absorption and eventually, photosynthetic yield of plants.[43] Soon after, it was reported that TiO_2-NPs target linear phosphorylation and cycling to augment the supply of photosynthates (improves photosynthetic yield) and, hence, improve plant quality. TiO_2-NPs have been further indicated to stimulate photoreduction activity.[44] ZnO-NPs fostered resistance to drought stress in corn via improvement in the photosynthetic rate and chlorophyll content.[45] In conclusion, NPs exert a comprehensive effect on all aspects of photosynthesis such as quantum yield of photosystems, water splitting, oxygen evolution, ETC (electron transport chain) and associated electron carriers, light absorption (enhanced via induction of plasmon resonance effect), and more.

9.3.3.2 Physiological indices

After several imperative evaluations, NPs have been indicated to induce structural and ultrastructural modifications to the plant physiology to upgrade its quality. NPs immensely alter pant metabolome that physically manifest as physiological traits that can serve as quality indices for plants. Few noteworthy experimental evidences have been summarized as follows. ZnO application has resulted in an increase in the number and weight of cotton bolls. In addition, the quality, strength, and uniformity of the fibers has been observed to substantially improve.[46] Upon in-depth investigation, it was revealed that NPs induce starch synthesis to increase the cellular content of soluble carbohydrates and enhance nutrient uptake (particularly that of zinc and phosphorus in tomato plants) thus, crop yield and quality.[42,47] Similarly, Fe_3O_4-NPs increase cellular concentration of proteins, nutrients, and carbohydrates to improve plant biomass (target crops: corn, wheat).[47–50] Enhancement in biomass accretion by a whopping 60% in spinach plants was credited to TiO_2-NPs and improvement of stem length and fruit weight was ascribed to varying dosages of CeO_2-NPs.[51,52] SiO_2-NPs targeted the leaching process to impede the loss of vital nutrients and enhanced phosphorus and nitrogen accumulation, simultaneously, in order to diminish intrinsic Na content and alleviate salt stress. Physiologically, SiO_2-NPs improved elasticity and promoted cell wall expansion to acclimatize to unfavorable environmental conditions.[53] It can be concluded that the mechanistic action of NPs fundamentally includes improvement of nutrient uptake, restoring nutritional imbalances and sustaining the biomolecule pool in order to invigorate diverse metabolic processes eventually resulting in high yield, enhanced seed germination, photosynthetic productivity, and superior physiology (root and shoot).

9.3.3.3 Defense mechanisms and enzymes

Inorganic cofactors, especially metallic compounds, are pivotal for optimal catalytic efficiency of enzymes. Thus, metal-derived NPs greatly boost activity of enzymes participating in metabolism and/or defense pathways.[54] NPs exhibit the ability to modulate the concentration of heavy metal ions and enhance the activity of antioxidant enzymes to curb adverse repercussions of abiotic stressors. They actively participate in the synthesis of organic acids, root exudates, and phytochelatin, which assist in improving defense responses.[55] Interestingly, besides bolstering enzymatic efficacy, certain NPs ($nCeO_2$, nFe_2O_3, nCo_3O_4, nCuO, nAu, nPt, etc.) exhibit enzyme-like properties. nAu has been indicated to mimic peroxidase, fullerene mimics superoxide dismutase, $nCeO_2$ mimics catalase, and many more NPs possessing mimetic activities have been recorded.[56] A study confirmed elevated activity of SOD (superoxide dismutase), CAT (catalase), APOX (ascorbate peroxidase), and GPOX (guaiacol peroxidase) in spinach upon application of TiO_2-A-NPs.[57] A dearth of extensive research limits our comprehension of effect of NPs on modulation of antioxidant defense system, associated enzymes, and ROS (reactive oxygen species) or oxidative stress with which plant system evades toxicity of free radical stress.[58]

9.3.3.4 Regulation at the molecular level

Genomic investigations have revealed NP interference with regulatory (central dogma) processes such as transcription initiation and elongation, translation, RNA processing, posttranscriptional processes, and protein degradation.[59] In certain investigations, NPs have demonstrated the ability to influence the expression of genes in the face of stress. For instance, NPs downregulate genes coding for heavy metal ion transporter proteins to alleviate heavy metal stress.[55] ZnO-NPs gave been indicated to upregulate 660 genes and downregulate 826 genes in plant model *Arabidopsis thaliana*. Likewise, MWCNT (multiwalled carbon nanotube) application on tomato resulted in upregulation to stress-associated genes.[60] A study employing RT-PCR (reverse-transcriptase polymerase chain reaction) to evaluate molecular responses generated by plants revealed that the expression of a vast array of antioxidant enzyme encoding genes is regulated by NPs.[61] Post Ag-NP application on *Arabidopsis* plants, 286 genes (including those associated with metal stress, oxidative injury, vacuolar proton exchangers, SOD, cytochrome P450-dependent oxidase and POX) were identified to be overexpressed through whole-genome cDNA expression microarray. In addition, 81 defense genes (auxin related, ethylene signaling related and systemic acquired resistance, SAR-related) were detected to be downregulated. This proves the effect of NP activity on plant genome yet suggests that molecular cross-talk of NPs and nucleic acids does not necessarily manifest favorable effects.[59] Recently, inclination toward molecular as well as omics analyzes is considerably growing as physiological traits, enzymatic functionality, synthesis of metabolites and photosynthates can be efficiently monitored by molecular phenomena. Such investigations yield comprehensive outcomes in a small amount of time. Furthermore, molecular reports also offer a bird's eye view of the modifications in biological pathways (citric acid cycle, glycolysis, secondary metabolite synthesis, fatty acid synthesis and oxidation, gluconeogenesis, etc.) fostered by NPs.[62]

9.3.4 Prospects and pitfalls: navigating the green highway

Understanding the transport and mechanisms governing NP journey within plant tissues unveils a realm where nanoscience intersects with plant physiology, presenting opportunities for sustainable agriculture, environmental remediation, and biomedical advancements. The transport and mechanism of biogenic nanoparticles in plant tissues transcend fundamental research, holding profound implications for diverse applications. Beyond the agricultural domain, the ability of plants to accumulate and translocate nanoparticles presents opportunities for environmental remediation, phytoremediation, and green synthesis of nanomaterials. Harnessing the unique properties of biogenic nanoparticles holds promise for pollutant sequestration, soil remediation, and water purification, mitigating the impact of anthropogenic activities on ecosystems. Additionally, studying the transport mechanisms of nanoparticles in plants is crucial for advancing agricultural and environmental sciences. Understanding how nanoparticles move through plant tissues informs the development of novel strategies for enhancing nutrient delivery, improving pest control, and remediating contaminated soils. By elucidating the pathways and processes involved in nanoparticle uptake, translocation, and accumulation, researchers can tailor nanoparticles for specific functions, such as targeted delivery of fertilizers or therapeutic agents. Additionally, this knowledge helps assess the potential risks associated with nanoparticle exposure, ensuring that their use does not adversely affect plant health or ecosystems. The insights gained from such studies can also guide the design of environmentally friendly nanomaterials, ultimately contributing to sustainable agricultural practices and environmental management.

Undertaking research pertaining to transport and mechanism of nanoparticles in plants poses several challenges, ranging from technical limitations to conceptual complexities. The current pool of scientific literature is deemed to be insufficient for transformative advancements to develop. In the light of same, absorption, transportation, and distribution of natural polymer NPs within plants remains untapped. Accurately characterizing nanoparticles in complex biological matrices remains a formidable task. Nanoparticle size, shape, surface chemistry, and aggregation state can influence their behavior within plant tissues. Owning to their unique chemical and mechanical attributes, NPs interact with multiple cellular and molecular targets, making it challenging for scientists and researchers to undertake explorations concerning uptake, transport, and mechanisms of NPs.[4,6] Visualizing nanoparticle uptake and distribution within plant tissues requires advanced imaging techniques with high spatial and temporal resolution. Live-cell imaging techniques offer real-time insights but face challenges in penetrating thick plant tissues and differentiating nanoparticles from endogenous structures.[63] Moreover, bioaccumulation of nanoparticles in edible plant parts raises concerns regarding food safety and human health, necessitating rigorous assessment of nanoparticle toxicity and bioavailability. Assessing nanoparticle toxicity, bioaccumulation, and long-term effects on ecosystems requires comprehensive risk assessments and adherence to ethical guidelines.[64–68] Balancing scientific exploration with environmental stewardship and societal responsibility is essential to ensure the responsible development and application of nanotechnology in plant science. Achieving consistency and reproducibility in nanoparticle uptake studies across different plant species, growth conditions, and experimental setups is challenging. Standardizing experimental protocols, reporting guidelines, and data analysis procedures enhances comparability between studies and facilitates the accumulation of robust scientific evidence.

9.4 Concluding remarks

The investigation into the transport and mechanism of biogenic nanoparticles within plant tissues reveals significant insights into the dynamic interactions between nanoparticles and plant systems. This chapter highlights the complex pathways through which biogenic nanoparticles traverse plant tissues, emphasizing the role of physiological and biochemical factors in influencing their movement and accumulation. The research demonstrates that biogenic nanoparticles, synthesized from natural sources, exhibit distinct transport behaviors compared to synthetic nanoparticles. Their interaction with plant tissues is largely mediated by both passive diffusion and active transport mechanisms, which are facilitated by the plant's vascular and cellular structures. The uptake of these nanoparticles is influenced by their size, surface charge, and the plant's metabolic activity, suggesting a nuanced relationship between nanoparticle properties and plant tissue permeability. Furthermore, the study reveals that biogenic nanoparticles can be systematically transported through plant tissues, with notable accumulation in specific cellular compartments. This transport mechanism appears to be modulated by the plant's intrinsic regulatory systems, which could potentially be harnessed to enhance nutrient delivery or bioremediation processes. These mechanisms underscore the potential of biogenic nanoparticles in agricultural and environmental applications, provided that their behavior within plant systems is thoroughly understood and managed.

Overall, the insights gained from this work contribute to a deeper understanding of nanoparticle–plant interactions and pave the way for future research aimed at optimizing the use of biogenic nanoparticles in various applications. The uptake and transport mechanisms of biogenic nanoparticles in plant tissues represents a frontier of interdisciplinary research, bridging the realms of nanoscience, plant biology, and environmental science. Through a comprehensive and sophisticated approach, informed by cutting-edge scientific methodologies and theoretical frameworks, we can unravel the intricacies of nanoparticle–plant interactions and harness their potential for advancing sustainable agriculture and environmental stewardship in the anthropocene era.

Acknowledgment

Higher education and science committee MESCS Republic of Armenia provided the 24FP-4C013 grant to support KG.

References

1. Sharma, V. K.; Yngard, R. A.; Lin, Y. Silver nanoparticles: green synthesis and their antimicrobial activities. *Adv. Colloid Interface Sci.* **2009**, *145*, 83–96.
2. Daniel, M. C.; Astruc, D. Gold Nanoparticles: assembly, supramolecular chemistry, quantum-size-related properties, and applications toward biology, catalysis, and nanotechnology. *Chem. Rev.* **2004**, *104*, 293–346.
3. Singh, J.; Dutta, T.; Kim, K. H.; Rawat, M.; Samddar, P.; Kumar, P. "Green" synthesis of metals and their oxide nanoparticles: applications for environmental remediation. *J. Nanobiotechnol.* **2018**, *16*, 1–24.
4. Dhand, C.; Dwivedi, N.; Loh, X. J.; Jie Ying, A. N.; Verma, N. K.; Beuerman, R. W.; Lakshminarayanan, R.; Ramakrishna, S. Methods and strategies for the synthesis of diverse nanoparticles and their applications: a comprehensive overview. *RSC Adv.* **2015**, *5*, 105003–105037.
5. Mabey, T.; Andrea Cristaldi, D.; Oyston, P.; Lymer, K. P.; Stulz, E.; Wilks, S.; William Keevil, C.; Zhang, X. Bacteria and nanosilver: the quest for optimal production. *Crit. Rev. Biotechnol.* **2019**, *39*, 272–287.
6. Mughal, B.; Zaidi, S. Z.; Zhang, X.; Hassan, S. U. Biogenic nanoparticles: synthesis, characterisation and applications. *Appl. Sci.* **2021**, *11* (6), 2598.
7. Mathivanan, K.; Selva, R.; Chandirika, J.U.; Govindarajan, R.K.; Srinivasan, R.; Annadurai, G.; Duc, P.A. Biologically synthesized silver nanoparticles against pathogenic bacteria: Synthesis, calcination and characterization. Biocatal. Agric. Biotechnol.
8. Markus, J.; Mathiyalagan, R.; Kim, Y. J.; Abbai, R.; Singh, P.; Ahn, S.; Perez, Z. E. J.; Hurh, J.; Yang, D. C. Intracellular synthesis of gold nanoparticles with antioxidant activity by probiotic Lactobacillus kimchicus DCY51T isolated from Korean kimchi. *Enzym. Microb. Technol.* **2016**, *95*, 85–93.

9. Sharma, A.; Sharma, S.; Sharma, K.; Chetri, S. P. K.; Vashishtha, A.; Singh, P.; Kumar, R.; Rathi, B.; Agrawal, V. Algae as crucial organisms in advancing nanotechnology: a systematic review. *J. Appl. Phycol.* **2016**, *28*, 1759−1774.
10. Yang, C.; Powell, C. A.; Duan, Y.; Shatters, R.; Zhang, M. Antimicrobial nanoemulsion formulation with improved penetration of foliar spray through citrus leaf cuticles to control citrus huanglongbing. *PLoS ONE* **2015**, *10*, e0133826.
11. Perez-de-Luque, A. Interaction of nanomaterials with plants: what do we need for real applications in agriculture? *Front. Environ. Sci.* **2017**, *5*, 12.
12. Eichert, T.; Kurtz, A.; Steiner, U.; Goldbach, H. E. Size exclusion limits and lateral heterogeneity of the stomatal foliar uptake pathway for aqueous solutes and water-suspended nanoparticles. *Physiol. Plant.* **2008**, *134*, 151−160.
13. Banerjee, K.; Pramanik, P.; Maity, A.; Joshi, D. C.; Wani, S. H.; Krishnan, P. Chapter 4—methods of using nanomaterials to plant systems and their delivery to plants (mode of entry, uptake, translocation, accumulation, biotransformation and barriers). In *Advances in Phytonanotechnology;* Ghorbanpour, M., Wani, S. H., Eds.; Academic Press: Cambridge, MA, USA, 2019; pp 123−152.
14. Li, L.; Luo, Y.; Li, R.; Zhou, Q.; Peijnenburg, W. J. G. M.; Yin, N.; Yang, J.; Tu, C.; Zhang, Y. Effective uptake of submicrometre plastics by crop plants via a crack-entry mode. *Nat. Sustain.* **2020**, *3*, 929−937.
15. Zhu, J.; Li, J.; Shen, Y.; Liu, S.; Zeng, N.; Zhan, X.; White, J. C.; Gardea-Torresdey, J.; Xing, B. Mechanism of zinc oxide nanoparticle entry into wheat seedling leaves. *Environ. Sci.-Nano* **2020**, *7*, 3901−3913.
16. Avellan, A.; Yun, J.; Zhang, Y.; Spielman-Sun, E.; Unrine, J. M.; Thieme, J.; Li, J.; Lombi, E.; Bland, G.; Lowry, G. V. Nanoparticle size and coating chemistry control foliar uptake pathways, translocation, and leaf-to-rhizosphere transport in wheat. *ACS Nano* **2019**, *13*, 5291−5305.
17. El-Shetehy, M.; Moradi, A.; Maceroni, M.; Reinhardt, D.; Petri-Fink, A.; Rothen-Rutishauser, B.; Mauch, F.; Schwab, F. Silica nanoparticles enhance disease resistance in Arabidopsis plants. *Nat. Nanotechnol.* **2021**, *16*, 344−353.
18. Lian, J.; Liu, W.; Meng, L.; Wu, J.; Chao, L.; Zeb, A.; Sun, Y. Foliar-applied polystyrene nanoplastics (PSNPs) reduce the growth and nutritional quality of lettuce (*Lactuca sativa* L.). *Environ. Pollut.* **2021**, *280*, 116978.
19. Sun, H.; Lei, C.; Xu, J.; Li, R. Foliar uptake and leaf-to-root translocation of nanoplastics with different coating charge in maize plants. *J. Hazard. Mater.* **2021**, *416*, 125854.
20. Zhu, J.; Wang, J.; Zhan, X.; Li, A.; White, J. C.; Gardea-Torresdey, J. L.; Xing, B. Role of charge and size in the translocation and distribution of zinc oxide particles in wheat cells. *Acs Sustain. Chem. Eng.* **2021**, *9*, 11556−11564.
21. Zhang, H.; Goh, N. S.; Wang, J. W.; Pinals, R. L.; González-Grandío, E.; Demirer, G. S.; Butrus, S.; Fakra, S. C.; Del Rio Flores, A.; Zhai, R., et al. Nanoparticle cellular internalization is not required for RNA delivery to mature plant leaves. *Nat. Nanotechnol.* **2022**, *17*, 197−205.
22. Schwabe, F.; Schulin, R.; Limbach, L. K.; Stark, W.; Bürge, D.; Nowack, B. Influence of two types of organic matter on interaction of CeO2 nanoparticles with plants in hydroponic culture. *Chemosphere* **2013**, *91*, 512−520.
23. Rajput, V.; Minkina, T.; Mazarji, M.; Shende, S.; Sushkova, S.; Mandzhieva, S.; Burachevskaya, M.; Chaplygin, V.; Singh, A.; Jatav, H. Accumulation of nanoparticles in the soil-plant systems and their effects on human health. *Ann. Agric. Sci.* **2020**, *65*, 137−143.

24. Hu, P.; An, J.; Faulkner, M. M.; Wu, H.; Li, Z.; Tian, X.; Giraldo, J. P. Nanoparticle charge and size control foliar delivery efficiency to plant cells and organelles. *ACS Nano* **2020**, *14*, 7970–7986.
25. Fan, X.; Cao, X.; Zhou, H.; Hao, L.; Dong, W.; He, C.; Xu, M.; Wu, H.; Wang, L.; Chang, Z., et al. Carbon dioxide fertilization effect on plant growth under soil water stress associates with changes in stomatal traits, leaf photosynthesis, and foliar nitrogen of bell pepper (*Capsicum annuum* L.). *Environ. Exp. Bot.* **2020**, *179*, 104203.
26. Bosker, T.; Bouwman, L. J.; Brun, N. R.; Behrens, P.; Vijver, M. G. Microplastics accumulate on pores in seed capsule and delay germination and root growth of the terrestrial vascular plant *Lepidium sativum*. *Chemosphere* **2019**, *226*, 774–781.
27. Parkinson, S. J.; Tungsirisurp, S.; Joshi, C.; Richmond, B. L.; Gifford, M. L.; Sikder, A.; Lynch, I.; O'Reilly, R. K.; Napier, R. M. Polymer nanoparticles pass the plant interface. *Nat. Commun.* **2022**, *13*, 7385.
28. Peng, C.; Duan, D.; Xu, C.; Chen, Y.; Sun, L.; Zhang, H.; Yuan, X.; Zheng, L.; Yang, Y.; Yang, J., et al. Translocation and biotransformation of CuO nanoparticles in rice (*Oryza sativa* L.) plants. *Environ. Pollut.* **2015**, *197*, 99–107.
29. Péret, B.; De Rybel, B.; Casimiro, I.; Benková, E.; Swarup, R.; Laplaze, L.; Beeckman, T.; Bennett, M. J. Arabidopsis lateral root development: an emerging story. *Trends Plant. Sci.* **2009**, *14*, 399–408.
30. Ali, S.; Mehmood, A.; Khan, N. Uptake, translocation, and consequences of nanomaterials on plant growth and stress adaptation. *J. Nanomater.* **2021**, *2021*, 6677616.
31. Ma, X.; Yan, J. Plant uptake and accumulation of engineered metallic nanoparticles from lab to field conditions. *Curr. Opin. Environ. Sci. Health* **2018**, *6*, 16–20.
32. Slomberg, D. L.; Schoenfisch, M. H. Silica nanoparticle phytotoxicity to *Arabidopsis thaliana*. *Environ. Sci. Technol.* **2012**, *46*, 10247–10254.
33. Chen, R.; Ratnikova, T. A.; Stone, M. B.; Lin, S.; Lard, M.; Huang, G.; Hudson, J. S.; Ke, P. C. Differential uptake of carbon nanoparticles by plant and mammalian cells. *Small* **2010**, *6*, 612–617.
34. Cornu, J. Y.; Bussiere, S.; Coriou, C.; Robert, T.; Maucourt, M.; Deborde, C.; Moing, A.; Nguyen, C. Changes in plant growth, Cd partitioning and xylem sap composition in two sunflower cultivars exposed to low Cd concentrations in hydroponics. *Ecotoxicol. Environ. Saf.* **2020**, *205*, 111145.
35. Raliya, R.; Saharan, V.; Dimkpa, C.; Biswas, P. Nanofertilizer for precision and sustainable agriculture: current state and future perspectives. *J. Agric. Food Chem.* **2018**, *66*, 6487–6503.
36. Bussières, P. Estimating the number and size of phloem sieve plate pores using longitudinal views and geometric reconstruction. *Sci. Rep.* **2014**, *4*, 4929.
37. Abdel-Aziz, H. M. M.; Hasaneen, M. N. A.-G.; Omer, A. M. Foliar application of nano chitosan NPK fertilizer improves the yield of wheat plants grown on two different soils. *Egypt. J. Exp. Biol.* **2018**, *14*, 63–72.
38. Azeem, I.; Adeel, M.; Ahmad, M. A.; Shakoor, N.; Jiangcuo, G. D.; Azeem, K.; Ishfaq, M.; Shakoor, A.; Ayaz, M.; Xu, M., et al. Uptake and accumulation of nano/microplastics in plants: a critical review. *Nanomaterials* **2021**, *11*, 2935.
39. Su, Y.; Ashworth, V.; Kim, C.; Adeleye, A. S.; Rolshausen, P.; Roper, C.; White, J.; Jassby, D. Delivery, uptake, fate, and transport of engineered nanoparticles in plants: a critical review and data analysis. *Environ. Sci.-Nano* **2019**, *6*, 2311–2331.

40. Tripathi, D. K.; Shweta.; Singh, S.; Singh, S.; Pandey, R.; Singh, V. P.; Sharma, N. C.; Prasad, S. M.; Dubey, N. K.; Chauhan, D. K. An overview on manufactured nanoparticles in plants: uptake, translocation, accumulation and phytotoxicity. *Plant. Physiol. Biochem.* **2017**, *110*, 2–12.
41. Schwab, F.; Zhai, G.; Kern, M.; Turner, A.; Schnoor, J. L.; Wiesner, M. R. Barriers, pathways and processes for uptake, translocation and accumulation of nanomaterials in plants—critical review. *Nanotoxicology* **2016**, *10*, 257–278.
42. Bala, R.; Kalia, A.; Dhaliwal, S. S. Evaluation of efficacy of ZnO nanoparticles as remedial zinc nanofertilizer for rice. *J. Soil. Sci. Plant. Nutr.* **2019**, *19*, 379–389.
43. Su, M.; Liu, H.; Liu, C.; Qu, C.; Zheng, L.; Hong, F. Promotion of nano-anatase TiO2 on the spectral responses and photochemical activities of D1/D2/Cyt b559 complex of spinach. Spectrochim. *Acta Part. A-Mol. Biomol. Spectrosc.* **2009**, *72*, 1112–1116.
44. Thabet, S. G.; Sallam, A.; Moursi, Y. S.; Karam, M. A.; Alqudah, A. M. Genetic factors controlling nTiO2 nanoparticles stress tolerance in barley (*Hordeum vulgare*) during seed germination and seedling development. *Funct. Plant. Biol.* **2021**, *48*, 1288–1301.
45. Sun, L.; Song, F.; Zhu, X.; Liu, S.; Liu, F.; Wang, Y.; Li, X. Nano-ZnO alleviates drought stress via modulating the plant water use and carbohydrate metabolism in maize. *Arch. Agron. Soil. Sci.* **2021**, *67*, 245–259.
46. Vaghar, M. S.; Sayfzadeh, S.; Zakerin, H. R.; Kobraee, S.; Valadabadi, S. A. Foliar application of iron, zinc, and manganese nanochelates improves physiological indicators and soybean yield under water deficit stress. *J. Plant. Nutr.* **2020**, *43*, 2740–2756.
47. Faizan, M.; Bhat, J. A.; Chen, C.; Alyemeni, M. N.; Wijaya, L.; Ahmad, P.; Yu, F. Zinc oxide nanoparticles (ZnO-NPs) induce salt tolerance by improving the antioxidant system and photosynthetic machinery in tomato. *Plant. Physiol. Biochem.* **2021**, *161*, 132–140.
48. Choukri, M.; Abouabdillah, A.; Bouabid, R.; Abd-Elkader, O. H.; Pacioglu, O.; Boufahja, F.; Bourioug, M. Zn application through seed priming improves productivity and grain nutritional quality of silage corn. *Saudi J. Biol. Sci.* **2022**, *29*, 103456.
49. Tondey, M.; Kalia, A.; Singh, A.; Dheri, G. S.; Taggar, M. S.; Nepovimova, E.; Krejcar, O.; Kuca, K. Seed Priming and coating by nano-scale zinc oxide particles improved vegetative growth, yield and quality of fodder maize (*Zea mays*). *Agronomy* **2021**, *11*, 729.
50. Armin, M.; Akbari, S.; Mashhadi, S. Effect of time and concentration of nano-Fe foliar application on yield and yield components of wheat. *Int. J. Biosci.* **2014**, *4*, 69–75.
51. Gao, G.; Tester, M. A.; Julkowska, M. M. The use of high-throughput phenotyping for assessment of heat stress-induced changes in Arabidopsis. *Plant. Phenomics* **2020**, *2020*, 3723916.
52. Nadeem, M.; Khan, R.; Afridi, K.; Nadhman, A.; Ullah, S.; Faisal, S.; Ul Mabood, Z.; Hano, C.; Abbasi, B. H. Green synthesis of cerium oxide nanoparticles (CeO2 NPs) and their antimicrobial applications: a review. *Int. J. Nanomed.* **2020**, *15*, 5951–5961.
53. Zhao, P.; Yuan, W.; Xu, C.; Li, F.; Cao, L.; Huang, Q. Enhancement of spirotetramat transfer in cucumber plant using mesoporous silica nanoparticles as carriers. *J. Agric. Food Chem.* **2018**, *66*, 11592–11600.
54. Hussein, M. M.; Abou-Baker, N. H. The contribution of nano-zinc to alleviate salinity stress on cotton plants. *R. Soc. Open. Sci.* **2018**, *5*, 171809.

55. Lian, J.; Zhao, L.; Wu, J.; Xiong, H.; Bao, Y.; Zeb, A.; Tang, J.; Liu, W. Foliar spray of TiO_2 nanoparticles prevails over root application in reducing Cd accumulation and mitigating Cd-induced phytotoxicity in maize (*Zea mays* L.). *Chemosphere* **2020**, *239*, 124794.
56. Wei, H.; Wang, E. Nanomaterials with enzyme-like characteristics (nanozymes): next-generation artificial enzymes. *Chem. Soc. Rev.* **2013**, *42*, 6060–6093.
57. Lei, Z.; Mingyu, S.; Xiao, W., et al. Antioxidant stress is promoted by nano-anatase in spinach chloroplasts under UV-B radiation. *Biol. Trace Elem. Res.* **2008**, *121*, 69–79.
58. Rico, C. M.; Peralta-Videa, J. R.; Gardea-Torresdey, J. L. Chemistry, biochemistry of nanoparticles, and their role in antioxidant defense system in plants. *Nanotechnology and Plant Sciences* **2015**, 1–7.
59. Abideen, Z.; Hanif, M.; Munir, N.; Nielsen, B. L. Impact of nanomaterials on the regulation of gene expression and metabolomics of plants under salt stress. *Plants* **2022**, *11* (5), 691.
60. Khodakovskaya, M.; Dervishi, E.; Mahmood, M.; Xu, Y.; Li, Z.; Watanabe, F.; Biris, A. S. Carbon nanotubes are able to penetrate plant seed coat and dramatically affect seed germination and plant growth. *ACS Nano* **2009**, *3*, 3221–3227.
61. Mazumdar, H.; Ahmed, G. Phytotoxicity effect of silver nanoparticles on *Oryza sativa*. *Int. J. ChemTech Res.* **2011**, *3*, 1494–1500.
62. Mhlongo, M. I.; Piater, L. A.; Madala, N. E.; Labuschagne, N.; Dubery, I. A. The chemistry of plant–microbe interactions in the rhizosphere and the potential for metabolomics to reveal signaling related to defense priming and induced systemic resistance. *Front. Plant. Sci.* **2018**, *9*, 112–121.
63. Avellan, A.; Schwab, F.; Masion, A.; Chaurand, P.; Borschneck, D.; Vidal, V.; Rose, J.; Santaella, C.; Levard, C. Nanoparticle uptake in plants: gold nanomaterial localized in roots of Arabidopsis thaliana by X-ray computed nanotomography and hyperspectral imaging. *Environ. Sci. & Technol.* **2017**, *51* (15), 8682–8691.
64. Wang, X.; Xie, H.; Wang, P.; Yin, H. Nanoparticles in plants: uptake, transport and physiological activity in leaf and root. *Materials* **2023**, *16* (8), 3097.
65. Ghazaryan, K.; Verma, A.; Rawat, S.; Rajput, P.; Minkina, T.; Mandzhieva, S., et al. Implementing sustainable nano-bioremediation for emerging pollutants. *Adv. Environ. Eng. green. Technol. book. Ser.* **2024**, 316–332. Available from: https://doi.org/10.4018/979-8-3693-1471-5.ch016.
66. Singh, A.; Rajput, V. D.; Agrawal, S.; Ghazaryan, K.; Minkina, T.; Tawaha, A. R. M. A., et al. Nanoparticles mediated salt stress resilience: a holistic exploration of physiological, biochemical, and nano-omics approaches. *Rev. Environ. Contamination Toxicol.* **2024**, *262*. Available from: https://doi.org/10.1007/s44169-024-00070-4.
67. Margaryan, G., et al. Unveiling the salinity tolerance potential of Armenian Dandur (*Portulaca oleracea* L.) genotypes: enhancing sustainable agriculture and food security. *J. King Saud. Univ. - Sci.* **2024**, *36*, 103332. Available from: https://doi.org/10.1016/j.jksus.2024.103332.
68. Singh, A.; Sharma, A.; Singh, O.; Rajput, V. D.; Movsesyan, H. S.; Minkina, T., et al. In-depth exploration of nanoparticles for enhanced nutrient use efficiency and abiotic stresses management: present insights and future horizons. *Plant. Stress.* **2024**, 100576. Available from: https://doi.org/10.1016/j.stress.2024.100576.

CHAPTER

10 Adaptation and resilience: plants and microorganism responding to biogenic nanomaterial

Morishwala Mohammad Farhan[1], Shreni Agrawal[1], Indrani Bhattacharya[1], Abhishek Singh[2], Anna Harutyunyan[2], Hasmik Movsesyan[2], Rupesh Kumar Singh[3], Francisco Roberto Quiroz-Figueroa[4], João Ricardo Sousa[3], Henrique Trindade[3], Mohamed S. Elshikh[5], Arunava Pradhan[6,7] and Karen Ghazaryan[2]

[1]*Department of Life Sciences, Parul Institute of Applied Sciences, Parul University, Waghodia, Gujarat, India*

[2]*Faculty of Biology, Yerevan State University, Yerevan, Armenia*

[3]*Centro de Investigação e Tecnologias Agroambientais e Biológicas (CITAB), UTAD, Vila Real, Portugal*

[4]*Laboratorio de Fitomejoramiento Molecular, Centro Interdisciplinario de Investigación para el Desarrollo Integral Regional-Unidad Sinaloa, Instituto Politécnico Nacional, Guasave, Sinaloa, Mexico*

[5]*Department of Botany and Microbiology, College of Science, King Saud University, Riyadh, Saudi Arabia*

[6]*Department of Biology, Centre of Molecular and Environmental Biology (CBMA), Aquatic Research Network (ARNET), Associate Laboratory, University of Minho, Campus of Gualtar, Braga, Portugal*

[7]*Institute for Science and Innovation for Bio-Sustainability (IB-S), University of Minho, Campus of Gualtar, Braga, Portugal*

10.1 Introduction

Biogenic nanomaterials (BNMs) are nano-sized materials produced by organisms or biological processes. These materials have unique properties that make them useful in a variety of applications including pharmaceuticals, electronics and environmental mitigation.[1] BNMs include metallic nanoparticles, magnetic nanoparticles, and carbon-based nanomaterials. Metallic nanoparticles, like gold and silver, have antibacterial, antiviral, and anti-inflammatory properties, useful for medical applications. Magnetic nanoparticles, like magneto tactic bacteria, are superparamagnetic, useful for drug delivery and magnetic separation techniques.[2] Biogenic

Biogenic Nanoparticles. DOI: https://doi.org/10.1016/B978-0-443-33819-9.00014-7

nanoparticles have gained attention in various fields such as health, agriculture, and environment due to their unique properties and potential applications but it is important to consider that surface capping can affect parameters such as particle morphology, aggregation, and dissolution rate. In order to determine characteristics, it is necessary to analyze the toxicity to different cell types and non-target organisms from the release of the products into the environment.[3]

The current state of climate change, which is producing broad consequences including droughts, temperature rises, and changes in precipitation patterns, highlights the significance of adaptation and resilience in plants and microbes. Changes in the environment can have an influence on agricultural output by changing the makeup and function of soil microbial communities.[4] The microbes and plants have developed defense mechanisms against environmental stresses like drought. Through chemical signals, plants may draw in "stress microbiomes" that help them become more drought-tolerant by increasing the amount of water they consume, the amount of water in their plants, the concentration of CO_2, the amount of chlorophyll, and on fruiting plants. Under subdrought circumstances, the prevalence of bacteria resistant to drought may also rise.[5]

Recent studies have shown that some soil microorganisms can adapt to drought independently of plant signals. These changes could help the benefits of plants introduced into previously uncultivated, drought-prone land, providing potential strategies for plant survival in drought-prone areas. Intensive soil management practices such as regular application of fertilizers and lime also play a role in how drought affects soil microbial communities as intensive management can lead to bacterial dominance of fungi after drought, which will determine plant diversity availability, biological tolerance to drought, and carbon storage.[4,6] Because biogenic nanomaterial contact holds the potential to improve nanoparticle functioning and performance in a variety of applications, it is very important. Biological systems, including fungus, bacteria, and plants, are used to create biogenic nanoparticles, which provide a more environmentally friendly and sustainable technique of synthesis than conventional chemical approaches.[7]

Researchers have indicated that distinct features, including size, shape, and composition, of biogenic nanoparticles may be adjusted by modifying the production circumstances or the biological system. In the context of nanomedicine, biosensors, and environmental remediation, in particular, these characteristics are crucial in defining how biogenic nanoparticles interact with other materials.[8] Additionally, waste materials and biomolecules can be used to synthesize biogenic nanoparticles, which can cut expenses and minimize the need for dangerous chemicals. This strategy was used in a recent work that created a unique biogenic synthesis technique for TiO_2 nanoparticles by employing an organic reducing agent and stabilizing agent in an aqueous extract of *Annona muricata* L. The resultant nanoparticles showed improved activity in the breakdown of organic dyes together with great biocompatibility and crystallinity.[9]

10.2 Adaptation mechanisms in plants

Adaptation strategies in plants may be elevated through the incorporation of BNMs. Nanoparticles, for example, metal–metal oxide nanoparticles, indicate promising outlooks in agriculture, proposing a potential remedy to lessen the adverse effects of abiotic stresses.[10] Employing a sustainable approach, particularly through green or biologically mediated synthesis of nanoparticles, facilitates an environmentally friendly production process. This, in turn, aids in modulating phytohormone signaling to bolster stress resistance mechanisms in plants. The utilization of these BNMs assumes a pivotal role in enhancing crop yield, promoting sustainability, optimizing resource utilization, and fostering environmental conservation.[11] The relief of the harmful effects of hyperreactive oxygen species ROS during abiotic stress situations may be achieved by using plant-based nanoparticles to manage redox balance and regulate oxidative stress.[12] Exploration of the dynamics of plant–nanoparticle interactions, as illustrated in a study released in the journal *Agronomy* in 2023, underlines the potential of such interactions to trigger various transcriptomic and proteomic modifications.[13] These alterations, in turn, influence various facets of plant physiology, ranging from growth and development to adaptation in intricate environmental settings. Ultimately, these interactions support enhanced plant fitness and resilience to environmental adversities.[14]

10.2.1 Nanoparticle uptake and transport

The absorption and transport of nanoparticles in plants is a multifaceted process that can impact several aspects of plant growth and development. Several mechanisms, including the root system and the leaf stomatal apertures, can be used by plants to absorb NPs. Through the xylem and phloem, nanoparticles can be translocated and delivered to various plant tissues, including the root system, stems, and leaves.[15] The amount, shape, charge, and surface chemistry of the NPs, together with the kind of plant and growing environment, all influence the mechanism of NP absorption and transport in plants. The physicochemical characteristics of the earth's soil and the rhizosphere, such as the pH, ionic strength, and organic matter content, might affect the absorption of Nanoparticles.[16] Vascular bundles and cell-to-cell movement pathways, including plasmodesmata and the endocytic pathway, can aid in the translocation and transport of nanoparticles (NPs) inside plant tissues. The kind of NPs and the species of plant can affect how they are distributed in the tissues of the plants.[15]

Plant growth and development may be impacted by the absorption and transport of NPs in a variety of ways. Higher stress tolerance, greater plant growth and output, and higher seed germination are only a few of the benefits of NP absorption and transport in plants. On the other hand, suppressed seed germination, decreased plant development and production, and compromised photosynthesis and respiration are some of the adverse impacts of NP absorption and transport in plant[16] (Fig. 10.1).

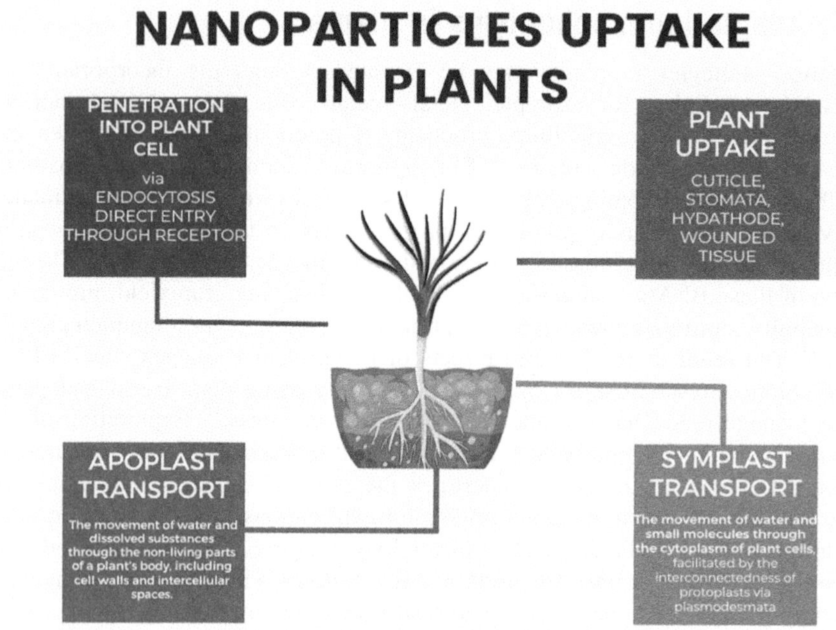

FIGURE 10.1

Various modes of nanoparticle uptake in plants.

10.2.2 Physiological responses to nanomaterial stress

The different alterations that transpire in plants upon exposure to nanoparticles are referred to as physiological responses to nanomaterial stress in plants. Depending on the kind and quantity of nanomaterial used as well as the type of plant, these reactions might be either positive or negative.[17] Oxidative stress is one of the main methods by which nanoparticles might impact plant physiology. This happens because of the nanoparticles' interactions with plant cells, which release reactive oxygen species ROS such hydroxyl, superoxide, and hydrogen peroxide. These ROS have the ability to harm plant cells and impair normal cell activity.[18] To combat oxidative stress, however, plants have evolved several defense strategies, including the synthesis of antioxidant enzymes like peroxidase, catalase, and superoxide dismutase. These enzymes aid in scavenging ROS and shielding plant cells from harm.[19]

Plant physiology can also be impacted by nanoparticles when they disrupt the absorption and transport of nutrients. Plant cell membrane permeability may be changed by nanoparticles, which may have an impact on the absorption of vital nutrients including potassium, phosphate, and nitrogen. Furthermore, nanoparticles have the ability to interact with a plant's root system and influence how nutrients are transferred from the roots to the shoots.[18] This may result in modifications to the plant's metabolic processes as well as growth and development. Additionally, plant hormone signaling pathways may be impacted by nanoparticles, which may have a major effect on the growth and development of plants. For instance, auxins, which are significant plant hormones involved in cell division, elongation, and differentiation, might be affected by nanoparticles in the production and transport processes. Additionally, abscisic acid, gibberellins, and cytokinins are examples of other plant hormones whose signaling pathways may be impacted by nanoparticles, potentially resulting in modifications to plant growth and development[17] (Fig. 10.2).

10.2.3 Genetic and epigenetic modifications

In plants, genetic and epigenetic modifications pertain to alterations in the expression and control of these genes, which can be achieved by structural or genetic modifications that do not impact DNA sequence.[18] Genetic engineering, sometimes referred to as genetic modification, is the process of changing the DNA sequence in plants to give them desired characteristics. This is accomplished by modifying, deleting, or adding new genes, which results in genetic alterations that can be inherited by subsequent generations. A lot of work has gone into using genetic engineering to increase crop output, nutritional content, and resilience to biotic and abiotic stress.[20] Conversely, heritable changes in gene expression that do not include modifications to the DNA sequence are known as epigenetic modifications. These alterations are brought about by chemical processes that alter the structure of DNA and make it accessible to transcription machinery, such as DNA methylation and histone modification.[21] It has been demonstrated that epigenetic changes significantly influence how plants react when exposed to nanoparticles. According to a study, multigenerational exposure to nanomaterials altered the global and gene-specific methylation patterns of DNA methylation in the model species *Enchytraeus crypticus*. Furthermore, there was a large regulation of the expression of genes associated to epigenetics, and these modifications were connected to endpoints that are extremely crucial to ecology, such as reproduction.[22] Environmental elements like temperature, light, and nutrient availability can cause epigenetic modifications, which in turn impact plant growth, development, and adaptation, according to another study on plant DNA methylation as an epigenetic mark in development, environmental interactions, and evolution. Genes linked to stress reaction growth, and development have been found to be regulated by DNA methylation, a common epigenetic mark. Variations in plant features can result from both genetic and epigenetic alterations.[21,23]

10.2.4 Morphological adaptations

When discussing BNMs, morphological adaptations in plants relate to the structural alterations that transpire because of plant exposure to nanoparticles. Plant

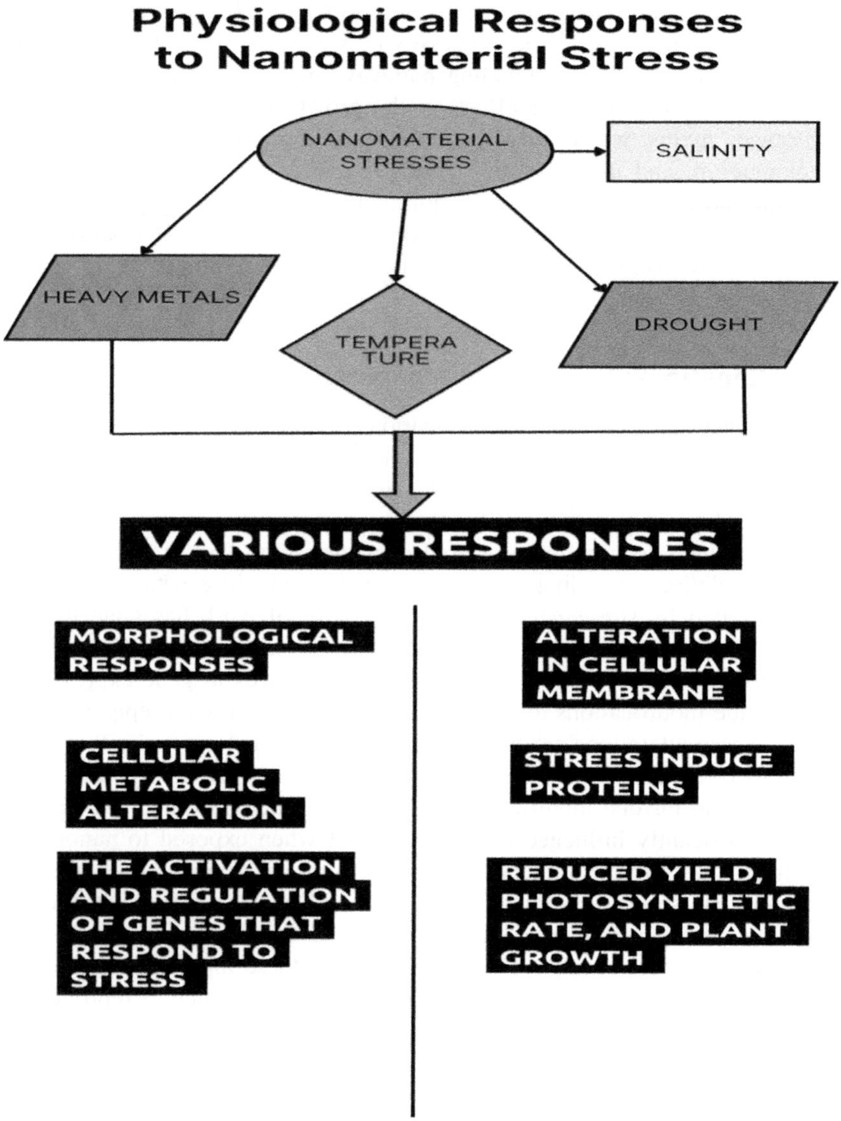

FIGURE 10.2

Types of nanomaterial stress and its related physiological and biochemical changes.

components such as roots, stems, leaves, and reproductive organs can all exhibit these adaptations.[3] The kind and size of the nanoparticles, the concentration utilized, the length of exposure, and the kind of plant are some of the variables that affect how they affect plant morphology. For example, while some researches have indicated that high concentrations of some nanoparticles, such titanium dioxide, gold, and silver, can restrict growth and lower biomass, other studies have discovered that exposure to these nanoparticles can boost plant growth and biomass.[17] In regard to morphological adaptations, plants exposed to nanoparticles may change the morphology of their roots, including their length, diameter, and branching. According to certain research, for instance, exposure to titanium dioxide nanoparticles can shorten roots and promote branching, but exposure to silver nanoparticles can boost both of these trait.[24] Similarly, exposure to nanoparticles may also have an impact on leaf morphology, including leaf size, shape, and quantity. According to some research, leaf size and quantity can increase when exposed to certain nanoparticles like zinc oxide, however leaf size can decrease when exposed to silver nanoparticles.[3] Moreover, exposure to nanoparticles might also have an impact on reproductive organs including fruits and flowers. According to certain research, for example, fruit output and quality can be decreased by exposure to silver nanoparticles, yet blooming and fruiting can be improved by exposure to gold nanoparticles.[17] Although the exact processes driving these morphological adaptations are unknown, it is thought that gene expression changes and modifications to plant hormone signaling pathways, including auxin, gibberellin, and abscisic acid, are involved. To clarify the precise processes and comprehend the long-term impacts of nanoparticle exposure on plant morphology and growth, more investigation is required.[24]

10.2.5 Resilience in plants

Plant growth and productivity can be significantly affected by temperature stress, whether it is high or low. These stresses can disrupt various processes within the plant at a biochemical, physiological, and molecular level, ultimately impacting the growth, development, and productivity of the plant.[25] Therefore, innovative strategies are required to address these challenges and promote sustainability. Nanotechnology emerges as a promising approach to enhance crop production through the utilization of nanoscale materials.[26]

The effectiveness of nanoparticles is influenced by factors such as their size, composition, method of application, environmental conditions, as well as the specific plant species and the environments they are exposed to.[11] Nanoparticles are able to easily penetrate plants due to their small size, solubility, and capability to traverse cellular barriers, making them increasingly popular in agriculture. Recent studies have shown that silver, silicon, titanium, and selenium nanoparticles can modify the physiological and biochemical responses of plants to mitigate the effects of temperature stress.[27]

Nanoparticles reach the targeted tissues through both apoplastic and symplastic pathways. Research conducted by various scientists has indicated that membrane carrier proteins facilitate the passage of nanoparticles through xylem channels. Moreover, nanoparticles must overcome barriers such as stomata and cuticle to access the internal system of leaves.[18]

10.3 Activation of antioxidant defense systems

Abiotic stress poses a significant challenge to plants in agriculture, prompting research efforts to identify compounds that can alleviate its effects. Biogenic particles, such as melatonin (MET), have emerged as key players in protecting plants against a range of abiotic stresses, including salinity, extreme temperatures, high light intensity, waterlogging, nutrient deficiencies, and combinations of stressors.[28] MET is known to modulate enzymatic and non-enzymatic antioxidant defense mechanisms, interacting with signaling molecules like ROS and nitric oxide (NO). NO, in particular, triggers S-nitrosylation and NO_2-Tyr modifications in crucial antioxidant-related proteins, thereby sustaining the antioxidant capacity of the AsA/GSH cycle in the face of nitro-oxidative stress.[29]

Moreover, MET has been found to regulate various plant responses, such as growth and stress tolerance, by collaborating with NO as a long-distance signaling molecule. Despite the progress made in understanding these interactions, further research is warranted to elucidate the precise mechanisms through which MET and NO influence plant responses to abiotic stresses.[29] Furthermore, the balance between ROS and antioxidant defense systems is pivotal for plants coping with abiotic stress. ROS are produced during different stress-related processes, and their excessive accumulation can lead to oxidative stress. Enzymatic and non-enzymatic antioxidants work in concert to maintain this equilibrium, ensuring plants' resilience under challenging environmental conditions.[30]

10.3.1 Hormonal regulation

Hormonal regulation is a critical factor in the resilience strategies of plants, allowing them to adjust to changing environmental conditions. Plant hormones, including auxins, gibberellins, cytokinins, ethylene, and abscisic acid, interact with biogenic particles to control various physiological processes like growth, development, and stress responses.[23] Biogenic particles, such as volatile organic compounds V Hormonal regulation is a critical factor in the resilience strategies of plants, allowing them to adjust to changing environmental conditions. Plant hormones, including auxins, gibberellins, cytokinin, ethylene, and abscisic acid, interact with biogenic particles to control various physiological processes like growth, development, and stress responses.[21] Biogenic particles, such as volatile organic compounds (VOCs), impact hormonal regulation in plants by signaling stress, regulating hormone production, and adjusting hormone signaling pathways.[22]

Auxins, like indole-3-acetic acid (IAA), work together with biogenic particles to regulate plant growth and development. For instance, terpenes, such as limonene, stimulate auxin production, which enhances root growth and development.[19] Ethylene, a biogenic particle, influences auxin signaling, affecting how plants respond to environmental cues like light and gravity. Gibberellins, such as gibberellic acid (GA), also collaborate with biogenic particles to regulate plant growth and development. Phenolics, like salicylic acid, trigger gibberellin production, promoting stem elongation and seed germination. Isoprene, a biogenic particle, adjusts gibberellin signaling, impacting how plants respond to environmental factors like temperature and light.[18]

Cytokinin, such as zeatin, interacts with biogenic particles to regulate plant growth and development. Monoterpenes, such as pinene, stimulate cytokinin production, which enhances shoot growth and leaf development. Ethylene, a biogenic particle, modulates cytokinin signaling, influencing plant responses to environmental stimuli like light and temperature. Ethylene, a crucial hormone in plant stress responses, collaborates with biogenic particles to regulate plant senescence and stress tolerance.[19] Methyl salicylate, a biogenic particle, triggers ethylene production, promoting fruit ripening and senescence. Isoprene, a biogenic particle, adjusts ethylene signaling, impacting how plants respond to environmental stimuli like mechanical stress. OCs impact hormonal regulation in plants by signaling stress, regulating hormone production, and adjusting hormone signaling pathways.[23]

cytokinin, such as zeatin, interacts with biogenic particles to regulate plant growth and development. Monoterpenes, such as pinene, stimulate cytokinin production, which enhances shoot growth and leaf development. Ethylene, a biogenic particle, modulates cytokinin signaling, influencing plant responses to environmental stimuli like light and temperature. Ethylene, a crucial hormone in plant stress responses, collaborates with biogenic particles to regulate plant senescence and stress tolerance.[5] Methyl salicylate, a biogenic particle, triggers ethylene production, promoting fruit ripening and senescence. Isoprene, a biogenic particle, adjusts ethylene signaling, impacting how plants respond to environmental stimuli like mechanical stress.[21]

10.3.2 Symbiotic relationships with microorganisms

Plant resilience strategies are significantly shaped by symbiotic interactions with microorganisms, which are critical for plant growth, development, and stress tolerance. Through the process of evolution, plants have established symbiotic relationships with various microorganisms like mycorrhizal fungi, rhizobia, and endophytes. These relationships are indispensable as they supply essential nutrients, enhance soil health, and boost plant defense mechanisms against pathogens and pests.[22] Furthermore, these microorganisms play a pivotal role in regulating plant hormone signaling, influencing growth and development, and fostering plant resilience to environmental stresses. For example, Mycorrhizal fungi establish mutually beneficial partnerships with plant roots by exchanging vital nutrients like phosphorus and nitrogen for carbohydrates. This symbiotic relationship not only boosts plant growth and soil quality but also fortifies plants against various environmental pressures. Additionally, mycorrhizal fungi play a crucial role in enhancing plant immunity by generating secondary metabolites that ward off harmful pathogens and pests.[10]

The practical applications of biogenic particle-mediated interactions between plants and microorganisms in agriculture are substantial. For example, biogenic particles have the capacity to enhance plant growth, resulting in improved crop yields. Additionally, biogenic particles can fortify plant resilience to environmental stressors, such as drought, salinity, and temperature extremes.[18] Furthermore, these interactions can decrease the requirement for synthetic fertilizers and pesticides, promoting sustainable agricultural practices. Finally, biogenic particles can be utilized for pest and disease control, reducing the need for chemical pesticides and fungicides.

10.3.3 Nutrient uptake enhancement

In plants, improving nutrient uptake is an essential resilience tactic, and biogenic particles are crucial to this process. When plants are exposed to temperature stress, they react differently to the nanoparticles that help them recover from the negative effects of both high and low temperatures and adjust to the new environment.[31] Numerous alterations, including physiological, morphological, anatomical, molecular, and biochemical ones, are what define this response. Increases in the rates of photosynthesis, respiration, and transpiration are examples of physiological alterations. Both a rise in their dry weight and the quantity of roots or leaves are examples of morphological alterations. Chloroplast and mitochondrial ultrastructure are two examples of anatomical modifications. Heat tolerance gene expression is a result of molecular alterations; biochemical alterations encompass responses to antioxidants and the generation of phytohormones.[4]

Plants use foliar uptake of nanoparticles in which the particles are first deposited on the leaf surface and then absorbed through the stomata or cuticle. Nanoparticles are mostly absorbed by foliar and root uptake by the xylem and phloem systems after being initially taken up by the plant's leaves or roots, and it takes time for plants to absorb and distribute them. The two most common channels for foliar uptake are stomatal and cuticular, the latter of which includes both lipophilic and hydrophilic uptake pathways.[21]

Improved crop yields can be attained by using biogenic particles, including nanoparticles, which can improve soil nutrient availability for plant development, as well as plant nutrient absorption and application reliability. The distinct characteristics of nanoparticles, such as their size, reactivity, and ability to transfer information precisely, offer a variety of options for reducing stress.[23] Nanoparticles have enormous potential to strengthen plants against a variety of environmental difficulties, including drought, salt, and metal toxicity. They can do this by enhancing nutrient absorption and inducing stress response pathways.[5]

To fully understand the intricate mechanisms underlying how diverse nanoparticle kinds interact with plant systems under varying environmental stress conditions, more research is needed. The potential of nanoparticles in enhancing plant stress tolerance must be realized through interdisciplinary techniques that combine plant biology, nanotechnology, and agricultural science. As a result, carefully used nanoparticles have the potential to transform farming methods and open the door to a future of greater plant resilience and food security[28] (Table 10.1).

Table 10.1 Effect of biostimulants in alleviating various abiotic stresses on different plant crops.

Species	Scientific name of plants	Biostimulants	Stress	Beneficial effect	References
Arabidopsis	*Arabidopsis thaliana*	Solid medium, liquid cultures, seaweed extracts and humic acid	Drought, salinity or cold stress and heat stress	Early growth cell death, chloroplast degradation and activate heat shock proteins	32–34
Corn	*Zea mays*	NOM-based biostimulant	Drought and salt stress	Root, shoot weights, and final yields	35
Jasmine	*Catharanthus roseus*	Chitosan nanoparticles (Ch-NPs)	Salinity stress	Higher alkaloid accumulation antioxidant capacity	36
Forages plants	*Urochloa brizantha* CV	Silicon	Water deficit	Nutrient efficiency. Root	37
Grapevine	*Vitis vinifera*	Kaolin, glycine betaine, chitosan, seaweed	Heat stress	Photosynthetic activity. Antioxidant capacity	37
Lettuce	*Lactuca sativa*	Protein hydrolysates	Salt and drought	Nitrogen metabolism and osmolytes	34
Maize	*Zea mays*	Humic substances HA, seed soaking	Salt and drought stress	Cell elongation in roots, antioxidants development	30,34
Mung bean	*Vigna radiata*	Zinc oxide	Heat stress	Final yield	38
Mustard	*Brassica nigra*	Foliar application of ZnONPs	Salt, drought stress, cadmium	Growth and antioxidants activities	39,40
Orange	*Citrus sinensis*	A nodosum seaweed extract	Drought stress	Weight, quality, and maturity	40
Olive	*Olea europaea*	Seaweed glycine betaine and kaolin	Heat and drought	Vegetative and leaf nutraceutical trait	40
Oilseed rape	*Brassica napus*	Application of selenium nanoparticles	Arsenic stress	Growth and development	41,42
Peanut	*Arachis hypogaen*	Biostimulants; MPB and MIRCEN (Bradyrhizobium)	Drought oxidative stress	ROS (reactive oxygen species); plant growth	43

(*Continued*)

Table 10.1 Effect of biostimulants in alleviating various abiotic stresses on different plant crops. *Continued*

Species	Scientific name of plants	Biostimulants	Stress	Beneficial effect	References
Sunflower	*Helianthus annuus*	Application of GABA metabolite	Drought stress, heat/chilling	Chlorophyll and sugar level development	44
Sugarcane	*Saccharum officinarum*	Seaweed extract (SWE)	Drought stress	Stalk and final yield	45,46
Soybean	*Glycine max*	NOM-based biostimulant	Drought and salt stress	Root, shoot weights, and final yields	35
Tomato	*Solanum lycopersicum*	Soil and foliar applied biostimulants (NPK)	Salt and drought stress	Pollen viability and photosynthetic protectants	45,46
Wheat	*Triticum aestivum*	NOM-based biostimulant, glutacetine	Drought and salt stress	Root, shoot weights, and spike	35

10.4 Microbial responses to biogenic nanomaterial

The ways in which microbes respond to biogenic nanomaterial are diverse and intricate, encompassing a range of mechanisms that impact the resilience of plants, the health of the soil, and interactions between microbes and plants.[21] Biogenic nanomaterials can be ingested by bacteria, fungus, and other microorganisms by a variety of pathways, one of which is endocytosis, in which the cell membrane invades and creates vesicles that engulf the nanoparticles. Microbes can also absorb nanoparticles by electrostatic interactions, in which the nanoparticles engage with the microbes through electrostatic forces, resulting in the ingestion of the particles, as well as through pores and channels in their cell membranes.[19] Microorganisms can undergo a variety of reactions when internalizing biogenic nanomaterial, such as modifications in metabolic activity, protein synthesis, and gene expression. Certain microorganisms have the potential to unregulated genes related to stress response, detoxification, and antioxidant synthesis in response to nanoparticle exposure. This can assist microbes in preserving cellular homeostasis and reducing the possible harmful effects of nanoparticles. On the other hand, some microbes could react to exposure to nanoparticles by suppressing genes related to growth and development, which would result in a decrease in cellular activity and growth.[10]

Additionally, it can affect the composition and activities of microbial communities, biogenic nanomaterial may influence the health of soil and the operation of ecosystems. For instance, certain microbial populations may be selectively inhibited or stimulated by nanoparticles, changing the diversity and makeup of the community.[28] On ecological processes including nitrogen cycling, decomposition, and plant–microbe interactions, this may have a domino effect. Additionally, BNMs may alter plant–microbe signaling, nutrition exchange, and defense mechanisms, among other aspects of microbial interactions with plants.[29] The variety and number of microorganisms present, as well as the size, shape, and surface chemistry of the nanoparticles, can all have an impact on how BNMs affect microorganisms. Microorganisms may be more receptive to smaller nanoparticles than to bigger ones, and specific surface chemistries may render certain nanoparticles more poisonous to them than to others. Furthermore, environmental elements like temperature, pH, and food availability might have an impact on how BNMs affect bacteria.[30]

10.4.1 Nanoparticle interaction with microbial cell structure

The intricate process of nanoparticle interaction with microbial cell structure comprises a number of different processes, such as chemical, biological, and physical interactions. The destiny of nanoparticles inside the cell is mostly determined by the composition of the microbial cell, which consists of the cytoplasm, cell wall, and cell membrane. Nanoparticles must pass through the cell wall, a tough layer made of peptidoglycan, teichoic acids, and other materials, in order to connect with the cell membrane.[10] The principal location of nanoparticle contacts and the semipermeable lipid bilayer that controls molecular transit into and out of the cell is the cell membrane. Numerous parameters, such as the size, shape, and surface chemistry of the nanoparticles, affect how they interact with the microbial cell membrane.[47] While bigger nanoparticles could be too big to go past the membrane and stay outside the cell, smaller ones—typically those with diameters less than 10 nm—can pass through and reach the cytoplasm. The way that nanoparticles interact with the cell membrane is also significantly influenced by their surface chemistry. While hydrophilic nanoparticles may interact with membrane proteins and receptors, hydrophobic nanoparticles, on the other hand, may interact with the lipid bilayer and disturb membrane structure.[48] Following their interaction with the cell membrane, nanoparticles can cause a variety of biological reactions, such as modifications to protein synthesis, gene expression, and metabolic activity. Nanoparticles and membrane proteins, such transporters and receptors, can interact to trigger signaling pathways that control physiological functions like antioxidant synthesis, detoxification, and stress response. Furthermore, nanoparticles have the ability to interfere with cellular shape and organization by interacting with the cytoskeleton, a network of protein filaments that gives cells structural support.[49]

The production of nanoparticle–microbe complexes, which can have profound effects on microbial physiology and ecology, can also result from the interaction of nanoparticles with microbial cell structure. These complexes have the ability to change the behavior, development, and survival of microorganisms as well as the movement and destiny of nanoparticles in the environment.[50] Moreover, the creation of extracellular polymeric substances (EPS), a complex combination of biomolecules essential to microbial adhesion, aggregation, and biofilm formation, might be influenced by the interaction between nanoparticles and microbial cell structure.[25] Aside from nanoparticles and microbial cell structure, environmental variables including pH, temperature, and nutrition availability can also have an impact. Temperature variations, for instance, might affect the cell membrane's permeability and fluidity, which can affect the absorption and interaction of nanoparticles. Analogously, variations in pH can modify the surface chemistry of nanoparticles, impacting their interplay with the microbial cell structure and cell membrane.[51] All things considered, the relationship between nanoparticles and microbial cell structure is a complicated one that is impacted by a number of variables, such as the characteristics of the nanoparticles, the physiology of the microbes, and the surrounding environment.[52]

10.4.2 Metabolic adaptations to nanomaterial stress

Metabolic adaptations to nanomaterial stress are an essential part of microbial responses to BNMs. Microorganisms have evolved diverse metabolic adaptations to respond to nanomaterial stress, which enable them to survive and thrive in the presence of BNMs. For instance, bacteria create antioxidants, such as glutathione, catalase, and superoxide dismutase, to neutralize ROS generated by nanoparticle exposure. They also maintain redox equilibrium by controlling the ratio of reduced to oxidized glutathione, which is crucial for antioxidant defencss.[53] Microorganisms adjust their energy metabolism to respond to nanomaterial stress, moving from aerobic to anaerobic respiration, which generates ATP more effectively in the presence of nanoparticles. They also change their carbon metabolism, increasing the generation of glycolytic intermediates, which can be used to create ATP and NADPH (1). Furthermore, microbes alter their amino acid metabolism, boosting the production of amino acids implicated in antioxidant defenses, such as cysteine and methionine.[13] Microorganisms also adjust their membrane composition and structure to adapt to nanomaterial stress, boosting the production of membrane lipids with antioxidant characteristics, such as cardiolipin. They create EPS, including as polysaccharides and proteins, to interact with BNMs, which can bind and sequester nanoparticles, limiting their bioavailability and toxicity. Additionally, microbes use quorum sensing to regulate their activity and respond to nanomaterial stress, controlling gene expression and coordinating the response to nanomaterial stress.[54]

10.4.3 Biofilm formation and quorum sensing

Biofilm production is an important feature of microbial reactions to BNMs, since it enables bacteria to survive and thrive in the presence of these materials. The process of biofilm formation involves numerous steps, commencing with initial attachment, where microorganisms connect to the surface of the biogenic nanomaterial, frequently helped by electrostatic interactions, van der Waals forces, and hydrophobic interactions.[55] This is followed by adhesion, where bacteria create adhesins, such as proteins and polysaccharides, that assist them attach to the surface. Next, microbes develop EPS, which offers a protective habitat for the microorganisms, allowing them to resist antimicrobial agents and other pressures.[56] The biofilm then grows as more microbes adhere and make EPS, generating a complex, three-dimensional structure. Finally, the biofilm disperses, releasing microorganisms into the surrounding environment. Biofilm development in response to biogenic nanoparticles gives various benefits to microorganisms. For example, the EPS matrix protects microorganisms from the harmful effects of BNMs. Additionally, biofilms can trap nutrients, making them available to microbes. Furthermore, biofilms enable microorganisms to attach to and colonize surfaces, allowing them to build a stable community. Overall, biofilm development is a crucial approach that bacteria adopt to adapt to the presence of BNMs.[42]

Quorum sensing is a cell-to-cell communication mechanism that enables microorganisms to coordinate their behavior in response to changes in their population density. In the context of microbial reactions to BNMs, quorum sensing plays a vital role in regulating biofilm formation and other adaptive responses. Quorum sensing involves the creation and detection of signaling molecules, such as autoinducers, which accumulate in the environment as the population density of microorganisms increases. When the concentration of autoinducers exceeds a threshold, it initiates a reaction in the microorganisms, leading to changes in gene expression and behavior.[57] In relation to BNMs, quorum sensing regulates several essential activities, including biofilm development, toxin generation, and antimicrobial resistance. For example, quorum sensing governs the generation of EPS and adhesins, which are important for biofilm formation. Additionally, quorum sensing limits the synthesis of poisons and other virulence factors, which can help microbes resist the damaging effects of BNMs. Furthermore, quorum sensing can modulate the expression of genes implicated in antimicrobial resistance, allowing microorganisms to resist the antimicrobial effects of BNMs. Overall, quorum sensing is a crucial process that enables bacteria to adapt to the presence of BNMs.[55] Several quorum sensing systems have been identified in microorganisms, including the LuxI/LuxR system, which is prevalent in Gram-negative bacteria and includes the creation of acyl-homoserine lactone (AHL) autoinducers. Other quorum sensing systems include the Agr system, which is prevalent in Gram-positive bacteria and includes the creation of autoinducing peptides (AIPs). Additionally, certain microbes use quorum sensing systems that include the synthesis of furanosyl borate diester (FB) autoinducers. Overall, quorum sensing is a complicated and versatile system that plays a vital role in controlling microbial responses to BNMs.[42]

10.4.4 Role of microbial communities in environmental resilience

Microbial communities play a critical role in environmental resilience, and their responses to BNMs (BNMs) can have substantial ramifications for ecosystem health and sustainability. Microbial communities are complex networks of microorganisms that interact with one other and their environment to conduct diverse ecosystem services. These functions include decomposition, nutrient cycling, carbon sequestration, and pollution degradation, among others. Microbial communities contribute to environmental resilience in numerous ways.[9] They break down organic debris, releasing nutrients that support plant development and ecological function. They can also store carbon in soils and sediments, limiting climate change. Additionally, microbes can breakdown contaminants, lowering their toxicity and environmental impact. Furthermore, microbial populations alter soil structure and fertility, affecting plant development and ecosystem function.[58] BNMs are nanomaterials derived from biological sources, such as plants, microbes, or animals. BNMs can interact with microbial communities, altering their structure, function, and ecosystem services. The consequences of BNMs on microbial communities can be severe, including toxicity, reduced nutrient availability, and alterations in surface interactions. For example, BNMs can be harmful to microorganisms, changing community makeup and function. They can also impact nutrient availability, influencing microbial development and activity.[59]

The impact of BNMs on environmental resilience can be far-reaching. Changes in microbial community composition and activity can affect ecosystem services, such as decomposition and nutrient cycling. BNMs can potentially affect carbon sequestration by modifying microbial populations and soil structure.

Furthermore, BNMs can increase pollutant toxicity by changing microbial breakdown pathways. Therefore, it is crucial to understand the interactions between BNMs and microbial populations to limit the possible hazards and optimize the advantages of these materials.[60]

10.5 Biogenic nanomaterials in ecosystem resilience

BNMs are nanomaterials derived from biological sources, such as plants, microbes, or animals. These materials offer unique qualities, such as tiny size, large surface area, and biocompatibility, making them suitable for numerous applications, including environmental remediation, agriculture, and medicine. However, the rising usage of BNMs raises concerns regarding their possible impact on ecosystem resilience. Ecosystem resilience refers to the ability of an ecosystem to tolerate and recover from disturbances, such as climate change, pollution, and invasive species.[55] Microbial communities serve a critical role in sustaining ecosystem resilience, as they are responsible for digesting organic matter, cycling nutrients, and degrading contaminants. The interactions between BNMs and microbial communities can have important implications for ecosystem resilience, as BNMs can modify microbial community composition, function, and ecosystem services.[57] The effects of BNMs on microbial populations can be both favorable and detrimental. For example, BNMs can boost microbial growth and activity, enhancing ecosystem functions such as decomposition and nutrient cycling. This can be performed by several processes, including the availability of nutrients, changing of pH, and modulation of redox conditions. Additionally, BNMs can also act as carriers for beneficial bacteria, increasing their colonization and activity in the environment. On the other side, BNMs can also be harmful to microorganisms, affecting community composition and function, and disrupting ecosystem functions. The toxicity of BNMs to microorganisms can be modified by parameters such as particle size, shape, and surface chemistry, as well as the type and concentration of BNMs. Several researches have studied the impact of BNMs on microbial populations and ecosystem resilience. A study demonstrated that silver nanoparticles, a kind of BNM, can modify the composition and function of soil microbial communities, resulting to alterations in ecosystem services such as decomposition and nutrient cycling.[61] The study found that silver nanoparticles can limit the growth of certain bacteria, while boosting the growth of others, resulting to changes in community makeup and function. Another study demonstrated that BNMs can be harmful to microorganisms, resulting to alterations in community composition and function. The study indicated that the toxicity of BNMs to microorganisms can be altered by parameters such as particle size and surface chemistry. The review underlines the necessity to analyze the intricate interactions between microorganisms, their habitat, and external factors such as climate change and pollution, in order to understand the impact of BNMs on ecosystem resilience.[10] The influence of BNMs on ecosystem resilience can have far-reaching consequences, including changes in ecosystem function, biodiversity, and ecosystem services. For example, changes in microbial community composition and function can disrupt nutrient cycling, leading to changes in plant growth and ecosystem function. Additionally, the toxicity of BNMs to microorganisms can lead to the accumulation of pollutants in the environment, causing dangers to human health and the environment.[1] Furthermore, the influence of BNMs on ecosystem resilience can also have cascading effects on higher trophic levels, such as plants and animals, leading to changes in ecosystem structure and function. To limit the possible hazards and enhance the advantages of BNMs, it is vital to understand the interactions between BNMs and microbial communities.[62] This can be achieved by more research on the effects of BNMs on microbial communities, as well as the development of measures to reduce the detrimental implications of BNMs on ecosystem resilience. For example, researchers can synthesize BNMs with lower toxicity to microorganisms, or design systems that minimize the exposure of microorganisms to BNMs.

10.5.1 Impact on soil health and nutrient cycling

The effect of BNMs on soil wellness and nutrient cycling is an important part of ecosystem resilience. Soil health is a vital component of ecosystem resilience, as it supports plant development, filters water, and stores carbon. BNMs might communicate with soil microorganisms, influencing their composition, function, and ecosystem services. For example, BNMs can boost microbe development and activity, enhancing ecosystem functions such as decomposition and nutrient cycling. However, BNMs can also be harmful to microorganisms, affecting community composition and function, and undermining ecosystem functions.[63] The soil micro biome is a fundamental component of soil health, encompassing a large array of microorganisms that play key roles in ecosystem activities such as decomposition, nitrogen cycling, and plant growth. BNMs can interact with the soil microbiome in numerous ways, including promotion of microbial growth and activity, toxicity to microorganisms, and changes in microbial community makeup. These interactions can have cascading consequences on ecosystem processes and services, including changes in nitrogen cycle, decomposition, and plant growth.[64] Nutrient cycling is a fundamental ecosystem process that involves the transformation and movement of nutrients through the ecosystem. BNMs can alter nutrient cycling in numerous ways, including improved nutrient availability, disruption of nutrient cycles, and changes in microbial-mediated nutrient transformations. For example, BNMs can boost the availability of nutrients such as nitrogen, phosphorus, and potassium, enhancing plant growth and ecosystem production. However, BNMs can also alter nutrient cycles, leading to changes in nutrient availability, plant growth, and ecosystem function.[51] The influence of BNMs on soil health and nitrogen cycling can have substantial implications for ecosystem resilience. Alterations in soil health and nutrient cycling can lead to changes in ecosystem function, including variations in decomposition rates, nutrient availability, and plant growth.[65] Disruptions to soil health and nutrient cycling can also contribute to the loss of ecosystem services, including carbon sequestration, water filtering, and climate regulation. Furthermore, changes in ecosystem function and services can have cascade impacts on ecosystem structure, including changes in plant and animal populations, and ecosystem biodiversity.[64] Despite the rising corpus of research on BNMs and ecosystem resilience, there are still considerable study gaps and future initiatives, including the effect of particle size, shape, and surface chemistry. Long-term studies are also needed to determine the impacts of BNMs on ecosystem resilience, including changes in ecosystem function, services, and structure. Additionally, research is needed to create measures to decrease the detrimental impacts of BNMs on ecosystem resilience, including the production of BNMs with reduced toxicity to microorganisms. By addressing these research gaps and future initiatives, we may better understand the influence of BNMs on soil health and nutrient cycling, and devise ways to limit the negative impacts and optimize the benefits of these materials for ecosystem resilience.[63]

10.5.2 Influences on plant–microbe interactions

One of the most important aspects of ecosystem resilience is the impact of BNMs on interactions between plants and microbes. Plant-derived nanomaterials, like plant-derived nanoparticles, can modify the interactions between microbes and plants, resulting in better soil health, stronger ecosystem resilience, and more robust plant growth. For example, it has been demonstrated that biogenic silver nanoparticles promote the development of advantageous bacteria in the rhizosphere, improving plant nutrition and health.[10] Additionally, BNMs have the power to modify the diversity and makeup of soil microbial populations, which can impact how microbes and plants interact. For instance, a study discovered that biogenic zinc oxide nanoparticles can boost plant development and health by increasing the number of advantageous bacteria in the soil, such as *Bacillus* and *Pseudomonas*.[2] Plant growth-promoting chemicals like hormones and antibiotics can also be produced by microorganisms in conjunction with BNMs. For example, it has been demonstrated that biogenic copper oxide nanoparticles can increase the production of the plant growth hormone indole-3-acetic acid by advantageous bacteria, which in turn promotes better plant growth and development.[54] Moreover, BNMs have the ability to alter the expression of genes in plants that are involved in interactions with microbes, changing the plant's defense mechanisms and susceptibility to infections. For instance, a study discovered that biogenic silica nanoparticles can enhance the expression of plant genes involved in defense reactions, improving the plants' ability to fend against infections.[28]

10.5.3 Contribution to biodiversity maintenance

The role that BNMs play in maintaining biodiversity is an essential component of ecosystem resilience. BNMs, including plant-derived nanoparticles, can influence a range of ecological processes and interactions, which can be important in preserving biodiversity.[66] For example, BNMs have the ability to alter soil microbial populations, which is crucial for preserving ecosystem biodiversity. Among their many jobs, microbial communities break down organic matter, fix nitrogen, and solubilize minerals. BNMs can affect nutrient cycling, plant growth promotion, and decomposition processes by changing the makeup and diversity of microbial communities. This can ultimately result in increased ecosystem biodiversity.[65] Enhancing plant diversity is one of the key ways that BNMs support the preservation of biodiversity. Since plants are the principal producers in ecosystems, preserving the diversity of these ecosystems is crucial to their resilience. By giving vital nutrients like nitrogen and phosphorus and encouraging the growth of helpful microbes, BNMs can enhance plant growth and development. For instance, it has been demonstrated that biogenic zinc oxide nanoparticles promote the growth of wheat plants, resulting in higher crop yields and enhanced biodiversity within ecosystems.[64] Similarly, it has been discovered that biogenic silver nanoparticles promote the growth of advantageous bacteria involved in nutrient cycling and plant growth promotion, such as Bacillus and Pseudomonas. BNMs can boost plant diversity and create more resilient and diversified ecosystems by encouraging plant growth and development.[63] BNMs have an impact on microbial populations, plant diversity, and pollinator health, all of which are critical to preserving ecosystem biodiversity. Pollinators, including bees and butterflies, are essential to the reproduction of plants and the health of ecosystems. It has been demonstrated that BNMs, such as biogenic copper oxide nanoparticles, enhance the fertility and health of honey bees, resulting in an increase in pollination services and biodiversity within ecosystems.[57] BNMs can preserve ecosystem biodiversity and guarantee the long-term viability of ecosystems by supporting pollinator health. Moreover, decomposition and nutrient cycling—two processes vital to preserving ecosystem biodiversity—can be impacted by BNMs. BNMs have the potential to modify ecosystem processes and biodiversity by influencing the activity of microbes engaged in decomposition and nutrient cycling. This can affect the availability of nutrients for plants and microorganisms.[67] The methods through which BNMs contribute to the preservation of biodiversity are intricate and multidimensional. BNMs have the ability to interact with plants, bacteria, and other creatures in ecosystems, changing biodiversity and ecological processes. BNMs, for instance, have the ability to encourage the growth of advantageous microbes, which in turn can aid in the growth and development of plants.[10] Additionally, by influencing the expression of genes involved in stress response and defense in plants, BNMs can alter the diversity of plants and the resilience of ecosystems. Moreover, BNMs have the power to alter ecosystem structure and function, which can modify biodiversity and ecosystem processes. We may create plans for utilizing BNMs' potential to increase ecosystem resilience and biodiversity by comprehending the mechanisms behind their contribution to biodiversity maintenance.[64] BNMs have a wide range of possible uses in maintaining biodiversity. Utilizing BNMs, sustainable farming methods like organic farming and precision agriculture can be developed. These methods can increase the resilience and biodiversity of ecosystems.[65] Additionally, degraded habitats, such as contaminated soils and damaged wetlands, can be restored with the use of BNMs, which may prevent biodiversity loss. Additionally, the development of innovative conservation techniques like assisted colonization and ecological restoration, which support ecosystem resilience and biodiversity, can be facilitated by the use of BNMs. Through the utilization of BNMs, creative approaches to preserving ecosystem resilience and biodiversity can be created, even in the face of environmental disruption.[56]

10.5.4 Implications for sustainable agriculture and environmental remediation

BNMs have a wide range of effects on sustainable agriculture. Enhancing nutrient uptake and utilization in plants is one of the biggest benefits of BNMs. For instance, it has been demonstrated that biogenic iron oxide nanoparticles improve plant iron uptake, which lessens the demand for artificial iron fertilizers. This lessens the negative effects of agricultural operations on the environment while simultaneously increasing crop yields. Furthermore, plants can respond to environmental challenges including salt, drought, and high temperatures by activating stress response mechanisms in response to BNMs.[67] Moreover, tailored and long-lasting methods of managing diseases and pests can be created with the help of BNMs. For example, it has been demonstrated that biogenic copper oxide nanoparticles have antibacterial activity against plant infections, which lessens the requirement for chemical pesticides.[68] This strategy fosters a more robust ecosystem while simultaneously lessening the negative effects of agricultural activities on the environment. Additionally, contaminated soils can be remedied with BNMs, restoring the fertility and health of the soil. For instance, it has been discovered that biogenic iron oxide nanoparticles can eliminate heavy metals from polluted soil, allowing it to be used for farming.[63] BNMs have demonstrated great promise for environmental clean-up. The capacity of BNMs to extract heavy metals from contaminated soil and water is one of its biggest benefits. For instance, it has been demonstrated that biogenic iron oxide nanoparticles may successfully remove cadmium and lead from contaminated soil.[53] Biogenic nanoparticles can also be utilized to break down organic contaminants found in soil and water. For example, it has been discovered that in contaminated soil, biogenic silver nanoparticles break down polycyclic aromatic hydrocarbons (PAHs). Additionally, the application of biogenic nanoparticles can improve the effectiveness of phytoremediation—a technique that employs plants to extract contaminants from contaminated soil and water.[4] BNMs may have advantages, but there are also a number of issues and worries that need to be resolved. The possible toxicity and environmental effects of BNMs are among the main worries. Although BNMs are usually thought to be less harmful than conventional nanoparticles, it is still unclear how they may ultimately affect the environment and public health. To be feasible for large-scale agricultural and environmental remediation applications, biogenic nanomaterial synthesis and application must also be made more cost- and scalarly efficient. It is also necessary to create regulatory frameworks to guarantee the sustainable and safe application of BNMs.[67]

10.6 Challenges and future directions

The potential of biogenic nanoparticles for environmentally remediation and sustainable agriculture must be taken into account, along with the difficulties and potential paths forward in relation to plant and microbe adaptation and resilience to these novel materials.[68–70]

10.6.1 Comprehending the interactions between plants, microbes, and nanomaterials

one of the key hurdles in this subject understands the complicated interactions between plants, microbes, and BNMs. While biogenic nanoparticles have showed promise in improving plant growth and clean-up, the processes behind these interactions are still not fully understood.[71] Future study should focus on deciphering the molecular and physiological responses of plants and microbes to BNMs, including the function of signaling networks, gene expression, and metabolomics. Acquiring this understanding will allow for the creation of more efficient and focused uses of BNMs in agriculture and environmental clean-up.

10.6.2 Microorganisms' ability to adapt and demonstrate resilience

Microorganisms play a significant role in soil ecosystems, and their adaptation and resilience to BNMs are critical for preserving soil health and fertility. However, the influence of BNMs on microbial communities is still largely understood. Future research should investigate the effects of BNMs on microbial diversity, community structure, and functional potential. This understanding will enable the production of BNMs that are compatible with microbial communities and support soil health.[72,73]

10.6.3 Plant tolerance and resistance to biogenic nanomaterial

Plants have evolved strategies to endure and resist diverse environmental stressors, including those caused by BNMs.[74,75] However, the genetic and molecular basis of plant tolerance and resistance to BNMs is still not fully known. Future research should focus on identifying the essential genes and signaling pathways involved in plant responses to BNMs, as well as generating crop types that are more tolerant to these materials. This understanding will enable the development of more effective and sustainable applications of BNMs in agriculture.

10.6.4 Ecotoxicological risks and environmental impacts

While BNMs are generally believed to be safer than regular nanoparticles, their ecotoxicological dangers and environmental implications are still not fully known. Future research should investigate the possible dangers of BNMs to nontarget creatures, including plants, microbes, and animals. This understanding will enable the development of more ecologically friendly and sustainable applications of BNMs.

10.6.5 Scalability and cost-effectiveness

One of the key obstacles in the creation of BNMs is scalability and cost-effectiveness. While BNMs have showed promise in small-scale applications, their large-scale manufacture and application are still constrained by high prices and technological obstacles. Future research should focus on creating more efficient and cost-effective ways for generating and deploying biogenic nanomaterials, as well as examining their potential for large-scale agricultural and environmental remediation applications.

10.6.6 Regulatory frameworks and public perception

Finally, the research and implementation of BNMs require regulatory frameworks and public perception that are supportive of these novel materials. Future study should investigate the regulatory frameworks and public perception of BNMs, as well as develop ways for raising public understanding and acceptance of these materials. This understanding will enable the development of more effective and sustainable applications of BNMs in agriculture and environmental remediation. The adaptation and resistance of plants and microorganisms to BNMs are crucial for the development of more effective and sustainable applications of these novel materials in agriculture and environmental remediation. Future research should focus on understanding the intricate interactions between plants, microbes, and BNMs, as well as resolving the obstacles and limits connected with the creation and deployment of these materials.

10.7 Conclusion

The ability of plants and microorganisms to adapt and withstand biogenic nanoparticles is essential for the advancement of sustainable and ecofriendly applications in agriculture and environmental cleanup. BNMs have potential in augmenting plant development, enhancing soil health, and remedying polluted settings. Nevertheless, the comprehensive understanding of the ecological impact of these substances over time and the possible hazards they pose to unintended organisms remains incomplete. Understanding the adaptation and resilience of plants and microorganisms to biogenic nanoparticles is vital for assuring environmental safety, maximizing applications, and encouraging sustainable development. By understanding how plants and microorganisms respond to BNMs, we can limit the potential dangers to ecosystems and non-target creatures. Acquiring this information will facilitate the creation of more efficient and enduring applications in agriculture and environmental remediation, so contributing to a more sustainable future.

The potential for future study and applications in comprehending and utilizing biogenic nanoparticles is immense. Further research is needed to clarify the intricate interactions between plants, microbes, and BNMs. Furthermore, it is imperative to prioritize research efforts towards the creation of specialized applications for BNMs that are designed to address specific environmental obstacles. Enhancing the ability to handle larger workloads and achieving optimal cost-efficiency is equally important for applications operating on a big scale. Furthermore, it is crucial to consider and tackle regulatory frameworks and public perception in order to guarantee the responsible advancement and utilization of biogenic nanomaterials. As we continue to explore the potential of biogenic nanomaterials, it is crucial that we do so in a responsible and sustainable manner. It is imperative that we give priority to ensuring environmental safety, maximize the use of applications, and actively encourage sustainable growth. We call for continuing investigation and careful application of biogenic nanomaterials for sustainable environmental management. Through collaboration, we may utilize biogenic nanomaterials to establish a sustainable future for future generations.

Acknowledgments

KG is supported by the Higher education and science committee MESCS Republic of Armenia provided the grant number 21AG-4C075 AS is supported by the 23PostDoc-4D007 grant provided by the Higher education and science committee MESCS Republic of Armenia.

References

1. Rai, M., et al. Biogenic silver nanoparticles: what we know and what do we need to know? *Nanomaterials* **2021**, *11* (11), 2901. Available from: https://doi.org/10.3390/nano11112901 *DOI.org (Crossref)*.
2. Patil, S.; Chandrasekaran, R. Biogenic nanoparticles: a comprehensive perspective in synthesis, characterization, application and its challenges. *J. Genet. Eng. Biotechnol.* **2020**, *18* (1), 67. Available from: https://doi.org/10.1186/s43141-020-00081-3 *DOI.org (Crossref)*.

3. Mughal, B., et al. Biogenic nanoparticles: synthesis, characterisation and applications. *Appl. Sci.* **2021**, *11* (6), 2598. Available from: https://doi.org/10.3390/app11062598 *DOI.org (Crossref).*
4. Chakraborty, S., et al. Biogenic nanoparticles and generation of abiotic stress-resilient plants: a new approach for sustainable agriculture. *Plant. Stress.* **2022**, *6*100117. Available from: https://doi.org/10.1016/j.stress.2022.100117 *DOI.org (Crossref).*
5. Barnes, E. M.; Tringe, S. G. Exploring the roles of microbes in facilitating plant adaptation to climate change. *Biochemical J.* **2022**, *479* (3), 327–335. Available from: https://doi.org/10.1042/BCJ20210793 *DOI.org (Crossref).*
6. Husen, A., Ed. *Harsh environment and plant resilience: molecular and functional aspects;* Springer International Publishing, 2021. Available from: https://doi.org/10.1007/978-3-030-65912-7 *DOI.org (Crossref).*
7. Kumar, A., et al. Biogenic metallic nanoparticles: biomedical, analytical, food preservation, and applications in other consumable products. *Front. Nanotechnol.* **2023**, *5*1175149. Available from: https://doi.org/10.3389/fnano.2023.1175149 *DOI.org (Crossref).*
8. Kulkarni, D., et al. Biofabrication of nanoparticles: sources, synthesis, and biomedical applications. *Front. Bioeng. Biotechnol.* **2023**, *11*1159193. Available from: https://doi.org/10.3389/fbioe.2023.1159193 *DOI.org (Crossref).*
9. Asmat-Campos, D., et al. Toward sustainable nanomaterials: an innovative ecological approach for biogenic synthesis of TiO_2 nanoparticles with potential photocatalytic activity. *Clean. Eng. Technol.* **2023**, *17*100702. Available from: https://doi.org/10.1016/j.clet.2023.100702 *DOI.org (Crossref).*
10. Abasi, F., et al. Biogenic silver nanoparticles as a stress alleviator in plants: a mechanistic overview. *Molecules* **2022**, *27* (11), 3378. Available from: https://doi.org/10.3390/molecules27113378 *DOI.org (Crossref).*
11. Aguirre-Becerra, H., et al. Nanomaterials as an alternative to increase plant resistance to abiotic stresses. *Front. Plant. Sci.* **2022**, *13*1023636. Available from: https://doi.org/10.3389/fpls.2022.1023636 *DOI.org (Crossref).*
12. Munir, N., et al. Plant–nanoparticle interactions: transcriptomic and proteomic insights. *Agronomy* **2023**, *13* (8), 2112. Available from: https://doi.org/10.3390/agronomy13082112 *DOI.org (Crossref).*
13. Maqsood, M. F., et al. Biogenic nanoparticles application in agriculture for ROS mitigation and abiotic stress tolerance: a review. *Plant. Stress.* **2023**, *10*100281. Available from: https://doi.org/10.1016/j.stress.2023.100281 *DOI.org (Crossref).*
14. Avila-Quezada, G. D., et al. Strategic applications of nano-fertilizers for sustainable agriculture: benefits and bottlenecks. *Nanotechnol. Rev.* **2022**, *11* (1), 2123–2140. Available from: https://doi.org/10.1515/ntrev-2022-0126 *DOI.org (Crossref).*
15. Wang, X., et al. Nanoparticles in plants: uptake, transport and physiological activity in leaf and root. *Materials* **2023**, *16* (8), 3097. Available from: https://doi.org/10.3390/ma16083097 *DOI.org (Crossref).*
16. Gao, M., et al. Advances in transport and toxicity of nanoparticles in plants. *J. Nanobiotechnology* **2023**, *21* (1), 75. Available from: https://doi.org/10.1186/s12951-023-01830-5 *DOI.org (Crossref).*
17. Rasheed, A., et al. The role of nanoparticles in plant biochemical, physiological, and molecular responses under drought stress: a review. *Front. Plant. Sci.* **2022**, *13*976179. Available from: https://doi.org/10.3389/fpls.2022.976179 *DOI.org (Crossref).*
18. Hossain, Z., et al. Plant responses to nanoparticle stress. *Int. J. Mol. Sci.* **2015**, *16* (11), 26644–26653. Available from: https://doi.org/10.3390/ijms161125980 *DOI.org (Crossref).*

19. Dilnawaz, F., et al. Involvement of nanoparticles in mitigating plant's abiotic stress. *Plant. Stress.* **2023**, *10*100280. Available from: https://doi.org/10.1016/j.stress.2023.100280 *DOI.org (Crossref).*
20. Abideen, Z., et al. Impact of nanomaterials on the regulation of gene expression and metabolomics of plants under salt stress. *Plants* **2022**, *11* (5), 691. Available from: https://doi.org/10.3390/plants11050691 *DOI.org (Crossref).*
21. Brukhin, V.; Albertini, E. Epigenetic modifications in plant development and reproduction. *Epigenomes* **2021**, *5* (4), 25. Available from: https://doi.org/10.3390/epigenomes5040025 *DOI.org (Crossref).*
22. Bicho, R. C., et al. Epigenetic effects of (nano)materials in environmental species—Cu case study in enchytraeus crypticus. *Environ. Int.* **2020**, *136*105447. Available from: https://doi.org/10.1016/j.envint.2019.105447 *DOI.org (Crossref).*
23. Panpatte, D. G., et al. *Nanoparticles: the next generation technology for sustainable agriculture.* In *Microbial Inoculants in Sustainable Agricultural Productivity;* Singh, D. P., et al., Eds.; Springer: India, 2016; pp 289–300. Available from: https://doi.org/10.1007/978-81-322-2644-4_18. *DOI.org (Crossref).*
24. Ali, S., et al. Uptake, translocation, and consequences of nanomaterials on plant growth and stress adaptation. *J. Nanomaterials* **2021**, *2021*, 1–17. Available from: https://doi.org/10.1155/2021/6677616 *DOI.org (Crossref).*
25. Sidhu, A. K., et al. Nanomaterial strategies for enhancing plant resilience in the face of temperature stress. *CABI Agriculture Biosci.* **2024**, *5* (1), 60. Available from: https://doi.org/10.1186/s43170-024-00255-w *DOI.org (Crossref).*
26. Schwab, F., et al. Barriers, Pathways and processes for uptake, translocation and accumulation of nanomaterials in plants—critical review. *Nanotoxicology* **2016**, *10* (3), 257–278. Available from: https://doi.org/10.3109/17435390.2015.1048326 *DOI.org (Crossref).*
27. Wu, H.; Li, Z. Nano-enabled agriculture: how do nanoparticles cross barriers in plants? *Plant. Commun.* **2022**, *3* (6), 100346. Available from: https://doi.org/10.1016/j.xplc.2022.100346 *DOI.org (Crossref).*
28. Guidi, L.; Tattini, M. Antioxidant defenses in plants: a dated topic of current interest. *Antioxidants* **2021**, *10* (6), 855. Available from: https://doi.org/10.3390/antiox10060855 *DOI.org (Crossref).*
29. Pardo-Hernández, M., et al. ROS and NO regulation by melatonin under abiotic stress in plants. *Antioxidants* **2020**, *9* (11), 1078. Available from: https://doi.org/10.3390/antiox9111078 *DOI.org (Crossref).*
30. Hasanuzzaman, M., et al. Reactive oxygen species and antioxidant defense in plants under abiotic stress: revisiting the crucial role of a universal defense regulator. *Antioxidants* **2020**, *9* (8), 681. Available from: https://doi.org/10.3390/antiox9080681 *DOI.org (Crossref).*
31. Dada, L., et al. Role of sesquiterpenes in biogenic new particle formation. *Sci. Adv.* **2023**, *9* (36), eadi5297. Available from: https://doi.org/10.1126/sciadv.adi5297 *DOI.org (Crossref).*
32. Cha, J.-Y., et al. Humic acid enhances heat stress tolerance via transcriptional activation of heat-shock proteins in Arabidopsis. *Sci. Rep.* **2020**, *10* (1).15042. Available from: https://doi.org/10.1038/s41598-020-71701-8 *DOI.org (Crossref).*
33. Benito, P., et al. A fast method to evaluate in a combinatorial manner the synergistic effect of different biostimulants for promoting growth or tolerance against abiotic stress. *Plant. Methods* **2022**, *18* (1), 111. Available from: https://doi.org/10.1186/s13007-022-00943-6 *DOI.org (Crossref).*

34. Nephali, L., et al. Biostimulants for plant growth and mitigation of abiotic stresses: a metabolomics perspective. *Metabolites* **2020**, *10* (12), 505. Available from: https://doi.org/10.3390/metabo10120505 *DOI.org (Crossref).*
35. Sleighter, R. L., et al. Abiotic stress mitigation: a case study from 21 trials using a natural organic matter based biostimulant across multiple geographies. *Agronomy* **2023**, *13* (3), 728. Available from: https://doi.org/10.3390/agronomy13030728 *DOI.org (Crossref).*
36. Hassan, F. E., et al. Effective priming techniques to enhance ghaf (*Prosopis cineraria* L. Druce) seed germination for mass planting. *Horticulturae* **2023**, *9* (5), 542. Available from: https://doi.org/10.3390/horticulturae9050542 *DOI.org (Crossref).*
37. Monteiro, E., et al. The role of biostimulants as alleviators of biotic and abiotic stresses in grapevine: a review. *Plants* **2022**, *11* (3), 396. Available from: https://doi.org/10.3390/plants11030396 *DOI.org (Crossref).*
38. Eftekhari, A., et al. Differences in the drought stress response of *DREB2* and *CAT1* genes and evaluation of related physiological parameters in some bread wheat cultivars. *Biotechnol. & Biotechnological Equip.* **2017**, 1–8. Available from: https://doi.org/10.1080/13102818.2017.1316214 *DOI.org (Crossref).*
39. Adrees, M., et al. Foliar Exposure of zinc oxide nanoparticles improved the growth of wheat (*Triticum aestivum* L.) and decreased cadmium concentration in grains under simultaneous Cd and water deficient stress. *Ecotoxicol. Environ. Saf.* **2021**, *208*111627. Available from: https://doi.org/10.1016/j.ecoenv.2020.111627 *DOI.org (Crossref).*
40. Graziani, G., et al. Biostimulants improve plant growth and bioactive compounds of young olive trees under abiotic stress conditions. *Agriculture* **2022**, *12* (2), 227. Available from: https://doi.org/10.3390/agriculture12020227 *DOI.org (Crossref).*
41. El-Badri, A. M., et al. Comparative efficacy of bio-selenium nanoparticles and sodium selenite on morpho-physiochemical attributes under normal and salt stress conditions, besides selenium detoxification pathways in *Brassica napus* L. *J. Nanobiotechnology* **2022**, *20* (1), 163. Available from: https://doi.org/10.1186/s12951-022-01370-4 *DOI.org (Crossref).*
42. Hu, C., et al. Nanomaterials regulate bacterial quorum sensing: applications, mechanisms, and optimization strategies. *Adv. Sci.* **2024**, *11* (15).2306070. Available from: https://doi.org/10.1002/advs.202306070 *DOI.org (Crossref).*
43. Furlan, A., et al. *Combined application of microbial and non-microbial biostimulants to improve growth of peanut plants exposed to abiotic stresses.* In *Microbial probiotics for agricultural systems;* Zúñiga-Dávila, D., et al., Eds.; Springer International Publishing, 2019; pp 239–256. Available from: https://doi.org/10.1007/978-3-030-17597-9_17. *DOI.org (Crossref).*
44. Ali, A. Y. A., et al. Exogenous jasmonic acid and humic acid increased salinity tolerance of sorghum. *Agron. J.* **2020**, *112* (2), 871–884. Available from: https://doi.org/10.1002/agj2.20072 *DOI.org (Crossref).*
45. Koleška, I., et al. Biostimulant prevents yield loss and reduces oxidative damage in tomato plants grown on reduced NPK nutrition. *J. Plant. Interact.* **2017**, *12* (1), 209–218. Available from: https://doi.org/10.1080/17429145.2017.1319503 *DOI.org (Crossref).*
46. Francesca, S., et al. Phenotyping to dissect the biostimulant action of a protein hydrolysate in tomato plants under combined abiotic stress. *Plant. Physiol. Biochem.* **2022**, *179*, 32–43. Available from: https://doi.org/10.1016/j.plaphy.2022.03.012 *DOI.org (Crossref).*

47. Waadt, R., et al. Plant hormone regulation of abiotic stress responses. *Nat. Rev. Mol. Cell Biol.* **2022**, *23* (10), 680–694. Available from: https://doi.org/10.1038/s41580-022-00479-6 *DOI.org (Crossref)*.
48. De Jonge, R., et al. How filamentous pathogens co-opt plants: the ins and outs of fungal effectors. *Curr. Opin. Plant. Biol.* **2011**, *14* (4), 400–406. Available from: https://doi.org/10.1016/j.pbi.2011.03.005 *DOI.org (Crossref)*.
49. Baslam, M. Advances and new perspectives in plant-microbe interactions. *Int. J. Mol. Sci.* **2023**, *24* (6), 5143. Available from: https://doi.org/10.3390/ijms24065143 *DOI.org (Crossref)*.
50. Balestrini, R., et al. The hidden side of interaction: microbes and roots get together to improve plant resilience. *J. Plant. Interact.* **2024**, *19* (1).2323991. Available from: https://doi.org/10.1080/17429145.2024.2323991 *DOI.org (Crossref)*.
51. Thabet, S. G.; Alqudah, A. M. Unraveling the role of nanoparticles in improving plant resilience under environmental stress condition. *Plant. Soil.* **2024**. Available from: https://doi.org/10.1007/s11104-024-06581-2 *DOI.org (Crossref)*.
52. Shaikh, A., et al. Microbial nanotechnology for producing stress smart crops. *Plant. Nano Biol.* **2024**, 7100063. Available from: https://doi.org/10.1016/j.plana.2024.100063 *DOI.org (Crossref)*.
53. Mgadi, K., et al. Nanoparticle applications in agriculture: overview and response of plant-associated microorganisms. *Front. Microbiology* **2024**, *15*1354440. Available from: https://doi.org/10.3389/fmicb.2024.1354440 *DOI.org (Crossref)*.
54. Noman, M., et al. Nano-enabled crop resilience against pathogens: potential, mechanisms and strategies. *Crop. Health* **2023**, *1* (1), 15. Available from: https://doi.org/10.1007/s44297-023-00015-8 *DOI.org (Crossref)*.
55. Tortella, G., et al. Nanoparticles as a promising strategy to mitigate biotic stress in agriculture. *Antibiotics* **2023**, *12* (2), 338. Available from: https://doi.org/10.3390/antibiotics12020338 *DOI.org (Crossref)*.
56. Ammendolia, M. G.; De Berardis, B. Nanoparticle impact on the bacterial adaptation: focus on nano-titania. *Nanomaterials* **2022**, *12* (20), 3616. Available from: https://doi.org/10.3390/nano12203616 *DOI.org (Crossref)*.
57. Sahli, C., et al. Recent advances in nanotechnology for eradicating bacterial biofilm. *Theranostics* **2022**, *12* (5), 2383–2405. Available from: https://doi.org/10.7150/thno.67296 *DOI.org (Crossref)*.
58. Ma, J., et al. Toward a Spatiotemporal understanding of dolomite dissolution in sandstone by CO $_2$-enriched brine circulation. *Environ. Sci. & Technol.* **2019**, *53* (21), 12458–12466. Available from: https://doi.org/10.1021/acs.est.9b04441 *DOI.org (Crossref)*.
59. Rajput, V. D., et al. Can nanomaterials improve the soil microbiome and crop productivity? *Agriculture* **2023**, *13* (2), 231. Available from: https://doi.org/10.3390/agriculture13020231 *DOI.org (Crossref)*.
60. Bourges, F., et al. Hydrogeological control on carbon dioxide input into the atmosphere of the Chauvet-Pont d'Arc Cave. *Sci. Total. Environ.* **2020**, *716*136844. Available from: https://doi.org/10.1016/j.scitotenv.2020.136844 *DOI.org (Crossref)*.
61. Thill, A., et al. Cytotoxicity of CeO_2 nanoparticles for *Escherichia coli*. physico-chemical insight of the cytotoxicity mechanism. *Environ. Sci. Technol.* **2006**, *40* (19), 6151–6156. Available from: https://doi.org/10.1021/es060999b *DOI.org (Crossref)*.
62. Nel, A. E., et al. Understanding biophysicochemical interactions at the nano–bio interface. *Nat. Mater.* **2009**, *8* (7), 543–557. Available from: https://doi.org/10.1038/nmat2442 *DOI.org (Crossref)*.

63. Fierer, N. Embracing the unknown: disentangling the complexities of the soil microbiome. *Nat. Rev. Microbiology* **2017**, *15* (10), 579–590. Available from: https://doi.org/10.1038/nrmicro.2017.87 *DOI.org (Crossref)*.
64. Khan, S. T., et al. Engineered nanomaterials in soil: their impact on soil microbiome and plant health. *Plants* **2021**, *11* (1), 109. Available from: https://doi.org/10.3390/plants11010109 *DOI.org (Crossref)*.
65. Mittal, D., et al. Nanoparticle-based sustainable agriculture and food science: recent advances and future outlook. *Front. Nanotechnol.* **2020**, 2579954. Available from: https://doi.org/10.3389/fnano.2020.579954 *DOI.org (Crossref)*.
66. Ur Rahim, H., et al. Nano-enable materials promoting sustainability and resilience in modern agriculture. *Nanomaterials* **2021**, *11* (8), 2068. Available from: https://doi.org/10.3390/nano11082068 *DOI.org (Crossref)*.
67. Singh, P., et al., Eds. *Plant-microbes-engineered nano-particles (PM-ENPs) Nexus in agro-ecosystems: understanding the interaction of plant, microbes and engineered nano-particles (ENPS);* Springer International Publishing, 2021. Available from: https://doi.org/10.1007/978-3-030-66956-0 *DOI.org (Crossref)*.
68. Asif, A., et al. Enhancing crop resilience by harnessing the synergistic effects of biostimulants against abiotic stress. *Front. Plant. Sci.* **2023**, *14*1276117. Available from: https://doi.org/10.3389/fpls.2023.1276117 *DOI.org (Crossref)*.
69. Margaryan, G., et al. Unveiling the salinity tolerance potential of Armenian Dandur (*Portulaca oleracea* L.) genotypes: enhancing sustainable agriculture and food security. *J. King Saud. Univ.—Sci.* **2024**, *36*.103332. Available from: https://doi.org/10.1016/j.jksus.2024.103332.
70. Sharma, A., et al. Revolutionizing agricultural sustainability and food security and management to achieve SDGs goals via nanotechnology. *Adv. Environ. Eng. green. Technol. book. Ser.* **2024**, 276–288. Available from: https://doi.org/10.4018/979-8-3693-1471-5.ch013.
71. Ghazaryan, K.; Verma, A.; Rawat, S.; Rajput, P.; Minkina, T.; Mandzhieva, S., et al. Implementing sustainable nano-bioremediation for emerging pollutants. *Adv. Environ. Eng. green. Technol. book. Ser.* **2024**, 316–332. Available from: https://doi.org/10.4018/979-8-3693-1471-5.ch016.
72. Verma, A.; Singh, A.; Rawat, S.; Rajput, P.; Minkina, T.; Mandzhieva, S., et al. Unveiling nanomaterial-induced toxicity. *Adv. Environ. Eng. green. Technol. book. Ser.* **2024**, 236–253. Available from: https://doi.org/10.4018/979-8-3693-1471-5.ch011.
73. Singh, A.; Sharma, A.; Singh, O.; Rajput, V. D.; Movsesyan, H. S.; Minkina, T., et al. In-depth exploration of nanoparticles for enhanced nutrient use efficiency and abiotic stresses management: present insights and future horizons. *Plant. Stress.* **2024**.100576. Available from: https://doi.org/10.1016/j.stress.2024.100576.
74. Upadhyay, P., et al. Nano-omics-based abiotic and biotic stresses management. *Adv. Environ. Eng. Green. Technol. book. Ser.* **2024**, 348–371. Available from: https://doi.org/10.4018/979-8-3693-1890-4.ch017.
75. Fathi, A.; Shiade, S. R. G.; Parmoon, G.; Yaghoubian, Y.; Pirdashti, H.; Rajput, V. D., et al. *Bioremediation of heavy metals contaminated soils using nanotechnology.* Elsevier eBooks, 2024. 611–628. Available from: https://doi.org/10.1016/b978-0-443-21610-7.00004-5.

CHAPTER

Role of fungi in biopesticides and biogenic synthesis of nanofertilizers and nanopesticides

11

Rupjyoti Bharali[1], Parismita Sarma[2], Indrani Barman[3], Sarvesh Rustagi[4], Sagar Mondal[5], Sumira Malik[5] and Jutishna Bora[5]

[1]*Department of Botany, Biswanath College, Biswanath Chariali, Assam, India*
[2]*Department of Botany, Rangia College, Rangia, India*
[3]*Program of Biochemistry, Faculty of Science, Assam Down Town University, Guwahati, Assam, India*
[4]*School of Applied and Life Sciences, Uttaranchal University, Dehradun, Uttarakhand, India*
[5]*Amity Institute of Biotechnology, Amity University Jharkhand, Ranchi, Jharkhand, India*

11.1 Introduction

The use of pesticides over many years have exerted negative impacts on ecosystem thus emphasizing on the need to develop alternatives that are efficient in managing pests and are environmentally safe.[1] Considering the multiple hazards associated with the application of agrochemicals, the attention has been shifted toward microbial agents for biogenic synthesis of pesticides and fertilizers. Microbial enzymes are crucial in developing potential biopesticides that can alter the cell wall of plant pathogens[2,3] using nanoparticle-based formulations.[4]

Biogenic nanoparticles have shown efficacy in release of pesticides and fertilizers in a safer way.[5] Detrimental effects have been exerted on ecosystem because of unplanned use of pesticides and fertilizers in agro-environments leading to the production of biogenic agriinputs. The domains of nanotechnology and microbial biotechnology are combined in the process of microbially assisted nanoparticle production.[6] Fungi is preferable for large-scale production of nanomaterials over other microbes and plant as they yield various metabolites, enzymes of marketable value—in their capacity to grow on substrates that are inexpensive and easily available[7,8] and their large surface area.[9] Fungal biosystem is very flexible as it can synthesize metal nanomaterials by intracellular as well as extracellular mechanisms[10] and are considered potential bio-nanafactories.[11] Fungal species, namely, *Aspergillus*, *Fusarium*, *Pencillium*, *Trichoderma*, *Alternaria*, and *Verticillium*, have been employed for fabrication of nanoparticles.[6,12,13]

Fungi especially endophytic fungi due to its role as plant growth promoter, secretion of antimicrobial metabolites, and phytotoxins have been utilized in

Biogenic Nanoparticles. DOI: https://doi.org/10.1016/B978-0-443-33819-9.00015-9

development of biopesticides.[14] Approximately, 90 genera are categorized as entomopathogenic fungi that infects and kills pests.[15,16] Utilizing the benefits of microorganisms to promote sustainable agriculture by replacing chemical pesticides is an easy and inexpensive way. The present chapter reviews the activity of fungi in synthesis of biopesticides, and nanomaterials as nanopesticides and nanofertilizers.

11.2 Role of fungi in biopesticide

Fungi-based biopesticides are very suitable in case of integrated pest management practices since they are biodegradable, have high specificity, high reproductive rate, with minimal chances of developing resistance in pests and there are very few controllable health risks.[17,18] Fungal-derived biopesticides constitutes only 10% of the total available biopesticides; rest are derived from other microbes.[19]

11.2.1 Role of entomopathogenic fungi

Entomopathogenic fungal genera, namely, *Beauveria*, *Metarhizium*, and *Trichoderma,* are most used in commercial biopesticides.[20–26] Entomopathogenic fungi consisting of more than 750 species are able to kill insects by entering through their integuments. They secrete enzymes, namely, upases, quitobiases, proteases, chitinases and lipoxygenases to microorganisms are degrade insect's cuticle through which they develop hyphae inside the insect's tissues thereby killing the host insect (Fig. 4.2).[27] Biopesticides derived from also known as biological control agents[27] and the most widely used agent is *Trichoderma* spp.-based commercial biopesticides that dominate 60% of the total fungi-derived biocontrol agents in the global market of biopesticides.[28] The popularity of the genus *Trichoderma* in formulations of biopesticides is due to its wide range of mechanisms for biocontrol such as colonization, competition, antibiosis, direct mycoparasitism and so on against foliar, soil, and postharvest phytopathogens.[19] In literature, frequently mentioned species of *Trichoderma* as active biocontrol agents are *T. harzianum*, *T. atroviridae*, *T. asperellum*, *T. gamsii*, and *T. polysporum* of which *T. harzianum* is most marketed globally. The demand for an entomopathogenic fungi, namely, *Beauveria bassiana,* is very high as an active ingredient in biopesticides and is recommended for use in crop species due to its antiinsecticidal properties capable of suppressing insect populations such as whitefly (*Bemisia tabaci*), weevil (*Gonipterus scutellatus*), borer beetle (*Hypothenemus hampei*), and mite (*Tetranychus urticae*).[17] Most of the brands having global sales do not share the vital informations of the fungi used in the formulation of biopesticides.[17]

Entomopathogenic fungi sp., namely, *Penicillium* sp. (Nlg 1), *Fusarium* sp. (khr4), and *Beauveria* sp. (deb4), showed efficiency against the aphid *Aphis gossypii* by producing and germinating spores on the surface of the aphis cuticle and thereby killing the aphid.[29] *Aphis gossypii* is the causal agent of Papaya Ring Spot Virus

(PRSV). Most of the entomopathogenic fungi belongs to Zygomycetes, Hyphomycetes, Laboulbeniales, and Pyrenomycetes and are easily extracted from cadavers of insects and soil. EPF, namely, *Aschersonia aleyrodis*, *Clonostachysrosea*, *Verticillium*, and *Isaria*, are known to control the infestation by whitefly.[30] *Beauveria bassiana*, an EPF, controls the infestation rate by *Curculio sayi*, a pest of chestnut tree either as natural repellent or by systematic fungus infections.[31] In another study, *A. flavus*, an EPF, was used as effective biocontrol agent against populations of seed bug (*E. pallens*) and on the mango seed weevil (*Sternochetus mangiferae*).[32]

11.2.2 Role of endophytic fungi

Endophytic fungi residing inside the host plant tissues can be investigated for new products for developing biopesticides to be used in the agroecosystem. Fungal edophytes promote growth in host plants, boost tolerance against any kind of abiotic or biotic stress, secrete various secondary metabolites such as isocoumarins, terpenoids, alkaloids, peptides, steroids, quinones, and benzopyranones that provide protection to host plant against numerous phytopathogenic microorganisms and pests (Fig. 4.1).[33,34] For instance, antifungal compounds secreted by species of *Trichoderma* act against the growth of phytopathogens, namely, *Colletotrichum musae*, *Fusarium verticillioides*, and *Lasiodiplodia theobromae*[35] and species of *Xylaria* against *Aspergillus niger* and *Fusarium avenaceum*,[36,37] antifungal activities of *Penicillium griseofulvum* against phytopathogens such as *Alternaria solani*, *Botrytis cinerea*, *Cladosporium gloeosporioides*, *Cytospora* sp., and *Fusarium solani*,[38] antibacterial and antifungal activity of species of *Diaporthe* against *Trichothecium roseum*, *F. avenaceum* and *A. alternata*,[39] *F. oxysporum* and *Colletotrichum* sp.,[40] *F. solani* and *Didymella bryoniae*.[41] Antibacterial activity was shown by *Diaporthe phragmitis* against *Pseudomonas syringae* pv. *actinidiae*.[42] Considering these beneficial roles that endophytic fungi play makes them suitable and sustainable agents for development of novel biopesticides exploiting their numerous agricultural bioactivities like antimicrobial, anti-insecticidal, etc. There are several mechanisms through which endophytes can control the pathogens like competition for space and nutrition,[43,44] secretion of extracellular hydrolytic enzymes,[45] secretion of metabolites that are inhibitory in action, that is, antibiosis,[44,46,47] induction of defensive response for protection from phytopathogens,[44,45] removal of toxicity of virulence factors.[45]

Xu et al. (2002)[34] reviewed 132 metabolites of endophytic fungi of which some of the metabolites are potent alternatives for chemical pesticides. Griseofulvin, a secondary metabolite belonging to class polyketide with more than 400 analogs, produced by many members of ascomycetes fungi, namely, species of *Nigrospora*, *Penicillium*, *Aspergillus*, and *Xylaria*, possesses antifungal properties against phytopathogenic fungi[48–50] and shows the potential of this compound for synthesis of biopesticides.[34] Other such metabolite with possibilities for development of biopesticide are trichothecenes belonging to class sesquiterpenes secreted by *Trichoderma* sp., *Spicellum* sp., *Myrothecium* sp.,

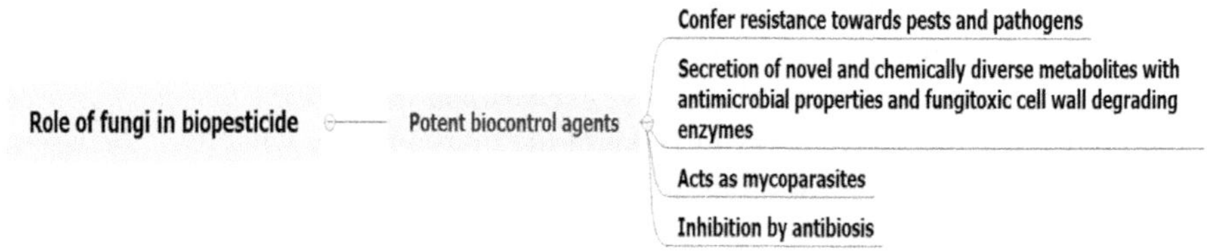

FIGURE 11.1

Benefits of using fungi in developing biopesticide.

Stachybotrys sp., *Trichothecium* sp., *Fusarium* sp., and *Cephalosporium* sp. and cytochalasin analogs, possessing antifungal properties produced and isolated from endophytes belonging to ascomycetes and basidiomycetes members.[51]

The application of commercial biopesticides obtained from fungal metabolites is still not usual in large scale and are limited to in vitro conditions as there are some research gaps, which needs to be addressed such as enhancing the production of metabolites by optimizing conditions in production, evaluation of the ingredients for developing commercial formulations, testing the efficacy of the formulations in field conditions and lastly commercialization of the bioproducts developed. Moreover, the farmers are not sure of the positive outcomes of utilizing these products in place of synthetic pesticides as the bioproducts can only minimize the phytopathogenic effects but not eradicate the pathogen.[52] Better efficacy of these fungi-mediated bioproducts over chemical pesticides has been documented when only the metabolites were used excluding the organism as the metabolites are not much influenced by the external conditions.[38,53] Companies for commercialization of these bioproducts derived from fungi must associate with research institutions to make necessary technological advancement for large-scale applications[52] (Fig. 11.1).

11.3 Fungi: an efficient biosystem for synthesis of nanoparticles

In recent years, interest in the manufacture of metal nanaoparticles by various microbes and plant has grown. Fungi in comparison to the other microbes and plants are more preferred for synthesis of metal nanoparticles.[6,54–56] Since fungi can be readily and economically cultivated on a big scale, in a short amount of time, they are an efficient material for synthesizing nanoparticles,[56] also for their capacity to fabricate nanoparticles by intracellular and extracellular ways,[6] their high tolerance for metal concentrations,[56] they secrete enzymes and biomolecules in large quantities that play the role as reducing agents; that is, they reduce the metal ions into their metallic NPs by and as stabilizing agents of the NPs by preventing their accumulation through adsorption around them, their biosystem is resistant in bioreactors toward any environmental conditions in the biosynthesis process allowing high yield, nanoparticles with specific desired properties can be synthesized as culture factors of fungus can be altered.[56] Furthermore, the NPs mycosynthesized extracellulary find application in different fields.[56–58]

Compared to other microbes, fungi are more competent for synthesis of nanoparticles as they can produce enormous bioactive substances[59] that subsequently find application as nanofertilizers and nanopesticides. These bioactive substances functions as reducing and capping agents during nanoparticle synthesis.[60] Although the exact mechanism of biogenic synthesis is not properly understood but it is known that fungi-mediated biosynthesis of nanoparticles takes place both by intracellular and extracellular pathways. Kashyap et al. (2013)[61] elucidated the theoretical method of intracellular synthesis of nanoparticles as trapping of Ag + by enzymes at the surface of the fungal cell wall followed by enzymatic reduction of the associated metal ions. This causes metal ions to build up to form nanoparticles. While extracellular mechanism is more feasible involving the association of metal ions and release of enzyme reductase that yields nanoparticles in solution. In extracellular mechanism unlike intracellular way, breakdown of fungal cell and purification of the nanoparticles is not required.[62]

11.4 As reducing agents

Fungal biosystem finds application as stabilizing and reducing agents because of their traits like internalization and bioaccumulation of metals and tolerance to heavy metals (Fig. 4.2). For example, *Fusarium oxysporum* was utilized to produce silver nanoparticles, and the results showed that anthraquinones and enzyme nitrate reductase were associated with the reduction of silver ions[63,64] and showed that extracellular NADPH-dependent nitrate reductase enzymes and quinones aided in the nanoparticles production using *Fusarium oxysporum*. Furthermore, they are feasible to manufacture desired quality of nanoparticles.[65,66] Secretion of enzymes and extracellular metabolites by fungi catalyzes metal ions reduction to metallic nanoparticles.[67]

The process of reduction of metal ions is described for some metals only and considered as same for the other metals in fungi-mediated synthesis. In case of extracellular mycosynthesis, the metals ions reduction to metallic nanoparticles is carried out by catalytic effect as a result of exposure of fungal enzymes and extracellular metabolites to varied stress conditions.[67] In intracellular mycosynthesis, the reduction takes place in two steps, namely, attachment of metal ions to the surface of fungal cell[68] and reduction of the metal ions to metallic NPs by the enzymes from the cell membrane as well as by sugars of the cell wall.[69]

11.5 Capping of nanoparticles

Capping of nanaoparticles during synthesis is done by the biomolecules of the organism used for production of nanoparticles to provide stability to the nanoparticles, also it prevents aggregation of particles and there may be some biological activity associated with it.[70,71] No extra steps are involved to coat the nanomaterials.[72] The negative carboxyl groups in the enzymes of cell wall in fungi provides stability.[73,74] Anchoring of drugs for transportation into cells can be achieved by these protein cappings or coatings.[75]

There is paucity of literature regarding cappings on biogenic NPs. Microscopic analysis showed the presence of capping on mycosynthesized silver NPs.[76] Using scanning electron microscopy for detection of capping and SDS-Page for fungal proteins charecterization,[72] confirmed the presence of cappings on fungi-mediated biogenic synthesis of silver NPs (Fig. 4.2). Guilger-Casagrande et al.[77] reported that the hydrolytic enzymes of *Trichoderma harzianum* acts as antipathogens against phytopathogens, and these enzymes were found to be present in the capped NPs, which were mycosynthesized using *T. harzianum*. For the uncapped nanoparticle samples in their study no activity was recorded. The antipathogenic activity of these enzymes in the capped NPs was seen in its ability to inhibit the growth of phytopathogen *Sclerotinia sclerotiorum* to a greater extent compared to the uncapped NPs. They suggested that further studies must analyze for the proteins and enzymes present in capped nanoparticles as capping enhances the effects of the nanoparticles against agricultural pests.

11.6 Potent entomopathogenic agents

These are group of fungal species that are parasitic in nature and finds application in agriculture by limiting the infestation by insects on crop plants naturally.[78] The host range of EPF is larger than other entomopathogenic microbes and they are host-specific thus infecting pests found in the aerial region as well as in the soil acting as specialist's pathogen. They can act as generalist pathogen also by infecting a range of insect hosts.[79,80] Fungi over other microbes are considered better entomopathogens as they can cause infection by utilizing the nutrients from the host body (insects), whereas other pathogenic microbes must be ingested or by other means must enter the host body (Fig. 4.2).[32] Among several other reasons, the loss in total agricultural output globally is due to the agricultural insect pests.[79,81] Although the use of chemical pesticides is considered efficient, but it has developed resistance in insect pests toward the insecticides.[82,83] EPF are more efficient than chemical insecticides.[84,85] *Beauveria*, *Metarhizium* and *Trichoderma* are few EPF species that infects and kills arthropods. EPFs are easily isolated from the insects infected and from soil and cultured in vitro.[30]

Secretions in the genome of entomopathogenic fungi is more compared to other fungal genome. The secondary metabolites become active and gain pathogenicity after contact with insect tissues or exposure to any environmental stress.[80] The metabolites from *Isaria fumosorosea* used against *Plutella xylostella*, a well-known pest of several crops showed mortality in the pest.[86]

11.7 Biogenic synthesis of metal nanoparticles by entomopathogenic fungi

Metal nanoparticles are known as efficient insecticidal agent and hence their synthesis is catching attention as safer way of controlling population of insect pests although this research is still in its developmental stage. Entomopathogenic fungi-mediated synthesis of nanosilver at low doses has shown efficiency as nanopesticide in limiting the incidence of infestation by a larger range of pest species. Microbes are capable of synthesizing nanaoparticles extracellularly (generation outside the cell) and intracellularly (inside the cell) depending on the site of synthesis.[80] Entomomycosynthesized silver nanoparticles may be used in agricultural crops against pests, but extensive studies on the toxicity of these particles must be carried out for its application in agriculture.[87,88]

11.8 Endophytic fungi-potent biofactories for biogenic synthesis of nanoparticles

Endophytic fungi are also used for mass production of nanaoparticles as they show higher tolerance and binding capacity of metal salts for relieving stress and toxicity in the host plant enhancing their competitive ability as well as adaptation,[89] cost-effective, and effective downstream process. Using both intracellular and extracellular mechanisms, endophyte's ability to detoxify metals can be channelized for production of metal-based NPs. However, extracellularly synthesized nanoparticles from endophytic fungi provides long-time stability as a result of slower kinetics in these group of fungi.[90] For example, extracellular synthesis of ZnONPs by zinc-tolerant endophyte *Cochliobolus geniculatus*, CoONPs by the CoO tolerant *Aspergillus nidulans*.[91]

Mycosynthesis of AgNPs using endophytic fungi has antimicrobial effects on phytopathogenic strains of fungi and bacteria[61,92–95] and growth in plants.[96] There is paucity of literature in the role of endophytes in the synthesis of AgNPs as potential nanopesticide and nanofertilizer.

Fungal endophytes are rich in bioactive metabolites capable of synthesizing nanoparticles of desired shape and size.[6,97] Apart from mycosynthesized metal NPs, metal oxide NPs also shows antimicrobial properties especially MnO NPs.[98–100] Mycosynthesis of MnO nanoparticles using metabolites from *Trichoderma* spp. exhibited antifungal effects against *Aspergillus favus*, *Aspergillus niger*, *Fusarium solani* as well as antibacterial effects against *Erwinia amylovora*, *Acetobacter pasteurianus*, *Erwinia persicina* indicating its possible use as nanopesticides[101] (Fig. 11.2).

11.9 Agro-based applications of diverse nanoparticles synthesized by fungi

Strong insecticidal activity was exhibited against *Lipaphis erysimi* (mustard aphid) by biogenic silver nanoparticles synthesized from *Beauveria bassiana* which is an

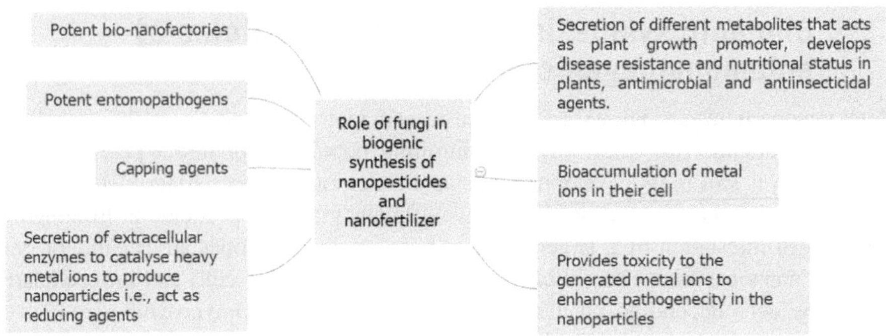

FIGURE 11.2

Applications of fungi in formulation of nanopesticides and nanofertilizers.

entomopathogenic fungi.[102] Titanium dioxide nanoparticles synthesized by *Trichoderma viride* were effective against growth of *Helicoverpa armigera* in crops.[4] In biogenic synthesis of nanoconjugate, fungal pesticidal metabolites are added with metallic nanoparticles for obtaining eco-friendly biopecsticidal agent against insect pests in agriculture. Karthick Raja Namasivayam et al.[103] synthesized nanoconjugate with silver nanoparticles and metabolites extracted from fungus *Nomuraea rileyi* (AgNp–FM) and analyzed their pesticidal activity against *Spodoptera litura* (Fab.) and the results were highly effective. Titanium nanoparticles derived from *Metarhizium anisopliae* showed strong insecticidal and larvicidal activity against *Galleria mellonella.*[104] Using 16 isolates of entomopathogenic fungi silver nanoparticles were synthesized and the efficacy was tested on a model insect, namely, *Tenebrio molitor,* and it was found that *B. bassiana*, *M. anisopliae*, and *I. fumosorosea* isolates were effective against the test insect showing their pesticidal potential.[105] Larvicidal activity against white grubs (*Holotrichia* sp), a sugarcane pest, was shown by silver nanoparticles synthesized from *Fusarium pallidoroseum.*[106] *Beauveria bassiana*-derived zinc oxide nanoparticles revealed insecticidal activities against whitefly *Trialeurodes vaporarioru.*[107] *Isaria fumosorosea*-based iron nanoparticles were acive against growth and survival of silverleaf whitefly *Bemisia tabaci*, an important agricultural pest.[108] *Beauveria brongniartii*-based iron nanoparticles showed activity against cotton leafworm *Spodoptera litura.*[109] Using *Trichoderma atroviride,* selenium nanospheres were synthesized, which showed antifungal activity on phytopatogenic fungi *Phytophthora infestans.*[110] In another study, SeNPs were synthesized using *T. harzianum* and its biocontrol efficacy was tested on pest *Spodoptera litura,* which showed larvicidal activities.[111] In a study by Helar and Chavan (2015),[112] it was reported that biogenic synthesis of gold nanoparticles by *Fusarium oxysporum* enhances the seed viability, which is very important agriculture sector. These studies show the potential of entomopathogenic fungi-mediated synthesis of nanoparticles especially silver nanaoparticles for developing nanopesticides for utilization in large-scale after being tested for field application.

Different structures of fungi-mediated synthesis of nanaomaterials are now used as biocontrol agents for protecting plants from a wide range of pests, insects, and plant diseases.[113,114] Mycosynthesized nanoparticles are also reported to decompose pesticides that are harmful like decomposition of chlorpyrifos pesticide by *Penicillium pinophilum*-derived silver nanoparticles.[115] Antifungal activity against rotting plant pathogens was shown by mycosynthesized copper nanoparticles using *Aspergillus versicolor* extract.[9] CuNPs synthesized using *Aspergillus flavus* showed antifungal activity against fungal crop pathogens, namely, *Alternaria alternata*, *Aspergillus niger*, and *Fusarium oxysporum.*[116] Mycosynthesized silica NPs derived from *Fusarium oxysporum* develops disease resistance in plant cells[117] and gold nanoparticles from the same fungi enhanced the germinating capacity of *Helianthus annuus* seeds.[112] *Fusarium solani*-derived AgNPs revealed antifungal effects against several species of fungal pathogens that infect agricultural crops.[118] Growth of phytopathogenic fungi, namely, *Fusarium moniliforme*, *Fusarium verticillioi-des*, *Helminthosporium oryzae*, *Penicillium brevicompac-tum,* and *Pyricularia grisea,* was restrained by mycosynthesized silver nanoparticles produced extracellularly by *Trichoderma longibrachiatum.*[67] *Penicillium chrysogenum*-mediated synthesis of CuO-NPs showed antimicrobial activities against phyto pathogenic fungi and bacteria.[114]

Mycosynthesized ZnONPs from *Fusarium solani* IOR 825 enhanced crop productivity in maize and have antipathogenic abilities.[119] Production of beneficial secondary metabolites by plant growth promoting fungi makes them suitable for biogenic synthesis of biofertilizer. For instance, mycosynthesis of zinc oxide nanoparticles (ZnO NPs) was carried out by coculturing few *Trichoderma* spp. and by monoculture of each species, and their antibacterial activity against *Xanthomonas oryzae* pv. Oryzae that causes bacterial leaf blight in rice was recorded. ZnO NPs using cocultured *Trichoderma* spp. showed more antipathogenic activity and their ability to synthesize more new secondary metabolites than monocultured species. The authors suggested that this NPs can serve as nanofertilizers.[120]

Nurfalah[121] reported that species of *Trichoderma* are nontoxic to environment and are safe to use as biofertilizer. Lee et al.[122] reported that the species of *Trichoderma* can be used as biocontrol agents in agriculture including soil-borne pathogens and can enhance crop productivity. These studies show that fungi are very beneficial for biogenic synthesis of nanoparticles, which can be utilized as nanofertilizer as well as nanopesticides.

Mycosynthesized TcAgNPs and $TcTiO_2$ NPs using *Trichoderma citrinoviride* extract showed enhanced germination and growth parameters in *Solanum lycopersicum* at lower concentrations of NPs and its higher concentrations were found to be toxic.[123] In another study by Albalawi et al. (2022),[124] it was observed that mycosynthesized SiO2-NPs from *Aspergillus niger* showed antifungal activity against the phytopathogen *Alternaria solani* and it also showed growth promotory effect on the host eggplant both in infected and in healthy plants. These studies suggest and indicate that at optimum concentrations, these biogenically synthesized NPs can be used as nanofertilizers to enhance crop productivity in an eco-friendly manner. ZnO nanoparticles were mycosynthesized using *Ganoderma lucidum* extracts and its efficacy as a nanofertilizer was tested on *Lepidium sativum* and the results showed that ZnONPs enhanced the growth parameters of the test plant.[125] In Table 11.1, further elucidation of recent findings on fungi-mediated nanoparticles and their activities was listed.

Table 11.1 Fungal species involved in the biogenic synthesis of nanoparticles with agro- applications.

S. no.	Fungal species	Nanoparticle synthesized	Effect on	Activity	References
1	*A. flavus* TFR 7	TiO_2 NPs	Mung bean	Plant nutrient fertilizer	126
2	*Aspergillus fumigatus* TFR-8	ZnO nanoparticles	Rhizosphere of mung bean	Nanofertilizer; mobilization of native phosphorous in rhizosphere of mung bean	113
3	*Aspergillus flavus*	CuNPs	*Aspergillus niger* *Fusarium oxysporum* *Alternaria alternata*	Antifungal	116
4	*Aspergillus niger*	Silver nanoparticles (AgNPs)	*Holotrichia* sp.	Pesticidal	106
5	*Aspergillus niger*	silver nanoparticles (AgNPs)	*Phytophthora infestans*	Potential fungicides	127
6	*Aspergillus niger*	ZNONPs	*Alternaria solani*	Potential fungicides	128
7	*Aspergillus niger* AH1	SiO_2-NPs	*Alternaria solani*	Biofungicide	124
			Egg plant	Plant growth-promoting agent	
8	*Aspergillus terreus*	Co3O4NPs, CuONPs, Fe3O4NPs, NiONPs, and ZnONPs	*Fusarium wilt*	Antifungal	129
9	*Aspergillus versicolor*	Silver nanoparticles	*Sclerotinia sclerotiorum*	Antipathogenic	76
10	*Aspergillus versicolor*	Copper nanoparticles	Against rotting plant pathogens	Antifungal activity	9
11	*Beauveria brongniartii*	FeO nanoparticles	*Spodoptera litura*	Bioinsecticidal activity	109
12	*Beauveria bassiana*	Silver nanoparticles	*Aedes aegypti*	Potential larvicides	130

13	*Epicoccum nigrum*	Silver nanoparticles	*Aspergillus flavus* *Aspergillus fumigatus* *Cryptococcus neoformans* *C. albicans* *Fusarium solani* *Sporothrix schenckii*	Antipathogenic	95
14	*F. oxysporum*	Silica NPs		Disease resistance in plant cells	117
15	*Fusarium oxysporum*	Nanoiron particles	*Vicia faba*	Nanobiofertilizer	
16	*Fusarium solani*	Silver nanoparticles	Grain borne fungi species	Antipathogenic	131
17	*Fusarium solani* IOR 825	AgNPs	Plant fungal pathogens	Antifungal	119
		ZnONPs	Maize	Promotes growth	
18	*Ganoderma lucidum*	ZnO nanoparticles	*Lepidium sativum*	Nanofertilizers properties on Lepidium sativum	132
19	*Guignardia mangiferae*	Silver nanoparticles	*Colletotrichum* sp. *Curvularia lunata* *Rhizoctonia solani*	Antipathogenic	133
20	*Isaria fumosorosea*	Silver nanoparticles	*Aedes aegypti* *Culex quinquefasciatus*	potential larvicides	134
21	*Mucor hiemalis*	Ag-NPs		Antifungal	135
22	*Penicillium chrysogenum*	CuO-NPs	*Aspergillus niger* *Alternaria solani* *Erwinia amylovora* *Fusarium oxysporum* *Ralstonia solanacearum*	Antifungal; antibacterial; agriculture	114

(*Continued*)

Table 11.1 Fungal species involved in the biogenic synthesis of nanoparticles with agro- applications. *Continued*

S. no.	Fungal species	Nanoparticle synthesized	Effect on	Activity	References
23	*Setosphaeria rostrata*	Silver nanoparticles	*Aspergillus niger* *Fusarium graminearum* *Fusarium udum* *Rhizoctonia solani*	Nano-based fungicides	136
24	*Trichoderma asperellum*	SeNPs	*Pearl millet plants*	Growth stimulant	137
			Sclerospora graminicola	Sioression of disease	
25	*Trichoderma atroviride*	SeNPs	*Phytopthora infestans* *Tomato plants*	Bio-nano plant growth stimulants Disease resistance in plant cells	100
26	*T. atroviride*	Selenium nanospheres	*Phytophthora infestans*	Antifungal	110
27	*Trichoderma citrinoviride*	TcAgNPs and $TcTiO_2NPs$	*Solanum lycopersicum*	Promotes growth	123
28	*Trichoderma harzianum*	Silver nanoparticles	*Aedes aegypti*	Potential larvicides	138
29	*Trichoderma harzianum*	Silver nanoparticles	*Alternaria alternata*	Phytopathogenic	139
			Fusarium oxysporum *Trichoderma harzianum*		
30	*Trichoderma harzianum*	Silver nanoparticles	*Sclerotinia sclerotiorum*	Antifungal	140
			Sclerotium rolfsii		
31	*Trichoderma harzianum* SYA.F4		*A. alternate*	Antipathogenic	141
			Botrytis sp. *Helminthosporium* sp. *P. arenaria*		

32	*T. harzianum, T. reesei*	ZnO NPs	*Xanthomonas oryzae* pv. *Oryzae*	Antibacterial	120
				Nanobiofertilizer	
33	*Trichoderma longibrachiatum*	Silver nanoparticles	*Fusarium moniliforme*	Phytopathogenic	67
			Fusarium verticillioides		
			Helminthosporium oryzae		
			Penicillium brevicompactum		
			Pyricularia grisea		
34	*Trichoderma virens* strain EG92	MnO NPs	*Acetobacter pasteurianus*	Nanobiopesticide	101
			Alternaria alternate		
			Aspergillus favus		
			Aspergillus niger		
			Clavibacter michiganensis		
			Erwinia amylovora		
			Erwinia carotovora		
			Erwinia persicina		
			Fusarium moniliforme		
			Fusarium solani		
			Helminthosporium sp.		
			Phytophthora arenaria		

11.10 Conclusion

Uncontrolled and unmindful utilization of agrochemicals has caused severe damage to the ecosystem for which green approach of designing biological pesticides and incorporation of biogenic nanotechnology is gaining interest. Fungal biosystem is enriched with novel bioactive metabolites, enzymes with numerous applications in various research areas like pharmaceutical, industrial, agricultural, and so on. The preference of fungal cultures over other microbial cultures in development of biopesticides, nanopesticides, and nanofertilizers is due to their ability to form large biomass with ease, high metal tolerance and binding capacity, inexpensive, and are ecofriendly. Furthermore, several fungal species especially endophytes and entomopathogens possess antimicrobial, antiinsecticidal, plant-growth promoting properties. Fungi are potential bionanofactories of diverse range of nanoparticles,which needs further exploration to be utilized as agri-inputs in agroecosystems. Due to limited ideas on environmental safety issues, the adoption of nanofertilizers and nanopesticides in agriculture is challenging. Further work on biosafety regarding use of nanopesticides and nanofertilizers is required and policies must be designed for use in agriculture at a large scale.[142–145] Advancement in field-based application and commercialization of the developed biogenic agriproducts must be made.

References

1. Bale, J.; Van Lenteren, J.; Bigler, F. Biological control and sustainable food production. *Philos. Trans. R. Soc. B: Biol. Sci.* **2008**, *363* (1492), 761–776.
2. Roberti, R., et al. Antagonistic fungi producing hydrolytic enzymes, active in degrading the cell wall of some foot rot pathogens (*Fusarium* spp.) of wheat/Antagonistische Pilze, die hydrolytische Enzyme erzeugen, welche zellwandabbauende Aktivität an einigen Erregern der Halmbasiskrankheiten (*Fusarium* spp.) des Weizens zeigen. *Z. für Pflanzenkrankheiten und Pflanzenschutz/Journal Plant. Dis. Prot.* **2002**, 101–108.
3. Mishra, P.; Mishra, J.; Dwivedi, S.; Arora, N. K. Microbial enzymes in biocontrol of phytopathogens. *Microb. enzymes: roles Appl. industries* **2020**, 259–285.
4. Chinnaperumal, K., et al. Bio-pesticidal effects of *Trichoderma viride* formulated titanium dioxide nanoparticle and their physiological and biochemical changes on *Helicoverpa armigera* (Hub.). *Pesticide Biochem. Physiol.* **2018**, *149*, 26–36.
5. Rai, M.; Ingle, A. Role of nanotechnology in agriculture with special reference to management of insect pests. *Appl. microbiology Biotechnol.* **2012**, *94*, 287–293.
6. Alghuthaymi, M. A., et al. Myconanoparticles: synthesis and their role in phytopathogens management. *Biotechnol. & Biotechnological Equip.* **2015**, *29* (2), 221–236.
7. Dhillon, G. S.; Brar, S. K.; Kaur, S.; Verma, M. Green approach for nanoparticle biosynthesis by fungi: current trends and applications. *Crit. Rev. Biotechnol.* **2012**, *32* (1), 49–73.

8. Sastry, R. K.; Rashmi, H.; Rao, N.; Ilyas, S. Integrating nanotechnology into agri-food systems research in India: a conceptual framework. *Technol. Forecast. Soc. Change* **2010**, *77* (4), 639−648.
9. Ammar, H. A.; Rabie, G. H.; Mohamed, E. Novel fabrication of gelatin-encapsulated copper nanoparticles using *Aspergillus versicolor*and their application in controlling of rotting plant pathogens. *Bioprocess. Biosyst. Eng.* **2019**, *42*, 1947−1961.
10. Rai, M., et al. Fusarium as a novel fungus for the synthesis of nanoparticles: mechanism and applications. *J. Fungi* **2021**, *7* (2), 139.
11. Loshchinina, E. A.; Vetchinkina, E. P.; Kupryashina, M. A. Diversity of biogenic nanoparticles obtained by the fungi-mediated synthesis: a review. *Biomimetics* **2022**, *8* (1), 1.
12. Mouhammed, H. A. Potential characterization and antimicrobial applications of newly bio-synthesized silver and copper nanoparticles using the novel marine-derived fungus *Alternaria tenuissima* KM651985. *Scimago* **2016**.
13. Naqvi, S. T. Q., et al. Characterization and biological studies of copper nanoparticles synthesized by *Aspergillus niger*. *J. Bionanoscience* **2017**, *11* (2), 136−140.
14. Hilário, S.; Gonçalves, M. F. Endophytic Diaporthe as promising leads for the development of biopesticides and biofertilizers for a sustainable agriculture. *Microorganisms* **2022**, *10* (12), 2453.
15. Moorhouse, E.; Gillespie, A.; Sellers, E.; Charnley, A. Influence of fungicides and insecticides on the entomogenous fungus *Metarhizium anisopliae* a pathogen of the vine weevil, *Otiorhynchus sulcatus*. *Biocontrol Sci. Technol.* **1992**, *2* (1), 49−58.
16. Hajek, A.; Leger, R., St Interactions between fungal pathogens and insect hosts. *Annu. Rev. entomology* **1994**, *39* (1), 293−322.
17. de la Cruz Quiroz, R., et al. Fungi-based biopesticides: shelf-life preservation technologies used in commercial products. *J. Pest. Sci.* **2019**, *92*, 1003−1015.
18. Thambugala, K. M., et al. Fungi vs. fungi in biocontrol: an overview of fungal antagonists applied against fungal plant pathogens. *Front. Cell. Infect. microbiology* **2020**, *10*, 604923.
19. Meher, J., et al. Trichoderma: a globally dominant commercial biofungicide. *Trichoderma: Agric. Appl. beyond* **2020**, 195−208.
20. Kakvan, N., et al. Development of new bioformulations using Trichoderma and Talaromyces fungal antagonists for biological control of sugar beet damping-off disease. *Crop. Prot.* **2013**, *53*, 80−84.
21. Rossoni, C., et al. Selectivity of *Metarhizium anisopliae* and *Beauveria bassiana* (Hypocreales: Clavicipitaceae) on adults of *Cotesia flavipes* (Hymenoptera: Braconidae). *Folia Biologica (Kraków)* **2014**, *62* (3), 269−275.
22. Woo, S. L., et al. Trichoderma-based products and their widespread use in agriculture. *Open. Mycology J.* **2014**, *8* (1).
23. Javanshir Javid, K.; Mahdian, S.; Behboudi, K.; Alizadeh, H. Biological control of *Fusarium oxysporum* f. sp. radicis-cucumerinum by some *Trichoderma harzianum* isolates. *Arch. Phytopathology Plant. Prot.* **2016**, *49* (17-18), 471−484.
24. Muñiz-Paredes, F.; Miranda-Hernández, F.; Loera, O. Production of conidia by entomopathogenic fungi: from inoculants to final quality tests. *World J. Microbiology Biotechnol.* **2017**, *33*, 1−9.
25. do Nascimento, M. O., et al. Antagonism of Trichoderma isolates against *Leucoagaricus gongylophorus*(singer) Möller. *J. basic. microbiology* **2017**, *57* (8), 699−704.

26. Ríos-Moreno, A.; Quesada-Moraga, E.; Garrido-Jurado, I. Treatments with *Metarhizium brunneum*BIPESCO5 and EAMa 01/58-Su strains (Ascomycota: Hypocreales) are low risk for the generalist predator *Chrysoperla carnea*. *J. pest. Sci.* **2018**, *91*, 385–394.
27. Kachhawa, D. Microorganisms as a biopesticides. *J. Entomol. Zool. Stud.* **2017**, *5* (3), 468–473.
28. Keswani, C.; Singh, S. P.; Singh, H. A superstar in biocontrol enterprise: *Trichoderma*spp. *Biotech. Today* **2013**, *3* (2), 27–30.
29. Mukherjee, A.; Debnath, P.; Ghosh, S. K.; Medda, P. K. Biological control of papaya aphid (*Aphis gossypii*Glover) using entomopathogenic fungi. *Vegetos* **2020**, *33*, 1–10.
30. Sani, I., et al. A review of the biology and control of whitefly, *Bemisia tabaci* (Hemiptera: Aleyrodidae), with special reference to biological control using entomopathogenic fungi. *Insects* **2020**, *11* (9), 619.
31. Filgueiras, C. C.; Willett, D. S. The lesser chestnut weevil (*Curculio sayi*): damage and management with biological control using entomopathogenic fungi and entomopathogenic nematodes. *Insects* **2022**, *13* (12), 1097.
32. Umaru, F. F.; Simarani, K. Efficacy of entomopathogenic fungal formulations against *Elasmolomus pallens* (Dallas) (Hemiptera: Rhyparochromidae) and their extracellular enzymatic activities. *Toxins* **2022**, *14* (9), 584.
33. Gouda, S., et al. Endophytes: a treasure house of bioactive compounds of medicinal importance. *Front. Microbiology* **2016**, *7*, 1538.
34. Xu, K.; Li, X.-Q.; Zhao, D.-L.; Zhang, P. Antifungal secondary metabolites produced by the fungal endophytes: chemical diversity and potential use in the development of biopesticides. *Front. Microbiology* **2021**, *12*, 689527.
35. Sangeetha, G.; Usharani, S.; Muthukumar, A. Biocontrol with Trichoderma species for the management of postharvest crown rot of banana. *Phytopathologia Mediterranea* **2009**, *48* (2), 214–225.
36. Deshmukh, S. K.; Gupta, M. K.; Prakash, V.; Saxena, S. Endophytic fungi: a source of potential antifungal compounds. *J. Fungi* **2018**, *4* (3), 77.
37. Wu, S.-H., et al. Guaiane sesquiterpenes and isopimarane diterpenes from an endophytic fungus *Xylaria* sp. *Phytochemistry* **2014**, *105*, 197–204.
38. Bai, Y.-B., et al. Antifungal activity of griseofulvin derivatives against phytopathogenic fungi in vitro and in vivo and three-dimensional quantitative structure–activity relationship analysis. *J. Agric. food Chem.* **2019**, *67* (22), 6125–6132.
39. Abramczyk, B., et al. Biocontrol potential and catabolic profile of endophytic *Diaporthe eres* strain 1420S from *Prunus domestica* L. in Poland—a preliminary study. *Agronomy* **2022**, *12* (1), 165.
40. da Silva Ribeiro, A., et al. Bioprospection of culturable endophytic fungi associated with the ornamental plant *Pachystachys lutea*. *Curr. Microbiology* **2018**, *75*, 588–596.
41. Polonio, J., et al. Biotechnological prospecting of foliar endophytic fungi of guaco (*Mikania glomerata* Spreng.) with antibacterial and antagonistic activity against phytopathogens. *Genet. Mol. Res.* **2015**, *14* (3), 7297–7309.
42. Yu, J.-J., et al. Secondary metabolites from cultures of the kiwi-associated fungus *Diaporthe phragmitis* and their antibacterial activity assessment. *Phytochemistry Lett.* **2021**, *46*, 143–148.

43. Landum, M. C., et al. Antagonistic activity of fungi of *Olea europaea* L. against *Colletotrichum acutatum. Microbiological Res.* **2016**, *183*, 100–108.
44. Jaber, L. R.; Ownley, B. H. Can we use entomopathogenic fungi as endophytes for dual biological control of insect pests and plant pathogens? *Biol. Control.* **2018**, *116*, 36–45.
45. Deketelaere, S.; Tyvaert, L.; França, S. C.; Höfte, M. Desirable traits of a good biocontrol agent against Verticillium wilt. *Front. microbiology* **2017**, *8*, 1186.
46. Rakshith, D.; Santosh, P.; Satish, S. Isolation and characterization of antimicrobial metabolite producing endophytic *Phomopsis* sp. from *Ficus pumila* Linn. (Moraceae). *Int. J. Chem. Anal. Sci.* **2013**, *4* (3), 156–160.
47. Santos, C. Md, et al. Enzymatic and antagonist activity of endophytic fungi from *Sapindus saponaria* L. (Sapindaceae). *Acta biológica colombiana* **2019**, *24* (2), 322–330.
48. Zhao, J., et al. Bioactive secondary metabolites from *Nigrospora* sp. LLGLM003, an endophytic fungus of the medicinal plant *Moringa oleifera* Lam. *World J. Microbiology Biotechnol.* **2012**, *28*, 2107–2112.
49. Petersen, A. B.; Rønnest, M. H.; Larsen, T. O.; Clausen, M. H. The chemistry of griseofulvin. *Chem. Rev.* **2014**, *114* (24), 12088–12107.
50. Tang, H.-Y.; Zhang, Q.; Li, H.; Gao, J.-M. Antimicrobial and allelopathic metabolites produced by *Penicillium brasilianum. Nat. Product. Res.* **2015**, *29* (4), 345–348.
51. Wei, G., et al. Flavichalasines A–M, cytochalasan alkaloids from *Aspergillus flavipes. Sci. Rep.* **2017**, *7* (1).42434.
52. Rodrigo, S.; García-Latorre, C.; Santamaria, O. Metabolites produced by fungi against fungal phytopathogens: review, implementation and perspectives. *Plants* **2021**, *11* (1), 81.
53. Kartsev, V., et al. Griseofulvin derivatives: synthesis, molecular docking and biological evaluation. *Curr. Top. Medicinal Chem.* **2019**, *19* (13), 1145–1161.
54. Guilger-Casagrande, M.; Lima, Rd Synthesis of silver nanoparticles mediated by fungi: a review. *Front. Bioeng. Biotechnol.* **2019**, *7*, 287.
55. Mohanta, S., et al. An overview of mycogenic nanoparticles: synthesis and their bioprospective applications. *Nanotechnol. Biol. Med.* **2019**, 87–97.
56. Almoneafy, A. A., et al. *Role of fungi-mediated nanoparticles in mitigation of biotic and abiotic stresses in plants. Fungal Cell Factories for Sustainable Nanomaterials Productions and Agricultural Applications;* Elsevier, 2023601–633.
57. Mukherjee, P., et al. Green synthesis of highly stabilized nanocrystalline silver particles by a non-pathogenic and agriculturally important fungus *T. asperellum. Nanotechnology* **2008**, *19* (7), 075103.
58. Gaikwad, S., et al. Antiviral activity of mycosynthesized silver nanoparticles against herpes simplex virus and human parainfluenza virus type 3. *Int. J. Nanomed.* **2013**, 4303–4314.
59. Neethu, S.; Radhakrishnan, E.; Jyothis, M. Biofabrication of nanoparticles using fungi. *Nanotechnol. Agriculture: Adv. Sustain. Agriculture* **2019**, 53–73.
60. Verma, V. C.; Kharwar, R. N.; Gange, A. C. Biosynthesis of antimicrobial silver nanoparticles by the endophytic fungus *Aspergillus clavatus. Nanomedicine* **2010**, *5* (1), 33–40.
61. Kashyap, P. L.; Kumar, S.; Srivastava, A. K.; Sharma, A. K. Myconanotechnology in agriculture: a perspective. *World J. Microbiology Biotechnol.* **2013**, *29*, 191–207.

62. Gade, A., et al. Exploitation of *Aspergillus niger* for synthesis of silver nanoparticles. *J. Biobased Mater. Bioenergy* **2008**, *2* (3), 243–247.
63. Durán, N., et al. Mechanistic aspects of biosynthesis of silver nanoparticles by several *Fusarium oxysporum* strains. *J. nanobiotechnology* **2005**, *3*, 1–7.
64. Kumar, S. A.; Ansary, A. A.; Ahmad, A.; Khan, M. Extracellular biosynthesis of CdSe quantum dots by the fungus, *Fusarium oxysporum. J. Biomed. Nanotechnol.* **2007**, *3* (2), 190–194.
65. Keat, C. L.; Aziz, A.; Eid, A. M.; Elmarzugi, N. A. Biosynthesis of nanoparticles and silver nanoparticles. *Bioresour. Bioprocess.* **2015**, *2*, 1–11.
66. Berdy, J. Bioactive microbial metabolites. *J. antibiotics* **2005**, *58* (1), 1–26.
67. Elamawi, R. M.; Al-Harbi, R. E.; Hendi, A. A. Biosynthesis and characterization of silver nanoparticles using *Trichoderma longibrachiatum* and their effect on phytopathogenic fungi. *Egypt. J. Biol. pest. Control.* **2018**, *28* (1), 1–11.
68. Riddin, T.; Gericke, M.; Whiteley, C. Analysis of the inter-and extracellular formation of platinum nanoparticles by *Fusarium oxysporum* f. sp. lycopersici using response surface methodology. *Nanotechnology* **2006**, *17* (14), 3482.
69. Mukherjee, P., et al. Bioreduction of AuCl4-ions by the fungus, *Verticillium* sp. and surface trapping of the gold nanoparticles formed. *Angew. Chemie-International Ed.* **2001**, *40* (19), 3585–3588.
70. Ballottin, D., et al. Elucidating protein involvement in the stabilization of the biogenic silver nanoparticles. *Nanoscale Res. Lett.* **2016**, *11*, 1–9.
71. Gopinath, V., et al. Biosynthesis of silver nanoparticles from Tribulus terrestris and its antimicrobial activity: a novel biological approach. *Colloids Surf. B: biointerfaces* **2012**, *96*, 69–74.
72. Chowdhury, S.; Basu, A.; Kundu, S. Green synthesis of protein capped silver nanoparticles from phytopathogenic fungus *Macrophomina phaseolina* (Tassi) Goid with antimicrobial properties against multidrug-resistant bacteria. *Nanoscale Res. Lett.* **2014**, *9*, 1–11.
73. Gole, A., et al. Pepsin – gold colloid conjugates: preparation, characterization, and enzymatic activity. *Langmuir* **2001**, *17* (5), 1674–1679.
74. Husseiny, S. M.; Salah, T. A.; Anter, H. A. Biosynthesis of size controlled silver nanoparticles by *Fusarium oxysporum*, their antibacterial and antitumor activities. *Beni-Suef Univ. J. Basic. Appl. Sci.* **2015**, *4* (3), 225–231.
75. Zhang, X.-F.; Liu, Z.-G.; Shen, W.; Gurunathan, S. Silver nanoparticles: synthesis, characterization, properties, applications, and therapeutic approaches. *Int. J. Mol. Sci.* **2016**, *17* (9), 1534.
76. Elgorban, A. M., et al. Extracellular synthesis of silver nanoparticles using *Aspergillus versicolor*and evaluation of their activity on plant pathogenic fungi. *Mycosphere* **2016**, *7* (6), 844–852.
77. Guilger-Casagrande, M., et al. Influence of the capping of biogenic silver nanoparticles on their toxicity and mechanism of action towards Sclerotinia sclerotiorum. *J. Nanobiotechnology* **2021**, *19*, 1–18.
78. Tanzini, M.R.; Alves, S.B.; Setten, A.; Augusto, N.T. *Compatibilidad de agentes tensoactivos con Beauveria bassiana y Metarhizium anisopliae*. 2001.
79. Mantzoukas, S.; Eliopoulos, P. A. Endophytic entomopathogenic fungi: a valuable biological control tool against plant pests. *Appl. Sci.* **2020**, *10* (1), 360.

80. Bihal, R., et al. Entomopathogenic fungi: an eco-friendly synthesis of sustainable nanoparticles and their nanopesticide properties. *Microorganisms* **2023**, *11* (6), 1617.
81. Dhaliwal, G.; Jindal, V.; Dhawan, A. Insect pest problems and crop losses: changing trends. *Indian. J. Ecol.* **2010**, *37* (1), 1–7.
82. Sharma, A.; Sood, K.; Kaur, J.; Khatri, M. Agrochemical loaded biocompatible chitosan nanoparticles for insect pest management. *Biocatalysis Agric. Biotechnol.* **2019**, *18*, 101079.
83. Deka, B.; Baruah, C.; Babu, A. Entomopathogenic microorganisms: their role in insect pest management. *Egypt. J. Biol. Pest. Control.* **2021**, *31*, 1–8.
84. Jaihan, P.; Sangdee, K.; Sangdee, A. Selection of entomopathogenic fungus for biological control of chili anthracnose disease caused by *Colletotrichum* spp. *Eur. J. plant. Pathol.* **2016**, *146*, 551–564.
85. Ríos-Moreno, A., et al. Destruxin A production by *Metarhizium brunneum* strains during transient endophytic colonisation of *Solanum tuberosum*. *Biocontrol Sci. Technol.* **2016**, *26* (11), 1574–1585.
86. Freed, S., et al. Toxicity of proteins secreted by entomopathogenic fungi against *Plutella xylostella* (Lepidoptera: Plutellidae). *Int. J. agriculture Biol.* **2012**, *14* (2).
87. Athanassiou, C., et al. Nanoparticles for pest control: current status and future perspectives. *J. Pest. Sci.* **2018**, *91*, 1–15.
88. Benelli, G. Mode of action of nanoparticles against insects. *Environ. Sci. Pollut. Res.* **2018**, *25* (13), 12329–12341.
89. Bogas, A. C., et al. Endophytic microorganisms from the tropics as biofactories for the synthesis of metal-based nanoparticles: healthcare applications. *Front. Nanotechnol.* **2022**, *4*, 823236.
90. Musoni, M., et al. Bioreactor design and implementation strategies for the cultivation of filamentous fungi and the production of fungal metabolites: from traditional methods to engineered systems. *BASE* **2015**.
91. Vijayanandan, A. S.; Balakrishnan, R. M. Biosynthesis of cobalt oxide nanoparticles using endophytic fungus *Aspergillus nidulans*. *J. Environ. Manag.* **2018**, *218*, 442–450.
92. Musarrat, J., et al. Production of antimicrobial silver nanoparticles in water extracts of the fungus *Amylomyces rouxii*strain KSU-09. *Bioresour. Technol.* **2010**, *101* (22), 8772–8776.
93. Raheman, F., et al. Silver nanoparticles: novel antimicrobial agent synthesized from an endophytic fungus *Pestalotia* sp. isolated from leaves of *Syzygium cumini* (L). *Nano Biomed. Eng.* **2011**, *3* (3), 174–178.
94. Bala, M.; Arya, V. Biological synthesis of silver nanoparticles from aqueous extract of endophytic fungus *Aspergillus fumigatus* and its antibacterial action. *Int. J. Nanomaterials Biostructures* **2013**, *3* (2), 37–41.
95. Qian, Y., et al. Biosynthesis of silver nanoparticles by the endophytic fungus *Epicoccum nigrum*and their activity against pathogenic fungi. *Bioprocess. Biosyst. Eng.* **2013**, *36*, 1613–1619.
96. Singh, A. K.; Rathod, V.; Singh, D.; Kulkarni, P. Effect of silver nanoparticles (AgNPs) produced by an endophytic fungus *Fusarium semitectum* isolated from a medicinal plant *Withania somnifera* (Ashwagandha) on seed germination. *Int. J. Res. Stud. Agric. Sci.* **2016**, *2* (5), 6–12.

97. Makvandi, P., et al. Metal-based nanomaterials in biomedical applications: antimicrobial activity and cytotoxicity aspects. *Adv. Funct. Mater.* **2020**, *30* (22). 1910021.
98. Sandhu, S. S.; Shukla, H.; Shukla, S. Biosynthesis of silver nanoparticles by endophytic fungi: its mechanism, characterization techniques and antimicrobial potential. *Afr. J. Biotechnol.* **2017**, *16* (14), 683–698.
99. Rajwade, J. M.; Chikte, R.; Paknikar, K. Nanomaterials: new weapons in a crusade against phytopathogens. *Appl. microbiology Biotechnol.* **2020**, *104* (4), 1437–1461.
100. Joshi, N. C.; Joshi, E.; Singh, A. Biological synthesis, characterisations and antimicrobial activities of manganese dioxide (MnO2) nanoparticles. *Res. J. Pharm. Technol.* **2020**, *13* (1), 135–140.
101. El-Moslamy, S. H.; Yahia, I.; Zahran, H.; Kamoun, E. A. Novel biosynthesis of MnO NPs using Mycoendophyte: industrial bioprocessing strategies and scaling-up production with its evaluation as anti-phytopathogenic agents. *Sci. Rep.* **2023**, *13* (1), 2052.
102. Kamil, D., et al., *Green synthesis of silver nanoparticles by entomopathogenic fungus Beauveria bassiana and their bioefficacy against mustard aphid (Lipaphis erysimi Kalt.)*. 2017.
103. karthick raja namasivayam, s.; arvind bharani, r. biocompatible silver nanoparticles-loaded fungal metabolites nanoconjugate (agnp–fm) preparation for the noteworthy pesticidal activity. *Natl Acad. Sci. Lett.* **2021**, 1–7.
104. Yosri, M.; Abdel-Aziz, M. M.; Sayed, R. Larvicidal potential of irradiated mycoinsecticide from *Metarhizium anisopliae* and larvicidal synergistic effect with its mycosynthesized titanium nanoparticles (TiNPs). *J. Radiat. Res. Appl. Sci.* **2018**, *11* (4), 328–334.
105. Santos, T. S., et al. Entomopathogenic fungi biomass production and extracellular biosynthesis of silver nanoparticles for bioinsecticide action. *Appl. Sci.* **2021**, *11* (6), 2465.
106. Shukla, G.; Gaurav, S. S.; Singh, A.; Rani, P. Synthesis of mycogenic silver nanoparticles by *Fusarium pallidoroseum* and evaluation of its larvicidal effect against white grubs (*Holotrichia* sp.). *Mater. Today: Proc.* **2022**, *49*, 3517–3527.
107. Khooshe-Bast, Z.; Sahebzadeh, N.; Ghaffari-Moghaddam, M.; Mirshekar, A. Insecticidal effects of zinc oxide nanoparticles and *Beauveria bassiana* TS11 on *Trialeurodes vaporariorum*(Westwood, 1856)(Hemiptera: Aleyrodidae). *Acta Agriculturae Slovenica* **2016**, *107* (2), 299–309.
108. Wang XiaoShuang, W.X., et al. *Isaria fumosorosea-based zero-valent iron nanoparticles affect the growth and survival of sweet potato whitefly, Bemisia tabaci (Gennadius)*. 2019.
109. Xu, J., et al. Toxicity and biological effects of *Beauveria brongniartii* Fe0 nanoparticles against *Spodoptera litura* (Fabricius). *Insects* **2020**, *11* (12), 895.
110. Joshi, S. M.; De Britto, S.; Jogaiah, S.; Ito, S.-i Mycogenic selenium nanoparticles as potential new generation broad spectrum antifungal molecules. *Biomolecules* **2019**, *9* (9), 419.
111. Arunthirumeni, M.; Veerammal, V.; Shivakumar, M. S. Biocontrol efficacy of mycosynthesized selenium nanoparticle using *Trichoderma* sp. on insect pest *Spodoptera litura*. *J. Clust. Sci.* **2022**, *33* (4), 1645–1653.

112. Helar, G.; Chavan, A. Synthesis, characterization and stability of gold nanoparticles using the fungus *Fusarium oxysporum* and its impact on seed. *Int. J. Recent. Sci. Res.* **2015**, *6*, 3181–3318.
113. Raliya, R.; Tarafdar, J. C.; Biswas, P. Enhancing the mobilization of native phosphorus in the mung bean rhizosphere using ZnO nanoparticles synthesized by soil fungi. *J. Agric. food Chem.* **2016**, *64* (16), 3111–3118.
114. El-Batal, A. I.; El-Sayyad, G. S.; Mosallam, F. M.; Fathy, R. M. *Penicillium chrysogenum*-mediated mycogenic synthesis of copper oxide nanoparticles using gamma rays for in vitro antimicrobial activity against some plant pathogens. *J. Clust. Sci.* **2020**, *31*, 79–90.
115. Deka, A. C.; Sinha, S. K. Mycogenic silver nanoparticle biosynthesis and its pesticide degradation potentials. *Int. J. Technol. Enhanc. Emerg. Eng. Res.* **2015**, *3*, 108–113.
116. Shende, S., et al. Myco-fabrication of copper nanoparticles and its effect on crop pathogenic fungi. *IEEE Trans. Nanobioscience* **2021**, *20* (2), 146–153.
117. Bansal, V., et al. Fungus-mediated biosynthesis of silica and titania particles. *J. Mater. Chem.* **2005**, *15* (26), 2583–2589.
118. Vigneshwaran, N., et al. Biological synthesis of silver nanoparticles using the fungus *Aspergillus flavus*. *Mater. Lett.* **2007**, *61* (6), 1413–1418.
119. Trzcińska-Wencel, J., et al. Biofabrication of novel silver and zinc oxide nanoparticles from *Fusarium solani*IOR 825 and their potential application in agriculture as biocontrol agents of phytopathogens, and seed germination and seedling growth promoters. *Front. Chem.* **2023**, *11*, 1235437.
120. Shobha, B., et al. Mycosynthesis of ZnO nanoparticles using *Trichoderma* spp. isolated from rhizosphere soils and its synergistic antibacterial effect against *Xanthomonas oryzae* pv. oryzae. *J. Fungi* **2020**, *6* (3), 181.
121. Nurfalah, A.; Ayuningrum, N.; Affandi, M.R.; Hastuti, L.D.S. *Promoting growth of tomato (Solanum lycopersicum L.) by using trichoderma–compost–rice bran based biofertilizer*. In *IOP conference Series: Earth and Environmental Science*. 2019. IOP Publishing.
122. Lee, S., et al. Volatile organic compounds emitted by Trichoderma species mediate plant growth. *Fungal Biol. Biotechnol.* **2016**, *3*, 1–14.
123. Sonawane, H.; Arya, S.; Math, S.; Shelke, D. Myco-synthesized silver and titanium oxide nanoparticles as seed priming agents to promote seed germination and seedling growth of *Solanum lycopersicum*: a comparative study. *Int. Nano Lett.* **2021**, *11* (4), 371–379.
124. Albalawi, M. A., et al. Mycosynthesis of silica nanoparticles using aspergillus Niger: control of *Alternaria solani* causing early blight disease, induction of innate immunity and reducing of oxidative stress in eggplant. *Antioxidants* **2022**, *11* (12), 2323.
125. Sedefoglu, N.; Zalaoglu, Y.; Bozok, F. Green synthesized ZnO nanoparticles using *Ganoderma lucidum*: characterization and in vitro nanofertilizer effects. *J. Alloy. Compd.* **2022**, *918*, 165695.
126. Raliya, R.; Biswas, P.; Tarafdar, J. TiO2 nanoparticle biosynthesis and its physiological effect on mung bean (*Vigna radiata* L.). *Biotechnol. Rep.* **2015**, *5*, 22–26.
127. Singh, A.; Gaurav, S. S.; Shukla, G.; Rani, P. Evaluation of mycosilver nanofungicides as potential control agent against phytophthora infestans. *Plant. Cell Biotechnol. Mol. Biol.* **2021**, *22*, 157–168.

128. Singh, A.; Gaurav, S. S.; Shukla, G.; Rani, P. Assessment of mycogenic zinc nano-fungicides against pathogenic early blight (*Alternaria solani*) of potato (*Solanum tuberosum* L.). *Mater. Today: Proc.* **2022**, *49*, 3528–3537.
129. El-Sayed, E.-S. R., et al. Bifunctional role of some biogenic nanoparticles in controlling wilt disease and promoting growth of common bean. *AMB. Express* **2023**, *13* (1), 41.
130. Najitha Banu, A.; Balasubramanian, C.; Moorthi, P. V. Biosynthesis of silver nanoparticles using *Bacillus thuringiensis* against dengue vector, *Aedes aegypti* (Diptera: Culicidae). *Parasitology Res.* **2014**, *113*, 311–316.
131. Abd El-Aziz, A.; Al-Othman, M.; Mahmoud, M.; Metwaly, H. Biosynthesis of silver nanoparticles using *Fusarium solani* and its impact on grain borne fungi. *Dig. J. Nanomaterials Biostructures* **2015**, *10* (2), 655–662.
132. Bairwa, P.; Kumar, N.; Devra, V.; Abd-Elsalam, K. A. Nano-biofertilizers synthesis and applications in agroecosystems. *Agrochemicals* **2023**, *2* (1), 118–134.
133. Balakumaran, M.; Ramachandran, R.; Kalaichelvan, P. Exploitation of endophytic fungus, *Guignardia mangiferae* for extracellular synthesis of silver nanoparticles and their in vitro biological activities. *Microbiological Res.* **2015**, *178*, 9–17.
134. Banu, A. N.; Balasubramanian, C. Optimization and synthesis of silver nanoparticles using *Isaria fumosorosea* against human vector mosquitoes. *Parasitology Res.* **2014**, *113*, 3843–3851.
135. Aziz, N.; Pandey, R.; Barman, I.; Prasad, R. Leveraging the attributes of *Mucor hiemalis*-derived silver nanoparticles for a synergistic broad-spectrum antimicrobial platform. *Front. Microbiology* **2016**, *7*, 1984.
136. Akther, T.; Hemalatha, S. Mycosilver nanoparticles: synthesis, characterization and its efficacy against plant pathogenic fungi. *Bio. Nano. Sci.* **2019**, *9* (2), 296–301.
137. Nandini, B., et al. Trichogenic-selenium nanoparticles enhance disease suppressive ability of *Trichoderma* against downy mildew disease caused by *Sclerospora graminicola* in pearl millet. *Sci. Rep.* **2017**, *7* (1), 2612.
138. Sundaravadivelan, C.; Padmanabhan, M. N. Effect of mycosynthesized silver nanoparticles from filtrate of *Trichoderma harzianum* against larvae and pupa of dengue vector *Aedes aegypti* L. *Environ. Sci. Pollut. Res.* **2014**, *21*, 4624–4633.
139. Al-Abboud, M.A., Fungal biosynthesis of silver nanoparticles and their role in control of Fusarium wilt of sweet pepper and soil-borne fungi in vitro. 2018.
140. El-Ashmony, R. M., et al. The biogenically efficient synthesis of silver nanoparticles using the fungus Trichoderma harzianum and their antifungal efficacy against *Sclerotinia sclerotiorum* and *Sclerotium rolfsii*. *J. Fungi* **2022**, *8* (6), 597.
141. El-Moslamy, S. H.; Elkady, M. F.; Rezk, A. H.; Abdel-Fattah, Y. R. Applying Taguchi design and large-scale strategy for mycosynthesis of nano-silver from endophytic *Trichoderma harzianum* SYA. F4 and its application against phytopathogens. *Sci. Rep.* **2017**, *7* (1), 45297.
142. Wu, H.; Li, Z. Recent advances in nano-enabled agriculture for improving plant performance. *Crop. J.* **2022**, *10* (1), 1–12. Available from: https://doi.org/10.1016/j.jksus.2024.103332.
143. Sharma, A., et al. Revolutionizing agricultural sustainability and food security and management to achieve SDGs goals via nanotechnology. *Adv. Environ. Eng. green. Technol. book. Ser.* **2024**, 276–288. Available from: https://doi.org/10.4018/979-8-3693-1471-5.ch013.

144. Ghazaryan, K.; Verma, A.; Rawat, S.; Rajput, P.; Minkina, T.; Mandzhieva, S., et al. Implementing Sustainable nano-bioremediation for emerging pollutants. *Adv. Environ. Eng. green. Technol. book. Ser.* **2024**, 316–332. Available from: https://doi.org/10.4018/979-8-3693-1471-5.ch016.
145. Singh, A.; Sharma, A.; Singh, O.; Rajput, V. D.; Movsesyan, H. S.; Minkina, T., et al. In-depth exploration of nanoparticles for enhanced nutrient use efficiency and abiotic stresses management: present insights and future horizons. *Plant. Stress.* **2024**.100576. Available from: https://doi.org/10.1016/j.stress.2024.100576.

SECTION

III

Human health effects of biogenic nanomaterials

CHAPTER

12 Exploring the combined role of nanoparticles and soil microorganisms in plant growth promotion and bioremediation

Jaagriti Tyagi[1,2], Riya Sharma[1], Upasana Bhagawati[1] and M.S. Smitha[1]

[1]*Amity Institute of Microbial Technology, Amity University, Noida, Uttar Pradesh, India*

[2]*Biotechnology Department, Dr. MPS Group of Institutions, Agra, Uttar Pradesh, India*

12.1 Introduction

In the modern era, the proliferation of chemical pollutants like dyes, pesticides, xenobiotics, and heavy metals has severely impacted environmental quality. The accumulation of these contaminants poses a significant threat to ecological equilibrium, necessitating urgent attention and innovative solutions from the scientific community. In response to these challenges, scientists have embarked on a quest to devise ecofriendly strategies that not only mitigate pollution but also align with broader sustainability objectives.[1,2] The convergence of nanotechnology and microbial technology has emerged as a promising approach in this endeavor. Nanotechnology, with its capacity to manipulate matter at the nanoscale, offers distinct advantages for environmental remediation. Nanoparticles, due to their small size, exhibit enhanced mobility and stability, enabling targeted delivery to contaminated sites. Various forms of nanomaterials hold immense potential for augmenting the efficiency and efficacy of microbial-based bioremediation processes. Microbial technology harnesses the metabolic capabilities of microorganisms to degrade pollutants and restore environmental balance. By leveraging the synergistic interplay between microorganisms and nanoparticles, novel strategies can be devised to tackle complex pollution challenges.

One innovative concept gaining traction is nanobioremediation, which integrates biological strategies involving microbes with nanoparticle-based technologies.[3,4] This holistic approach enhances plant growth-promoting (PGP) activities and addresses major pollution issues, fostering sustainable environmental remediation. The integration of nanoparticles and microbial strategies opens new avenues for remediation activities, ranging from soil and water remediation to air purification.[5,6] Furthermore, nano-based bioremediation holds promise for enhancing the productivity of

Biogenic Nanoparticles. DOI: https://doi.org/10.1016/B978-0-443-33819-9.00016-0

microbe-based biofertilizers and other bioremediation agents. Nanoparticles interact with plant growth-promoting microbes (PGPMs) and other microbial consortia, influencing both microbial communities and crucial soil processes. PGPMs play a vital role in various ecological processes, such as nitrogen fixation, mineralization, and pollutant degradation. Integrating nanoparticles with PGPRs enhances plant growth-promoting activities and offers a sustainable approach to bioremediation challenges.[7]

Bioremediation techniques involve bioaccumulation, biotransformation, biosorption, and biological stabilization using microbes, plants, enzymes, or their combinations. The integration of nanoparticles and microbial technology provides effective solutions for addressing various challenges in agriculture and environmental sustainability.[8]

Studies have shown that nanoparticles efficiently treat soil pollution while minimizing secondary environmental impacts. Additionally, nanoparticles combined with microbial processes offer effective remediation solutions for cleaning up contaminated sites, including those contaminated with heavy metals, pesticides, and organic compounds. Traditional agricultural practices often lead to adverse environmental effects such as soil degradation, water pollution, and biodiversity loss.[9] Nanotechnology offers more sustainable approaches to agriculture, minimizing environmental harm and enabling more precise resource management. Furthermore, nanoparticles and microbial technology help enhance plant resilience to environmental stressors such as climate change, drought, salinity, and pest infestations. They also offer solutions for disease diagnostics, prevention, and control, reducing crop losses and improving food security. Inadequate postharvest management practices often result in significant food losses. Nanotechnology provides solutions for improving food storage, processing, and shelf life, ultimately reducing postharvest losses in agriculture. Integrating nanotechnology with bioremediation techniques offers innovative solutions to complex environmental challenges, necessitating adherence to strict regulatory standards for human and environmental safety.[10]

The synthesis of nanomaterials has shifted toward sustainable methods utilizing microorganisms and plant extracts.[11,12] These biomaterials interact with microorganisms to remove toxins from polluted locations, contributing to environmental remediation efforts. Microbes like *Azospirillum*, *Pseudomonas*, and *Bacillus*, in consortium with nanoparticles, enhance soil micronutrient availability and create favorable environment for plant growth. They also increase soil nutrient availability, improve resistance against pathogens, and contribute to sustainable agricultural practices and enhanced crop productivity.[13]

In summary, soil microorganisms play a crucial role in promoting plant growth and facilitating bioremediation, making them essential for agricultural ecosystems and environmental cleanup.[14–17] The primary objective of this chapter is to explore the intricate relationship between nanoparticles, soil microorganisms, plant growth promotion, and bioremediation, providing insights into how nanotechnology and microbial processes can be harnessed to address agricultural challenges and environmental pollution effectively.

12.2 Nanoparticles in plant growth promotion

12.2.1 Types of nanoparticles used in agriculture

Various types of nanomaterials and nanoparticles have been implemented in research studies for promoting plant growth in agricultural fields. Combined formulations of nanoparticles and microbes have been successfully applied to enhance absorption and effectiveness in promoting plant growth. Nanoparticles have emerged as valuable tools in modern agriculture, offering a range of applications aimed at improving crop yields, enhancing soil fertility, and mitigating environmental impacts. Metal nanoparticles, including silver, gold, copper, zinc, and titanium dioxide, exhibit antimicrobial properties useful for controlling plant pathogens and promoting growth. Carbon-based nanoparticles such as carbon nanotubes and graphene contribute to soil remediation efforts and facilitate nutrient delivery to plants.[18,19] Metal oxide nanoparticles like zinc oxide and iron oxide find utility in nutrient delivery systems and pest management strategies. Silica nanoparticles play a role in improving soil structure and water retention, while polymeric nanoparticles enable controlled release of fertilizers and pesticides, minimizing environmental harm.[20] Chitosan nanoparticles derived from chitin show promise in enhancing plant defenses and delivering bioactive compounds efficiently.[21] Lipid-based nanoparticles offer targeted delivery of agrochemicals, further enhancing efficacy. Additionally, nanoemulsions and nanosuspensions improve stability and efficacy of agrochemicals.[22,23] ongoing exploration of nanoparticles in agriculture underscores their potential to revolutionize farming practices, promoting sustainable and efficient crop production in the face of evolving agricultural challenges (Fig. 12.1).

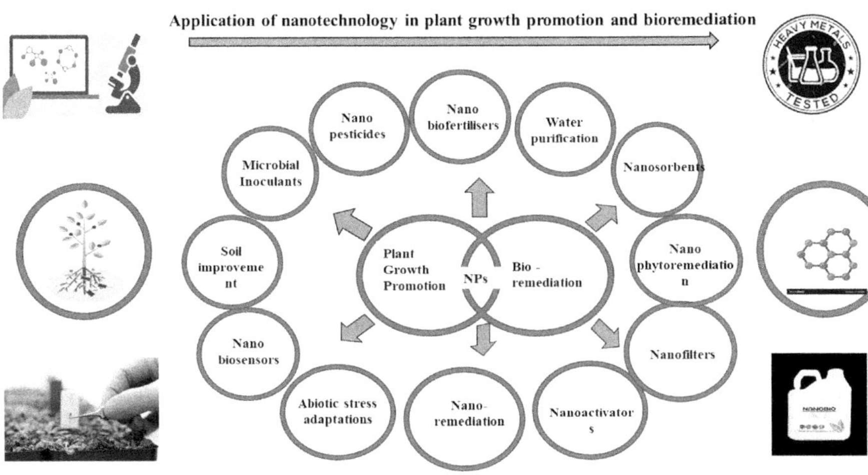

FIGURE 12.1

Different roles of nanoparticles in plant growth promotion and bioremediation.

12.2.2 Mechanism of action of nanoparticles in promoting plant growth

Understanding plant transport pathways is essential for comprehending the mechanism of action of nanoparticles in promoting plant growth. These pathways serve as conduits for nanoparticles to enter plants and their surrounding environment, enabling them to exert their beneficial effects.[24,25] Nanoparticles can penetrate plant cells through various mechanisms, including engulfing, endosome formation, or sieving, with their entry influenced by factors such as plant species and nanoparticle properties.[26]

In the context of promoting plant growth, nanoparticles leverage these transport pathways to deliver essential nutrients or growth-promoting substances directly into plant cells. By traversing the cell wall and membrane, nanoparticles can access vascular bundles and distribute throughout the plant, facilitating nutrient uptake and enhancing physiological processes.[27] For example, nanoparticles may carry micronutrients or growth hormones, aiding in plant development, stress tolerance, and overall yield improvement. Furthermore, nanoparticles can interact with cellular components, modulating gene expression or enzyme activity to optimize plant growth and productivity.[28] Therefore, understanding how nanoparticles utilize plant transport pathways elucidates their mechanism of action in promoting plant growth. By harnessing these pathways, nanoparticles can efficiently deliver bioactive compounds, nutrients, or regulatory signals, enhancing plant health, resilience, and productivity.

12.2.3 Effects of nanoparticles on nutrient uptake, stress tolerance, and yield enhancement

Nanoparticles have a significant impact on the nutrient content of crops, particularly through the use of nanofertilizers. These nanofertilizers, which encapsulate nutrients in nanoparticles, have been shown to enhance the efficiency of nutrient uptake by plants, leading to improved crop growth and development.[29,30] Additionally, nanoparticles can help combat nutrient deficiencies, promote stress tolerance, and enhance the yield and quality of crops by improving nutrient uptake efficiency[19,31] demonstrated that nanoparticles, when used in agriculture, can positively influence plant physiological activity by increasing nutrient content and improving photosynthetic processes, regulate phytohormone activities.[32,33] Nanoparticles play a crucial role in enhancing plant growth and quality by increasing nutrient levels and promoting stress tolerance in crops.[34,35] As per the reports consortium of iron oxide nanoparticles in combination with *Bacillus subtilis S4* helped in alleviating Arsenic (As) heavy metal stress in Pumpkin plant showed eminent results increased activity of antioxidative enzymes like POD (Peroxidase), enhanced stress mitigating compounds resulted in combatting stress, increase yield and nutrition uptake[36] (Table 12.1).

Table 12.1 Combined role of NPS and microbes in plant growth promotion.

S. no.	Nanoparticle and microbe	Heavy metal	Host	Response	References
1	Nanoscale zero-valent Iron (nZVI) + *Pseudomonas fluorescens*	Antimony (Sb)	White clover (*Trifolium repens*)	Increased accumulation capacity (PGPR + 500 mg/kg nZVI) of *T. repens* for Sb, improved physiological responses, plant photosynthesis Successful phytoremediation of Sb from soil, Improved accumulation of sb in shoots, phytoremediation activity found with nZVI (1000 mg/kg)	Zand et al.[37] Zand et al.[38]
2	Ag-nanoparticles + *Pseudomonas stutzeri* strain Kx574858	Manganese (Mn) and Iron (Fe)	Bitter gourd *(Momordica charantia L.)*	Increased level of carotenoids, proline, antioxidant enzyme SOD (superoxide dismutase) and CAT (catalase), alleviated oxidative stress, higher germination rate, enhanced ROS production, enhancement in PGPR-mediated bioremediation, reduction in metal stress	Tariq et al.[39]
3	Green molybdenum nanoparticles (MoNPs) + *Bacillus* sp. ZH16	Arsenic (As)	Wheat (*Triticum aestivum* L.)	Remodeled ionic/nutrient profiles under As stress, reduction in metal toxicity, promoted indole-3-acetic acid synthesis, phosphate solubilization and ACC (1-aminocyclopropane-1-carboxylic acid) deaminase activity	Ahmed et al.[40]
4	Titanium dioxide (TiO_2) + *Bacillus mycoides PM35*	Cadmium (Cd)	Barley *(Hordeum vulgare L.)*	Alleviated Cd toxicity, oxidative stress, reactive oxygen level (ROS), improved plant growth and hence, balanced exudation of organic acids and secondary metabolites (ascorbic acid, phenolic, flavonoids)	Ma et al.[41]

(*Continued*)

Table 12.1 Combined role of NPS and microbes in plant growth promotion. *Continued*

S. no.	Nanoparticle and microbe	Heavy metal	Host	Response	References
5	Titanium dioxide zinc oxide nanoparticles (ZnO NPs) + *Bacillus pumilus*	Cadmium (Cd)	Maize (*Zea mays L.)*	Improved plant growth metabolic activity, reduced Cd toxicity, lipid peroxidation, Cd accumulation in tissues, enhanced stress metabolism, defense responses Increased water content, photosynthetic pigments, proline, sugars, and proteins, antioxidant enzyme levels.	Shafiq et al.[42]
6	Copper nanoparticles (CuONPs) + *Bacillus* sp.	Chromium (Cr)	Spinach (*Spinacia oleracea* L.)	Alleviated metal-induced toxicity, increased chlorophyll, plant growth, yield, improved photosynthesis, carotenoid, total protein and nonenzymatic antioxidant compounds Reduction in oxidative stress.	Ali et al.[20]
7	Zinc oxide (Zno) + *Bacillus fortis* IAGS 223	Cadmium (Cd)	Muskmelon (*Cucumis melo* L.)	Mitigated Cd stress, by modulation of antioxidant enzymes CAT (catalase), SOD (Superoxide dismutase) and POD (Peroxidase) system, Improved growth, Decreased level of stress markers like H_2O_2, and MDA (malondialdehyde).	Shah et al.[43]
8	Iron oxide nanoparticles + *Bacillus subtilis* S4	Arsenic (As)	Pumpkin (*Cucurbita moschata)*	Increased activity of antioxidative enzymes peroxidase (POD) and superoxide dismutase (SOD), reduced levels of hydrogen peroxide (H_2O_2), malondialdehyde (MDA), and electrolyte leakage (EL), enhanced the biosynthesis of stress-mitigating polyamines, spermidine and putrescine.	Mushtaq et al.[36]

9	Ag NPs + (HAS31) rhizobacteria	Chromium (Cr)	Barley (*Hordeum vulgare* L.)	Alleviated Cr stress, regulated antioxidant activity increased plant growth and biomass, improved photosynthetic apparatus, mineral uptake, diminished the exudation of organic acids and oxidative stress.	Zhu et al.[44]
10	Cerium (IV) oxide (CeO_2) nanoparticles + *Staphylococcus aureus*	Chromium (Cr)	Sunflower (*Helianthus annuus*)	Alleviated metal toxicity, due to decline leakage of electrolyte, Increased chlorophyll a and b, enhanced enzymatic activities, enhancement in plant growth and biomass production.	Ma et al.,[45]
11	Zinc oxide nanoparticles (ZnO NPs) + *Bacillus cereus*and *Lysinibacillus macroides*	Lead (Pb) and copper (Cu)	Rice *(Oryza sativa L)*	Reduced heavy metal toxicity, uptake of heavy metals, bioaccumulation Index and metallothionine (MTs) content. Enhanced plant growth, tolerance index, inhibited the passage of heavy metals to plant tissues by improving the plant tolerance index	Akhtar et al.[46]
12	Zero-valent iron nanoparticles (nZVI) + *Oscillatoria* sp., arbuscular mycorrhizal fungi (*Glomus Mossie* and *Glomus intraradices*) (CCAN25)	Chromium (Cr VI and Cr III) and nickel (Ni)	Princess tree (*Paulownia elongata S. Y. Hu)*	Increased growth, total dry mass, root length, stem diameter, and leaf area, total dry mass, root length, stem diameter, and leaf area, phytoaccumulation of Cr and Ni, bioconcentration factor and bioaccumulation coefficient values being less than 1.0 for both metals, enhanced growth in heavy metal contaminated environments.	Khoshyomn et al.[47]

(*Continued*)

Table 12.1 Combined role of NPS and microbes in plant growth promotion. *Continued*

S. no.	Nanoparticle and microbe	Heavy metal	Host	Response	References
13	Nano zero-valent iron (nZVI) + *Pseudomonas aeruginosa* CKQ9	Chromium (Cr)	Aloe Vera (*Aloe barbadensis* L.)	Improved plant height, root length, and leaf development, enhanced activities of antioxidants such as catalase (CAT) by 124%, superoxide dismutase (SOD) by 87%, and ascorbate peroxidase by 36%, increased proline content by 34%, enhancements in phytoremediation activity of the host plant in soil (50%), roots (77%), and leaves (84%).	Komal et al.[48]
14	Iron oxide nanoparticles (FeO-NPs) + *Providencia vermicola*	Arsenic (As)	Ajwain (*Trachyspermum ammi* L.)	Decreased As concentration in plant tissues, reduction in reactive oxygen species (ROS) and oxidative stress, restoration of structure and cellular function, enhanced activities of antioxidants and nutrients (phenolic, flavonoid, ascorbic acid, and anthocyanin).	Sun et al.[49]
15	Zero-valent iron nanoparticles (Navi) + *Rhizophagus irregularis* (AMF) + *Pseudomonas fluorescens*	Copper (Cu), lead (Pb), cadmium (Cd)	White willow (*Salix alba*)	Low dose showed positive impact on plant growth, biochemical, or physiological responses, increased root length, leaf area, Cd accumulation in roots, and bioconcentration factors (BCF) value of Cd, neutralized toxicity effects of high dose of nZVI.	Mokarram-Kashtiban et al.[50]

12.2.4 Environmental implications and safety concerns of nanoparticle use

Nanoparticles, utilized in diverse fields like agriculture, raise pertinent environmental and safety concerns. Physicochemical characteristics such as size, shape, and solubility influence their potential health risks, emphasizing the need for meticulous handling and adherence to regulatory guidelines. Despite lacking specific federal regulations, cautious management of nanomaterials during transportation, storage, and disposal is imperative. Environmental impact, particularly through air or wastewater release, underscores the necessity for comprehensive studies and proper disposal methods to mitigate contamination risks. Effective risk assessment before market introduction remains paramount, necessitating further research to understand nanoparticle behavior and impacts in different environments. In conclusion, while nanoparticles offer substantial benefits, addressing environmental and safety concerns demands diligent research, regulatory measures, and responsible handling practices.

12.3 Soil microorganisms in plant growth promotion

12.3.1 Diversity and physical role of soil microorganisms in plant–microbe interactions

Soil is a flourishing ecosystem teeming with a wide variety of microorganisms that play an important role in plant–microbe interactions. These interactions are complex and multifaceted, ranging from symbiotic associations that benefit partners to antagonistic ones in which microbes behave as plant diseases.[51] Understanding the diversity and roles of soil microorganisms is critical for improving plant health, production, and sustainability in both agricultural and natural systems. One of the most well-known examples of beneficial plant–microbe interactions is the symbiosis between leguminous plants and nitrogen-fixing bacteria such as *Rhizobium* and *Bradyrhizobium* genera.[52] These bacteria generate root nodules and convert atmospheric nitrogen into a form that the plant can use, making them an important supply of nitrogen. The symbiotic plant microbe relationship not only benefits the plant but also helps improve soil fertility, which further benefits other plant species in the ecosystem.

Arbuscular mycorrhizal fungi (AMF) are another remarkable group of microorganisms that are involved in plant–microbe interaction.[53,54] These fungi establish symbiotic relationships with the roots of terrestrial plant species, extending their hyphal networks into the soil and promoting nutrient and water intake by the plant. In exchange, the plant provides fungi with carbohydrates obtained from photosynthesis.[55] AMF plays an important role in plant growth, nitrogen uptake and stress tolerance, especially in nutrient-deficient or drought-prone areas. Furthermore, arbuscular mycorrhizal fungi can improve plant tolerance to heavy metals and facilitate phytoremediation, making soil microorganisms of contaminated sites beneficial for bioremediation.[56]

Plant growth-promoting rhizobacteria (PGPR) are a varied group of bacteria that colonize the rhizosphere and help plants grow and develop.[57] These bacteria can promote plant growth by a variety of processes, which includes production of phytohormones such as auxins and cytokinin, increased nutrient availability, inhibition of plant pathogens and the induction of systemic tolerance to biotic and abiotic stresses.[58,59] PGPR includes species from the genera *Pseudomonas, Bacillus, Azospirillum*, and *Rhizobium*.[60] PGPR can produce biosurfactants and siderophores, which increases the bioavailability and mobility of contaminants, making them more susceptible for degradation by other microorganisms.[61] Soil microbes also play an important role in organic matter decomposition and nutrient cycling, which are necessary processes for soil fertility and plant growth.[62] Saprophytic fungi and some ectomycorrhizal fungi can break down organic molecules like lignin and cellulose, releasing nutrients that plants can utilize. Bacteria from *Bacillus* and *Pseudomonas* genera contribute to organic matter degradation and nutrient cycling.[22,23]

12.3.2 Mechanisms of plant growth promotion by soil microorganisms

Plant growth-promoting microbes (PGPM) constitute a vital cohort of beneficial microorganisms inhabiting soil ecosystems, exerting profound effects on plant physiology and development. These microorganisms play pivotal roles in augmenting nutrient acquisition, bolstering resistance against phytopathogens, and enhancing tolerance to abiotic stressors, including drought and high temperatures.[63–65] PGPM operate via a multifaceted modus operandi to address environmental challenges, exhibiting mechanisms not only to mitigate biotic and abiotic stresses but also to deploy antiphyto pathological strategies against pathogens and diseases.[66,67]

The intricate interplay between plants and microbes within the rhizosphere underscores a delicate ecological balance. Soil reservoirs harbor essential nutrients such as nitrogen, iron, phosphate, and assorted minerals, often sequestered in complex, recalcitrant forms. This necessitates the intervention of specialized microbial communities, notably Plant Growth-Promoting Rhizobacteria (PGPRs), adept at enzymatic transformations to render these nutrients bioavailable. Prominent examples include *Azotobacter*, *Bacillus* spp., *Pseudomonas*, *Trichoderma*, and *Rhizobium*, all intricately associated with root systems, orchestrating symbiotic interactions essential for plant vigor and productivity.[68–71]

12.3.2.1 Direct mechanism

12.3.2.1.1 Nitrogen fixation

In soils, nitrogen stands out as a critical but largely inaccessible nutrient for plants. However, a specialized group of microbes known as nitrogen-fixing PGPRs (Plant Growth-Promoting Rhizobacteria) play a crucial role in facilitating nitrogen acquisition. These PGPRs form symbiotic associations with plant roots, utilizing signalling molecules such as flavonoids and nod factors to initiate nodulation, a process essential for nitrogen fixation.[72–74] Unlike other nitrogen fixation methods involving natural phenomena like lightning or industrial processes, PGPR-mediated nitrogen fixation relies on the enzymatic activity of nitrogenase. This enzyme catalyzes the conversion of atmospheric nitrogen (N_2) into ammonia (NH_3), a form readily utilized by plants. Moreover, PGPRs can augment nitrogen uptake in plants, thereby improving nitrogen use efficiency.

While the precise mechanisms underlying nitrogen fixation by PGPRs remain incompletely understood, certain strains exhibit the ability to produce growth-promoting substances like auxins and gibberellins, which stimulate plant growth and modulate root architecture. Furthermore, PGPRs can synthesize cytokinins, leading to increased root exudation by plants, thereby fostering a conducive environment for PGPR colonization.[75,76] Moreover, PGPRs play a role in delaying plant senescence and continuously supplying fixed nitrogen to plants, thereby sustaining high grain yield and protein content. This multifaceted interaction underscores the significance of PGPRs in optimizing plant growth and productivity in nutrient-limited environments.

12.3.2.1.2 Siderophore production

Plant growth-promoting rhizobacteria (PGPR) can positively influence plant growth, although the precise mechanism remains somewhat unclear. Some proposed mechanisms include direct iron provision in which siderophores facilitate the uptake of iron from the environment, making it available to plants, Indirect iron provision: By sequestering iron, siderophores limit the availability of iron to plant pathogens, thereby protecting plants from infection.[77] Improved nutrient utilization can occur as, siderophores might indirectly support plant growth by improving nutrient cycling and mobilizing micronutrients like zinc and copper, altered root architecture; siderophores may induce changes in root morphology, leading to increased nutrient absorption capacity, stress mitigation in which siderophores can protect plants from oxidative stress induced by heavy metals and other abiotic factors.[26,78,79] These mechanisms likely act synergistically to promote plant growth, but further research is needed to fully understand the intricate relationships among siderophores, plants, and the environment.[80] Siderophores produced by plant growth-promoting rhizobacteria (PGPR) help plants acquire iron through the following general mechanisms:

1. Chelating iron: Siderophores bind to ferric iron (Fe^3) with high affinity, creating stable complexes that can be taken up by plants.
2. Reduction of $Fe^3 \rightarrow Fe^2$: When siderophore-bound Fe^3 reaches the plant root, it can be reduced to Fe^2, which is more easily absorbed by plants.
3. Transport across the plasma membrane: Fe^2 enters the plant via specific transporter proteins, such as iron-regulated transporters (IRTs).[81–83]

The exact details of how plants recognize and absorb siderophore-bound iron remain uncertain, but it is believed that plants may utilize receptor-mediated transport systems for Fe acquisition from siderophores. Additionally, siderophores can benefit plants by preventing the growth of pathogens and improving overall nutrient cycling.

12.3.2.1.3 Solubilization of other minerals

Microbes like *Erwinia* and *Enterobacter* help in solubilization of various minerals lead to plant growth. Phosphorous is important but found in complex form plants can only absorb in monobasic and dibasic forms therefore through release of organic acids, phosphatase and phytase enzyme played by phosphate solubilizing microbes leads to bioavailability of phosphorus to plants.[61] Uptake of root exudates, lysates result in positive interaction between plants and microbes, which help in release certain volatile compounds, antibiotic, organic acids, vitamins, and enzymes and help in plant uptake nutrient and efficient growth.

12.3.2.2 In direct mechanism

12.3.2.2.1 Antibiosis biocontrol mechanism

The antibiosis biocontrol mechanism facilitates the suppression of pathogens and diseases through microbial antagonism, inducing systemic resistance, and the production of antifungal metabolites such as chitinase and cellulases, thereby enhancing resistance and elimination of fungal pathogens.[84,85]

Plant Growth-Promoting Rhizobacteria (PGPR) exert their effects through the production of phytohormones such as auxins, which stimulate root initiation and growth, and play critical roles in plant development and stress mitigation.[57,86] Additionally, PGPR produces cytokinin, which regulates cell division and differentiation, promoting plant growth and delaying senescence. Abscisic acid (ABA), another hormone produced by PGPR, mediates stress responses, regulates seed dormancy, and controls stomatal closure under adverse conditions. Ethylene, also synthesized by PGPR, participates in various physiological processes including fruit ripening, senescence, and responses to both biotic and abiotic stresses.[87]

12.3.2.2.2 Phytohormones production

PGPR can directly promote plant growth and development, improve mineral nutrition, and enhance plant resistance to biotic and abiotic stresses. Jasmonic acid (JA) plays a central role in Induced Systemic Resistance (ISR) mediated by Plant Growth-Promoting Rhizobacteria (PGPR).[88] It is a signaling molecule that is synthesized in response to wounding and/or insect feeding, and it causes the transcriptional activation of genes encoding proteinase inhibitors (PIs) and enzymes involved in the production of secondary compounds such as nicotine and numerous phenolics, and other defense-related compounds.[89] Plant growth promoting rhizobacteria (PGPR) can produce Jasmonic acid (JA), which triggers Induced Systemic Resistance (ISR) in plants by enhancing their defenses against pests and diseases, independently of salicylic acid (SA). SA, on the other hand, is crucial for Systemic Acquired Resistance (SAR), a lasting defense mechanism against pathogens.[90,91] It activates the expression of pathogenesis-related (PR) genes, bolstering plant immunity. Additionally, SA aids in mitigating various abiotic stresses like salt, metal toxicity, and oxygen deficiency. In PGPR interactions, SA primarily facilitates SAR induction, contributing to long-term plant resilience against pathogens.

In the context of plant growth promoting rhizobacteria (PGPR), the role of SA is primarily linked to SAR and the activation of plant defense responses against pathogens. The induction of SAR is generally associated with SA signaling, and it contributes to the long-term resistance of plants to a wide range of pathogens.[92]

12.4 Nanoparticle and soil microorganisms in bioremediation

12.4.1 Overview of bioremediation and its importance

Environmental contamination is increasing rapidly with increase in density of population, industrialization, urbanization, and even requirement of higher agricultural output levels.[93] The pollutants discharged have the potential to degrade soil, cause deforestation, reduce biodiversity, contaminate air, water, and cause harm to human health. These pollutants include sulfur dioxide, particulates, hydrocarbons, nitrogen oxides, organic compounds (volatile organic compounds and dioxins), hydrocarbons, heavy metals (arsenic, chromium, lead, cadmium, mercury, and zinc), and carbon monoxide (CO).[94] The contamination caused by organic/inorganic composites has increasingly expanded due to rising anthropological encroachment, which is the primary challenge for environmental safety.[95]

Many aspects are carefully considered when choosing a technology to remediate a contaminated site with these contaminants.[96] Few of the criteria include costs, efficiency, technology complexity, recovery time, and availability of physical and human resources for implementation. There are various physical and chemical techniques available for remediation, which are quite expensive, and it may not be effective, efficient, or sustainable.[97,98] In order to overcome the drawbacks of each technology and reduce the cost of the process, combined technologies—which use two or more technologies sequentially—must be developed and used. Thus, besides the physical and chemical techniques for remediating the contaminants, biological treatments are becoming more important because of their affordability, versatility, efficiency and also sustainable in use and such technique is termed bioremediation.[99]

Bioremediation is the process of employing microorganisms and their byproducts to detoxify a wide range of common environmental contaminants, such as hazardous metal compounds, organochlorine insecticides, poly-chlorinated hydrocarbons (PCH), and chlorinated organic solvents. The process of bioremediation includes techniques such as biological stabilization, bioaccumulation, biotransformation, and biosorption.[100] These technologies make use of various microorganisms (bacteria, fungi, algae) and plants, as well as mixtures of these microbes. Proper temperature, nutrition, and food combinations are necessary for bioremediation. The removal of pollutants could take longer if certain components are missing. Environmental "amendments" like molasses, vegetable oil, or simply air might be added to enhance unfavourable conditions for bioremediation.[101]

12.4.2 Role of nanoparticles in enhancing bioremediation efficiency

In recent years, nanotechnology has emerged as a subject of extensive research, which involves all forms of life and intersects with numerous other fields of science.[102] It plays crucial role in production of new products, substitution of existing production equipment and the reformulation of chemicals and materials with enhanced performance, all of which reduce energy and material consumption, minimize environmental damage, and require less environmental remediation. In a way that is both economical and efficient, nanotechnology can address some of the problems associated with challenges faced during remediation and improve the overall efficiency of the remediation process[22,23] (Fig. 12.2) (Table 12.2).

Nanotechnology includes the synthesis and application of a wide range of nanomaterials (NMs), comprising devices and structures sized between 1 and 100 nm that have unique features, which are not present in bulk materials. They are ultrafine particles that have excellent selectivity, crystallinity, regulated shape, and size that vary based on their morphology. Due to such properties, nanomaterials are used for extensive applications in biomedicine and pharmaceuticals,

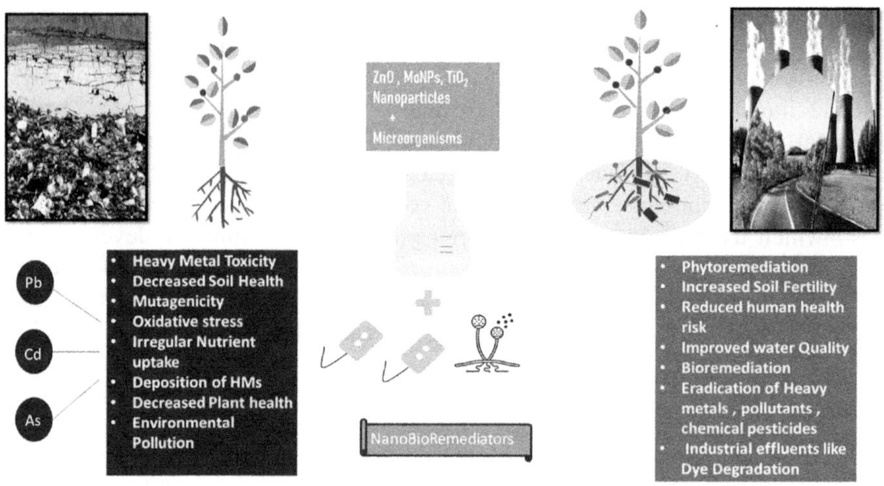

FIGURE 12.2

Response of nanoparticles and microbes in plant growth promotion and bioremediation.

Table 12.2 Combined effect of microbes and NPs in bioremediation.

S. no.	Nanoparticle	Microorganism	Heavy metal /pollutants	Response	References
1	Iron oxide nanoparticles (IONPs) Fe_2O_3NPs and Fe_3O_4NPs	*Methylobacterium currus DSD (Meb)*	Cadmium (Cd)	Enhanced resilience, absorption capacity of bacteria under Cd-induced stress conditions, Adjustment of surface chemical carboxyl and hydroxyl groups, Control of extracellular polymeric substance (EPS), Consortium led to increase protein levels, polysaccharide levels mitigation of toxicity and Cd accumulation in crops.	Ding et al.[103]
2	Magnesium oxide nanoparticles (MgO-NPs)	*Aspergillus niger* strain F1	Tannery Effluent and heavy Metals Cr, Co, Pb, Cd, and Ni	Demonstrated antimicrobial activity, decreased levels of total suspended solids (TSS), total dissolved solids (TDS), and chemical oxygen demand (COD), toxicity, removal of heavy metal Cr, Co, Pb, Cd, and Ni from both textile and tannery effluents, Reduced Toxicity of effluents increased corn, bean seeds germination.	Fouda et al.[104]
3	Magnesium oxide nanoparticles (MgO-NPs)	*Aspergillus terreus S1*	Chromium ions	Exhibited Antimicrobial activity, reduced levels of total suspended solids (TSS), total dissolved solids (TDS), and chemical oxygen demand (COD), removed 97.5% Chromium ions from tanning effluent.	Saied et al.[105]
4	Magnetite (Fe3O4) nanoparticles	*Enterobacter cloacae and Pseudomonas otitidis*	Oily industrial wastewater	Enhanced removal efficiencies of contaminants, removal efficiencies 96% for total suspended solids (TSS), 65.4% for chemical oxygen demand (COD), 83.9% for oil and grease (O&G), and 85% for total petroleum hydrocarbons (TPH), highest coating efficiency, glycine modified-Fe_3O_4 exhibited high ability to immobilize bacteria.	Bestawy et al.[106]

(Continued)

Table 12.2 Combined effect of microbes and NPs in bioremediation. *Continued*

S. no.	Nanoparticle	Microorganism	Heavy metal /pollutants	Response	References
5	Copper oxide nanoparticles (CuO-NPs)	*Fusarium oxysporum OSF18*	Heavy metals, dye	Exhibited superior microbial disinfection, heavy metal removed 93% for Pb, 55% for Cr, and 30% for Ni, achieved 90% dye decolorization, resistance activity to copper, antimicrobial activity showed high efficiency eliminating textile pollutants like dye and heavy metals.	Darwesh et al.[107]
6	Copper oxide nanoparticles (CuO-NPs)	*Stenotrophomonas sp.*	Azo Dyes RB5/ DR23	Improved decolorization efficiency DR23 dye: 97% removal at 20 g/L concentration using 0.04 g, RB5 dye: 94% removal using 0.08 g nanoparticles within 60 and 80 minutes, Np absorption activity for waste water treatment.	Kamalian et al.[108]
7	Zero-valent iron (nZVI) with a carbon/ phosphate matrix (nZVI@C/P)	*Leclercia adecarboxylata*	Lead (Pb)	Enhanced phosphate solubilizing bacteria + Nps enhanced ABC transporters, siderophore, sulfur, phosphorus metabolism, Improved soil enzymatic activities (alkaline phosphatase, catalase and urease), betterment of the soil health led to conversion labile lead into stable form 70.58% from $Pb_5(PO_4)_3Cl$, $PbSO_4$ and $3PbCO_3{\cdot}2Pb(OH)_2{\cdot}H_2O$ in soil.	Teng et al.[109]
8	Iron oxide nanoparticles (FeO Np)	*Corynebacterium xerosis and Lysinibacillus fusiformis 5B*	Diesel, crude oil	Reduction in diesel and crude oil over 20 days, enhanced microbial enhanced activity remediation of Petroleum-contaminated environments, *Corynebacterium*sp. utilize (91%) with isolate count 192 x 10^6 CFU/mL, diesel, *Lysinbacillus fusiformis 5B* utilized (37.95%) crude oil with 3.8 x 10^5 CFU/mL.	Mordecai et al.[110]
9	Silver bio nanoparticles (AgNPs)	*Bacillus thuringiensis israelensis*	Chlorpyrifos pesticide biodegradation	Facilitated biodegradation of chlorpyrifos pesticide from contaminated water (93.4%) with less harmful by products such as 5-octadecene, 7,10- hexadecadienal, 2,13-octadecadien-1-ol with −36.33 mV zeta potential.	Al-Zubaidi et al.[111]

10	Silver gold nanoparticles Drp-Au-AgNPs	*Deinococcus radiodurans*	Malachite green dye degradation	Degradation of malachite green dye (83.68%) with degraded by product,4-(dimethylamino) benzophenone Bimetallic degradation showed reusability after 2 cycles (64%) dye degradation	Weng et al.[112]
11	Ceria nanoparticles (CeO_2 NPs)	*Spirulina platensis*	Methylene blue dye degradation	Methylene blue photodegradation, pure ceria showed (76.4%), green CeO_2@Sp 2:1 (86.2 %), 1: 1 (88.8%) (pH = 11, contact time = 90 min, catalyst dose = 0.3 g/L and MB dye initial concentration = 100 mg/L), expeditious degradation of MB dye under UV irradiation for wastewater treatment	Ali et al.[113]
12	Silver nanoparticles (Ag)	*Glutamicibacter uratoxydans VRAK 24*	Lead (Pb)	Exhibited antibiofilm activity against *Staphylococcus aureus* recorded −5.37 mV zeta potential stability having 0.376 size contributed to lead removal efficiency 87.69%.	Vishnupriya et al.[114]
13	Zinc oxide nanoparticles (ZnO NPs)	*Fusarium proliferatum WC416*	Anthracene	Photocatalytic remediation of Anthracene, a polycyclic aromatic hydrocarbon (PAH), resulted in the generation of 1,2 benzene carboxylic acid and salicylic acid, exhibited removal rates of Anthracene at 65 ± 0.03% for 100 ppm, 50.7 ± 0.02% for 250 ppm, and 40.83 ± 0.03% for 500 ppm concentrations.	Gupta et al.[115]
14	Titanium dioxide nanoparticles (TiO_2 NPs)	*Bacillus subtilis* MTCC 8322	Dyes methylene blue and orange G	Removed 98.85% methylene blue, 80.43% of Orange G dye, synthesis of (TiO_2 NPs) from microbe therefore, photocatalytic, antibacterial activity led to pollutant removal	Rathore et al.[116]
15	Ferrous sulfide nanoparticles (nFeS)	*Geobacter sulfurreducens (SRB)*	Lead (Pb), cadmium (Cd), copper (Cu), nickel Ni(II), Mn(II), Zn(II)	Removed heavy metal Pb(II) (92.6%), Cd(II) (78.7%), Cu(II) (76.0%), Ni(II) (62.5%), Mn(II) (62.2%), and Zn(II) (88.5%) from acid mine drainage, enhanced heavy metal sequestration (nFeS and SRB), improved stability, and sulfur reducing bacteria (SRB)-mediated conversion of SO_4^{2-} to S^{2-} facilitated nFeS synthesis, ensuring stability.	Chen et al.[117]

electronics and optoelectronics, environmental detection, cosmetics and clothing industries.[118] In comparison to the conventional materials, the magnetic, visual, and electrical properties present in nanomaterials varies a lot. Nanomaterials play crucial role in remediation of pollutants from soil and water ecosystem. It includes carbon nanotubes, zero-valent nanoparticles, metal oxide nanoparticles, and nanocomposites.[119] It enables the use of various high-tech materials such as membranes, adsorption materials, functionalized surfaces, coatings, and reagents for more efficient remediation procedure. Thus, nanomaterial-based technologies have shown great potential as low-cost, highly efficient substitutes for existing water treatment methods while meeting the increasing stringent water quality standards.[120]

12.4.3 Interaction between nanoparticles, soil microorganisms, and contaminants in soil

The interaction between nanomaterials, microbes, and pollutants in soil is a complex and dynamic process that has a significant impact on soil health and environmental functioning. Various natural processes such as atmospheric deposition or human activities like applications of biosolids are administered for introducing nanoparticles into soil and play role in environmental remediation. In soil environment, these nanoparticles interact with soil microorganisms and contaminants aiding transport of pollutants into soil and carrying out various microbial activity.[121,122]

Soil microorganisms that include bacteria, fungi, archaea, protist etc. constitute diverse components of soil ecosystems. They play an important role in biogeochemical cycling, organic matter breakdown and pollutant degradation.[123] Soil microorganisms interact with nanoparticles and pollutants in a variety of ways. For example, nanoparticles can affect microbial community composition, diversity and function either by physical interaction or indirectly by changing the physiochemical properties parameters in soil.[18,124] Furthermore, microorganisms can colonize nanoparticle surfaces and use them as growth and metabolic substrates.[125]

The interaction of nanoparticles, soil microorganisms, and pollutants has major significance for degrading contaminants and carrying out remediation process. Soil microorganisms have various metabolic capacities that allow them to break down a wide range of toxins, including organic pollutants, heavy metals, and other hazardous substances.[126] Nanoparticles can impact microbial degradation processes in a variety of ways. Nanoparticles, such as zero-valent iron nanoparticles, can operate as electron shuttles or carriers for microbial enzymes, allowing for more efficient pollutant degradation processes. Similarly, metal nanoparticles, including silver nanoparticles, can exhibit catalytic activity, enabling the transformation and detoxification of organic contaminants in soil settings.[127]

Moreover, metal-based nanoparticles such as copper, zinc, and titanium oxides interact with soil organic matter, influencing its cycle and availability to microbes.[128] As nanoparticles consists of higher surface area to volume ratio compared to bulk particles results in having increased reactivity and solubility. This releases soluble metals, which act as both pollutants that stress microorganisms and required trace nutrients.[129] Nanoparticles can adhere to bacterial cell wall, leading to membrane stress and preventing respiration and food intake. However, at low concentrations, some nanoparticles promote microbial development by supplying necessary metals. Furthermore, due to their larger surface area to volume ratio, nanoparticles provide numerous sites for the adsorption of various contaminants, including heavy metals, organic pollutants, and new contaminants such as medicines and nanoplastics.[130] For example, iron oxide nanoparticles have been demonstrated to adsorb heavy metals such as lead and cadmium, limiting their mobility and bioavailability in soil.[131] Similarly, carbon-based nanoparticles such as graphene and carbon nanotubes can absorb organic contaminants such as pesticides and hydrocarbons, reducing their effects on soil ecosystems.[132]

Another example of adsorption in which pollutants and nanoparticles attach to each other's surfaces is through use of copper oxide nanoparticle.[133] The nanoparticle copper oxide binds well to triclosan, an antibacterial chemical commonly found in biosolids used to treat agricultural soils. The sorption of nanoparticles with triclosan inhibits the breakdown of the organic pollutant by soil bacterial populations such as *Pseudomonas*.[134] This may lead to the accumulation of triclosan in surface soils. Meanwhile, soil fungal mortality rises because fungi are more susceptible than bacteria. The metals produced by nanoparticles are also absorbed by microorganisms, resulting in oxidative stress via reactive oxygen species.[135]

In some circumstances, nanoparticles can transform organic pollutants into less harmful products via increased redox processes, thus facilitating bioremediation downstream. For example, nanoscale zero-valent iron (nZVI) nanoparticles can enhance breakdown of several chlorinated herbicides.[136] The breakdown intermediates are more accessible to soil microbial communities and are capable of complete mineralization. However, inadequate degradation of some persistent organic pollutants such as DDT (dichloro-diphenyl-trichloroethane) can result in even more toxic product that harm sensitive nitrifying bacterial populations.[137] Metals in certain nanoparticles supply vital trace nutrients to bacteria while emitting harmful dissolved ions.[58,59] Selenium nanoparticles at low concentrations improved microbial respiration and nitrogen mineralization rates in agricultural soils containing legumes such as lucerne and soybeans. However, with larger concentrations, selenium nanoparticles reduced microbial activity and soil enzyme expression because of oxidative stress.[138]

The antibacterial activities of nanoparticles such as nanosilver disrupt the integrity of microbial cell walls and membrane transport channels. This may reduce microbial proliferation and extracellular enzyme production, which are required to break down soil organic matter, hence limiting nutrient cycling.[139]

However, data suggests that soil microbial populations can develop resistance mechanisms to reduce toxicity. For example, genetic and enzymatic alterations allowed *Pseudomonas putida* populations that had been chronically exposed to silver nanoparticles to aggregate the nanoparticles on their surface, limiting internalization and antibacterial effects.[140]

Furthermore, the interaction of nanoparticles, soil microbes, and pollutants in soil may have larger environmental consequences beyond soil health and ecosystem function.[141] For example, nanoparticles and their byproducts may seep into groundwater or surface water bodies, posing a threat to aquatic ecosystems and human health. The fate of nanoparticles and its transport in soil habitats may be modified by soil features, environmental variables, and microbial activity, impacting their permanence and bioavailability in terrestrial and aquatic ecosystems.[142]

12.4.4 Case studies and examples of nanoparticle-assisted bioremediation approaches

A major branch of nanotechnology deals with removal of various contaminants, which includes organic and inorganic pollutants, dye, and heavy metals from contaminated locations by use of nanoparticles synthesized by microbes.[143] Nanoremediation is a promising approach that provides an environment friendly, sustainable, and feasible alternative for removal of contaminants. Unlike the traditional approaches, biosynthesized nanoparticles have unique features that can effectively catalyze and degrade organic contaminants. High efficiency of biosynthesized nanoparticles is due to their large surface area at nanoscale.[144]

Dyes are used for various applications. But when discharged in water causes adverse effect on human health and environment. Toxic dyes effluents from textile and dyeing units affects aquatic plant photosynthesis by reducing light penetration and increasing chemical and biological oxygen demand (BOD) in water bodies.[145] Metallic nanoparticles, which have unique physicochemical, biological, and catalytic properties, have recently been demonstrated to be a useful tool for wastewater treatment. Thus, catalytic degradation is thought to be the most effective strategy for dye degradation. For example, titanium dioxide nanoparticles act as photocatalyst for degradation of methylene blue dye and organic dye.[146] Also, it was seen that bacterial biomass of *Erwinia herbicola* formed from selenium nanoparticles have shown a high ability to absorb methylene blue from wastewater via photocatalytic degradation.[147]

The expansion of urban, agricultural industries, and man-made activity has resulted in significant increase in organic pollutant contamination of water. Organic pollutants include xenobiotics such as pesticides, insecticides, synthetic dyes, pharmaceuticals, aromatic hydrocarbons, halogenated hydrocarbons and phenol.[148] These pollutants enhance water turbidity, reducing sunlight penetration and negatively impacting aquatic and marine ecosystems. Nano-enabled technology eliminates the organic pollutants by employing unique novel products such as nanoadsorbent, nanocatalyst, and nanofilms.[149] Various nanoparticles and its derivative are used for degradation of organic pollutants as they exhibit high level of degrading efficiency for refractory and microorganic contaminants. For example, iron nanoparticles can help to degrade organic compounds such as chlorinated chemicals like trichloroethylene (TCE) by serving as an electron donor for microorganisms' dehalorespiration.[150] A study found that addition of iron nanoparticles to TCE contaminated groundwater stimulated growth of dehalorespiring bacteria and accelerated TCE degradation rate.[151] Moreover, the biosynthesis of cadmium sulfide (CdS) on the cell surface of *Geobacter sulfurreducens* PCA and *Shewanella oneidensis* MR-1 shows effective photocatalytic activity toward degradation of organic pollutants.[152] Also, TiO_2 based nanoparticles shows photocatalytic activity when activated by UV light for remediation of organic contaminants such as petroleum hydrocarbon. It produces reactive oxygen species that degrades the contaminants. A study showed that TiO_2 nanoparticles could decompose more than 95% of diesel oil in contaminated soil after 90 minutes of UV exposure.[153]

Heavy metal contamination is a severe environmental issue as they are nonbiodegradable, toxic, and carcinogenic.[154] Discharge of these heavy metals into water bodies is a severe cause of water pollution and poses a hazard to human health and environment. Lead (Pb), zinc (Zn), mercury (Hg), nickel (Ni), cadmium (Cd), copper (Cu), chromium (Cr), and arsenic (As) are the most commonly occurring heavy metals.[155] Use of nanovalent iron particles for remediation of heavy metals like arsenic, chromium, lead from soil and groundwater. The nanoparticles provide strong reducing power and can immobilize soluble heavy metals by reducing them into soluble heavy metals by reducing them into insoluble forms that precipitate out. A field study showed that injecting a suspension of nanozero-valent iron into groundwater contaminated with over 500 ppb arsenic reduced levels to below 50 ppb within 5 hours.[156] Moreover, bacterial species such as *Shewanella* and *Geobacter* species can synthesize biogenic palladium nanoparticle (Pd-NPS), which have capability for removal of strong Cr6 + .[128] A large concentration of arsenic, Fe (II) and Mn (II) are present in groundwater and surface waters, which have a negative impact on human health and the aquatic ecology. The bacterial species of *Pseudomonas* (*Pseudomonas QJX-1*) are capable of oxidizing manganese (II), which was extracted from the soil of manganese mines and is utilized in the production of biosynthesis of Fe–Mn oxides (BFMO).[157] It was found that these oxides could both oxidize and absorb arsenic. They are also capable of adsorption of Fe (II) and Mn (II), thus remediating three contaminants from ground water, which includes as (III or V), Fe (II), and Mn (II).[158]

Encapsulating bacterial cells or enzymes in silica nanoparticles allows them to breakdown pesticides and insecticides in water and soil. The encapsulation protects the cells and enzymes against denaturation.[159] A study employed nanocapsules carrying the enzyme organophosphorus hydrolase to break down organophosphate pesticides such as chlorpyrifos and parathion in polluted water. Within 24 hours, more than 95% of degradation had occurred.[160]

12.5 Synergistic effects of nanoparticles and soil microorganisms

12.5.1 Combined applications of nanoparticles and soil microorganisms in plant growth promotion and bioremediation

The combination of nanoparticles and beneficial soil microbes is a potential strategy for improving plant development and enabling bioremediation of contaminated soil. Nanoparticles such as zinc oxide, titanium dioxide, silver, copper, iron, and silica when utilized effectively improves properties such as seed germination, plant growth, yield, nutrient uptake, and stress tolerance.[161] On the other hand, plant growth promoting rhizobacteria (PGPR) and arbuscular mycorrhizal fungi (AMF) can stimulate plant growth, improve soil and degrade organic pollutants through the involvement of processes such as nitrogen fixation, phosphate solubilization, and phytohormone production. However, nanoparticles at high doses can be hazardous to plants.[162] The synergistic integration of nanoparticles and microorganisms provides an opportunity to harness the benefits to both while minimizing their potential negative impacts.[163]

There are several hypothesized mechanisms behind such synergistic interactions. Nanoparticles act as crucial mineral nutrient reservoirs, while microbes such as PGPR and AMF facilitate nutrient mobilization through acidification, chelation, ion exchange and reduction processes mediated by organic acids, siderophores etc.[164] Microbes further expand soil–root interfaces through the formation of extensive hyphal networks that improves nutrient availability. Furthermore, the vast surface area of nanoparticles provides binding sites for microbial accumulation and plant growth-promoting substances.[165] Another key mechanism is the regulation of soil enzyme activities and the increased degradation of organic contaminants.[166] Nanoparticles and coexisting soil bacteria can synergistically affect the activity of enzymes such as dehydrogenase, urease and catalase, which are involved in nutrient transformation and pollutant degradation. Microbes directly participate in reactions, serve as binding scaffolds, and generate cofactors and redox mediators that influence catalytic rates.[167] Nanoparticles promote microbial enzyme expression, protect against denaturation, and mimic active sites via surface electrostatic interactions. At the molecular level, synergistic effect of nanoparticles and soil microorganisms influence numerous plant signaling pathways, which results to be more effective than their individual actions such as those involving reactive oxygen species, phytohormones and pathogenesis related proteins.[168]

12.5.2 Application of synergistic effect

The combination of nanoparticles and soil microorganisms offers numerous possibilities for developing more effective plant growth enhancers and soil remediation agents. Several studies have shown that combining nanoparticles with beneficial soil microbes like PGPR and AMF improves plant performance and soil health than their individual exposure.[169]

Cerium oxide nanoparticles and an AMF species, *Funneliformis mosseae,* showed great effect in alleviating arsenic toxicity in maize.[170] Researchers treated maize seeds with cerium oxide nanoparticles and infected them in arsenic free and arsenic-contaminated soil. They observed that the combination of nanoparticle and *Funneliformis mosseae* resulted in dramatically greater tolerance to arsenic-induced oxidative damage, increased growth, elevated antioxidant enzyme activity as compared to their individual uses.[171] Cerium oxide nanoparticles uptake in plant tissues, but the fungus boosted nutritional intake and antioxidant response by signaling and priming defense systems against the metalloid.[172]

Similarly, scientists investigated the synergistic effects of selenium nanoparticles and the rhizobacterial strain *Pseudomonas fluorescens* on Indian mustard growing on regular soil. Coinoculation of selenium nanoparticles and bacteria significantly increased selenium accumulation in plant tissues, resulting in 2–3 times higher selenium content than their solo treatments.[173] Researchers attributed this synergistic effect to increased selenium mobility/bioavailability and plant selenium metabolism because of nanoparticle-activated bacterial signaling pathways. For instance, zinc oxide nanoparticles and bacterial species, *Azospirillum brasilense* were administered for combating salinity stress in cotton. They observed that coapplication of nanoparticles and bacteria resulted in significant improvements in many metrics such as seed germination, seedling fresh weight, leaf area, and root elongation compared to solo exposures in both normal and saline soils.[174] The synergistic process was attributed to nanoparticles acting as a zinc nutrition supply, whereas bacteria improved nutrient uptake through increased root surface area and phytohormone signaling.[175]

Moreover, the combination of nanoparticles and beneficial soil microorganisms such as plant growth-promoting bacteria and fungus has enormous potential as an emerging tool for more efficient bioremediation of organic and inorganic contaminants from polluted soil and wastewater.[176] For instance, scientists evaluated the individual and combined potential of zinc oxide nanoparticles and white-rot fungus *Phanerochaete chrysosporium* in eliminating the azo dye Acid Orange 7 from stimulated wastewater. They achieved a synergistic effect in dye removal efficiency employing fungal nanoparticle combination compared to their individual exposures.[177] Researchers suggested that dye adsorption on nanoparticles enabled easy destruction by fungal lignolytic enzymes, and nanoparticles also fragmented dye clumps, boosting accessibility for the fungus.[178] Arsenic is a harmful pollutant, which, when discharged to soil, causes various negative impact on human health and environment. To remove arsenic from contaminated soil, experts combined selenium nanoparticles with mixture of three dissimilatory arsenate-reducing bacterium strains: *Bacillus licheniformis, Bacillus subtilis*, and *Bacillus pumilus*.[129] They observed increased soil arsenic mobilization and transformation into less hazardous elemental form because of synergistic biotransformation when bacteria were assisted by nanoparticles rather than bacterial inoculation alone. Nanoparticles released selenium for bacterial metabolic activities, and bacteria affected bioavailability and nanoparticle dissolution kinetics via soil interactions.[179]

Another example is the use of nanoparticles, such as iron oxide nanoparticles (IONPs), when combined with certain soil microorganisms like *Rhizobium* spp. to remediate heavy metal-contaminated soils.[97,98] *Rhizobium* spp. are nitrogen-fixing bacteria that are typically found in leguminous plants, but they also have metal-tolerant properties and heavy metal detoxifying processes. When combined with IONPs, which have been proven to immobilize heavy metals via adsorption and precipitation process. *Rhizobium* spp. can improve heavy metal immobilization and sequestration in polluted soils.[180] The nanoparticles provide a surface for bacteria to colonize and interact with heavy metal ions, resulting in synergistic actions that reduce heavy metal bioavailability and toxicity in the environment.[181] Moreover, coexposure of selenium nanoparticles with the fungus *Trametes versicolor* resulted in enhanced degradation of anthracene and pyrene contaminating soil than fungal inoculation alone. The researchers attributed that this synergism led to increase in production of essential lignolytic enzymes such as laccase and manganese peroxidase in fungus, which was induced by nanoparticles.[182]

12.6 Challenges and opportunities for integrating nanoparticles and microbial approaches

The integration of nanoparticles and microbial methods in plant growth promotion and bioremediation generates both obstacles and opportunities that must be properly addressed. One of the primary challenges is the composition, size and surface qualities of nanoparticles that affect plants, microbes and other species in the environment leading to potential toxicity and biosafety issues.[183] It is critical to assess and mitigate the harmful effects of beneficial microbes utilized to promote plant growth or bioremediation. Understanding the complex relationships between nanoparticles and microbial cells is another key challenge since nanoparticles might have an impact on microbial physiology, gene expression and metabolic pathway, potentially affecting desirable functionality.[184,185]

Ensuring the long-term stability of nanoparticles and its persistence in soil habitat is of particular importance as factors such as pH, organic matter content, and microbial activity can have impact on their availability and effectiveness. Establishing effective delivery techniques in order to introduce nanoparticles and microbial formulations into the plant or soil environment involve obstacles such as nanoparticles dispersion, microbial viability and targeted administration to specific plant tissues or soil zone.[186] Moreover, determining the potential environmental fate and impact of these intervention such as bioaccumulation, biomagnification and long-term consequences on soil health and ecosystem is crucial for their safe and sustainable application.[187] Establishing proper regulatory frameworks and policies for the use of nanoparticles and microbial methods in plant growth promotion and bioremediation is additional problem, since concerns about potential dangers, environmental release, and public acceptance must be addressed.

On the other hand, the combination of nanoparticles with microbial methods offers tremendous opportunities. Nanoparticles can improve the distribution and efficacy of beneficial microorganisms, potentially boosting plant growth and bioremediation capacities.[37,38] Nanoparticles, for example, can serve as carriers for bacteria, insulating them from external challenges while facilitating targeted delivery to specific sites.[188] Furthermore, nanoparticles can have special features that promote plant development and bioremediation. Certain nanoparticles, for example, can boost nutrient uptake, water retention, and stress tolerance in plants, whilst others can help with pollutant breakdown or contaminant immobilization in soil.[189]

12.7 Conclusion and future prospectives

The integration of nanoparticles and beneficial soil microorganisms offers promising opportunities for sustainable agriculture and environmental remediation. Emerging trends includes the use of nanoparticles carriers for targeted nutrient delivery, growth regulators, and biocontrol agents to increase plant productivity. Synergistic combinations of nanoparticles and plant growth promoting microorganisms have shown to improve nutrient bioavailability and plant–microbe symbioses. In bioremediation, nanoparticles catalyze pollutant degradation while few microbial consortia metabolize pollutants resulting in a comprehensive remediation technique.

Looking ahead, optimizing nanoparticle production, delivery techniques and investigating novel microorganisms will help to enhance these technologies. Improved characterization and modeling techniques will help understand the intricate nanobiointeractions and provide appropriate solutions. The future integration of nanoparticles and soil microorganisms offers fascinating developments ongoing research will improve nanoparticle synthesis processes, investigate novel nanomaterials and identify microbial species/consortia for specific uses. Advanced characterization techniques and computational modeling will provide molecular-level details about nanoparticles, driving rotational system design. Thus, combining these complimentary technologies has enormous potential for sustainable food production, soil remediation and ecosystem restoration. By integrating nanoparticles with microorganisms, a more sustainable and resilient future in agriculture and environmental management can be developed.

References

1. Syed, S.; Tollamadugu, N. P. *Role of plant growth-promoting microorganisms as a tool for environmental sustainability. Recent Developments in Applied Microbiology and Biochemistry;* Academic Press, 2019, pp. 209–222.
2. Vandana, U. K.; Chopra, A.; Bhattacharjee, S.; Mazumder, P. B. Microbial biofertilizer: a potential tool for sustainable agriculture. *Microorganisms for Green Revolution: Volume 1: Microbes for Sustainable Crop Production* **2017**, 25–52.
3. Rajput, V. D.; Minkina, T.; Kimber, R. L.; Singh, V. K.; Shende, S.; Behal, A.; Sushkova, S.; Mandzhieva, S.; Lloyd, J. R. Insights into the biosynthesis of nanoparticles by the Genus *Shewanella. Appl. Environ. Microbiol.* **2021**, *87* (22), e0139021. Available from: https://doi.org/10.1128/AEM.01390-21. Epub 2021 Sep 8. PMID: 34495739.
4. Vázquez-Núñez, E.; Molina-Guerrero, C. E.; Peña-Castro, J. M.; Fernández-Luqueño, F.; de la Rosa-Álvarez, M. G. Use of nanotechnology for the bioremediation of contaminants: a review. *Processes* **2020**, *8* (7), 826.
5. Sharma, U.; Sharma, J. G. Nanotechnology for the bioremediation of heavy metals and metalloids. *J. Appl. Biol. Biotechnol.* **2022**, *10* (5), 34–44.
6. Tahat, M.; Alananbeh, M. K.; Othman, A. Y.; Leskovar, I. D. Soil health and sustainable agriculture. *Sustainability* **2020**, *12* (12), 4859.
7. Saba, I.; Wani, K.; Syed, A.; Rehman, S. Role of microbial nanotechnology in bioremediation of heavy metals. *Microbial Nanotechnol. Green Synth. Appl.* **2021**, 307–319.
8. González-González, R. B.; Flores-Contreras, E. A.; Parra-Saldívar, R.; Iqbal, H. M. Bio-removal of emerging pollutants by advanced bioremediation techniques. *Environ. Res.* **2022**, *214*, 113936.
9. Sharma, I. *Bioremediation techniques for polluted environment: concept, advantages, limitations, and prospects. Trace Metals in the Environment-new Approaches and Recent Advances;* IntechOpen, 2020.
10. Singh, D.; Gurjar, B. R. Nanotechnology for agricultural applications: facts, issues, knowledge gaps, and challenges in environmental risk assessment. *J. Environ. Manage.* **2022**, *322*, 116033.
11. Patil, A. G.; Kounaina, K.; Aishwarya, S.; Harshitha, N.; Satapathy, P.; Hudeda, S. P.; Reddy, K. R.; Alrafas, H.; Yadav, A. N.; Raghu, A. V., et al. Myco-nanotechnology for sustainable agriculture: challenges and opportunities. *Recent Trends Mycol. Res. Vol. 1 Agric. Med. Persp.* **2021**, 457–479.
12. Kumar, M.; Ahmad, S.; Singh, R. P. Plant growth promoting microbes: diverse roles for sustainable and ecofriendly agriculture. *Energy Nexus* **2022**, *7*, 100133.
13. Lopes, M. J. D. S.; Dias-Filho, M. B.; Gurgel, E. S. C. Successful plant growth-promoting microbes: inoculation methods and abiotic factors. *Front. Sustain. Food Syst.* **2021**, *5*, 606454.
14. Tyagi, J.; Bandyopadhyay, P.; Shrivastava, N.; Sharma, A. K.; Varma, A.; Pudake, R. N. Effect of inoculum concentrations of *Piriformospora indica (Serendipita indica)* on phenotypic and biochemical characteristics of finger millet plant under drought stress. *Biochem. Cell. Arch.* **2017**, *17* (2).
15. Tyagi, J.; Varma, A.; Pudake, R. N. Evaluation of comparative effects of arbuscular mycorrhiza (*Rhizophagus intraradices*) and endophyte (*Piriformospora indica*) association with finger millet (*Eleusine coracana*) under drought stress. *Eur. J. Soil Biol.* **2017**, *81*, 1–10.

16. Ghasemnezhad, A.; Ghorbanpour, M.; Sohrabi, O.; Ashnavar, M. A general overview on application of nanoparticles in agriculture and plant science. *Comprehen. Analyt. Chem.* **2019**, *87*, 85–110.
17. Hidangmayum, A.; Debnath, A.; Guru, A.; Singh, B. N.; Upadhyay, S. K.; Dwivedi, P. Mechanistic and recent updates in nano-bioremediation for developing green technology to alleviate agricultural contaminants. *Int. J. Environ. Sci. Technol.* **2023**, *20* (10), 11693–11718.
18. Xin, X.; Zhao, F.; Zhao, H.; Goodrich, S. L.; Hill, M. R.; Sumerlin, B. S.; Stoffella, P. J.; Wright, A. L.; He, Z. Comparative assessment of polymeric and other nanoparticles impacts on soil microbial and biochemical properties. *Geoderma* **2020**, *367*, 114278.
19. Zhao, L.; Lu, L.; Wang, A.; Zhang, H.; Huang, M.; Wu, H.; Xing, B.; Wang, Z.; Ji, R. Nano-biotechnology in agriculture: use of nanomaterials to promote plant growth and stress tolerance. *J. Agric. Food Chem.* **2020**, *68* (7), 1935–1947.
20. Ali, A.; Alghanem, S. M. S.; Al-Haithloul, H. A. S.; Muzammil, S.; Adrees, M.; Irfan, E.; Waseem, M.; Anayat, A.; Saqalein, M.; Rizwan, M., et al. Co-application of copper nanoparticles and metal tolerant *Bacillus* sp. for improving growth of spinach plants in chromium contaminated soil. *Chemosphere* **2023**, *345*, 140495.
21. Yadav, A.; Yadav, K.; Abd-Elsalam, K. A. Nanofertilizers: types, delivery and advantages in agricultural sustainability. *Agrochemicals* **2023**, *2* (2), 296–336.
22. Yadav, A. N.; Kour, D.; Kaur, T.; Devi, R.; Yadav, A.; Dikilitas, M.; Abdel-Azeem, A. M.; Ahluwalia, A. S.; Saxena, A. K. Biodiversity, and biotechnological contribution of beneficial soil microbiomes for nutrient cycling, plant growth improvement and nutrient uptake. *Biocatal. Agric. Biotechnol.* **2021**, *33*, 102009.
23. Yadav, N.; Garg, V. K.; Chhillar, A. K.; Rana, J. S. Detection and remediation of pollutants to maintain ecosustainability employing nanotechnology: a review. *Chemosphere* **2021**, *280*, 130792.
24. Das, P.; Barua, S.; Sarkar, S.; Karak, N.; Bhattacharyya, P.; Raza, N.; Kim, K. H.; Bhattacharya, S. S. Plant extract–mediated green silver nanoparticles: efficacy as soil conditioner and plant growth promoter. *J. Hazard. Mater.* **2018**, *346*, 62–72.
25. Ghorbanpour, M.; Bhargava, P.; Varma, A.; Choudhary, D. K., Eds. *Biogenic Nano-Particles and Their Use in Agro-Ecosystems;* Springer: Singapore, 2020; pp 1–606.
26. Gupta, K.; Dubey, N. K.; Singh, S. P.; Kheni, J. K.; Gupta, S.; Varshney, A. Plant growth-promoting rhizobacteria (PGPR): current and future prospects for crop improvement. *Current Trends in Microbial Biotechnology for Sustainable Agriculture* **2021**, 203–226.
27. Agrahari, S.; Dubey, A. *Nanoparticles in Plant Growth and Development;* Springer *Link*, 2020, pp. 9–37.
28. Ingle A.; Aziz H.M.; Sebasta M.; Ramakanth I. et al. Nanotechnology in Plant Growth Promotion and Protection: Recent Advances and Impacts. 2021, pp. 170–202.
29. Mathur, P.; Roy, S. Nanosilica facilitates silica uptake, growth and stress tolerance in plants. *Plant Physiol. Biochem.* **2020**, *157*, 114–127.
30. Ali, S.; Mehmood, A.; Khan, N. Uptake, translocation, and consequences of nanomaterials on plant growth and stress adaptation. *J. Nanomater.* **2021**, *2021*, 1–17.
31. Rameshraddy.; Pavithra, G. J.; Rajashekar Reddy, B. H.; Salimath, M.; Geetha, K. N.; Shankar, A. G. Zinc oxide nano particles increases Zn uptake, translocation in rice with positive effect on growth, yield and moisture stress tolerance. *Indian J. Plant Physiol.* **2017**, *22*, 287–294.

32. Al-Khayri, J. M.; Rashmi, R.; Surya Ulhas, R.; Sudheer, W. N.; Banadka, A.; Nagella, P.; Aldaej, M. I.; Rezk, A. A. S.; Shehata, W. F.; Almaghasla, M. I. The role of nanoparticles in response of plants to abiotic stress at physiological, biochemical, and molecular levels. *Plants* **2023**, *12* (2), 292.
33. Elemike, E. E.; Uzoh, I. M.; Onwudiwe, D. C.; Babalola, O. O. The role of nanotechnology in the fortification of plant nutrients and improvement of crop production. *Appl. Sci.* **2019**, *9* (3), 499.
34. Azameti, M. K.; Imoro, A. W. M. Nanotechnology: a promising field in enhancing abiotic stress tolerance in plants. *Crop Des.* **2023**, 100037.
35. Kalia, A.; Kaur, H. *Nanofertilizers: an innovation towards new generation fertilizers for improved nutrient-use efficacy and environmental sustainability. Nanoagroceuticals & Nanophytochemicals;* CRC Press, 2018, pp. 45–61.
36. Mushtaq, T.; Shah, A. A.; Akram, W.; Yasin, N. A. Synergistic ameliorative effect of iron oxide nanoparticles and *Bacillus subtilis* S4 against arsenic toxicity in *Cucurbita moschata*: polyamines, antioxidants, and physiochemical studies. *Int. J. Phytoremediation* **2020**, *22* (13), 1408–1419.
37. Zand, A. D.; Tabrizi, A. M.; Heir, A. V. The influence of association of plant growth-promoting rhizobacteria and zero-valent iron nanoparticles on removal of antimony from soil by *Trifolium repens*. *Environ. Sci. Pollut. Res.* **2020**, *27* (34), 42815–42829.
38. Zand, A. D.; Mikaeili Tabrizi, A.; Vaezi Heir, A. Application of titanium dioxide nanoparticles to promote phytoremediation of Cd-polluted soil: contribution of PGPR inoculation. *Bioremediat. J.* **2020**, *24* (2-3), 171–189.
39. Tariq, S.; Bano, A., et al. Role of PGPR and silver nanoparticles on the physiology of *Momordica charantia* L. irrigated with polluted water comprising high Fe and Mn. *Int. J.Phytoremediation* **2023**, *25* (12), 1643–1655.
40. Ahmed, T.; Noman, M.; Rizwan, M.; Ali, S.; Ijaz, U.; Nazir, M. M.; ALHaithloul, H. A. S.; Alghanem, S. M.; Abdulmajeed, A. M.; Li, B. Green molybdenum nanoparticles-mediated bio-stimulation of *Bacillus* sp. strain ZH16 improved the wheat growth by managing in planta nutrients supply, ionic homeostasis and arsenic accumulation. *J. Hazard. Mater.* **2022**, *423*, 127024.
41. Ma, J.; Li, Y.; Chen, F.; Sun, Y.; Zhu, Y.; Wang, L. Bacillus mycoides PM35 in combination with titanium dioxide (TiO2) nanoparticles enhanced morpho-physio-biochemical attributes in Barley (*Hordeum vulgare* L.) under cadmium stress. *Chemosphere* **2023**, *323*, 138224.
42. Shafiq, T.; Yasmin, H.; Shah, Z. A.; Nosheen, A.; Ahmad, P.; Kaushik, P.; Ahmad, A. Titanium oxide and zinc oxide nanoparticles in combination with cadmium tolerant bacillus pumilus ameliorates the cadmium toxicity in maize. *Antioxidants* **2022**, *11* (11), 2156.
43. Shah, A. A.; Aslam, S.; Akbar, M.; Ahmad, A.; Khan, W. U.; Yasin, N. A.; Ali, B.; Rizwan, M.; Ali, S. Combined effect of *Bacillus fortis* IAGS 223 and zinc oxide nanoparticles to alleviate cadmium phytotoxicity in *Cucumis melo. Plant Physiol. Biochem.* **2021**, *158*, 1–12.
44. Zhu, Y.; Wang, L.; Ma, J.; Li, Y.; Chen, F.; Peijnenburg, W. Comparative physiological and metabolomics analyses using Ag NPs and HAS31 (PGPR) to alleviate Cr stress in barley (*Hordeum vulgare* L). *Environ. Pollut.* **2023**, *333*, 122010.
45. Ma, J.; Alshaya, H.; Alwasel, Y. A.; Chen, F.; Adrees, M.; Hussain, A.; Hameed, S. Application of cerium dioxide nanoparticles and chromium-resistant bacteria reduced chromium toxicity in sunflower plants. *Front. Plant Sci.* **2022**, *13*, 876119.

46. Akhtar, N.; Khan, S.; Rehman, S. U.; Rehman, Z. U.; Khatoon, A.; Rha, E. S.; Jamil, M. Synergistic effects of zinc oxide nanoparticles and bacteria reduce heavy metals toxicity in rice (*Oryza sativa* L.) plant. *Toxics* **2021**, *9* (5), 113.
47. Khoshyomn, S.; Heidari, A.; Farzam, M.; Shariatmadari, Z.; Karimian, Z. Integrated approaches for heavy metal–contaminated soil remediation: harnessing the potential of *Paulownia elongata* SY Hu, Oscillatoria sp., arbuscular mycorrhizal fungi (*Glomus mosseae* and *Glomus intraradices*), and iron nanoparticles. *Environ. Sci. Pollut. Res.* **2024**, 1–20.
48. Komal.; Shabaan, M.; Ali, Q.; Asghar, H. N.; Zahir, Z. A.; Yousaf, K.; Aslam, N.; Zulfiqar, U.; Ejaz, M.; Alwahibi, M. S., et al. Exploring the synergistic effect of chromium (Cr) tolerant Pseudomonas aeruginosa and nano zero valent iron (nZVI) for suppressing Cr uptake in Aloe Vera. *Int. J. Phytoremediation* **2024**, 1–12.
49. Sun, Y.; Ma, J.; Chen, F. Combined application of plant growth-promoting bacteria and iron oxide nanoparticles ameliorates the toxic effects of arsenic in Ajwain (*Trachyspermum ammi* L.). *Front. Plant Sci.* **2022**, *13*, 1098755.
50. Mokarram-Kashtiban, S.; Hosseini, S. M.; Tabari Kouchaksaraei, M.; Younesi, H. The impact of nanoparticles zero-valent iron (nZVI) and rhizosphere microorganisms on the phytoremediation ability of white willow and its response. *Environ. Sci. Pollut. Res.* **2019**, *26*, 10776–10789.
51. De Bellis, T.; Laforest-Lapointe, I.; Solarik, K. A.; Gravel, D.; Kembel, S. W. Regional variation drives differences in microbial communities associated with sugar maple across a latitudinal range. *Ecology* **2022**, *103* (8), e3727.
52. Mahmud, K.; Makaju, S.; Ibrahim, R.; Missaoui, A. Current progress in nitrogen fixing plants and microbiome research. *Plants* **2020**, *9* (1), 97.
53. Sangwan, S.; Prasanna, R. Mycorrhizae helper bacteria:unlocking their potential as bioenhancers of plant–arbuscular mycorrhizal fungal associations. *Microb. Ecol.* **2022**, *84* (1), 1–10.
54. Tyagi, J.; Sultan, E.; Mishra, A.; Kumari, M.; Pudake, R. N. *The impact of amf symbiosis in alleviating drought tolerance in field crops*. In *Mycorrhiza—Nutrient Uptake, Biocontrol, Ecorestoration;* Varma, A., Prasad, R., Tuteja, N., Eds.; Springer: Cham, 2017. Available from: https://doi.org/10.1007/978-3-319-68867-1_11.
55. Khaliq, A.; Perveen, S.; Alamer, K. H.; Zia Ul Haq, M.; Rafique, Z.; Alsudays, I. M.; Althobaiti, A. T.; Saleh, M. A.; Hussain, S.; Attia, H. Arbuscular mycorrhizal fungi symbiosis to enhance plant–soil interaction. *Sustainability* **2022**, *14* (13), 7840.
56. Riaz, M.; Kamran, M.; Fang, Y.; Wang, Q.; Cao, H.; Yang, G.; Deng, L.; Wang, Y.; Zhou, Y.; Anastopoulos, I., et al. Arbuscular mycorrhizal fungi-induced mitigation of heavy metal phytotoxicity in metal contaminated soils: a critical review. *J. Hazard. Mater.* **2021**, *402*, 123919.
57. Santoyo, G.; Urtis-Flores, C. A.; Loeza-Lara, P. D.; Orozco-Mosqueda, M. D. C.; Glick, B. R. Rhizosphere colonization determinants by plant growth-promoting rhizobacteria (PGPR). *Biology* **2021**, *10* (6), 475.
58. Khan, M. R.; Fromm, K. M.; Rizvi, T. F.; Giese, B.; Ahamad, F.; Turner, R. J.; Füeg, M.; Marsili, E. Metal nanoparticle–microbe interactions: synthesis and antimicrobial effects. *Particle Particle Syst. Character.* **2020**, *37* (5), 1900419.
59. Khan, N.; Bano, A.; Ali, S.; Babar, M. A. Crosstalk amongst phytohormones from planta and PGPR under biotic and abiotic stresses. *Plant Growth Regul.* **2020**, *90*, 189–203.

60. Verma, M.; Mishra, J.; Arora, N. K. Plant growth-promoting rhizobacteria: diversity and applications. *Environ. Biotechnol. Sustain. Future* **2019**, 129–173.
61. Vocciante, M.; Grifoni, M.; Fusini, D.; Petruzzelli, G.; Franchi, E. The role of plant growth-promoting rhizobacteria (PGPR) in mitigating plant's environmental stresses. *Appl. Sci.* **2022**, *12* (3), 1231.
62. Basu, S.; Kumar, G.; Chhabra, S.; Prasad, R. *Role of soil microbes in biogeochemical cycle for enhancing soil fertility. New and Future Developments in Microbial Biotechnology and Bioengineering;* Elsevier, 2021, pp. 149–157.
63. Sharma, N.; Dabral, S.; Tyagi, J.; Yadav, G.; Aggarwal, H.; Joshi, N. C.; Varma, A.; Koul, M.; Choudhary, D. K.; Mishra, A. Interaction studies of *Serendipita indica* and *Zhihengliuella* sp. ISTPL4 and their synergistic role in growth promotion in rice. *Front. Plant Sci.* **2023**, *14*, 1155715.
64. Tyagi, J.; Chaudhary, P.; Jyotsana, U. B.; Bhandari, G.; Chaudhary, A. Impact of endophytic fungi in biotic stress management. *Plant Protect. Chem. Biologicals* **2022**, 447–462.
65. Tyagi, J.; Chaudhary, P.; Mishra, A.; Khatwani, M.; Dey, S.; Varma, A. Role of endophytes in abiotic stress tolerance: with special emphasis on *Serendipita indica. Int. J. Environ. Res.* **2022**, *16* (4), 62.
66. Mazumdar, S.; Singhal, S.; Tyagi, J.; Mishra, A.; Joshi, N. C.; Pudake, R. N.; Farooq, M.; Varma, A. *Understanding of belowground biochemical communication in millets through metabolomics. Millet Rhizosphere;* Springer Nature Singapore: Singapore, 2023, pp. 259–274.
67. Jeyanthi, V.; Kanimozhi, S. Plant growth promoting rhizobacteria (PGPR)-prospective and mechanisms: a review. *J. Pure Appl. Microbiol.* **2018**, *12* (2), 733–749.
68. Saeed, Q.; Xiukang, W.; Haider, F. U.; Kučerik, J.; Mumtaz, M. Z.; Holatko, J.; Naseem, M.; Kintl, A.; Ejaz, M.; Naveed, M., et al. Rhizosphere bacteria in plant growth promotion, biocontrol, and bioremediation of contaminated sites: a comprehensive review of effects and mechanisms. *Int. J. Molecular Sci.* **2021**, *22* (19), 10529.
69. Cao, M.; Narayanan, M.; Shi, X.; Chen, X.; Li, Z.; Ma, Y. Optimistic contributions of plant growth-promoting bacteria for sustainable agriculture and climate stress alleviation. *Environ. Res.* **2023**, *217*, 114924.
70. El-Saadony, M. T.; Saad, A. M.; Soliman, S. M.; Ahmed, A. I.; El-Tahan, A. M.; El-Mageed, A.; Taia, A.; Elrys, A. S.; El-Tarabily, K. A.; AbuQamar, S. F. Plant growth-promoting microorganisms as biocontrol agents of plant diseases: mechanisms, challenges and future perspectives. *Front. Plant Sci.* **2022**, *13*, 923880.
71. Kushwaha, M.; Mishra, A.; Shankar, S.; Goel, D.; Joshi, S.; Ram, S. Plant growth-promoting rhizobacteria for sustainable agriculture: recent progress and challenges. *RoleGreen Chem. Ecosyst. Restoration Achieve Environ. Sustain.* **2024**, 333–342.
72. Alves, A. R.; Yin, Q.; Oliveira, R. S.; Silva, E. F.; Novo, L. A. Plant growth-promoting bacteria in phytoremediation of metal-polluted soils: current knowledge and future directions. *Sci. Total Environ.* **2022**, *838*, 156435.
73. Utami, D. S.; Setiawati, M. R.; Simarmata, T. Synergistic and competence of native plant growth promoting rhizobacteria on direct mechanism on growth plants: a review. *Int. J. Life Sci. Agricult. Res.* **2024**, *3* (2), 103–111.
74. Hakim, S.; Naqqash, T.; Nawaz, M. S.; Laraib, I.; Siddique, M. J.; Zia, R.; Mirza, M. S.; Imran, A. Rhizosphere engineering with plant growth-promoting microorganisms for agriculture and ecological sustainability. *Front. Sustain. Food Syst.* **2021**, *5*, 617157.

75. Oleńska, E.; Małek, W.; Wójcik, M.; Swiecicka, I.; Thijs, S.; Vangronsveld, J. Beneficial features of plant growth-promoting rhizobacteria for improving plant growth and health in challenging conditions: a methodical review. *Sci. Total Environ.* **2020**, *743*, 140682.
76. Zhu, Y.; Wang, Y.; He, X.; Li, B.; Du, S. Plant growth-promoting rhizobacteria: A good companion for heavy metal phytoremediation. *Chemosphere* **2023**, 139475.
77. Backer, R.; Rokem, J. S.; Ilangumaran, G.; Lamont, J.; Praslickova, D.; Ricci, E.; Subramanian, S.; Smith, D. L. Plant growth-promoting rhizobacteria: context, mechanisms of action, and roadmap to commercialization of biostimulants for sustainable agriculture. *Front. Plant Sci.* **2018**, *9*, 402666.
78. Bagmare, R. R.; Ismail, S.; Ingole, A. J.; Bagmare, P. A. Siderophore production by plant growth promoting microorganisms. *J. Pharmacogn. Phytochem.* **2019**, *8* (4), 1450–1452.
79. Borah, P.; Gogoi, N.; Asad, S. A.; Rabha, A. J.; Farooq, M. An insight into plant growth-promoting rhizobacteria-mediated mitigation of stresses in plant. *J. Plant Growth Regul.* **2023**, *42* (5), 3229–3256.
80. Chouhan, G. K.; Singh, S.; Mukherjee, A.; Gaurav, A. K.; Lepcha, A.; Kumari, S.; Verma, J. P. *Exploration of plant growth-promoting rhizobacteria (PGPR) for improving productivity and soil fertility under sustainable agricultural practices. Re-visiting the Rhizosphere Eco-System for Agricultural Sustainability;* Springer Nature Singapore: Singapore, 2022, pp. 245–269.
81. Timofeeva, A. M.; Galyamova, M. R.; Sedykh, S. E. Bacterial siderophores: classification, biosynthesis, perspectives of use in agriculture. *Plants* **2022**, *11* (22), 3065.
82. Giannelli, G.; Potestio, S.; Visioli, G. The contribution of PGPR in salt stress tolerance in crops: unravelling the molecular mechanisms of cross-talk between plant and bacteria. *Plants* **2023**, *12* (11), 2197.
83. Khoshru, B.; Mitra, D.; Mahakur, B.; Sarikhani, M. R.; Mondal, R.; Verma, D.; Pant, K. Role of soil rhizobacteria in utilization of an indispensable micronutrient zinc for plant growth promotion. *J. Crit. Rev* **2020**, *21*, 4644–4654.
84. Orozco-Mosqueda, M. D. C.; Flores, A.; Rojas-Sánchez, B.; Urtis-Flores, C. A.; Morales-Cedeño, L. R.; Valencia-Marin, M. F.; Chávez-Avila, S.; Rojas-Solis, D.; Santoyo, G. Plant growth-promoting bacteria as bioinoculants: attributes and challenges for sustainable crop improvement. *Agronomy* **2021**, *11* (6), 1167.
85. Rani, A. A.; Basha, S. M.; Darsha, K. D.; Christy, C. A.; Nagaiah, H. P.; Kasthuri, T.; Pandian, S. K. *Plant growth promoting Rhizobacteria and their biofilms in promoting sustainable agriculture and soil health. Understanding Microbial Biofilms;* Academic Press, 2023, pp. 629–647.
86. Rosier, A.; Medeiros, F. H.; Bais, H. P. Defining plant growth promoting rhizobacteria molecular and biochemical networks in beneficial plant-microbe interactions. *Plant Soil* **2018**, *428*, 35–55.
87. Gupta, S.; Pandey, S. Plant growth promoting rhizobacteria to mitigate biotic and abiotic stress in plants. *Sustain. Agricult. Rev. 60 Microb. Processes Agricult.* **2023**, 47–68.
88. Raj, M.; Kumar, R.; Lal, K.; Sirisha, L.; Chaudhary, R.; Patel, S. K. Dynamic role of plant growth promoting rhizobacteria (PGPR) in agriculture. *Int. J. Chem. Stud* **2020**, *8*, 105–110.

89. Adedayo, A. A.; Babalola, O. O.; Prigent-Combaret, C.; Cruz, C.; Stefan, M.; Kutu, F.; Glick, B. R. The application of plant growth-promoting rhizobacteria in Solanum lycopersicum production in the agricultural system: a review. *PeerJ* **2022**, *10*, e13405.
90. Mekonnen, H.; Kibret, M. The roles of plant growth promoting rhizobacteria in sustainable vegetable production in Ethiopia. *Chem. Biological Technol. Agricult.* **2021**, *8*, 1−11.
91. Lobo, L. L. B.; Da Silva, M. S. R. D. A.; Carvalho, R. F.; Rigobelo, E. C. The negative effect of coinoculation of plant growth-promoting bacteria is not related to indole-3-acetic acid synthesis. *J. Plant Growth Regul.* **2023**, *42* (4), 2317−2326.
92. Tanveer, F.; Iqrar, I.; Shinwari, Z. K.; Gul, I. *Microbial biostimulants and their role in environmental bioremediation. Microbial Biostimulants for Sustainable Agriculture and Environmental Bioremediation;* CRC Press, 2022, pp. 101−112.
93. Proshad, R.; Kormoker, T.; Mursheed, N.; Islam, M. M.; Bhuyan, M. I.; Islam, M. S.; Mithu, T. N. Heavy metal toxicity in agricultural soil due to rapid industrialization in Bangladesh: a review. *Int. J. Adv. Geosci.* **2018**, *6* (1), 83−88.
94. Borah, P.; Kumar, M.; Devi, P. *Types of inorganic pollutants: metals/metalloids, acids, and organic forms. Inorganic Pollutants in Water;* Elsevier, 2020, pp. 17−31.
95. Pemawat, G., 2022. Ayushi Bhatnagar1, Dr. Rama Kanwar Khangarot1, Kiran Meena1, and Dr. Current and Future Perspectives of Environmental Pollution and its Remediation, 1.
96. Liu, L.; Li, W.; Song, W.; Guo, M. Remediation techniques for heavy metal-contaminated soils: principles and applicability. *Sci. Total Environ.* **2018**, *633*, 206−219.
97. Ahmed, S. F.; Mofijur, M.; Nuzhat, S.; Chowdhury, A. T.; Rafa, N.; Uddin, M. A.; Inayat, A.; Mahlia, T. M. I.; Ong, H. C.; Chia, W. Y., et al. Recent developments in physical, biological, chemical, and hybrid treatment techniques for removing emerging contaminants from wastewater. *J. Hazard. Mater.* **2021**, *416*, 125912.
98. Ahmed, T.; Noman, M.; Manzoor, N.; Shahid, M.; Abdullah, M.; Ali, L.; Wang, G.; Hashem, A.; Al-Arjani, A. B. F.; Alqarawi, A. A., et al. Nanoparticle-based amelioration of drought stress and cadmium toxicity in rice via triggering the stress responsive genetic mechanisms and nutrient acquisition. *Ecotoxicol. Environ. Saf.* **2021**, *209*, 111829.
99. Sharma, J.; Goutam, J.; Dhuriya, Y. K.; Sharma, D. Bioremediation of Industrial pollutants. *Microb. Rejuven. Polluted Environ.* **2021**, *2*, 1−31.
100. Saeed, M. U.; Hussain, N.; Sumrin, A.; Shahbaz, A.; Noor, S.; Bilal, M.; Aleya, L.; Iqbal, H. M. Microbial bioremediation strategies with wastewater treatment potentialities—a review. *Sci. Total Environ.* **2022**, *818*, 151754.
101. Tegene, B. G. Mode of action, mechanism and role of microbes in bioremediation service for environmental pollution management. *J. Biotechnol. Bioinformat. Res.* **2020**, *2* (3), 1−18.
102. Hussain, C. M.; Mishra, A. K. *Nanotechnology in Environmental Science, 2 volumes*, 1. John Wiley & Sons, 2018.
103. Ding, S.; Hu, Z.; Liu, Y.; Li, L.; Zeng, Y.; Jin, D.; Chen, A.; Shao, J.; Su, P.; Luo, S. Enhanced cadmium absorption and tolerance of rice epiphytic microbes by iron oxide nanoparticles. *Int. Biodeterior. Biodegradation* **2024**, *190*, 105770.
104. Fouda, A.; Hassan, S. E. D.; Saied, E.; Hamza, M. F. Photocatalytic degradation of real textile and tannery effluent using biosynthesized magnesium oxide nanoparticles

(MgO-NPs), heavy metal adsorption, phytotoxicity, and antimicrobial activity. *J. Environ. Chem. Eng.* **2021**, *9* (4), 105346.

105. Saied, E.; Eid, A. M.; Hassan, S. E. D.; Salem, S. S.; Radwan, A. A.; Halawa, M.; Saleh, F. M.; Saad, H. A.; Saied, E. M.; Fouda, A. The catalytic activity of biosynthesized magnesium oxide nanoparticles (MgO-NPs) for inhibiting the growth of pathogenic microbes, tanning effluent treatment, and chromium ion removal. *Catalysts* **2021**, *11* (7), 821.
106. Bestawy, E. E.; El-Shatby, B. F.; Eltaweil, A. S. Integration between bacterial consortium and magnetite (Fe3O4) nanoparticles for the treatment of oily industrial wastewater. *World J. Microbiol. Biotechnol.* **2020**, *36* (9), 141.
107. Darwesh, O. M.; Li, H.; Matter, I. A. Nano-bioremediation of textile industry wastewater using immobilized CuO-NPs myco-synthesized by a novel Cu-resistant *Fusarium oxysporum* OSF18. *Environ. Sci. Pollut. Res.* **2023**, *30* (6), 16694–16706.
108. Kamalian, M.; Shahnavaz, B.; Karrabi, M. Biogenic synthesis of copper oxide nanoparticles using a cold tolerant bacterium for RB5/DR23 Azo dye adsorption. *Biomass Conversion Biorefinery* **2024**, 1–18.
109. Teng, Z.; Zhao, X.; Jia, B.; Ye, L.; Tian, S.; Guo, H.; Guo, Y.; Ji, X.; Li, T.; Li, M. Bioremediation system consisted with Leclercia adecarboxylata and nZVI@ carbon/phosphate for lead immobilization: the passivation mechanisms of chemical reaction and biological metabolism in soil. *J. Environ. Manage.* **2023**, *340*, 117888.
110. Mordecai, G. A. N. A.; Olusanya, C. S.; Shaba, A. D.; Adamu, B. B.; Idris, A. D. Biodegradation of diesel and crude oil using *Corynebacterium* sp and *Lysinobacillus fusiformis* 5B stimulated with biosurfactant, biochar and iron oxide nanoparticles. *Covenant J. Physical and Life Sciences* **2021**.
111. Al-Zubaidi, L. A.; Alnuaimi, M. T.; Aljanabi, Z. Z.; Adel, M. M. 2021. Biodegradation of chlorpyrifos pesticide using silver bio-nanoparticles *Bacillus thuringiensis israelensis* extracts. In IOP Conference Series: Earth and Environmental Science (Vol. 779, No. 1, pp. 012113). IOP Publishing.
112. Weng, Y.; Li, J.; Ding, X.; Wang, B.; Dai, S.; Zhou, Y.; Pang, R.; Zhao, Y.; Xu, H.; Tian, B., et al. Functionalized gold and silver bimetallic nanoparticles using *Deinococcus radiodurans* protein extract mediate degradation of toxic dye malachite green. *Int. J. Nanomedicine* **2020**, 1823–1835.
113. Ali, M. H.; Abdelkarim, M. S.; Al-Afify, A. D. Characterization and photodegradation of methylene blue dye using bio-synthesized cerium oxide nanoparticles with *Spirulina platensis* extract. *Discover Applied Sci.* **2024**, *6* (3), 1–17.
114. Vishnupriya, C.; Mohamedrizwan, K.; Arya, P. R.; Vijayakumar, S.; Kavitha, R. Evaluation of heavy metal removal and antibiofilm efficiency of biologically synthesized chitosan-silver nano-bio composite by a soil actinobacterium *Glutamicibacter uratoxydans* VRAK 24. *Int. J. Biol. Macromol.* **2024**, *255*, 128032.
115. Gupta, S.; Ravi, R. K.; Pathak, B. Photocatalytic removal of anthracene using zinc oxide nanoparticles synthesized by *Fusarium proliferatum* WC416. *Geomicrobiol. J.* **2024**, *41* (1), 72–81.
116. Rathore, C.; Yadav, V. K.; Amari, A.; Meena, A.; Chinedu Egbosiuba, T.; Verma, R. K.; Mahdhi, N.; Choudhary, N.; Sahoo, D. K.; Chundawat, R. S., et al. Synthesis and characterization of titanium dioxide nanoparticles from *Bacillus subtilis* MTCC 8322 and its application for the removal of methylene blue and orange G dyes under UV light and visible light. *Front. Bioeng. Biotechnol.* **2024**, *11*, 1323249.

117. Chen, J.; Gan, L.; Han, Y.; Owens, G.; Chen, Z. Ferrous sulfide nanoparticles can be biosynthesized by sulfate-reducing bacteria: synthesis, characterization and removal of heavy metals from acid mine drainage. *J. Hazard. Mater.* **2024**, *466*, 133622.
118. Prajitha, N.; Athira, S. S.; Mohanan, P. V. Bio-interactions and risks of engineered nanoparticles. *Environ. Res.* **2019**, *172*, 98–108.
119. Dhasmana, A.; Uniyal, S.; Kumar, V.; Gupta, S.; Kesari, K. K.; Haque, S.; Lohani, M.; Pandey, J. Scope of nanoparticles in environmental toxicant remediation. *Environ. Biotechnol. Sustain. Future* **2019**, 31–44.
120. Mazive, P. A.; Hu, B.; Zhu, H.; He, W.; Mazive, A. M. Graphene oxide; adsorption; organic, inorganic; biological; radiological contaminants. *J. Nanotechnol. Res.* **2020**, *2* (4), 60–91.
121. Pudake, R. N.; Mittal, J.; Tripathi, R. M.; Tyagi, J.; Mohanta, T. K. Biochemical responses of maize seedlings exposed to SnNPs. *Micro Nano Lett.* **2019**, *14* (6), 645–649.
122. Thanigaivel, S.; Priya, A. K.; Gnanasekaran, L.; Hoang, T. K.; Rajendran, S.; Soto-Moscoso, M. Sustainable applicability and environmental impact of wastewater treatment by emerging nanobiotechnological approach: future strategy for efficient removal of contaminants and water purification. *Sustain. Energy Technol. Assessments* **2022**, *53*, 102484.
123. Hu, W.; Liu, D.; Su, S.; Ren, L.; Ren, H.; Wei, L.; Yue, S.; Xie, Q.; Zhang, Z.; Wang, Z., et al. Photochemical degradation of organic matter in the atmosphere. *Adv. Sustain. Syst.* **2021**, *5* (11), 2100027.
124. Tyagi, J.; Mishra, A.; Kumari, S.; Singh, S.; Agarwal, H.; Pudake, R. N.; Varma, A.; Joshi, N. C. Deploying a microbial consortium of *Serendipita indica*, *Rhizophagus intraradices*, and *Azotobacter chroococcum* to boost drought tolerance in maize. *Environ. Exp. Bot.* **2023**, *206*, 105142.
125. Mi, G.; Shi, D.; Wang, M.; Webster, T. J. Reducing bacterial infections and biofilm formation using nanoparticles and nanostructured antibacterial surfaces. *Adv. Healthc. Mater.* **2018**, *7* (13), 1800103.
126. Sterzenbach, T.; Helbig, R.; Hannig, C.; Hannig, M. Bioadhesion in the oral cavity and approaches for biofilm management by surface modifications. *Clin. Oral Investig.* **2020**, *24*, 4237–4260.
127. Bhavya, G.; Belorkar, S. A.; Mythili, R.; Geetha, N.; Shetty, H. S.; Udikeri, S. S.; Jogaiah, S. Remediation of emerging environmental pollutants: a review based on advances in the uses of eco-friendly biofabricated nanomaterials. *Chemosphere* **2021**, *275*, 129975.
128. Rajput, V. D.; Singh, A.; Singh, V. K.; Minkina, T. M.; Sushkova, S. *Impact of nanoparticles on soil resource. Nanomaterials for Soil Remediation;* Elsevier, 2021, pp. 65–85.
129. Mishra, S.; Bharagava, R. N.; More, N.; Yadav, A.; Zainith, S.; Mani, S.; Chowdhary, P. Heavy metal contamination: an alarming threat to environment and human health. *Environ. Biotechnol. Sustain. Future* **2019**, 103–125.
130. Yu, F.; Yang, C.; Zhu, Z.; Bai, X.; Ma, J. Adsorption behavior of organic pollutants and metals on micro/nanoplastics in the aquatic environment. *Sci. Total Environ.* **2019**, *694*, 133643.
131. Vasarevičius, S.; Danila, V.; Januševičius, T. Immobilisation of cadmium, copper, lead, and nickel in soil using nano zerovalent iron particles: ageing effect on heavy metal retention. *Water Air Soil Pollut.* **2020**, *231*, 1–11.

132. Borišev, I.; Borišev, M.; Jović, D.; Župunski, M.; Arsenov, D.; Pajević, S.; Djordjevic, A. *Nanotechnology and remediation of agrochemicals. Agrochemicals Detection, Treatment and Remediation;* Butterworth-Heinemann, 2020, pp. 487−533.
133. Naseem, T.; Durrani, T. The role of some important metal oxide nanoparticles for wastewater and antibacterial applications: a review. *Environ. Chem. Ecotoxicol.* **2021**, *3*, 59−75.
134. Bolan, S.; Padhye, L. P.; Mulligan, C. N.; Alonso, E. R.; Saint-Fort, R.; Jasemizad, T.; Wang, C.; Zhang, T.; Rinklebe, J.; Wang, H., et al. Surfactant-enhanced mobilization of persistent organic pollutants: potential for soil and sediment remediation and unintended consequences. *J. Hazard. Mater.* **2023**, *443*, 130189.
135. Horie, M.; Tabei, Y. Role of oxidative stress in nanoparticles toxicity. *Free Radic. Res.* **2021**, *55* (4), 331−342.
136. Phenrat, T.; Le, T. S. T.; Naknakorn, B.; Lowry, G. V. Chemical reduction and oxidation of organic contaminants by nanoscale zerovalent iron. *Nanosc. Zeroval. Iron Part. Environ. Restoration Fundam. Sci. Field Scale Eng. Appl.* **2019**, 97−155.
137. Nayak, S.; Sahoo, A.; Kolanthasamy, E.; Rao, K. Role of pesticide application in environmental degradation and its remediation strategies. *Environ. Degrad. Causes Remediation Strategies* **2020**, *1*, 36.
138. Hussein, H. A. A.; Darwesh, O. M.; Mekki, B. B. Environmentally friendly nano-selenium to improve antioxidant system and growth of groundnut cultivars under sandy soil conditions. *Biocatalysis Agricult. Biotechnol.* **2019**, *18*, 101080.
139. Coonan, E. C.; Kirkby, C. A.; Kirkegaard, J. A.; Amidy, M. R.; Strong, C. L.; Richardson, A. E. Microorganisms and nutrient stoichiometry as mediators of soil organic matter dynamics. *Nutrient Cycling Agroecosyst.* **2020**, *117* (3), 273−298.
140. Chatterjee, P.; Chauhan, N.; Jain, U. Confronting antibiotic-resistant pathogens: the drug delivery potential of nanoparticle swords. *Microb. Pathog.* **2023**, 106499.
141. Hao, Y.; Ma, C.; Zhang, Z.; Song, Y.; Cao, W.; Guo, J.; Zhou, G.; Rui, Y.; Liu, L.; Xing, B. Carbon nanomaterials alter plant physiology and soil bacterial community composition in a rice-soil-bacterial ecosystem. *Environ. Pollut.* **2018**, *232*, 123−136.
142. Wang, F.; Wang, Q.; Adams, C. A.; Sun, Y.; Zhang, S. Effects of microplastics on soil properties: current knowledge and future perspectives. *J. Hazard. Mater.* **2022**, *424*, 127531.
143. Kumari, S.; Tyagi, M.; Jagadevan, S. Mechanistic removal of environmental contaminants using biogenic nano-materials. *Int. J. Environ. Sci. Technol.* **2019**, *16*, 7591−7606.
144. Maity, D.; Gupta, U.; Saha, S. Biosynthesized metal oxide nanoparticles for sustainable agriculture: next-generation nanotechnology for crop production, protection and management. *Nanoscale* **2022**, *14* (38), 13950−13989.
145. Akter, T.; Protity, A. T.; Shaha, M.; Al Mamun, M.; Hashem, A. *The impact of textile dyes on the environment. Nanohybrid Materials for Treatment of Textiles Dyes;* Springer Nature Singapore: Singapore, 2023, pp. 401−431.
146. Kiwaan, H. A.; Atwee, T. M.; Azab, E. A.; El-Bindary, A. A. Photocatalytic degradation of organic dyes in the presence of nanostructured titanium dioxide. *J. Mol. Struct.* **2020**, *1200*, 127115.
147. Islam, T.; Ali, I.; Naz, I.; Peng, C.; Khan, Z. M.; Amjed, M. A. *Bionanocomposites for wastewater treatment. Aquananotechnology;* Elsevier, 2021, pp. 249−272.

148. Singh, G.; Singh, A.; Singh, P.; Shukla, R.; Tripathi, S.; Mishra, V. K. The fate of organic pollutants and their microbial degradation in water bodies. *Pollut. Water Manage. Resources Strategies Scarcity* **2021**, 210–240.
149. Younas, F.; Mustafa, A.; Farooqi, Z. U. R.; Wang, X.; Younas, S.; Mohy-Ud-Din, W.; Ashir Hameed, M.; Mohsin Abrar, M.; Maitlo, A. A.; Noreen, S., et al. Current and emerging adsorbent technologies for wastewater treatment: trends, limitations, and environmental implications. *Water* **2021**, *13* (2), 215.
150. Duhl, T.R., 2020. Microbially-mediated transformations of chlorinated solvents and engineered nanomaterials in the subsurface: implications for bioremediation, natural attenuation, fate and transport (Doctoral dissertation, Tufts University).
151. Dong, H.; Li, L.; Lu, Y.; Cheng, Y.; Wang, Y.; Ning, Q.; Wang, B.; Zhang, L.; Zeng, G. Integration of nanoscale zero-valent iron and functional anaerobic bacteria for groundwater remediation: a review. *Environ. Int.* **2019**, *124*, 265–277.
152. Gahlawat, G.; Choudhury, A. R. A review on the biosynthesis of metal and metal salt nanoparticles by microbes. *RSC Adv.* **2019**, *9* (23), 12944–12967.
153. Adamu, H. *Titanium dioxide—a missing photo-responsive material for solar-driven oil spill remediation. Titanium Dioxide-Adv. Appl.;* IntechOpen, 2021.
154. Zhang, P.; Yang, M.; Lan, J.; Huang, Y.; Zhang, J.; Huang, S.; Yang, Y.; Ru, J. Water quality degradation due to heavy metal contamination: health impacts and eco-friendly approaches for heavy metal remediation. *Toxics* **2023**, *11* (10), 828.
155. Hussain, S.; Rengel, Z.; Qaswar, M.; Amir, M.; Zafar-ul-Hye, M. Arsenic and heavy metal (cadmium, lead, mercury and nickel) contamination in plant-based foods. *Plant Human Health* **2019**, *2*, 447–490. *Phytochemistry and Molecular Aspects.*
156. Chapman, E. E. V.; Moore, C.; Campbell, L. M. Evaluation of a nanoscale zero-valent iron amendment as a potential tool to reduce mobility, toxicity, and bioaccumulation of arsenic and mercury from wetland sediments. *Environ. Sci. Pollut. Res.* **2020**, *27*, 18757–18772.
157. Ramezani, M.; Rad, F. A.; Ghahari, S.; Ghahari, S.; Ramezani, M. Nano-bioremediation application for environment contamination by microorganism. *Microb. Rejuvenation Polluted Environ.* **2021**, *2*, 349–378.
158. Okonji, S. O.; Yu, L.; Dominic, J. A.; Pernitsky, D.; Achari, G. Adsorption by granular activated carbon and nano zerovalent iron from wastewater: a study on removal of selenomethionine and selenocysteine. *Water* **2020**, *13* (1), 23.
159. Sabry, A. K. H. Role of nanotechnology applications in plant-parasitic nematode control. *Nanobiotechnol. Appl. Plant Protect.* **2019**, *2*, 223–240.
160. Kong, X. P.; Zhang, B. H.; Wang, J. Multiple roles of mesoporous silica in safe pesticide application by nanotechnology: a review. *J. Agric. Food Chem.* **2021**, *69* (24), 6735–6754.
161. Jampílek, J.; Kráľová, K. Beneficial effects of metal-and metalloid-based nanoparticles on crop production. *Nanotechnol. Agricult. Adv. Sustain. Agricult.* **2019**, 161–219.
162. Begum, N.; Wang, L.; Ahmad, H.; Akhtar, K.; Roy, R.; Khan, M. I.; Zhao, T. Co-inoculation of arbuscular mycorrhizal fungi and the plant growth-promoting rhizobacteria improve growth and photosynthesis in tobacco under drought stress by up-regulating antioxidant and mineral nutrition metabolism. *Microb. Ecol.* **2022**, 1–18.
163. Karnwal, A.; Dohroo, A.; Malik, T. Unveiling the potential of bioinoculants and nanoparticles in sustainable agriculture for enhanced plant growth and food security. *BioMed Res. Int.* **2023**, *2023*.

164. Khatoon, K.; Arfi, N.; Malik, A. Microbes-mediated facilitation of micronutrients uptake by plants from soil especially zinc. *Microb. Biofert. Micronutrient Availability Role Zinc Agricult. Human Health* **2022**, 331–359.
165. de Moraes, A. C. P.; Ribeiro, L. D. S.; de Camargo, E. R.; Lacava, P. T. The potential of nanomaterials associated with plant growth-promoting bacteria in agriculture. *3 Biotech* **2021**, *11* (7), 318.
166. Aponte, H.; Medina, J.; Butler, B.; Meier, S.; Cornejo, P.; Kuzyakov, Y. Soil quality indices for metal (loid) contamination: an enzymatic perspective. *Land Degrad. Dev.* **2020**, *31* (17), 2700–2719.
167. Jiang, Y.; Zhang, X.; Yuan, H.; Huang, D.; Wang, R.; Liu, H.; Wang, T. Research progress and the biotechnological applications of multienzyme complex. *Appl. Microbiol. Biotechnol.* **2021**, *105*, 1759–1777.
168. Raza, A.; Salehi, H.; Rahman, M. A.; Zahid, Z.; Madadkar Haghjou, M.; Najafi-Kakavand, S.; Charagh, S.; Osman, H. S.; Albaqami, M.; Zhuang, Y., et al. Plant hormones and neurotransmitter interactions mediate antioxidant defenses under induced oxidative stress in plants. *Front. Plant Sci.* **2022**, *13*, 961872.
169. Ameen, F.; Alsamhary, K.; Alabdullatif, J. A.; ALNadhari, S. A review on metal-based nanoparticles and their toxicity to beneficial soil bacteria and fungi. *Ecotoxicol. Environ. Saf.* **2021**, *213*, 112027.
170. Hua, T.; Zhang, R.; Sun, H.; Liu, C. Alleviation of boron toxicity in plants: Mechanisms and approaches. *Crit. Rev. Environ. Sci. Technol.* **2021**, *51* (24), 2975–3015.
171. Zhou, H. Y.; Nian, F. Z.; Chen, B. D.; Zhu, Y. G.; Yue, X. R.; Zhang, N. M.; Xia, Y. S. Synergistic reduction of arsenic uptake and alleviation of leaf arsenic toxicity in maize (*Zea mays* L.) by arbuscular mycorrhizal fungi (AMF) and exogenous iron through antioxidant activity. *J. Fungi* **2023**, *9* (6), 677.
172. Vasudevan, V.; Arumugam, V. Impact of metal oxide nanoparticles against salinity stress in plants. *Nanometal Oxides Horticult. Agronomy* **2023**, 141–163.
173. Ehis-Eriakha, C. B.; Adetunji, C. O. Recent advances in application of biostimulants derived from beneficial microorganisms: agriculture and environmental perspective. *Agricult. Biotechnol.* **2022**, 47–88.
174. Saharan, B. S.; Brar, B.; Duhan, J. S.; Kumar, R.; Marwaha, S.; Rajput, V. D.; Minkina, T. Molecular and physiological mechanisms to mitigate abiotic stress conditions in plants. *Life* **2022**, *12* (10), 1634.
175. Gauba, A.; Hari, S. K.; Ramamoorthy, V.; Vellasamy, S.; Govindan, G.; Arasu, M. V. The versatility of green synthesized zinc oxide nanoparticles in sustainable agriculture: a review on metal-microbe interaction that rewards agriculture. *Physiol. Molecular Plant Pathol.* **2023**, 102023.
176. Bhatt, P.; Pandey, S. C.; Joshi, S.; Chaudhary, P.; Pathak, V. M.; Huang, Y.; Wu, X.; Zhou, Z.; Chen, S. Nanobioremediation: a sustainable approach for the removal of toxic pollutants from the environment. *J. Hazard. Mater.* **2022**, *427*, 128033.
177. Gomaa, H.; Emran, M. Y.; El-Gammal, M. A. *Biodegradation of azo dye pollutants using microorganisms. Handbook of Biodegradable Materials;* Springer International Publishing: Cham, 2023, pp. 781–809.
178. Negi, B. B.; Das, C. Mycoremediation of wastewater, challenges, and current status: a review. *Bioresource Technol. Rep.* **2023**, 101409.
179. Escobar-Ramírez, M. C.; Castañeda-Ovando, A.; Pérez-Escalante, E.; Rodríguez-Serrano, G. M.; Ramírez-Moreno, E.; Quintero-Lira, A.; Contreras-López, E.;

Añorve-Morga, J.; Jaimez-Ordaz, J.; González-Olivares, L. G. Antimicrobial activity of Se-nanoparticles from bacterial biotransformation. *Fermentation* **2021**, *7* (3), 130.

180. Rana, A.; Sindhu, M.; Kumar, A.; Dhaka, R. K.; Chahar, M.; Singh, S.; Nain, L. Restoration of heavy metal-contaminated soil and water through biosorbents: a review of current understanding and future challenges. *Physiol. Plant.* **2021**, *173* (1), 394–417.
181. Rahman, Z.; Singh, V. P. Bioremediation of toxic heavy metals (THMs) contaminated sites: concepts, applications and challenges. *Environ. Sci. Pollut. Res.* **2020**, *27* (22), 27563–27581.
182. Punetha, A.; Saraswat, S.; Rai, J. P. N. An insight on microbial degradation of benzo [a] pyrene: current status and advances in research. *World J. Microbiol. Biotechnol.* **2022**, *38* (4), 61.
183. Fatehbasharzad, P.; Fatehbasharzad, P.; Sillanpää, M.; Shamsi, Z. Investigation of bioimpacts of metallic and metallic oxide nanostructured materials: size, shape, chemical composition, and surface functionality: A review. *Particle Particle Syst. Character.* **2021**, *38* (10), 2100112.
184. Mansor, M.; Xu, J. Benefits at the nanoscale: a review of nanoparticle-enabled processes favouring microbial growth and functionality. *Environ. Microbiol.* **2020**, *22* (9), 3633–3649.
185. Chaudhary, P.; Chaudhary, A.; Khati, P.; Kumar, G.; Tyagi, J.; Behera, M. *Nanotechnology and omics approach in agrobiotechnology*. In *Agricultural and Environmental Nanotechnology. Interdisciplinary Biotechnological Advances;* Fernandez-Luqueno, F., Patra, J. K., Eds.; Springer: Singapore, 2023. Available from: https://doi.org/10.1007/978-981-19-5454-2_12.
186. Lang, C.; Mission, E. G.; Fuaad, A. A. H. A.; Shaalan, M. Nanoparticle tools to improve and advance precision practices in the Agrifoods Sector towards sustainability—a review. *J. Cleaner Prod.* **2021**, *293*, 126063.
187. Saxena, G.; Purchase, D.; Mulla, S. I.; Saratale, G. D.; Bharagava, R. N. Phytoremediation of heavy metal-contaminated sites: eco-environmental concerns, field studies, sustainability issues, and future prospects. *Rev. Environ. Contam. Toxicol.* **2020**, *249*, 71–131.
188. Raj, S.; Khurana, S.; Choudhari, R.; Kesari, K. K.; Kamal, M. A.; Garg, N.; Ruokolainen, J.; Das, B. C.; Kumar, D. *Specific targeting cancer cells with nanoparticles and drug delivery in cancer therapy. Seminars in Cancer Biology.*, 69. Academic Press, 2021, pp. 166–177.
189. Rajput, V. D.; Minkina, T.; Kumari, A.; Shende Sudhir, S.; Ranjan, A.; Faizan, M.; Barakvov, A.; Gromovik, A.; Gorbunova, N.; Rajput, P., et al. A review on nanobioremediation approaches for restoration of contaminated soil. *Eurasian J. Soil Sci.* **2022**, *11* (1), 43–60.

Further reading

Ali, E. M.; El-Shehawy, A. S. *Metallic nanoparticles and bioremediation for wastewater treatment. Advanced Application of Nanotechnology to Industrial Wastewater;* Springer Nature Singapore: Singapore, 2023, pp. 215–239.

Al-Tammar, F. K.; Khalifa, A. Y. Z. Plant growth promoting bacteria drive food security. *Brazilian J. Biol.* **2022**, *82*, e267257.

Harman, G. E.; Uphoff, N. Symbiotic root-endophytic soil microbes improve crop productivity and provide environmental benefits. *Scientifica* **2019**, *2019*.

Paulami, D. A. M.; Paret, M. L.; Mondal, R.; Mandal, A. K. Advancement of noble metallic nanoparticles in agriculture: a promising future. *Pedosphere* **2023**, *33* (1), 116–128.

Thulasisingh, A.; Venkataramakrishnan, A.; Clement, P. R.; Kannaiyan, S. *The role of nanotechnology in bioremediation of pollutants. Industrial Wastewater Reuse: Applications, Prospects and Challenges;* Springer Nature Singapore: Singapore, 2023, pp. 133–163.

CHAPTER

Nanoshields: harnessing biogenic nanomaterials to combat the adverse effects of heavy metal stress 13

Simran Choudhury[1], Nishi Patel[2], Shreni Agrawal[2], Indrani Bhattacharya[2], Abhishek Singh[3], Armine Chakhmakhchyan[3], Ani Hayrapetyan[3], Rupesh Kumar Singh[4], Francisco Roberto Quiroz-Figueroa[5], João Ricardo Sousa[4], Henrique Trindade[4], Mohamed S. Elshikh[6], Arunava Pradhan[7,8], Priyadarshani Rajput[9] and Karen Ghazaryan[3]

[1]*School of Life Sciences, Central University of Gujarat, Vadodara, Gujarat, India*

[2]*Department of Life Sciences, Parul Institute of Applied Sciences, Parul University, Waghodia, Gujarat, India*

[3]*Faculty of Biology, Yerevan State University, Yerevan, Armenia*

[4]*Centro de Investigação e Tecnologias Agroambientais e Biológicas (CITAB), UTAD, Vila Real, Portugal*

[5]*Laboratorio de Fitomejoramiento Molecular, Centro Interdisciplinario de Investigación para el Desarrollo Integral Regional-Unidad Sinaloa, Instituto Politécnico Nacional, Guasave, Sinaloa, Mexico*

[6]*Department of Botany and Microbiology, College of Science, King Saud University, Riyadh, Saudi Arabia*

[7]*Department of Biology, Centre of Molecular and Environmental Biology (CBMA), Aquatic Research Network (ARNET), Associate Laboratory, University of Minho, Campus of Gualtar, Braga, Portugal*

[8]*Institute for Science and Innovation for Bio-Sustainability (IB-S), University of Minho, Campus of Gualtar, Braga, Portugal*

[9]*Academy of Biology and Biotechnology, Southern Federal University, Rostov-on-Don, Russia*

13.1 Introduction

Despite being the primary engine of cultural and economic growth, industrialization throws up many harmful agents into the ecosystems through generation and accumulation thereby posing danger to human beings, flora and fauna. Heavy metals (HMs) are a class of pollutants that require considerable attention among others because of their toxicity and persistence in ecosystems.[1] These HMs exhibit metallic properties with atomic density exceeding 3 g/cm^3 and atomic number greater than 20, which includes various semimetals, actinides as well as

Biogenic Nanoparticles. DOI: https://doi.org/10.1016/B978-0-443-33819-9.00017-2

lanthanides. Even though heavy metals are plentiful in the earth's surface, humankind is barely able to reach them and so do plants in their soil forms, but anthropogenic operations like fertilizers munificence, smelting of metals, industrial waste disposal, and sewage sludge putrefaction among others have increased their availabilities to animals. HMs are severely toxic to both plants and animals, as cadmium (Cd), lead (Pb), arsenic (As), and nickel (Ni) clearly show in many cases (1). Also, because of their ability to persist in the environment and their innate resistance to biological decomposition HMs can easily get access to the food chain. A significant risk to human health comes with these strong properties, and thus we should be extremely worried. Accordingly, other than functioning as elements of matter in the body part where they are required in small amounts, many HMs turn into poison to life at less quantitatively levels.[2]

The security of agricultural production has been severely threatened over the past ten years by the increase in arable land degradation with harmful heavy metals. Studies have shown that high concentrations of heavy metals (HM) can have detrimental effects on crops due to a number of factors, including biomacromolecule-caused hindrance, ion balance disruption, excessive generation of reactive oxygen species, toxic radicals that cause oxidative stress, and finally the destruction of photosynthetic apparatus.[3] Several methods have been developed to minimize the effects of HMs on crops including marker-assisted breeding, conventional breeding, and transgenic crop engineering. Moreover, over the recent past, nano-based techniques have appeared as very promising modes used to control stress (conditions) caused by the environment, improve crop production and growth, eliminate deficiency in nutrients, as well as changing biological systems. In addition, there has been considerable interest in the use of nanoparticles (NPs) to deliver micronutrients in a targeted way that is both affordable and eco-friendly, unlike chemical fertilizers. In recent times, several methods have been utilized for the synthesis of metal NPs including chemical ones, like coprecipitation, microemulsions, sol–gel processes, and hydrothermal reactions; physical ways, such as ion sputtering and scattering procedures as well as gamma radiation treatment, electroexplosion technique, ion sputtering methods, pulsed laser ablation, or even mechanical ball millings; and biological methods, such as green syntheses utilizing microorganisms and plant extracts.[4]

In regard to latest research, external application of NPs improves plant response capabilities to HMs through modulation of morphophysiological and molecular mechanisms. Thus, promoting plant adaptation to HM toxicity using NPs in sustainable agriculture is a dependable, environmentally friendly, and persistent approach as opposed to the use of traditional means.[1]

Nanotechnology is broadly recognized for its positive contributions towards increasing crop quality as well as quantity. Some of these applications are its use in mitigating abiotic stressors and adapting to the climate change phenomenon. Some of the main nano-based technologies which are useful in enhancing crop yield and growth if various ways in this case include; using of nanosensors when monitoring changes in plant nutrients as well as their administration via foliar spray or soil irrigation, coating seeds with them and using nanotechnology while advancing photosynthetic efficiency through altering genes responsible for photosynthesis or making them resistant to pathogens that may cause diseases.[5]

13.2 What are biogenic nanomaterials?

As we know, nanomaterials (NMs) are materials of nanometer range, and have their properties modified at the scale of atoms and molecules, which bestows upon them novel characteristics at macrolevel. There are methods of synthesizing nanomaterials which can be largely classified as physical, chemical, and biological. The physical method is also known as "top-down synthesis," while chemical method as "bottom-up synthesis." The approaches in synthesis include aerosol method, ball milling method, planetary ball milling method, sol–gel method, chemical vapor deposition (CVD) method, gamma irradiation method, etc., which are all physical and chemical approaches.

However, the nanomaterials synthesized using above mentioned are not as stable in biological systems and have less biocompatibility as compared to those synthesized using biological approach. The biological or "biogenic" approach is known to produce NMs with enhanced biological activity, better biocompatibility, ecofriendly nature, less ecologic and biologic toxicity, along with the process being cost-effective. A distinguishing feature of biogenic NMs is that they have remarkable stability and biological activity within living systems, due to the presence of "capping." Capping refers to the characteristic of biogenic NMs due to the activity of the proteins and enzymes produced by their source organism.

Biogenic NMs can be synthesized from various sources, like algae, bacteria, fungi, etc., and can be composed of various substances. Currently, the most prominent ones are biogenic metallic NMs, biogenic metal oxide NMs, biogenic lipid NMs, and biogenic carbon-based NMs. Of these, biogenic metal NMs and biogenic metal oxide NMs are the most common ones.[6]

Green nanotechnology describes the field of research that focuses on novel NP synthesis techniques applying a framework borrowed from the sphere of green chemistry. Techniques used in producing NPs are drawn from the biological approach where they can be produced using naturally occurring reducing and stabilizing agents. It therefore provides an economically viable way forward when compared against chemical and physical methods not to mention that no energy is required while shunning toxic chemical use.[7]

Utilizing a bottom-up methodology, NPs are biologically manufactured from simple unicellular to complex multicellular biological organisms, including bacteria, fungus, actinomycetes, yeast, algae, and plant materials. Using microbes to aid in the synthesis of nanoparticles is another method of producing them. The microbial culture filtrates are used in this specific synthesis approach as a reductive medium to produce nanoparticles. In general, bacteria, fungi, yeasts, and actinomycetes have the ability to tolerate metals, and they live in extremely severe environments.[8] This will deter microbes from all the agents that may lead to makes metals, to take up these metals and change them into their ions. For example, *Bacillus subtilis* is the microbe that engineered the first bacterial gold nanoparticles.[9]

13.3 What is heavy metal stress?

Plants are naturally facing various harsh environmental factors such as biotic and abiotic stress. It is unfortunate that heavy metal stress is considered to be among the most significant of all other stresses influencing growth and crop yield.[10] This is because there is evidence showing that there are unfavorable effects of heavy metal stress upon growth and crop productivity in the agriculture sector as a whole. The responses differ from biochemical responses to crop yield among plants at large under heavy metal stress. The term "heavy metals" refers to any metallic element possessed with high density that would lead to toxic poisoning or death at low concentration.[11] Heavy metals, a collective name usually given, refers to a group of metals and metalloids with atomic density greater than 4 g/cm^3, five times or more higher than that of water. However, when compared to their mass, it is the chemical properties of a heavy metal that matter most.[12]

Other heavy metals found in rock dispersion include lead, cadmium nickel iron, zinc chromium iron, arsenic, silver, and platinum group elements. Urbanization and industrialization are two examples of human activities that are increasing the anthropogenic load of heavy metals in the environment.[13] The available levels are the largest for heavy metal in the environment of both soil and water, however, they are found in lesser amounts as particulate or vapor in the air. Being essential elements for plants development, a lot of heavy metals cause different intensity of toxicity among different plant types, differentiating with respect to types of metal, concentrations thereof, chemical forms, soil compositions and pH being slightly acidic or near-neutral, though some tend to grow in acid soils.[14] Some metals are heavy; copper and zinc, for example, are used as activators or cofactors for enzymes reactions. They may form a complex between enzyme/substrate by acting as enzyme itself.[15] They may also act in such a way by becoming prosthetic group in metalloprotein as well acting as catalyst. Some metals are heavy; copper and zinc, for example, are used as activators or cofactors for enzymes reactions. They may form a complex between enzyme/substrate by acting as enzyme itself. They may also act in such a way by becoming prosthetic group in metalloprotein as well acting as catalyst.[16]

According to an alternate categorization of metals, which considers heavy metals to have coordination chemistry as highly poisonous nonessential trace elements like mercury, silver, and lead, heavy metals are categorized as class B metals. There are heavy metals that do not decompose in the environment and are not readily metabolized in living organisms. These metals bio accumulate along food chains: starting with absorption at primary producers' levels and continuing through consumption by other organisms.

Particles that settle on leaves' foliar surfaces allow heavy metals to be directly absorbed by the leaves. Since plants remain stationary in their permanent habitat, their roots function as the main points of interaction for heavy metal ions. The entire body of plants in aquatic systems, including leaves, fruits, and flowers, are exposed to these ions when they come into contact with soil, water, or air currents.[17]

13.4 How are heavy metals Introduced?

Since the Earth's formation, metals have always been present in its crust. The overuse of heavy metals has resulted in an increase in metallic materials in both the terrestrial and marine environments. The different human activities have contributed to the increase in the pollution caused by heavy metals. The main sources of pollution are mining, smelting, foundries, and other metal-based businesses. Leaching of metals from other sources, including garbage dumps, landfills, excretion, livestock and chicken dung, runoffs, and road construction, is also a factor. The use of fertilizers, insecticides, and pesticides in the agriculture industry is the secondary reason. The natural sources such as volcanic eruptions, soil erosion, metal corrosion, geological weathering serve as the reason for rise in heavy metal pollution.[18]

13.4.1 Sediment deposition

Excessive metallic particle pollution in sediments is a huge environmental woe that portends potential danger to human life and marine life. Important factors in the natural environment that affect the adsorption, desorption, and post-concentration of heavy metals in sediments include temperature, hydrodynamics, redox status, organic materials, microbes, and particle size (Hazrat Ali et al., n.d.). One of the most frequent historical sources of heavy metal(loid) contamination of alluvial soils is sediment from metalliferous mines. During river flooding events that exceed their banks, a substantial number of suspended particles carried by the rivers that contain heavy metals (loids) settle out onto these soils. With the exception of finely crushed ore, gangue minerals, and country rock, the majority of the metallic ore extracted from the subterranean was removed after treatment. The term "tailings" refers to the sheer particles of mine waste that were once dumped into water bodies. These were typically piled up within yards of streams, and after the ban was passed, the tailings gradually washed away by the weather and poisoned the rivers for almost 130 years. The remnants of the current mining operations are kept in sizable ponds beneath the dams. But sometimes, these dams break, and large sections of soil and water get contaminated with thick layers of "tailings mud" buried deep down, like what happened at Aznalcollar Mine (Southern Spain) among others in 1998.[19]

13.4.2 Agricultural sources of heavy metals

Often referred to as growth and yield improvement substances, both organic and inorganic manures are important sources of heavy metals in agricultural soil. These soil types also contain heavy metals from liming, sewage sludge, irrigation waters, and hashish chemicals. Fungicides, inorganic fertilizers, and phosphate fertilizers are examples of farm inputs. Depending on where they come from, these materials have varying concentrations of Cd, Cr, Ni, Pb, and Zn. Because it can be ingested by humans and animals, cadmium poses a serious hazard to plants due to its ability to accumulate in very high concentrations in leaves.[20]

13.4.3 Sewage sludges

Sewage treatment creates a waste product called sewage sludge, or "biosolids." Biosolids are usually reused as fertilizer on farm fields because they have valuable nutrients, for example, N, P, and organic matter.[19] The sources of the heavy metal(loid)s found in sewage sludges are human waste matter, household "waste" water coming from showers, bathtubs and dishwashing basins, roof and street runoff, and industrial effluents discharged into municipal sewers and taken care of in sewage treatment plants (STPs).

13.4.4 Inorganic fertilizers

Heavy metal (loid) is present in almost all agricultural soils due to the input of nitrogenous fertilizers, nitrogenous fertilizers have nitrogen (among other macronutrients), which help in plant development and production of different products, that is, food from such plants Fertilizer management techniques and crop type determine the amount of nitrogenous fertilizer should be applied to the soil. In particular, plant growth and higher yields require the addition of nitrogen (N), phosphorus (P), and potassium (K), which can be achieved by applying the aforementioned nutrients.[19] Many inorganic compounds found in macronutrient fertilizers are contaminated with high levels of metalloids and heavy metals most of which sometimes may be useful when the amount of feeding is in bulk, because minerals like zinc can be one of the trace elements used as an additive for them. The pollutants in fertilizers result in a notable accumulation of certain heavy metal(loid)s on top soils that used fertilizers with exceptionally huge amounts of their contents.

13.4.5 Military activity and sports shooting

Army training grounds, destroyed cities, and battlefields are contaminated largely by organic and heavy metal pollutants. The unexploded mines, cartridge boxes, leaking gasoline, burning buildings, shells, bullets, and explosives are the possible sources as pointed out by Moller. One use of depleted uranium metal is in shells that penetrate tanks. In the last two World Wars after all signs of damage have been cleared and places developed in western Europe, soil in such places could remain toxic for a period exceeding a century. Dominant elements are Pb, Cu, Zn, Ni, Sb, POISONS, Hg, As, and Bi. Even after decreasing its highly radioactive 235U content, this substance may still have a high amount of radioactivity as well as chemical toxicity. Swiss military bullets have 95% Pb, 2% Sb, 3% Cu, and 0.5% Ni in the mix.[21]

13.4.6 Atmospheric deposition

One of the most hazardous pollutants that could have a negative effect on public health is heavy metal (HM) pollution, which is discharged into the atmosphere by human activity. Numerous trace metals with documented negative effects are released by different pollution processes, and these impacts vary depending on the chemical forms and concentration levels of the metals. In power plants, smelters, incinerators, cement kilns, residential furnaces, fireplaces, and automobiles carbon-containing materials usually burn and produce these particles. As things stand now, those pollutants are having an impact on most towns in which there are lots of cars and people as well as many businesses dedicated to selling products.[22] Even though it may not seem like such- every time we observe the movement of metal(loids) laden particles in the atmosphere that reach into soils or water bodies through either rains, dry winds or some other means, what scientists normally do is collect and measure them at once as one composite sample called "bulk deposition."[19]

13.4.7 Industrial sources

Mining, processing, and disposal of waste rock and tailings, ore transportation, smelting, metal finishing, and recycling are a few industrial sources of heavy metals. Different heavy metals are emitted over the course of mining operations depending mostly on the kind of mine. For example, coal mines are among the factors that contaminate the terrain around them with As, Cd, Fe, and other elements in more or less direct ways.[23] A high level of environmental pollution today occurs mainly because of mercury used during gold extraction and extraction of significantly enormous amounts of this substance from abandoned mines.[20] Metals are released in particulate or vapor form through high-temperature metal processing operations such as smelting and casting. When water and vaporized heavy metals such as As, Cd, Cu, Pb, Sn, and Zn mix, it results in aerosols being formed. Through equi-distribution by wind known as dry deposition or water bodies contamination due to rainfall recognized as wet deposition part of these get deposited in soil at different sites.[20]

During the transportation of crude ores, as mine waste corrodes, or metallic elements leach into the ground and soil water, dust is generated, resulting in heavy soil pollution, in addition to water pollution. Soil pollution from heavy metals results from refining activities, while other sources of pollution are in microelectronics, paper milling, textiles production, wood conservation and processing of plastics. Recent studies have shown that grazing animals get poisoned by the high concentration of copper contaminated plants that grow under electric power transmission line (Sherene, 2010). The main source of elevated metal levels in rivers and lakes is most likely these waste streams. The following can be found in domestic wastewaters: (a) materials that have gone through biological treatment plant filters; (b) raw or primary treated wastewaters; and (c) substances that have flowed through sewage discharge pipes to the receiving waters and are often found out at the coastal residential regions as sea sources.[24] Using detergents can be dangerous because they might interfere with water quality in that most enzyme detergents were found to contain some percentage of the following trace elements: Fe, Mn, Cr, Co, Zn, Sr, and B[25] (Fig. 13.1).

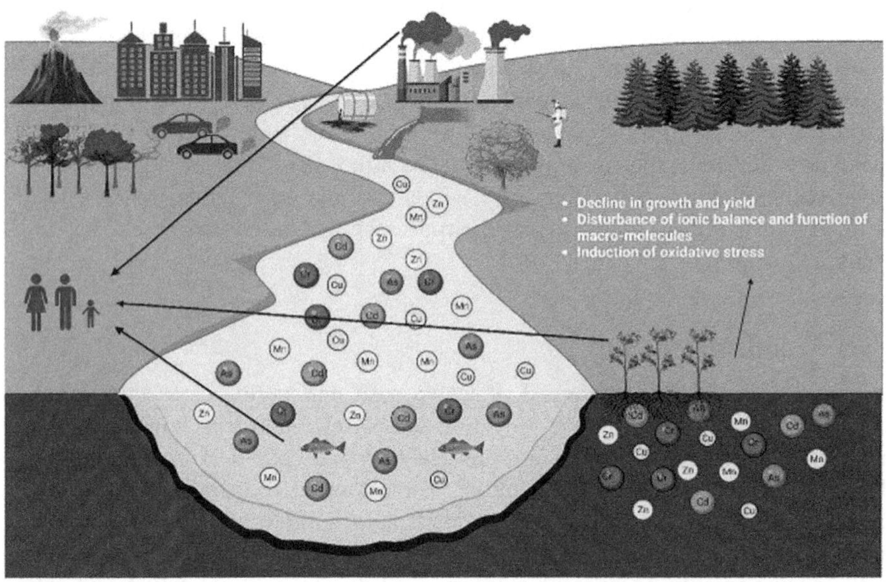

FIGURE 13.1

The beginning of heavy metal (HM) toxicity existence that is detrimental to plant growth and productivity as well as poses a major human health concern from food chain contamination by HMs.[20]

13.5 Mechanism of interaction

Soil contains a variety of metals in different forms at any one time. The physical, chemical, and biological characteristics of the soil affect the actual balance, but in general, the soil regulates the metal balance.

Since most compounds do not dissolve, it is hard for them to be taken up by plants except for a few of them that can easily pass through the root hairs of plant.[26] Microbes and root exudates are responsible for the availability and movements of metals in the rhizosphere while plants move them through mass flow. Metal ions can be transported into the cells of a plant through the root membranes.[27] The metals first enter through the roots' apoplast, a free intercellular area that is found on the side facing the xylem. The interconnection of the cortex and epidermis of plant roots allows for the apoplastic transfer of heavy metals. For the metals in root cells to reach the xylem, they must pass via the endodermis and casparian strip.[28] The endodermis' and the casparian strip's cell walls function as a barrier to prevent apoplastic diffusion into the circulatory system. Transpiration pumps likely power metal transporters that move metal ions from the root symplast into the xylem apoplast during the root–shoot translocation of metals.[29]

Plant cells' heavy metal levels are controlled by tonoplasts and plasma membrane metal transporters. Plants absorb metals via a range of membrane protein groups, but the molecular details of heavy metal transport across plant membranes are still uncertain. To improve the process of soil decontamination and remediation, transgenic plants that can accumulate particular metals must be created, which requires a thorough understanding of plant transport mechanisms[20] (Fig. 13.2).

13.5.1 Mode of action ols

Heavy metals exist in two forms: redox active (Fe, Cu, Cr, and Co) and nonredox active (Cd, Zn, Ni, Al, etc.). Redox-inactive redox-active heavy metals participate in Haber-Weiss and Fenton reactions, which produce $O_2{\bullet}-$ initially and H_2O_2 and •OH through later processes. This allows them to quickly facilitate redox reactions in cells. By such indirect processes, oxidative stress (Fig. 13.3) is also induced in plants by metals, which are redox inactive due to their interaction with an antioxidant protection system, blocking of electron donation and lipid peroxidation initiation.[10] Activated lipoxygenase (LOX) levels due to heavy metals may be a trigger for the first one. Toxicity of heavy metals is highly attributable to their strong bonding with oxygen, nitrogen, and sulfur atoms. This bonding affinity is related to the low solubility of heavy metal and ligand products as well as their free energy of production.[30] Such features enable heavy metals to bind on cysteine residues leading to enzyme inactivation. An example is proteins and enzymes with Cd attached to their SH groups which make them fold incorrectly and hence have lower enzyme activity or disrupt the redox regulation.[5] The enzyme's function requires additional compounds for some other heavy metal ions such as iron, magnesium or copper ions, and also includes organic ligands such as hemoglobin, biotin, NAD, FAD or coenzyme A. If a heavy metal ion is

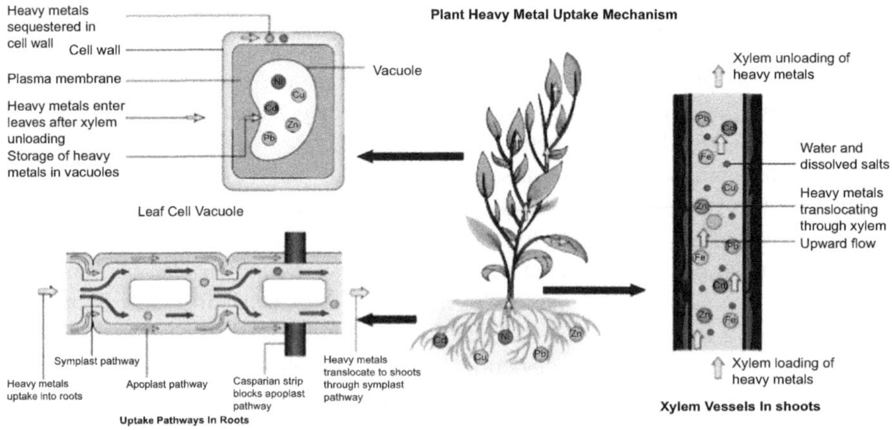

FIGURE 13.2

Heavy metals are taken to roots due to high solubility, and then they translocate into shoots through symplast and apoplast pathway consequently this process nowadays has been seen as one of the ways regarding detoxification of plants in relation to soil pollution. Conversely, the uptake and sequestration of heavy metals are done in three pathways. Another strategy for removing these heavy metals involved xylem loading before they reach leaves where they are stored in vacuoles.[20]

replaced with another, the enzyme activity gets reduced or inhibited.[31] In ribulose-1,5-bisphosphate carboxylase/oxygenase of (RuBisCO), Mg^{2+} is removed by Co^{2+}, Ni^{2+}, and Zn^{2+}, which are trivalent cations hence leading to the subsequent decrease in its activity.[32] Calmodulin is an important protein in cell signaling. Substitution of Ca^{2+} with Cd^{2+} in calmodulin suppressed radish calmodulin-dependent phosphodiesterase activity. Moreover heavy metals lead on to changes in membrane through various known ways.[33] This could be either as oxidizing proteins with sulfur containing amino acids or forming cross-links with different amino acids on protein molecules, blocking some essential membrane proteins such as H + -ATPase or altering the fluidity and components of membrane lipid.[34]

13.6 Environmental impacts of heavy metal stress

13.6.1 Effects on water

Increased human activities pose a threat to both aquatic environments and the environment because they discharge many dangerous chemicals into water bodies Heavy metal ions, which are extremely toxic, nonbiodegradable, and have a propensity to bioaccumulate and biomagnify, are the main cause of water pollution. Due to natural occurrences such as volcanic and seismic activities or storms and human activities like mining and agriculture that are not environmentally friendly,

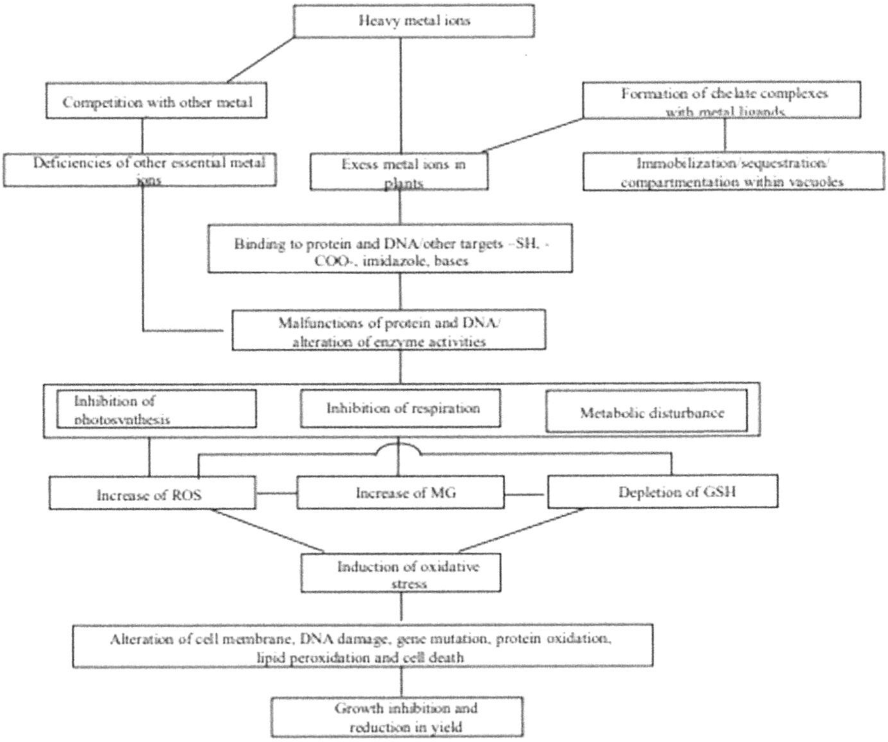

FIGURE 13.3

Potential chemical and molecular pathways which heavy metal exposure triggers reactive oxygen species leading to oxidative harm to higher plants growth.[10]

it is almost sure that sewages will have dangerous substances in them.[35] Environmental poisoning by heavy metals is currently one of the most terrifying threats. The safety of both people and animals depends on the wastewater's cleanliness. To remove heavy metal ions, a variety of procedures are employed, such as chemical precipitation, ion exchange, adsorption membrane filtration, and electrochemical treatment techniques. This is considered to be one of the metals that is most toxic to human health. Reverse osmosis, ion exchange, complexation/sequestration, coagulation/flocculation, and electrochemical approaches are a few popular ways for removing heavy metal ions from wastewater. These methods are not commercially viable and may result in harmful consequences.[18]

13.6.2 Effects on plants

Heavy metal pollution in the soil could lead to two major issues, which include losing soil value and posing a higher risk to people's health living close by. The use of any soil that has been poisoned by heavy metals will reduce at least some of it functions. The soils may even function properly despite having heavy metal concentrations lower than the permitted levels. Heavy metals attack in different places inside plants and, as a result, their overall visible toxic response varies. Leaf chlorosis is one of the components of reduced plant development; necrosis, turgor loss, decreased seed germination rates are others. One of the most common signs of heavy metal poisoning is also dysfunction of photosynthetic machinery due to senescence or plant death whenever it occurs.[36] Food contains varying levels of this element, which depend on where it came from, how it was kept, or what was done to it after harvest. They fall under a certain class of elements; peculiarities associated with this group of elements include the following: (1) lack of deterioration with age; (2) plants might need or benefit from them at some doses, while others exceed toxicity limits; and (3) they are always present in the natural environment in trace amounts, with their contribution to soils coming through paedogenesis and parent rock weathering. (4) Heavy metals are frequently found in soil as cations, which bond firmly to soil particles and become mobile in response to changes in the surrounding environment. Several studies show that anthropogenic activities are primarily responsible for heavy metal pollution in the environment. It has been demonstrated that soil bacteria can modify the mobility and bioavailability of heavy metals by releasing chelating agents, solubilizing metal phosphates, causing redox shifts, and acidifying the soil.[35]

Various examples of impact of heavy metals on plants:

1. **Inhibition of seed germination:** Lead (Pb), cadmium (Cd), and copper (Cu) are some of the heavy metals that can prevent the germination of seeds in plants. They have the capacity to interfere with different physiological processes during germination, for instance, water absorption, enzyme activation and mobilization of nutrients.
2. **Stunted growth and reduced biomass production:** Elevated levels of heavy metals in the soil may hinder the growth of the roots and prevent shoots from developing, which leads to retarded growth and lesser biomass production in plants.
3. **Disruption of nutrient uptake and transport:** Heavy metals are known to inhibit the absorption and transport of vital nutrients in plants, leading to nutrient deficiencies. For instance, in plants, an overabundance of zinc can hinder iron and manganese uptakes.[37]
4. **Oxidative stress and cellular damage:** Since heavy metals produce reactive oxygen species (ROS), which destroy cellular components like proteins, lipids, and DNA, they put plants through oxidative stress. Plant cells die as a result of this damaging cell processes.[38]
5. **Altered photosynthesis and chlorophyll degradation:** Heavy metals distort the shape and working of the chloroplast in plants thus impairing photosynthesis; they can halt chlorophyll production and amplify the speed of its breakdown, thus reducing photosynthesis efficiency. This sentence in a way conflates or incorrectly aligns ideas concerning various impacts of heavy metals on photosynthesis of plants.[36]

13.7 Impacts of heavy metals on human health

Several edible crops can accumulate heavy metals, even at extremely low amounts, which can be harmful since it can upset the food chain. These heavy metals pose a health concern to humans by entering the food chain and causing conditions like cancer and liver disease. Some vegetables are more likely to absorb heavy metals through their roots, which can then transport the metals to the edible sections of the plant. Examples of these plants are brinjal, gourd, spinach, coriander, tomato, and pumpkin.[4] Our exposure to 35 different metals, of which 23 are heavy metals, at home or at work: antimony, arsenic, bismuth, cadmium, cerium, chromium, cobalt, copper, gallium, gold, iron, lead, manganese, mercury, nickel, platinum, silver, tellurium, thallium, tin, uranium, vanadium, and zinc.[39]

These heavy metals are frequently found in food and the environment. While tiny amounts are necessary for health, larger levels can be fatal or harmful. Heavy metals in large quantities can ruin wellbeing and damage fundamental body systems like the heart, lungs, kidney, liver and all components of blood. Prolonged exposure can result in slowly developing neurological, musculoskeletal, and physical degenerative processes that mimic conditions including muscular dystrophy, multiple sclerosis, Parkinson's disease, and Alzheimer's disease. Long-term exposure to some metals and their compounds on a regular basis may even lead to cancer.[40]

13.7.1 Impact on enzymatic pathways

In the bodies of humans, competition among metal ions for protein-binding centres can result in challenges posed by specific molecular traffic and chemical reaction catalysis. Metal selectivity is associated with the composition of the immediate vicinity of the metal-binding site. What chiefly affects human biology is how metal ions promote interaction between proteins. To ensure that metallic elements distribute evenly within the cell, it employs peptides rich in cysteine (Cys) that bind metals, for example, glutathione (GSH), metallothioneins (MTs), or phytochelatins (PCs) in plants and bacteria.[38]

Since it doesn't decay, any living organism cannot break it down and accumulates. Thus, they continue to exist for even centuries or decades in the ecosystem. Consequently, it leads to diseases and disorders in neural, immunity, reproductive, and digestive aspects of organisms through a variety of ways. Melanin might shield tissues by taking away heavy metals from surrounding photoreceptor cells and neuronal retina. Just as retinal pigment epithelium does, the choroid plexus of the brain keeps off dangerous substances from entering it, absorbs lead.[41] Pregnant women may have potentially dangerous substances in their blood that may reach the foetus and endanger the health of the unborn child. Cadmium (Cd), lead (Pb), manganese (Mn), and mercury (Hg) have been extensively studied because of their widespread exposure. Humans and wildlife have experienced a range of issues as a result of exposure to heavy metals and other toxins, including teratogenic, mutagenic, and carcinogenic impacts.[35]

13.8 Natural methods of mitigating heavy metal stress

13.8.1 Nanotechnological and nanomedicinal approaches to treat heavy metal toxicity

Nanotechnology is used in analyzing and removing heavy metals from water sources, even from food sources. For this, a number of nanomaterials have been employed. These consist of carbon nanotubes, graphene, magnetic nanoparticles, and metal oxide nanoparticles. Nanotechnology offers far more benefits than the conventional ways hence why it is good at analysing off heavy metal in both food as well as water sources while on the other hand removing them too. These advantages include high sensitivity, fine selectivity, low detection and quantification limitations, and a broad linear range. Nanotechnology-based technologies require simple and safe field application.[37]

13.8.2 Metal oxide nanoparticles

Metal oxide nanoparticles have unique physical and chemical characteristics that make them highly effective at removing dangerous heavy metals from contaminated waterways. Recently, some innovative green chemical processes have emerged for the synthesis of magnetic nanoadsorbents using biological waste products and natural biopolymers. These technologies have a great affinity for metal capping, are easily accessible, affordable, and biodegradable.[42]

Nanoscale synthesis with various structural modifications of CuO has been proved to have good capacity of absorbing As^{3+}, As(V), Pb^{2+}, and Cr (VI). Changing the structure of CuO at the nanoscale creates more active sites for metal ions, leading to effective removal of heavy metals from the environment.

Using an organic surfactant to create SiO_2 mesoporous nanoparticles is one way to alleviate the environmental issues brought on by Cr^{3+} ions. Sunflower oil or n-dodecyl amine can be used as the template agents to generate organo-silica mesoporous materials containing cyano groups via a single step cocondensation of 2-cyanoethyltriethoxysilane and tetraethoxysilane in the ratios 1:4 or 1:9.[43]

13.8.3 Magnetic nanoparticles

Magnetic Fe_3O_4 nanoparticles with surface coatings have improved selectivity towards certain targets and are kept from oxidizing or aggregating. One described a core-shell structure comprising of Fe_3O_4, SiO_2, and polythiophene. Fe_3O_4 hybridized with polyaniline and MnO_2 (Fe_3O_4/PANI/MnO_2) is a cost-effective and environmentally friendly way to separate and enrich Hg^{2+} ions in various matrices. It also has a high capacity for heavy metal ion adsorption (including Pb^{2+}, Zn^{2+}, Cd^{2+}, and Cu^{2+}).[44]

Poly-(m-phenylenediamine), also known as PmPD, is an amino-conjugated polymer that has many amine and imine functional groups in this respect that impart to it an improved redox performance, chelating ability, and adsorption capacity; PmPD and PANI are identical in terms of functional properties and structure as well. It could be a theoretical consideration to introduce MnO_2 into PmPD.[45] The MnO_2/Fe_3O_4/PmPD core–shell hybrid is characterized by its high adsorption capacity as well as being paramagnetic in nature; these properties allow it to separate and eliminate various heavy metal pollutants by ion exchange, electrostatic attraction, and coordination interaction at large scales. Considering its great remedial capabilities and low cost of production.[46]

Nano-zerovalent iron (NZVI) is a possible future material. Such feat, it to be noted, is overshadowed when it comes to pollutions of inorganic and heavy metal ions by no other material so far. When exposed to oxygen, air, or water, NZVI becomes oxidized and reactive, reducing its reactivity. Instead of iron, one may use nickel ions, as they are more stable to retard oxidation. NaBH4 was used as reducing agent and ground was a simple reduction in preparation of nanoscale zero-valent nickel. It can be used as adsorbents for (VI) Cr.[37]

13.8.4 Carbon nanotubes

There is the increasing popularity of carbon nanoparticles due to their better physicochemical properties that help in efficient treatment of water contaminated by heavy metals. Carbon nanomaterials, including carbon nanotubes, graphene, fullerenes, graphene oxide, and activated carbon, are very effective at removing heavy metals from water due to their large surface area, nanoscale size, and availability of a variety of functionalities. They are also easier to recycle and chemically change. Multiwalled carbon tubes (MWCNTs) that are unmodified are useful in absorbing the toxin mercury (Hg^{2+}) out of waste water. In an ideal setting, all pollutants can be eliminated. Functional groups, which are bound to the surface of multiwalled carbon nanotubes, primarily contribute to their adsorption properties.[47] It is revealed that introduction of amino and thiol groups onto carbonaceous adsorbents and carbon nanotubes (CNTs) enhances cannot only the removal efficiency of heavy metals but also that of organic compounds.[48]

13.8.5 Microbial-assisted remediation of heavy metal

The practice of utilizing live microorganisms, such as bacteria, fungus, and archaea, to decompose and detoxify different types of harmful chemical and metallic pollutants from the environment is known as microbial remediation. Bioremediation stands for employing microorganisms on the contaminated surface directly for aiding pollution degradation while different techniques used in remediation such as bioaugmentation and biostimulation incorporate microorganisms.[49] Bioaugmentation involves the addition of microorganisms in a polluted area to help in the breakdown of toxins. Biostimulation on the other hand alters the environment so as to foster the growth and activities of microbes which are able to decompose pollutants. There are some drawbacks to conventional methods of restoring something back to its original condition like involving high costs, use of huge machinery, problems with transportation, because it can also be harmful to the environment in some cases. Bioremediation has been growing tremendously in this decade and has witnessed significant improvements in its use for the removal of soil and water contaminants.[50]

In mycoremediation, a type of bioremediation, fungi are used to treat pollutants. In soil and aquatic environments, fungi play a crucial role in the cleanup process. Different kinds of microorganisms can be encountered by individuals in different environments. Certain microorganism species, such as laccases and peroxidase, are capable of surviving in hostile environments and produce ligninolytic enzymes. These enzymes help fungi transform contaminants into less harmful forms. Extracellular enzymes have the ability to absorb pollutants.[38]

13.8.6 Rhizofiltration

Rhizofiltration is the removal of heavy metals through plant roots. Similar to phytoextraction, this approach involves plants removing toxins from wastewater or groundwater instead of soil. The roots of the plant take in pollutants in water recycling, underground water, as well as surface water. On the other hand, aquatic plant types use rhizofiltration as a way of reducing pollutants. Rhizofiltration is beneficial in eliminating Cd, Pb, and Cr from the roots. Sunflower, tobacco, and spinach show potential in eliminating lead from water.[38]

13.8.7 Phytostabilization

The roots of the plant take in pollutants in water recycling, underground water, as well as surface water. On the other hand, aquatic plant types use rhizofiltration as a way of reducing pollutants. Contaminants are captured on the root surface using transport proteins or secondary metabolites. t converts complicated chemical substances into easily soluble ones by mixing them with derivatives of proteins, amino acids and sugars. Black nightshade, sunflower, and cowpea are among plant species that use phytostabilization as an avoidance strategy for heavy metals pollution.[38]

13.9 Application of biogenic nanomaterials

13.9.1 Organic pollutant removal

An organic pollutant is any substance that contains nitrogen, sulfur, halogens, phosphorus, artificial colors, insecticides, pesticides, aromatic hydrocarbons, phenols, and halogenated hydrocarbons. Moreover, the release of such persistent organic pollutants from industrial wastewater to agricultural zones seriously threatens aquatic ecosystems. Nano-enabled technology has made it several more nanomaterials, such as nanofilms, nanocatalysts, and nanoadsorbents, possible. Some have been associated with allergies and cancer.[6,51–54]

13.9.2 Growth promotion

Iron nanoparticles increase the dry mass of leaves and pods through a process which leads to increased yields of peanut, wheat, and soybean seeds in the presence of FeNPs. CuNPs have contributed to the growth of shoots and roots of mung beans, despite the fact that CsNPs have led to an increase in peanut pod weights. Confirming that the ingestion of copper and iron nanoparticles improved wheat production through physiological effects, the amount of sugar and superoxide dismutase activity increased after proteomic analysis. Additionally, plants were positively influenced by TiO_2 NPs leading to improved cucumber photosynthetic activity and a better germination rate in spinach. Spinach and titanium dioxide had different effects on chlorophyll production levels, which were higher for rutile than for anatase. Conversely, TiO_2 nanoparticles enhanced ambient N_2 photoreduction for NH_3 synthesis thereby supporting spinach growth.

The carbon nanotube, a crucial part of nanotechnology increases the uptake of water encouraging plants grow and seeds sprout thus bringing about the need to develop more farming solutions such as nanofertilizers. Moreover, in agricultural regions, it decreases the variety of microorganisms and increases the amount of harmful metals found in farm produce by prompting cell death through oxidative damage (ROS formation).[6]

13.9.3 Antimicrobial activity

Antibiotic resistant bacteria pose a serious threat to the global community therefore it is extremely important to find a new compound that can effectively deal with these types. Since ancient times, metals like copper, silver, iron, gold, magnesium among others have been used in traditional medicine. Scientists are now using metallic/metal oxide nanoparticles (NPs) as an alternative to develop antimicrobial agents since these metals have natural antibacterial characteristics. Among eukaryotes, Gram-positive and Gram-negative bacteria, metallic nanoparticles of plant origin have been shown to exhibit many biocidal actions. Additionally, efficacious inhibitory activity of metallic nanoparticles (NPs) to against various resistant strains have been reported, for example, vancomycin-resistant *Staphylococcus aureus*, methicillin-resistant *Staphylococcus aureus* (MRSA), erythromycin-resistant *Streptococcus pyogenes*, ampicillin-resistant *Escherichia coli* and *Pseudomonas aeruginosa*.

13.9.4 Antiviral activity of metallic nanoparticles

One of the infectious agents that can cause a wide range of illnesses in people, animals, and plants is a virus. Particularly, the severity of viral illnesses in humans can be frightening and have negative consequences. Human life is particularly threatened by viruses such as the human immunodeficiency virus (HIV), hepatitis, herpes simplex virus (HSV), and influenza. Through engineered nanomedicines development, plant based nanoparticles have brought about a lot of new chances in the combat against viral diseases. Silver nanoparticles are most effective when used in the form of metallic nanoparticles aimed at treating viral infections because silver can be used to treat conjunctivitis when used in the form of an ocular solution. Subsequently, tests were conducted on silver nanoparticles produced chemically against the hepatitis, herpes, influenza, and HIV viruses. The development of NPs as antiviral medicines is made possible by these results, which were therapeutically significant.[1]

13.10 Current challenges and future prospects

Apart from numerous benefits there are still certain challenges faced which are as follows:

1. Production standardization: Most biogenic nanomaterials are produced using biological methods such as microbial synthesis or extraction from plants among other sources found in nature; thus, getting uniform quality in output quantity might prove cumbersome since organisms differ.
2. "Scaling up production" is possible in the laboratory but difficult at industrial scale. It is however possible to address such challenges like sustaining the most favorable biological system conditions and ensuring yield is sufficient.
3. Biogenic nanomaterials have to be biocompatible in order to be utilized for biomedical purposes and must also meticulously reviewed for possible toxicity problems to assure human and environmental safety.[27]

By introduction of NPs above fifty years, researchers have not yet clearly understood the perception of NPs. The green chemistry philosophy requires synthesis of NPs as an ecofriendly option for traditional methods of NPs' synthesis. The green chemistry principle was the basis on which NPs synthesis was founded; the principle suggested that the synthesis should be safe, made from nature, should not include hazardous chemicals, should not be poisonous, and should not cost much. Many studies have shown that plants, bacteria, fungi, yeast, and actinomycetes provide ideal materials for synthesizing metal/metal oxide nanoparticles. Plants are the most common source of metal oxide nanoparticles probably because they are of value in traditional medicine and contain active ingredients that can easily be obtained.[6]

13.11 Conclusion

In the field of pollution control, nanotechnology might introduce revolution. A sustainable way of making nanoparticles with aid of a living thing is synthesis. It implies that there is very little emission from biogenic nanoparticles synthesized through bacteria, algae, and fungi to mean that they can successfully degrade many pollutants in the most polluted water or soil areas as evidenced in various studies that have been done so far. Using biogenic nanoparticles for environmental cleanup is a cheaper, safer, and sustainable alternative to chemical treatment for various environmental problems including pesticides, dyes, halogenated compounds, heavy metals, and pharmaceuticals in wastewater. Biogenic nanoparticles effectively eliminate heavy metal from polluted soils. It can be a revolution in creating sustainable environment, or treating soil pollution if one opts to unify production of these nanoparticles.[27]

Acknowledgment

Internal grant of Yerevan State University (YSU)-2023 to support KG. Internal grant of Yerevan State University (YSU)- 2024 to support AS. Authors acknowledge the support by the Strategic Academic Leadership Program of the Southern Federal University ("Priority 2030").

References

1. Ghorbani, et al. Nano-enabled agrochemicals: mitigating heavy metal toxicity and enhancing crop adaptability for sustainable crop production. *J Nanobiotechnology.* **2024**, *22* (1), 91.
2. Rezapour, S., et al. Urbanization influences the distribution, enrichment, and ecological health risk of heavy metals in croplands. *Sci. Rep.* **2022**, *12* (1), 3868.
3. Rahman.; Shafeeq Ur., et al. State-of-the-art OMICS strategies against toxic effects of heavy metals in plants: a review. *Ecotoxicol. Environ. Saf.* **2022**, *242*, 113952.
4. Manzoor, N., et al. Recent advancements and development in nano-enabled agriculture for improving abiotic stress tolerance in plants. *Front. plant. Sci.* **2022**, *13*, 951752.
5. El-Saadony.; Mohamed, T., et al. Role of nanoparticles in enhancing crop tolerance to abiotic stress: a comprehensive review. *Front. plant. Sci.* **2022**, *13*, 946717.
6. Patil, S.; Chandrasekaran, R. Biogenic nanoparticles: a comprehensive perspective in synthesis, characterization, application and its challenges. *J. Genet. Eng. Biotechnol.* **2020**, *18* (1), 67.
7. Kalishwaralal, K., et al. Silver nanoparticles impede the biofilm formation by *Pseudomonas aeruginosa* and *Staphylococcus epidermidis*. *Colloids Surf. B: Biointerfaces* **2010**, *79* (2), 340–344.
8. Karunakaran, G., et al. Rapid biosynthesis of AgNPs using soil bacterium Azotobacter vinelandii with promising antioxidant and antibacterial activities for biomedical applications. *JOM* **2017**, *69*, 1206–1212.

9. Phanjom, P.; Ahmed, G. Effect of different physicochemical conditions on the synthesis of silver nanoparticles using fungal cell filtrate of *Aspergillus oryzae* (MTCC No. 1846) and their antibacterial effect. *Adv. Nat. Sci: Nanosci. Nanotechnol.* **2017**, *8*, 045016.
10. Mukti, G. Heavy metal stress in plants: a review. *Int. J. Adv. Res.* **2014**, *2* (6), 1043–1055.
11. Emamverdian, A., et al. Heavy metal stress and some mechanisms of plant defense response. *The Scientific World Journal* **2015**, *2015*, 756120. Available from: https://doi.org/10.1155/2015/756120.
12. Yadav, S. K. Heavy metals toxicity in plants: an overview on the role of glutathione and phytochelatins in heavy metal stress tolerance of plants. *South. Afr. J. botany* **2010**, *76* (2), 167–179.
13. Andrés, A.; García-Muse, T. Causes of genome instability. *Annu. Rev. Genet.* **2013**, *47* (1), 1–32.
14. Rucińska-Sobkowiak, R. Water relations in plants subjected to heavy metal stresses. *Acta Physiologiae Plant.* **2016**, *38*, 1–13.
15. Aery, N. C. Plant defence against heavy metal stress. *Plant. Defence: Biol. Control.* **2012**, 241–269.
16. Singh, A.; Agrawal, S.; Rajput, V. D., et al. A Biogenic nanoparticles for managing salinity stress-related crop and environmental risks: realistic applications and challenges. *Discover Sust* **2025**, *6*, 120. Available from: https://doi.org/10.1007/s43621-025-00855-0.
17. Shahid, M., et al. Heavy metal stress and crop productivity. *Crop. Prod. Glob. Environ. Issues* **2015**, 1–25.
18. Jessica, B.; Sinagra, E.; Blundell, R. Heavy metal pollution in the environment and their toxicological effects on humans. *Heliyon* **2020**, *6*, 9.
19. Ghazaryan, K.; Agrawal, S.; Margaryan, G., et al. Soil pollution: an agricultural and environmental problem with nanotechnological remediation opportunities and challenges. *Discover Sust* **2024**, *5* (453). Available from: https://doi.org/10.1007/s43621-024-00666-9.
20. Sandeep, G.; Vijayalatha, K. R.; Anitha, T. Heavy metals and its impact in vegetable crops. *Int. J. Chem. Stud.* **2019**, *7* (1), 1612–1621.
21. Duruibe, J.; Ogwuegbu, M. O. C.; Ogwuegbu.; Egwurugwu, J. N. Heavy metal pollution and human biotoxic effects. *Int. J. Phys. Sci.* **2007**, *2* (5), 112–118.
22. Hassanien, M. A. Atmospheric heavy metals pollution: exposure and prevention policies in mediterranean basin. Environmental heavy metal pollution and effects on child mental development: risk assessment and prevention strategies. Springer Netherlands, 2011.
23. Petruzzelli, G. *Heavy metals in compost and their effect on soil quality. The Science of Composting;* Springer Netherlands: Dordrecht, 1996, 213–223.
24. Sankhla, M. S.; Rajeev, K. Contaminant of heavy metals in groundwater & its toxic effects on human health & environment. *Available SSRN* **2019**, 3490718.
25. Mahmood, Q., et al. Current status of toxic metals addition to environment and its consequences. *The Plant Family Brassicaceae. Environmental Pollution.* **2012**, 35–69.
26. Wu, X., et al. A review of toxicity and mechanisms of individual and mixtures of heavy metals in the environment. *Environ. Sci. Pollut. Res.* **2016**, *23*, 8244–8259.
27. Kumari, S.; Tyagi, M.; Jagadevan, S. Mechanistic removal of environmental contaminants using biogenic nano-materials. *Int. J. Environ. Sci. Technol.* **2019**, *16* (11), 7591–7606.

28. Giri.; Ved Prakash., et al. A review of sustainable use of biogenic nanoscale agro-materials to enhance stress tolerance and nutritional value of plants. *Plants* **2023**, *12* (4), 815.
29. Khandel, P., et al. Biogenesis of metal nanoparticles and their pharmacological applications: present status and application prospects. *J. Nanostructure Chem.* **2018**, *8*, 217–254.
30. Mathur, S.; Deeksha, S.; Rajiv, R. Remediation of heavy metal (loid) contaminated soil through green nanotechnology. *Front. Sustain. Food Syst.* **2022**, *6*, 932424.
31. Maqsood.; Muhammad Faisal., et al. Plant-based nanoparticles in ROS mitigation and abiotic stress tolerance: a review. *Plant. Stress.* **2023**, 100281.
32. Sharma, U.; Sharma, J. G. *Nanotechnology for the bioremediation of heavy metals and metalloids. JABB.* **2022**, 34–44.
33. Sun, W., et al. Microbially mediated remediation of contaminated sediments by heavy metals: a critical review. *Curr. Pollut. Rep.* **2021**, *7*, 201–212.
34. Abd, E.; Manar, K., et al. Toxicity of heavy metals and recent advances in their removal: a review. *Toxics* **2023**, *11* (7), 580.
35. Khalef, R. N.; Hassan, A. I.; Saleh, H. M. *Heavy metal's environmental impact. Environmental Impact and Remediation of Heavy Metals;* IntechOpen, 2022.
36. Kumar, P., et al. Impact of heavy metal contamination on human health. *Eur. Chem. Bull.* **2023**, *2* (12), 1366–1379.
37. Mitra, S., et al. Impact of heavy metals on the environment and human health: Novel therapeutic insights to counter the toxicity. *J. King Saud. University-Science* **2022**, *34* (3), 101865.
38. Das, S., et al. Heavy metal pollution in the environment and its impact on health: exploring green technology for remediation. *Environ. health insights* **2023**, *17*, 11786302231201259.
39. Jaishankar, M., et al. Toxicity, mechanism and health effects of some heavy metals. *Interdiscip. Toxicol.* **2014**, *7* (2), 60–72.
40. Witkowska.; Danuta.; Joanna, S.; Chilicka, K. Heavy metals and human health: possible exposure pathways and the competition for protein binding sites. *Molecules* **2021**, *26* (19), 6060.
41. Engwa, G. A., et al. Mechanism and health effects of heavy metal toxicity in humans. *Poisoning in the Modern World—New Tricks for an Old Dog?IntechOpen* **2019**, *10*, 70–90.
42. Bayat.; Mehrnoosh.; Hossein Beyki, M.; Shemirani, F. One-step and biogenic synthesis of magnetic Fe3O4—Fir sawdust composite: application for selective preconcentration and determination of gold ions. *J. Ind. Eng. Chem.* **2015**, *21*, 912–919.
43. Kaushik, A.; Singh, R. K.; Tyagi, P. K. Green synthesized nanoparticle based drug delivery: recent trends and future prospects. *Precis. Nanomed.* **2023**, *6* (4), 1109–1131.
44. Gupta.; Vinod Kumar., et al. Removal of hexavalent chromium ions using CuO nanoparticles for water purification applications. *J. Colloid Interface Sci.* **2016**, *478*, 54–62.
45. Moosa, A. A.; Mousa Ridha, A.; Al-Kaser, M. Process parameters for green synthesis of silver nanoparticles using leaves extract of Aloe vera plant. *Int. J. Multi Curr. Res.* **2015**, *3*, 966–975.
46. Xiong, T., et al. Mechanistic insights into heavy metals affinity in magnetic MnO2@ Fe3O4/poly (m-phenylenediamine) core – shell adsorbent. *Ecotoxicol. Environ. Saf.* **2020**, *192*, 110326.

47. Rao, B.; Tang, R.-C. Green synthesis of silver nanoparticles with antibacterial activities using aqueous *Eriobotrya japonica* leaf extract. *Adv. Nat. Sci: Nanosci. Nanotechnol.* **2017**, *8* (1), 015014.
48. Dong, H., et al. Chromate removal by surface-modified nanoscale zero-valent iron: effect of different surface coatings and water chemistry. *J. colloid interface Sci.* **2016**, *471*, 7–13.
49. Ahmad, A., et al. Extracellular biosynthesis of monodisperse gold nanoparticles by a novel extremophilic actinomycete, *Thermomonospora* sp. *Langmuir* **2003**, *19* (8), 3550–3553.
50. Murugesan, S.; Bhuvaneswari, S. U. N. D. A. R. E. S. A. N.; Sivamurugan, V. Green synthesis, characterization of silver nanoparticles of a marine red alga *Spyridia fusiformis* and their antibacterial activity. *Int. J. Pharm. Pharm. Sci. 9.5* **2017**, 192–197.
51. Ghazaryan, K.; Verma, A.; Rawat, S.; Rajput, P.; Minkina, T.; Mandzhieva, S., et al. Implementing sustainable nano-bioremediation for emerging pollutants. *Nanotechnology Applications and Innovations for Improved Soil Health.* **2024**, 316–332. Available from: https://doi.org/10.4018/979-8-3693-1471-5.ch016.
52. Verma, A.; Singh, A.; Rawat, S.; Rajput, P.; Minkina, T.; Mandzhieva, S., et al. Unveiling nanomaterial-induced toxicity. Nanotechnology Applications and Innovations for Improved Soil Health. **2024**, 236–253. Available from: https://doi.org/10.4018/979-8-3693-1471-5.ch011.
53. Singh, A.; Sharma, A.; Singh, O.; Rajput, V. D.; Movsesyan, H. S.; Minkina, T., et al. In-depth exploration of nanoparticles for enhanced nutrient use efficiency and abiotic stresses management: present insights and future horizons. *Plant. Stress.* **2024**, 100576. Available from: https://doi.org/10.1016/j.stress.2024.100576.
54. Upadhyay, P., et al. Nano-omics-based abiotic and biotic stresses management. Harnessing NanoOmics and Nanozymes for Sustainable Agriculture. **2024**, 348–371. Available from: https://doi.org/10.4018/979-8-3693-1890-4.ch017.

CHAPTER

14 Navigating the nanofields: risks and sustainable solutions of nanoparticle accumulation in cereal crops

Rupjyoti Bharali[1], Ranju Chetri[2], Nayan Talukdar[3], Sarvesh Rustagi[4], Sagar Mondal[5], Sumira Malik[5] and Jutishna Bora[5]

[1]*Department of Botany, Biswanath College, Biswanath Chariali, Assam, India*

[2]*Department of Zoology, Tezpur College, Tezpur, Assam, India*

[3]*Program of Biotechnology, Faculty of Science, Assam Downtown University, Guwahati, Assam, India*

[4]*School of Applied and Life Sciences, Uttaranchal University, Dehradun, Uttarakhand, India*

[5]*Amity Institute of Biotechnology, Amity University Jharkhand, Ranchi, Jharkhand, India*

14.1 Introduction

A linear trend has been observed between the growing population and demand for agricultural products to uphold the food requirements of people.[1] To combat this situation, developments over the years have been made to produce fertilizers, and pesticides for obtaining large quantities of nutritive crops in a brief period.[2] However, overuse of these products for enhancing agricultural productivity has exerted negative impacts on soil health causing pollution. For the adverse effects of synthetic fertilizers such as increased rate of disease incidence, eutrophication of water bodies, plant roots weakening, soil acidification[3], and so on, attention has been shifted toward the development of cost-effective and ecofriendly agricultural inputs, namely, biofertilizers, organic fertilizers, and biostimulants.

Microorganisms residing inside plants act as agents promoting plant growth by supplying nutrients to the plants[4] and the use of these microorganisms as biofertilizers for enhancing growth in agricultural crop plants is an innovative green approach toward sustainable agriculture.[5] These strains of microbes to be used as biofertilizers must have growth-promoting action, colonizing capabilities, and traits like organic matter decomposition, secretion of phytohormones, nutrient-enhancing ability, and resistance toward varied stresses.[6] The biofertilizer market shall reach USD 1.88 billion by 2025.[7] Based on their activity, the types of biofertilizers are potassium mobilizing and solubilizing, nitrogen-fixing, phosphorous mobilizing and

Biogenic Nanoparticles. DOI: https://doi.org/10.1016/B978-0-443-33819-9.00018-4

solubilizing, micronutrient, and growth promoting.[8] For production of high-yielding varieties of crops, without deteriorating the soil health, biofertilizers are considered as a better choice replacing harmful chemical fertilizers.[8]

With the intention of minimizing the use of artificial fertilizers, search for green alternatives has paved the path toward development of biostimulants. This group are no fertilizers as they are not for providing any nutrients, however, they accelerate the processes that indirectly boosts plant growth.[9] Since there is no legal definition for biostimulants, a variety of substances may be categorized under it. However, certain categories, namely, Humic compounds, beneficial fungi and bacteria, seaweed extracts, biopolymers, protein hydrolysates represent biostimulants in global market.[10] These substances/microorganisms are added to crops to enhance the efficiency of the plants for nutrient uptake, provides resistance toward stress, promotes growth in plants by facilitating the metabolic processes in plants and soil.[11] The accessory products of industrial and agricultural sector are recycled and renewed to beneficial products and serves as important sources for formulation of biostimulants. The global market for biostimulants showed 10%–12% of compound annual growth from the year 2013–2018.[12]

Globally, per day a huge amount of waste are formed and threefold rise in this value is estimated by 2100[13] which needs to be handled in a sustainable way as leakage of nutrients from waste into environment threatens food security[14] and valorization of these waste into useful products such as organic fertilizers solve the issue to a greater extent. Organic fertilizers are carbon-based compounds or nutrient-rich materials derived from animal or plant sources that have a definite chemical composition and supply nutrients in available form to plants.[15] Utilizing these organic fertilizers ensures production of toxic free agricultural products by enhancing crop and soil productivity.[16]

This review summarizes the developments in the production of biofertilizers, biostimulants and organic fertilizers from various sources including the advances in the production technology. Furthermore, the transformation of potential sources or feedstocks into biostimulants and organic fertilizers has been discussed in detail.

14.2 What are biofertilizers?

Biofertilizer is a substance that has living microorganisms in it. They promote plant growth by providing essential nutrients.[17] Biofertilizers are mixtures of microbes that improves the nutrient uptake ability in plants and condition of the soil.[18]

Biofertilizers aggressively colonize the underground parts of plants, thereby facilitating the mobilization of vital nutrients and hormones boosting crop production by up to roughly 10%–40%. Biofertilizers, sometimes referred to as microbial inoculants, are organic substances that include certain microorganisms that are isolated from underground zones.[19] When these are applied onto the plants to encourage development, they colonize the environment.[20] They not only enrich the soil with nutrients to increase crop output and soil fertility, but provide shield to protect the plants from pests. They eradicate the dangerous compounds, and accelerate flowering.[3] The effect of these inoculants stays after its use for years, due to presence of parental inocula that multiplies further.[21]

Microbial inoculants have potential to replace chemical fertilizers and pesticides in agricultural systems and increase environmental sustainability, making their application of strategic significance.[22] The plant growth-promoting microorganisms have been characterized for their vital roles such as modulation of phytohormones, minimizing incidence of plant diseases, improving the availability of nutrients.[23] Under in vivo and in vitro environments, a single microbe also can promote crop production.[24]

14.3 Global biofertilizer market

In developed countries, rising concern toward development of sustainable agriculture has moved the global biofertilizer market to a certain range.[25] The current expected growth rate by 2025 is estimated to reach US$3.8 billion.[26] There are two broad forms, namely, the carrier-based, and the liquid biofertilizers in biofertilizer market. However, the other major beneficial categories such as potash mobilizing, nitrogen fixing, phosphate, sulfur, zinc, and boron solubilizing. The market comprises microbes, namely, AM fungi.[27] Earlier 80% of global demand for biofertilizers was met through N-fixing biofertilizers in the year 2012 and now attention has shifted toward phosphate solubilizing biofertilzers as they are cost-effective and due to their diverse roles as biocontrol agents.[28]

14.4 Technological advances in production of biofertilizers

14.4.1 Multistrain inoculants biofertilizer

Use of single strain inoculants for formulation of biofertilizers is a common practice, and coinoculation of different combinations of beneficial microorganisms have emerged as a new method in development of biofertilizers. Coinoculation has shown better yield in comparison to single strain biofertilizers. Combination of several microorganisms through their synergistic action shows numerous benefits such as enhancing the activities of other microorganisms that are beneficial, promotes plant growth and yield, better absorption of nutrients. However, there are certain limitations of this technology, namely, the interaction between the microorganisms is unknown and, in the fermenter, it becomes difficult to adjust the requirements for nutrition and growth for the different microorganisms used.[29] An example of successful multistrain biofertilizer developed for enhancing yield of rice is BioGro.[30]

14.4.2 Type of carrier material

The carriers are major component of biofertilizers that are used to preserve contents and growth of microbial inoculants during production and to deliver enough live microbes in good physiological condition to the field.[31] Utilization of cheap carrier materials derived from waste products, namely, peat and bagasse that are available in large quantities for biofertilizer production ensures sustainability in market making the products affordable to farmers.[32] Microbial biofilms as suitable carriers have shown success in formulation of inoculants. Microbes found in roots and soil can form biofilms that serves as an excellent source for development of biofertilizers.[33] Moreover, incorporation of nanobiotechnology in developing carrier-type biofertilizers has emerged as a novel approach.[34] For instance, macroscopic filters are used in the fermentation process to absorb bacterial cells for distribution in field.[33]

Strerilization of carrier material is very vital for maintaining the quality of biofertilizers produced which is done using the methods, namely, gamma irradiation and autoclaving.[35,36] Gamma irradiation is most preferred as it does not modify the properties of the final products enhancing quality of products. On other hand, inoculant's pure culture is obtained using autoclaving method which is a cost-effective method.[36]

14.4.3 Evaluation of inoculant quality

Viability of the microbial strain and contamination level determines the quality of biofertilizer and through the entire production process, physical factors are adjusted.[37,38] Conventional methods such as ELISA and nonconventional ones, namely, reporter gene techniques, RFLP, PCR-based techniques are used for quality control test.[37] For quality test of the bioinoculant, use of reporter genes that changes color under particular radiation or on a specific substrate is known.[39] Specifications for different categories of biofertilizers, namely, *Azotobacter*, *Azospirillum*, *Rhizobium*, mycorrhiza, phosphate solubilizing biofertilizer, and zinc solubilizing biofertilizer have been prescribed by the Indian Ministry of Agriculture and Cooperation as reviewed by Maan et al.[29]

14.5 Technological advancements in production of organic fertilizers

14.5.1 Composting and vermicomposting of organic waste

Composting is a technology that manages varied wastes of organic origin and develops commercially valued end-product known as organic fertilizers.[40] developed organic fertilizers by cocomposting wastes generated in food market and urban gardens with varied composting mixtures. The composts were ecofriendly and showed properties that were agronomically vital to be used as organic fertilizers.[40] Along with organic fertilizer production, composting also recovers other value-added products like compost tea.[41] Compost tea is formed by fermentation of compost soaked in water for some days which is extracted and developed as liquid fertilizer as they are enriched with nutrients and numerous beneficial microorganisms.[42] Similarly, vermicomposting tea is also extracted and has been developed as liquid fertilizer that provides resistance to plants against diseases[43]

Vermicomposting is another method where earthworms are nurtured in organic waste to convert the wastes into organic fertilizer.[44] A new technology that eliminates the need of manual extraction of earthworms in large-scale vermicomposting has been developed where earthworms are extracted by centrifugation with 84% worm recovery rate.[45] For production of quality fertilizers in less time and maintenance of worm growth, vermicompost reactors have been designed which are very efficient compared to the conventional containers marking another technological development.[46,47]

Thermophilic composting is a treatment for transforming solid organic wastes into useful organic fertilizer and it is performed under aerobic conditions. It is easily operated, inactivates pathogens from waste, has few solid residues and cost-effective.[48] Compared to this, hyperthermophilic composting has short period of composting, higher temperature, higher pathogen inactivating efficiency, better quality of compost. In HTC treatment, wastes are inoculated with hyperthermophilic microbes.[49] In Beijing, China, the HTC plant for sewage sludge treatment has the processing capacity of 600 t per day and the end-product is used in garden soil as fertilizer. Again, in Jaozuo, China, the cattle manure treatment using HTC is highly efficient as it produces 10 t of compost per day from 40 t cattle manure.[50] These treatments need further improvement before it is applied in a large scale.[49]

Organic carbon extraction using compost technology from solid waste of municipal origin produced 200 L of liquid fertilizer per 100 kg of dry compost at €1/L.[51] More work must be done to verify the quality of liquid fertilizer[52] and also in risks associated with extraction from solid wastes of municipality.[53]

14.5.2 Anerobic way of organic waste digestion

This results in formation of nutritive end-product called digestate, where pollutants and pathogens are removed as a result of presence of mesophilic or thermophilic phase in the process.[54] It is a technology where microbial consortia are used to decompose organic matter in oxygen-free conditions to obtain nutrient-rich fertilizer.[55] Production of liquid digestate using poultry waste as N source,[56] digestate from animal manures as N source[57] and as P source,[58] use of animal slurries in digestate production[59] is well documented that shows the fertilizer value of digestate. Every technology to produce organic fertilizer from different sources of organic waste has its own drawbacks hence to modify the digestate quality,[60] used both composting and anerobic digestion to treat wastes from rural slaughterhouses.

14.5.3 Pyrolysis of organic waste

Carbon-rich biochar production from pyrolysis of organic waste is used in agriculture as fertilizer to improve crop health and protects plants from pathogens.[61] In pyrolysis, waste is thermochemically degraded under anerobic condition and high temperature and based on these conditions there are types of it. The grade and amount of biochar vary due to variations in operating conditions.[62] Potential feedstocks for biochar production food waste, animal waste, waste from agriculture serves as vital feedstocks.[63] Biochar procured from animal manure are more nutrient rich compared to plant waste;[64] however, it needs further evaluation.[65]

14.5.4 Thermal drying of organic wastes

A cost-effective method of transforming organic wastes into fertilizers is thermal drying the wastes at optimum temperature[66] and this technology also removes the pathogens present.[67] Slaughterhouse wastes were recycled using tray drying method for 6–8 hours at 100–120°C to obtain fertilizer and its efficacy was tested on bell pepper and amaranth. Recently in India, an equipment for drying has been designed and patented for production of fertilizer on-site.[24] For production of mature organic fertilizer, it is very important to eradicate the pathogens which is possible only by combining different techniques of waste conversion.[60]

14.5.5 Subcritical water (SCW) treatment

Another technology used for fast bioconversion of any organic waste into mature compost at low temperature and pressure is subcritical water (SCW) treatment. It takes only seven days to complete the composting process to obtain mature compost from this treatment whereas traditional composting takes atleast 3 months. The compost obtained from treated organic waste like rice straw, chicken feather and banana waste with SCW showed its efficacy to be used an organic fertilizer after germination index test on mustard seeds which was found to be higher after dilution and nutrient analysis. This method is efficient in nutrient recovery as observed in the study with production of potassium from rapid hydrolysis of rice straw, nitrogen from banana and chicken feather waste and phosphorous from banana waste and in formulation of nutrient-rich organic fertilizer from these organic wastes.[68]

14.5.6 Hydrothermal carbonization (HTC)

Hydroachar, a soil amendment rich in carbon, is produced from organic wastes by hydrothermal carbonization (HTC).[69] This are slow-release fertilizers used in agriculture sector as it is rich in nutrients,[70] although it is feedstock dependent.[71] Manure-derived hydrochars are reported to be richer in nutrients in comparison to plant biomass-derived hydrochars.[72] Studies have shown that the application of hydrochar in agroecosystem enhances production and soil conditions.[71,73,74] However, there are reports of its negative impacts on few studied crops.[75–77] These variations in hydrochar activity is mainly attributed to the selection of type of feedstock, conditions during its production by hydrothermal carbonization, soil, crop, and environmental interactions.[78]

Although biochar and hydrochar are similar, but they vary in their technique of production, type of feedstocks. Since HTC for hydrochar production can use both wet and dry waste, the cost and energy involved is very less as compared to pyrolysis technique for biochar production where only dry biomass can be used for the later and pre-drying treatments multiplies the cost required.[72] Use of amendments and fertilizers for improving plant health and soil environments needs further research as both products cannot meet all the requirements of agriculture sector.[78]

14.5.7 Chemical decomposition

Digestion of degradable waste thermochemically by using HCl and KOH at 100°C in a prototype machine resulted in the quick and sustainable production of organic fertilizer with quality equivalent to conventional compost. The process for managing waste can be adopted in developing countries as it is economically feasible.[79]

14.5.8 Conventional solvent extraction of organic waste

The conventional extraction technology using alkaline solvent was found to be most suitable for mass production of organic liquid fertilizer from economical, technical, and environmental view compared to conventional extraction using water and microwave assisted extraction using water/alkaline solvent with ten times higher production of fertilizer and the cost involved is also four time less.[80] Extraction through microwave is an emerging green technology for nutrient recovery from organic wastes and its application requires optimization.[80]

14.5.9 Use of microbial inoculants in bioconversion of organic wastes

Using microbial inoculants for bioconverting organic waste into fertilizer or more accurately termed as bioorganic fertilizer is gaining interest and evolving. The techniques used in these bioconversions are fermentation (solid and liquid) and aerobic composting. Majority of these studies have used monoculture of microbes for bioconversion of organic wastes.[81] Different microorganisms are used to compost a variety of biowastes to speed up the decomposition process and improve quality of compost. *Aspergillus niger* was inoculated in municipal waste which shortened the cycle[82]; to degrade the various components in food waste, inoculation of a decomposing microbial agent is done[83]; inoculation of fungal consortium consisting of various strains of microbes to municipal solid waste showed increased mineralization of carbon and nitrogen and reduced the maturity time of compost.[84]

14.6 Potential sources of organic wastes for production of organic fertilizers

14.6.1 Sewage sludge

High nitrogen and phosphorus content found in sewage sludge degrades organic matter thus activating the soil factors making it useful to be utilized as organic fertilizer; however, it also composed of undesirable toxic substances.[85,86] Accumulation of such toxic materials in sewage sludge is minimized through pyrolysis making the matter suitable for use in agronomic sector.[87] Moreover, phosphorous is recoverable from sewage.[88]

Sewage sludge composting with phosphate creates more efficient fertilizer and it also affects the mobility of heavy metals thus making safer, ecofriendly compost to be utilized in agriculture (Fig. 14.1).[89] In a study, vermicomposting of wastewater sludge was done by nurturing earthworm species *Eisenia foetida* in biosolids obtained from digested wastewater sludge with flocculants in a reactor.[90]

Furthermore, ashes with high phosphorous content are produced after combustion of sewage sludge making them efficient organic fertilizer after recovering phosphorous by treating thermochemically,[91] lime water precipitation,[92] phosphoric acid reaction,[93] and acid or alkaline leaching.[94] The end-product is formed either as precipitate, liquid fertilizers, NP, PK, or NPK fertilizers.[95] Low-temperature

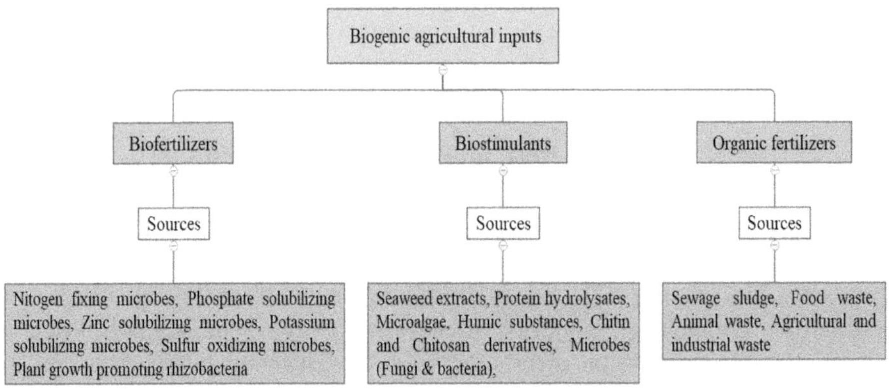

FIGURE 14.1

Production of biogenic agricultural inputs from various sources.

combustion treatment has great potential to recover and enrich phosphorous content from sewage sludge.[96]

In modern agriculture, the role of phosphorous being an essential element is very crucial and the fertilizer industry is focusing on finding alternative sources for recovering phosphorus for sustainable development.[94] An alternative would be sewage sludge ash as it is enriched with phosphorous compounds.[95,97] Cracow University of Technology developed a method which was patented with phosphorous recovery efficiency of 80%–96% applying acid extraction technique. Gorazda et al.[95,98] showed that extraction of sewage sludge ash with phosphoric acid produces organic fertilizer with the final products being either liquid fertilizer, NP, PK, NPK, or precipitates.

14.6.2 Food waste

Food industry is a potential source to be transformed into organic fertilizer as it is composed of carbohydrates, organic acids, proteins, lipids and are biodegradable.[86] A dynamic high-temperature aerobic fermentation (DHAF) process has been documented that accelerates the manufacture of organic fertilizers through food waste. In this process food waste undergoes continuous collisions and friction in a bioreactor due to rotation in mixers which facilitates the growth of microbes. Within 96 hours, the food waste is transformed into mature organic fertilizer.[99] Same study showed the germination of the cabbage seeds by using DHAF produced compost was found to be 88% after 16 hours which was greater than the GI of compost of traditional compost, that is, 82%. An organic fertilizer or compost is considered matured and nontoxic when its germination index is above 80%.[100]

14.6.3 Animal-based wastes

Untreated organic wastes are source of diseases, toxic compounds, greenhouse gases that affect human health and crops and cause environment pollution.[65] Increase in antibiotic-resistant bacteria is an issue due to animal waste that are not processed.[101] Hence waste recycling is very crucial in agriculture sector and there are technologies that converts these organic wastes into fertilizers that are safe and ecofriendly. The major challenge in development of organic fertilizer from these organic wastes depends on technologies that can eliminate heavy metals, minimize the effect of pathogens.[65] The animal excreta are rich in nutrients for production of fertilizer. For example, in cultivating vegetables belonging to family Solanaceae, animal wastes are sun dried and mixed which act as N source. Bhunia et al.[60] and Malancha Roy et al.[102] reported efficient development of bell pepper and amaranth after recycling slaughterhouse wastes using tray-drying method. For production of organic fertilizer, bone-based waste was treated with high-temperature pyrolysis to be used as a source of phosphorous.[103]

Since the fertilizer produced using animal waste by the process of vermicomposting is not guaranteed to be devoid of pathogens as heating phase is not present (Fig. 14.1),[104] inoculation of earthworms in the composting method was carried out after the thermophilic phase to eradicate the pathogens.[105] Moreover, it is reported that used earthworms have the capacity to suppress the bioavailability of metal ions thus reducing the metal ion mobility.[106]

14.6.4 Agricultural and industrial waste

Agrowaste materials are potential source of production of organic fertilizers with high efficiency as evident in the studies carried out in intensive horticultural system where yield and nutrient status of tomato crop was enhanced with their application. Carricondo-Martínez et al.[107] and Phibunwatthanawong et al.[108] formulated six liquid organic fertilizer from different ratios of agroindustry organic residues, namely, molasses, and leaves of sugarcane by the process of fermentation. The efficacy of these formulas was proved based on successful germination of lettuce seeds indicating the nontoxic character of the prepared formulas thus making it suitable for distribution commercially. The study found that fertilizer contains nitrogen fixing, phosphate and potassium solubilizing bacteria that assists in plant growth thus improving the product quality.

Carricondo-Martínez et al.[107] explored the activity of agrowaste-derived compost and vermicompost on nutrient status and yield of tomato and found that these can be used as an alternative to mineral fertilzers.

Recently for recycling agroindustrial waste a low-cost composting technology using black soldier fly larvae has been reported by Beesigamukama et al.[109] for conversion into nutrient-rich organic fertilizer. The experimental substrate for rearing BSF larvae were brewer's spent grains and sawdust, which were combined to produce different ratios of substrates. The maturity of the compost using BSF was attained at 5 weeks, which is very short duration compared to the conventional composting method.

A nutrient-rich organic fertilizer was manufactured utilizing animal and plant based solid waste, which was found to be efficient on the growth of *Catharanthus roseus*. Majee et al.[110,111] developed cost-effective, ecofriendly organic fertilizers from agricultural and industrial waste enriched with nitrogen, potassium and phosphorous. They also tested the efficiency of the fertilizer on *Tagetes* plant, which showed increase in growth in them. Wet blue leather, a waste from leather industry, contains heavy metal, removal of which yields a collagen material rich in nitrogen, which is used in organic fertilizer formulation.[112] The collagen obtained from wet blue leather when enriched with phosphorous and potassium results in formulation of nutrient-rich fertilizer and was tested on rice plants. The obtained results after application were better than the commercial N-P-K based fertilizers. The chromium levels in the rice grains, leaves and soil were within limits.[112] The chromium removal method is well documented and patented.[113]

14.7 What are biostimulants?

These are substances or microorganisms that are used on plants to broaden the horizon of their stress tolerance capacity with nutrient uptake efficiency (Fig. 14.1).[114] It was defined as materials that enhance growth in plants when applied in small amounts. Kauffman et al.[115] and Kumari et al.[116] defined biostimulants as substance(s) or microorganisms that stimulates nutrient uptake processes and crop quality They are not categorized under pesticides.[116] Similar definition of plant biostimulant was given by European union as a product that improves overall plant health conditions. Li et al.[117] defined biostimulants as a bioactive substance that improves nutrition, tolerance and nutrients availability in the soil. There are alternative terminologies used for biostumalants, namely, agricultural biostimulants, plant strengtheners,[118] organic biostimulants,[119] biogenic stimulators,[120] and phytostimulators.[121]

14.8 Categories of biostimulants

Depending on some parameters like the source of material, type of product, mode of action many authors have classified biostimulants which has been reviewed recently by Kumari et al.[116] Based on origin, biostimulants are broadly categorized as biological and nonbiological (physical and chemical factors)[122]; microbial (fungi, bacteria, microalgae) and nonmicrobial (plants, biopolymers, humic substances etc).[123,124] Nonmicrobial biostimulants were further divided into phytohormonal and nonphytohormonal (protein-based).[125]

14.9 Mass production of biostimulants from various sources

The growing popularity in the production and use of biostimulants in global context is due to the roles they play in promoting sustainable agriculture by enhancing performance of crops, nutrient uptake capacity, resistance toward stress thereby minimizing the dependence on synthetic fertilizers. In 2012 Europe was the biggest biostimulant market.[126] Recently, in the year 2022, a greater section of global market was occupied by European biostimulants market reaching \$1.5–2 billion.[127]

The various organic sources from which plant biostimulants are produced are discussed in the further sections.

14.9.1 Seaweed extracts

Seaweeds are large algae inhabiting benthic region in oceans.[128] They have been used as fertilizer[129] as they secrete plant growth hormones, proteins, minerals, antioxidants, vitamins, fatty acids, polysaccharides and pigments.[130] The species of brown alga which are used to produce commercial extracts as biofertilizers are *Laminaria*, *Sargassum*, *Ascophyllum*, *Fucus* spp., *Padina*, *Turbinaria*, *Ralfsia*[131–133] and among these *Ascophyllum nodosum* is the most popular.[134] Liquid extracts of seaweeds are reported to improve structure of soil and aeration in soil.[135] 33% is contributed by seaweed extracts to biostimulant market.[136] Among the various extraction technologies utilized to manufacture seaweed extracts, the green technology, namely, enzyme-assisted extraction provides better outcome although the limitation regarding the consistency in extraction needs improvement. As exploitation of macroalgae by industries is limited to abundant or invasive species, availability of biomass remains a challenge for the industries for mass production of seaweed extracts. Moreover, based on seaweed extracts market survey it was found that the high price of the products makes them less popular among farmers.[137] In this regard, adoption of bio-refinery concept in exploitation of macroalgae and manufacture of useful products as a suitable developed technology has been discussed by some authors. Furthermore, these authors also suggested for developing techniques for cultivation of macroalga in coastal regions itself to solve the issue of availability of biomass for mass production of seaweed extaracts biostimulants.[137]

14.9.2 Protein hydrolysates

These are combination of peptides and amino acids procured from either animal or plant protein sources through partial hydrolysis.[12] Commercially, in market PH biostimulants are found as powder, granules and liquid[138] and have various names worldwide like Macro Sorb foliar, Aminoplant or Siapton and ILSATOP.[12] Plant-derived protein sources like alfalfa residues, algal proteins, carob-germ protein and animal-derived sources such as epithelial tissues, elastin, animal tissue or fish waste are chemically, thermally or enzymatically hydrolyzed to manufacture protein hydrolysates. Sources of plant proteins are mostly treated enzymatically whereas animal sources are treated chemically or thermally.[12,138,139] However, for animal proteins like collagen and keratins enzymatic and fermentative processes has been suggested for commercial PH production as these sources shows partial insolubility and requires extra pretreatments.[140] Wastes from food processing industries are composed of high proteins content which are a potent source for development of PH biostimulant. It has been mentioned that the production process of PH biostimulant has not been updated for years.[141] Use of bioinformatics in screening of desirable substrates for production of PH is discussed by Moreno-Hernández et al.[142] To date, the enzymatic processing of protein sources is considered the promising approach for protein hydrolysates formation.

14.9.3 Microalgal and cyanobacterial biostimulants

Interest in use of microalgae and cyanobacteria as potent renewable resource to develop biostimulants is rising in agroindustries and farmers due to their nutrient composition that helps in promoting growth in crop plants.[12] Moreover, they produce large biomass for development of biostimulants at large-scale. Research on their activities as biostimulants however is conducted only in limited genera.[143] The microalgal species that are of global interest in context of commercial production are *Arthrospira* spp., *Aphanizomenon* spp., and others.[144] Two systems have been adopted in industries for production of large quantities of microalgal biomass, namely, raceway ponds, an open system and photobioreactor, a closed system to procure products which finds application in different sectors.[145] As these processes involves high-cost during cultivation of microalgae, an alternative cost and ecofriendly way has been suggested which is coproduction of microalgae with tomato using hydroponic system[146]; production of microalgae using nutrient-rich wastewater.[147] To develop microalgal or cyanobacterial based biostimulants, it is necessary to cut the cost of production by implementing other approaches, namely, biorefinery approach such as utilizing wastewater treatments for cultivation, by selecting thermotolerant strains, using Co2 from waste streams.[148] Along with wastewater, anerobic digestate also serves the role of nutrient source for microalgal biomass production, however, studies on its efficiency as biostimulants is limited and needs further research.[149] To grow microalgae, commonly utilized streams are domestic wastewater, piggery wastewater, agricultural runoff, industrial wastewater from varied industries[150] although there is paucity of literature in this area.[149] However, Hans et al.[151] revealed the efficiency of microalgal consortium mediated formulation of biostimulants. It has been suggested that the certain assessment tools must be expanded to scale-up the production of biostimulants in an ecofriendly way.[149]

It is suggested that instead of utilizing whole biomass, extraction of cell ensures higher recovery of bioactive compounds.[151] The cell-walls of microalgae contains various active compounds which can be released by applying high pressure break the cell wall; enzymatic hydrolysis of same[152] and by other processes also, namely, cryo-processing and drying, use of high-pressure piston homogenizer, supercritical fluid extraction as reviewed by Ronga et al.[153] Presence of various substances in microalgal extracts, namely, auxins, cytokinins, vitamins, polysaccharides betaines, amino acids, and polyamines has attracted industries to explore their biostimulant properties.[153] There is paucity of literature based on microalgal and cyanobacterial biostimulants which indicates that this sector in developing commercial biostimulants is in its initial stage in market.[154] Among other microalgal and cyanobacterial species, microalgae species, namely, *Chlorella* spp. and cyanobacterium species, namely, *Arthrospira platensis* are exploited in the global market to obtain products and for exporing biostimulants activities.[155] Only a small percentage in the biostimulant market is contributed by cyanobacterial based products specially from *Arthrospira* spp. with Spain being the highest producer of biostimulants based on cyanobacteria.[137] The microalgal species from where the registered products have been developed remains unspecified for most of the products as reviewed by Santini et al.[154] The algal biostimulants are marketed under the category of organic fertilizer as these are new products that has not been recognized as a separate category in Asia-Pacific countries.[149]

14.9.4 Humic substances

These are diverse organic molecules obtained from the soil microbes' metabolites and decomposed organic matter that makes the soil conditions nutritive for better growth of plants.[156,157] Humic like substances vary in their composition with diverse roles based on molecular weight as they are obtained from diverse sources like plant tissues, biowaste etc.[158,159] They have biostimulating effect on soil and plant activities.[116] These substances based on their molecular weight and solubility are classified as fulvic acids, humic acids and humins. Humic acid can be extracted easily from acidic media as they gets precipitated, while fulvic acid cannot be extracted easily from acidic and alkaline media as it is soluble in either case.[160,161] The biostimulant market is dominated by humic substances and extracts of seaweed extracts sharing more than half of the world market.

14.9.5 Chitin and chitosan derivatives

Chitin is a biopolymer and it is highly biodegradable.[162,163] In agriculture, interest in use of chitin as a biostimulant can be credited to its antimicrobial, antioxidant activities and other bioactivities.[164] Derivatives of chitin, namely, chitosan, are formed by chemical and enzymatic deacetylation and oligochitosan is produced by chemical and enzymatic hydrolysis of chitin.[163] The role of chitin and its derivatives as plant biostimulants is studied.[165] Source of chitin are exoskeleton crustaceans like shrimp, lobster, and crabs; fungal and algal cell wall; mollusks. Insects with year-round breeding abilities are good source of chitosan production.[41,165] For production of chitosan, either chemical or biological method is utilized and the steps involved in recovery of chitin is demineralization, deproteination, elimination of pigments and lipids, that is, decolorization followed by deacetylation.[166–168] The biological method is preferred over chemical as the former is a green approach and involves two ways, one is enzymatic process and the other is fermentation.[169] The enzymatic process is similar to the chemical method except enzymes being used in the former and acid is used in the later method alongwith some modification in enzymatic method.[170] Fermentation method utilizes microorganism in the recovery of chitosan.[171] Chitosan has limitations in terms of their solubility and in most cases, acids have shown to be the best solvents which reduces their chemical and mechanical strength. To improve this shortcoming, and widening its potential applications, chitosans are modified using techniques that generates chitosan derivatives using substitution reactions and depolymerization.[172] Derivatives of chitosan produced by such modifications finds applications in various sectors including agrotechnology.[173] Utilizing chitosan-based nanomaterials alongwith plant biostimulant properties and other bioactivities has been reviewed by Stasińska-Jakubas et al.[164] However, the authors suggested that for the large-scale application, these chitosan-based nanomaterials as biostimulants must be evaluated for their toxicity.

14.9.6 Microbial biostimulants

Microbes provides beneficial ecological services that enhances agricultural productivity.[174,175] Minimizing chemical fertilizers' use, to promote plant, soil health and fertility, attention is drawn toward beneficial soil microbes.[176] Potentiality of several species of fungi and bacteria having biostimulating properties have been explored of which the most reported groups are *Rhizobium* and rhizospheric PGPR.[176] There are 20 commercial PGPR-derived microbial plant biostimulants available Another group of microorganisms with biostimulating properties are arbuscular mycorrhizal fungi (AMF), the symbiotic partners of more than 80% of terrestrial species.[177] Commercial inocula derived from AMF belongs mostly to genera *Funneliformis* and *Rhizoglomus*.[178,179] Formulation of potential plant biostimulant by coinoculating AMF's with PGPR's and the benefits of this mixture has been reviewed by Adeleke et al.[180]. Furthermore, different formulations have been explored for their potential biostimulating effects like combinations containing microorganisms and biologically active matrices such as composts, extracts of compost, algal extracts, agrofood or industrial by-products.[181] Microbes of bacterial origin also supports nitrogen cycling, enhances soil nutrient and absorption of plant nutrient.[182] Documentation of plant growth promoting microbes of fungal origins are few such as mycorrhizal fungi, namely, *Funneliformis*, *Gigaspora* (Glomus).[183,184]

A variety of microorganisms such as *Trichoderma* spp. are successful as effective biostimulants apart from their biopesticidal activities.[185]

Rhizosphere engineering, a new approach which involves recovery of microbial groups associated with soil fertility by adding microbial inoculations to replenish the natural microbiome.[186] This treatment involves potential microbes, culturing the microbes followed by selection of the different components, compatibility assessment, analysis, developing techniques for manufacture and distribution, providing support to users.[187] Such treatments enhance plant growth by by activating physiological and environmental processes.[11,188]

14.10 Risks and sustainable solutions of nanoparticle accumulation in cereal crops

The unplanned utilization of both organic and inorganic nanomaterials in agriculture can display severe ill-effects on physiology of crop species based on their path of entry and dose.[189] The risks occur at gene and cell levels such as disruption of biochemistry, morphology and physiology of crops generating ROS, thereby causing cellular death.[189] However, the effects of various NPs depend also on the crop species.[190] Table 14.1 shows the negative impact of NPs accumulation in cereal crops. Once NPs enter plant body, they interact with all the environmental factors and their agglomeration in different parts or translocation can

Table 14.1 Toxicity of nanoparticles in cereal crops.

S. no.	Type of nanoparticles	Target crop species	Toxic effect	References
1	TiO_2 and ZnO NPs	*Triticum aestivum*	Biomass reduction	192
2	TiO_2 NPs	*Triticum aestivum*	Phytotoxic	193
3	TiO_2 NPs	*Triticum aestivum*	Phytotoxic	194
4	AZ NPs	*Triticum aestivum*	Genotoxic	195
5	Ag NPs	*Triticum aestivum*	Genotoxic	196
6	AgNPs	*Triticum aestivum*	Phytotoxic; genotoxic	197
7	CuO NPs	*Triticum aestivum*	Reduction in biomass	198
8	Cu NPs	*Triticum aestivum*	Reduction in growth rate	199
9	TiO_2 NPs	*Zea mays*	genotoxicity	200
10	ZnO NPs	*Zea mays*	Reduction in length of root and shoot	201
11	Ag NPs	*Zea mays*	Reduction in germination	202
12	ZnO NPs	*Zea mays*	Decrease in germination of seeds	203
13	Zn NPs	*Zea mays*	Reduction in root growth	204
14	Ag NPs	*Barley*	Reduction in length of root and shoot	205
15	ZnO NPs	*Lolium perenne*	Reduction in biomass	206
16	CuO NPs	*Lolium perenne*	DNA damage	207
17	ZnO NPs	*Lolium perenne*	Phytotoxic	206
18	CuO NPs	*Lolium perenne*	Reduction in root growth	208
19	CuO NPs	*Oryzasativa*	Reduced root growth by altering genes	207
20	CuO NPs	*Oryzasativa*	Inhibition of seedlings growth	209
21	CeO_2 NPs	*Oryzasativa*	Phytotoxic	210
22	TiO_2 NPs	*Oryza sativa*	Reduction in germination	211
23	ZnO NPs	*Oryza sativa*	Reduction into shoot and root elongation.	212
24	Ag NPs	*Oryza sativa*	Root growth inhibition	213
25	Ag NPs	*Oryza sativa*	Phytotoxic	214
26	Ag NPs	*Oryza sativa*	Phytotoxic	215
27	TiO_2 NPs	*Oryza sativa*	Phytotoxic	198

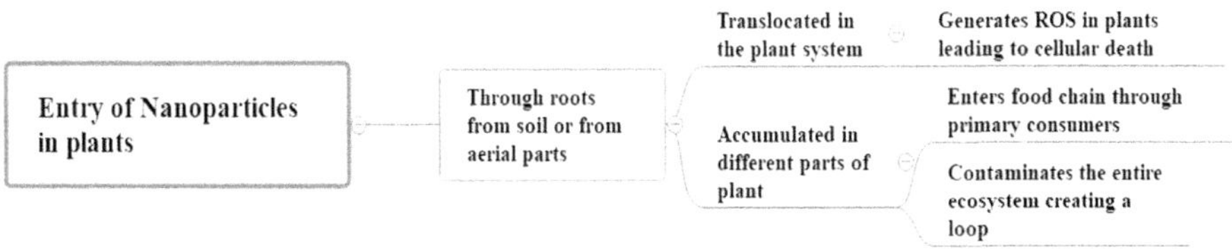

FIGURE 14.2

Path and outcome of nanoparticles in plant system.

cause harm (if present in higher concentration) to the host or it will enter and contaminate the food chain (Fig. 14.2).[191]

To overcome these challenges, new sustainable ways must be explored, like, adoption of biopolymer nanoparticles in agriculture to neutralize the effect of their association without disrupting the plant–soil health conditions.[216] Specific microbes or plants can be explored for nullifying the toxic effects of accumulated NPs. The NPs must be assessed thoroughly for any possible risks before delivering for use and these evaluation technologies are yet to develop.[217] Moreover, the nanoparticles must be manufactured in such a way that it exerts lesser effects on ecosystem.[218]

14.11 Conclusion

To meet the demands of growing population by ensuring food security is greatly influenced using promising tools such as biofertilizers, organic fertilizers, and biostimulants. These agricultural products developed through green technology play an important role in supplying nutrients to the plants and promote growth in plants, thereby enhancing yield of crops and improving soil fertility. Dependence on use of chemical fertilizers over decades has exerted negative impact on human health and has disturbed the ecological balance. Moreover, the high cost involved in the use of synthetic fertilizers is not affordable to farmers. Exploration of beneficial microorganisms inhabiting rhizosphere, plant tissues aid in development of novel agricultural inputs providing solutions to issues of agroecosystem and environment. Recycling of organic waste generated from agricultural and industrial sector by valorizing the wastes into beneficial agri-products has shown successful results in replacing synthetic products by producing bio-based products. Furthermore, these products developed from microbes, recycling waste are cost-effective, nontoxic, and ecofriendly facilitating production of bio-based agri-inputs in large scale. Although there are limitations associated in these processes, advancement can be made by exploring ways for developing preparations with broad spectrum of functions and by developing innovative technologies.

References

1. Batista, B. D., et al. Screening of tropically derived, multi-trait plant growth-promoting rhizobacteria and evaluation of corn and soybean colonization ability. *Microbiological Res.* **2018**, *206*, 33–42.
2. Egamberdieva, D.; Shrivastava, S.; Varma, A. *Plant-growth-promoting rhizobacteria (PGPR) and medicinal plants*, Vol. 42. Springer, 2015.
3. Youssef, M.; Eissa, M. Biofertilizers and their role in management of plant parasitic nematodes. A review. *J. Biotechnol. Pharm. Res.* **2014**, *5* (1), 1–6.
4. Kumar, M. S.; Reddy, G. C.; Phogat, M.; Korav, S. Role of bio-fertilizers towards sustainable agricultural development: a review. *J. Pharmacognosy Phytochemistry* **2018**, *7* (6), 1915–1921.
5. Fasusi, O. A.; Cruz, C.; Babalola, O. O. Agricultural sustainability: microbial biofertilizers in rhizosphere management. *Agriculture* **2021**, *11* (2), 163.
6. Zhang, J., et al. Harnessing the plant microbiome to promote the growth of agricultural crops. *Microbiological Res.* **2021**, *245*, 126690.
7. Raja, N. Biopesticides and biofertilizers: ecofriendly sources for sustainable agriculture. *J. Biofertil. Biopestici* **2013**, *4* (1), 1–2.
8. Daniel, A. I., et al. Biofertilizer: the future of food security and food safety. *Microorganisms* **2022**, *10* (6), 1220.
9. Drobek, M.; Frąc, M.; Cybulska, J. Plant biostimulants: importance of the quality and yield of horticultural crops and the improvement of plant tolerance to abiotic stress—a review. *Agronomy* **2019**, *9* (6), 335.
10. González-Pérez, B. K.; Rivas-Castillo, A. M.; Valdez-Calderón, A.; Gayosso-Morales, M. A. Microalgae as biostimulants: a new approach in agriculture. *World J. Microbiology Biotechnol.* **2022**, *38* (1), 4.
11. Van Oosten, M. J., et al. Root inoculation with *Azotobacter chroococcum* 76A enhances tomato plants adaptation to salt stress under low N conditions. *BMC plant. Biol.* **2018**, *18*, 1–12.
12. Calvo, P.; Nelson, L.; Kloepper, J. W. Agricultural uses of plant biostimulants. *Plant. soil.* **2014**, *383*, 3–41.
13. Hoornweg, D.; Bhada-Tata, P.; Kennedy, C. Environment: waste production must peak this century. *Nature* **2013**, *502* (7473), 615–617.
14. Stiles, W. A., et al. Using microalgae in the circular economy to valorise anaerobic digestate: challenges and opportunities. *Bioresour. Technol.* **2018**, *267*, 732–742.
15. Gupta, P. K. *Handbook of Soil, Fertilizer and Manure*. 2003: Agrobios (India).
16. Assefa, S.; Tadesse, S. The principal role of organic fertilizer on soil properties and agricultural productivity-a review. *Agri Res. Tech: Open. Access. J.* **2019**, *22* (2), 556192.
17. Sneha, S., et al. *Biofertilizer for Crop Production and Soil Fertility*, 2018.
18. Bhattacharjee, R.; Dey, U. Biofertilizer, a way towards organic agriculture: a review. *Afr. J. Microbiology Res.* **2014**, *8* (24), 2332–2343.
19. Stewart, W.; Roberts, T. Food security and the role of fertilizer in supporting it. *Procedia Eng.* **2012**, *46*, 76–82.
20. Nosheen, S.; Ajmal, I.; Song, Y. Microbes as biofertilizers, a potential approach for sustainable crop production. *Sustainability* **2021**, *13* (4), 1868.
21. Bumandalai, O.; Tserennadmid, R. Effect of *Chlorella vulgaris* as a biofertilizer on germination of tomato and cucumber seeds. *Int. J. Aquat. Biol.* **2019**, *7* (2), 95–99.

22. Mohanty, P., et al. Insight into the role of PGPR in sustainable agriculture and environment. *Front. Sustain. Food Syst.* **2021**, *5*, 667150.
23. Glick, B. R.; Gamalero, E. Recent developments in the study of plant microbiomes. *Microorganisms* **2021**, *9* (7), 1533.
24. Santoyo, G., et al. Plant growth stimulation by microbial consortia. *Agronomy* **2021**, *11* (2), 219.
25. Vosátka, M.; Albrechtová, J.; Patten, R. *The international market development for mycorrhizal technology. Mycorrhiza: state of the art, genetics and molecular biology, eco-function, biotechnology, eco-physiology, structure and systematics;* Springer, 2008419−438.
26. Kumawat, K.; Nagpal, S.; Sharma, P. *Present scenario of bio-fertilizer production and marketing around the globe. Biofertilizers;* Elsevier, 2021389−413.
27. Thomas, L.; Singh, I. Microbial biofertilizers: types and applications. *Biofertil. Sustain. agriculture Environ.* **2019**, 1−19.
28. Suyal, D. C.; Soni, R.; Sai, S.; Goel, R. Microbial inoculants as biofertilizer. *Microb. Inoculants Sustain. Agric. Productivity: Vol. 1: Res. Perspect.* **2016**, 311−318.
29. Maan, P. K.; Garcha, S. *Production technology, properties, and quality management. in* Biofertilizers*;* Elsevier, 202131−43.
30. Kecskés, M., et al. Effects of bacterial inoculant biofertilizers on growth, yield and nutrition of rice in Australia. *J. Plant. Nutr.* **2016**, *39* (3), 377−388.
31. Tabassam, T., et al. Suitability of different formulated carriers for sustaining microbial shelf life. *Pak. J. Agric. Res.* **2015**, *28* (2).
32. Raimi, A.; Roopnarain, A.; Adeleke, R. Biofertilizer production in Africa: current status, factors impeding adoption and strategies for success. *Sci. Afr.* **2021**, *11*, e00694.
33. Malusá, E.; Sas-Paszt, L.; Ciesielska, J. Technologies for beneficial microorganisms inocula used as biofertilizers. *Sci. World J.* **2012**, *2012*(1), 491206.
34. Duhan, J. S., et al. Nanotechnology: the new perspective in precision agriculture. *Biotechnol. Rep.* **2017**, *15*, 11−23.
35. Deaker, R., et al., *Practical methods for the quality control of inoculant biofertilisers*, 2011.
36. Abd El-Fattah, D. A.; Eweda, W. E.; Zayed, M. S.; Hassanein, M. K. Effect of carrier materials, sterilization method, and storage temperature on survival and biological activities of *Azotobacter chroococcum* inoculant. *Ann. Agric. Sci.* **2013**, *58* (2), 111−118.
37. Catroux, G.; Hartmann, A.; Revellin, C. Trends in rhizobial inoculant production and use. *Plant. soil.* **2001**, *230* (1), 21−30.
38. Herrmann, L.; Lesueur, D. Challenges of formulation and quality of biofertilizers for successful inoculation. *Appl. Microbiology Biotechnol.* **2013**, *97* (20), 8859−8873.
39. Gupta, R.; Kalia, A.; Kapoor, S. *Bioinoculants: A Step Towards Sustainable Agriculture;* New India Publishing, 2007.
40. Jara-Samaniego, J., et al. Development of organic fertilizers from food market waste and urban gardening by composting in Ecuador. *PLoS One* **2017**, *12* (7), e0181621.
41. Ma, J.; Faqir, Y.; Tan, C.; Khaliq, G. Terrestrial insects as a promising source of chitosan and recent developments in its application for various industries. *Food Chem.* **2022**, *373*, 131407.
42. Verrillo, M., et al. Bioactivity and antimicrobial properties of chemically characterized compost teas from different green composts. *Waste Manag.* **2021**, *120*, 98−107.

43. Yatoo, A. M.; Ali, M. N.; Baba, Z. A.; Hassan, B. Sustainable management of diseases and pests in crops by vermicompost and vermicompost tea. A review. *Agron. Sustain. Dev.* **2021**, *41* (1), 7.
44. Mahaly, M., et al. Vermicomposting of distillery sludge waste with tea leaf residues. *Sustain. Environ. Res.* **2018**, *28* (5), 223–227.
45. Walling, E.; Trémier, A.; Vaneeckhaute, C. A review of mathematical models for composting. *Waste Manag.* **2020**, *113*, 379–394.
46. Ramprasad, C.; Alekhya, D. Design and optimisation of high rate decentralised vermicomposting reactor for the household organic waste. *Int. J. Environ. waste Manag.* **2021**, *27* (4), 420–439.
47. Ghorbani, M.; Sabour, M. R. Global trends and characteristics of vermicompost research over the past 24 years. *Environ. Sci. Pollut. Res.* **2021**, *28*, 94–102.
48. Moreno, J., et al. Revisiting the succession of microbial populations throughout composting: a matter of thermotolerance. *Sci. Total. Environ.* **2021**, *773*, 145587.
49. Wang, S.; Wu, Y. Hyperthermophilic composting technology for organic solid waste treatment: recent research advances and trends. *Processes* **2021**, *9* (4), 675.
50. Liao, H., et al. Development of hperthermophinic aerobic composting and its engineering applications in organic solid wastes. *J. Fujian Agric. For. Univ.(Nat. Sci. Ed.)* **2017**, *46* (04), 439–444.
51. Camposeco-Negrete, C. Optimization of cutting parameters for minimizing energy consumption in turning of AISI 6061 T6 using Taguchi methodology and ANOVA. *J. Clean. Prod.* **2013**, *53*, 195–203.
52. Norouzi, O.; Dutta, A. The current status and future potential of biogas production from Canada's organic fraction municipal solid waste. *Energies* **2022**, *15* (2), 475.
53. Thanigaivel, S., et al. Exploration of effective biorefinery approach to obtain the commercial value-added products from algae. *Sustain. Energy Technol. Assess.* **2022**, *53*, 102450.
54. Nkoa, R. Agricultural benefits and environmental risks of soil fertilization with anaerobic digestates: a review. *Agron. Sustain. Dev.* **2014**, *34*, 473–492.
55. Chen, Y.; Cheng, J. J.; Creamer, K. S. Inhibition of anaerobic digestion process: a review. *Bioresour. Technol.* **2008**, *99* (10), 4044–4064.
56. Nkoa, R.; Coulombe, J.; Desjardins, Y.; Tremblay, N. Towards optimization of growth via nutrient supply phasing: nitrogen supply phasing increases broccoli (*Brassica oleracea* var. italica) growth and yield. *J. Exp. botany* **2001**, *52* (357), 821–827.
57. Loria, E. R., et al. Use of anaerobically digested swine manure as a nitrogen source in corn production. *Agron. J.* **2007**, *99* (4), 1119–1129.
58. Collins, H. P.; Kimura, E.; Frear, C. S.; Kruger, C. E. Phosphorus uptake by potato from fertilizers recovered from anaerobic digestion. *Agron. J.* **2016**, *108* (5), 2036–2049.
59. Möller, K.; Müller, T. Effects of anaerobic digestion on digestate nutrient availability and crop growth: a review. *Eng. life Sci.* **2012**, *12* (3), 242–257.
60. Bhunia, S., et al. Application of recycled slaughterhouse wastes as an organic fertilizer for successive cultivations of bell pepper and amaranth. *Sci. Horticulturae* **2021**, *280*, 109927.
61. Yadav, A., et al. Vacuum pyrolysed biochar for soil amendment. *Resource-Efficient Technol.* **2016**, *2*, S177–S185.

62. Zhang, P.; Zhang, X.; Li, Y.; Han, L. Influence of pyrolysis temperature on chemical speciation, leaching ability, and environmental risk of heavy metals in biochar derived from cow manure. *Bioresour. Technol.* **2020**, *302*, 122850.
63. Nanda, S.; Dalai, A. K.; Berruti, F.; Kozinski, J. A. Biochar as an exceptional bioresource for energy, agronomy, carbon sequestration, activated carbon and specialty materials. *Waste Biomass Valoriz.* **2016**, *7*, 201–235.
64. Shinogi, Y.; Kanri, Y. Pyrolysis of plant, animal and human waste: physical and chemical characterization of the pyrolytic products. *Bioresour. Technol.* **2003**, *90* (3), 241–247.
65. Bhunia, S.; Bhowmik, A.; Mallick, R.; Mukherjee, J. Agronomic efficiency of animal-derived organic fertilizers and their effects on biology and fertility of soil: a review. *Agronomy* **2021**, *11* (5), 823.
66. Bhunia, S.; Bhowmik, A.; Mukherjee, J. *Use of rural slaughterhouse wastes (SHWs) as fertilizer in agriculture: a review,. International Conference on Energy Management for Green Environment (UEMGREEN)*, 2019. IEEE, 2019.
67. Roy, M., et al. Organic cultivation of tomato in India with recycled slaughterhouse wastes: evaluation of fertilizer and fruit safety. *Agriculture* **2015**, *5* (3), 826–856.
68. Huzir, N., et al. *Rapid production of organic fertilizer using subcritical water treatment on waste biomass. IOP Conference Series: Earth and Environmental Science;* IOP Publishing, 2023.
69. Yu, S., et al. Biowaste to treasure: application of microbial-aged hydrochar in rice paddy could improve nitrogen use efficiency and rice grain free amino acids. *J. Clean. Prod.* **2019**, *240*, 118180.
70. Fei, Y.-h, et al. Feasibility of sewage sludge derived hydrochars for agricultural application: nutrients (N, P, K) and potentially toxic elements (Zn, Cu, Pb, Ni, Cd). *Chemosphere* **2019**, *236*, 124841.
71. de Jager, M.; Röhrdanz, M.; Giani, L. The influence of hydrochar from biogas digestate on soil improvement and plant growth aspects. *Biochar* **2020**, *2* (2), 177–194.
72. Kumar, A.; Saini, K.; Bhaskar, T. Hydochar and biochar: production, physicochemical properties and techno-economic analysis. *Bioresour. Technol.* **2020**, *310*, 123442.
73. Mau, V.; Arye, G.; Gross, A. Poultry litter hydrochar as an amendment for sandy soils. *J. Environ. Manag.* **2020**, *271*, 110959.
74. Hou, P., et al. Win-win: application of sawdust-derived hydrochar in low fertility soil improves rice yield and reduces greenhouse gas emissions from agricultural ecosystems. *Sci. Total. Environ.* **2020**, *748*, 142457.
75. Wagner, A.; Kaupenjohann, M. Suitability of biochars (pyro-and hydrochars) for metal immobilization on former sewage-field soils. *Eur. J. Soil. Sci.* **2014**, *65* (1), 139–148.
76. Rillig, M. C., et al. Material derived from hydrothermal carbonization: effects on plant growth and arbuscular mycorrhiza. *Appl. Soil. Ecol.* **2010**, *45* (3), 238–242.
77. Gajić, A.; Koch, H. J. Sugar beet (*Beta vulgaris* L.) growth reduction caused by hydrochar is related to nitrogen supply. *J. Environ. Qual.* **2012**, *41* (4), 1067–1075.
78. Khosravi, A., et al. Production and characterization of hydrochars and their application in soil improvement and environmental remediation. *Chem. Eng. J.* **2022**, *430*, 133142.
79. Sudharmaidevi, C.; Thampatti, K.; Saifudeen, N. Rapid production of organic fertilizer from degradable waste by thermochemical processing. *Int. J. Recycling Org. Waste Agriculture* **2017**, *6*, 1–11.

80. Fernández-Delgado, M., et al. Liquid fertilizer production from organic waste by conventional and microwave-assisted extraction technologies: Techno-economic and environmental assessment. *Sci. Total. Environ.* **2022**, *806*, 150904.
81. Kiruba N, J. M.; Saeid, A. An insight into microbial inoculants for bioconversion of waste biomass into sustainable "bio-organic" fertilizers: a bibliometric analysis and systematic literature review. *Int. J. Mol. Sci.* **2022**, *23* (21), 13049.
82. Heidarzadeh, M. H.; Amani, H.; Javadian, B. Improving municipal solid waste compost process by cycle time reduction through inoculation of *Aspergillus niger*. *J. Environ. Health Sci. Eng.* **2019**, *17*, 295–303.
83. Wang, C., et al. Bacterial dynamics and functions driven by a novel microbial agent to promote kitchen waste composting and reduce environmental burden. *J. Clean. Prod.* **2022**, *337*, 130491.
84. Awasthi, M. K., et al. Evaluation of thermophilic fungal consortium for organic municipal solid waste composting. *Bioresour. Technol.* **2014**, *168*, 214–221.
85. Andriamananjara, A.; Rabeharisoa, L.; Prud'homme, L.; Morel, C. Drivers of plant-availability of phosphorus from thermally conditioned sewage sludge as assessed by isotopic labeling. *Front. Nutr.* **2016**, *3*, 19.
86. Chew, K. W., et al. Food waste compost as an organic nutrient source for the cultivation of *Chlorella vulgaris*. *Bioresour. Technol.* **2018**, *267*, 356–362.
87. Frišták, V.; Pipíška, M.; Soja, G. Pyrolysis treatment of sewage sludge: a promising way to produce phosphorus fertilizer. *J. Clean. Prod.* **2018**, *172*, 1772–1778.
88. Becker, G., et al. Novel approach of phosphate-reclamation as struvite from sewage sludge by utilising hydrothermal carbonization. *J. Environ. Manag.* **2019**, *238*, 119–125.
89. Wang, X., et al. Effect of phosphate amendments on improving the fertilizer efficiency and reducing the mobility of heavy metals during sewage sludge composting. *J. Environ. Manag.* **2019**, *235*, 124–132.
90. Valdez-Perez, M., et al. Cultivation of beans (*Phaseolus vulgaris* L.) in limed or unlimed wastewater sludge, vermicompost or inorganic amended soil. *Sci. Horticulturae* **2011**, *128* (4), 380–387.
91. Mattenberger, H., et al. Sewage sludge ash to phosphorus fertiliser: variables influencing heavy metal removal during thermochemical treatment. *Waste Manag.* **2008**, *28* (12), 2709–2722.
92. Franz, M. Phosphate fertilizer from sewage sludge ash (SSA). *Waste Manag.* **2008**, *28* (10), 1809–1818.
93. Weigand, H., et al. RecoPhos: full-scale fertilizer production from sewage sludge ash. *Waste Manag.* **2013**, *33* (3), 540–544.
94. Petzet, S.; Peplinski, B.; Cornel, P. On wet chemical phosphorus recovery from sewage sludge ash by acidic or alkaline leaching and an optimized combination of both. *Water Res.* **2012**, *46* (12), 3769–3780.
95. Gorazda, K., et al. Fertilisers production from ashes after sewage sludge combustion—a strategy towards sustainable development. *Environ. Res.* **2017**, *154*, 171–180.
96. Meng, X., et al. Improved utilization of phosphorous from sewage sludge (as fertilizer) after treatment by low-temperature combustion. *Waste Manag.* **2018**, *80*, 349–358.
97. Herzel, H.; Krüger, O.; Hermann, L.; Adam, C. Sewage sludge ash—a promising secondary phosphorus source for fertilizer production. *Sci. Total. Environ.* **2016**, *542*, 1136–1143.

98. Gorazda, K.; Tarko, B.; Werle, S.; Wzorek, Z. Sewage sludge as a fuel and raw material for phosphorus recovery: combined process of gasification and P extraction. *Waste Manag.* **2018**, *73*, 404–415.
99. Jiang, Y., et al. Rapid production of organic fertilizer by dynamic high-temperature aerobic fermentation (DHAF) of food waste. *Bioresour. Technol.* **2015**, *197*, 7–14.
100. Nakhshiniev, B., et al. Evaluation of hydrothermal treatment in enhancing rice straw compost stability and maturity. *Bioresour. Technol.* **2014**, *151*, 306–313.
101. Urra, J.; Alkorta, I.; Garbisu, C. Potential benefits and risks for soil health derived from the use of organic amendments in agriculture. *Agronomy* **2019**, *9* (9), 542.
102. Malancha Roy, M. R., et al., *Conversion of rural abattoir wastes to an organic fertilizer and its application in the field cultivation of tomato in India*, 2016.
103. Zwetsloot, M. J.; Lehmann, J.; Solomon, D. Recycling slaughterhouse waste into fertilizer: how do pyrolysis temperature and biomass additions affect phosphorus availability and chemistry? *J. Sci. Food Agriculture* **2015**, *95* (2), 281–288.
104. Jayakumar Pathma, J. P.; Natarajan Sakthivel, N. S. *Microbial diversity of vermicompost bacteria that exhibit useful agricultural traits and waste management potential*, 2012.
105. Tognetti, C.; Laos, F.; Mazzarino, M.; Hernandez, M. Composting vs. vermicomposting: a comparison of end product quality. *Compost. Sci. & Utilization* **2005**, *13* (1), 6–13.
106. Swati, A.; Hait, S. Fate and bioavailability of heavy metals during vermicomposting of various organic wastes—a review. *Process. Saf. Environ. Prot.* **2017**, *109*, 30–45.
107. Carricondo-Martínez, I.; Berti, F.; Salas-Sanjuán, Md. C. Different organic fertilisation systems modify tomato quality: an opportunity for circular fertilisation in intensive horticulture. *Agronomy* **2022**, *12* (1), 174.
108. Phibunwatthanawong, T.; Riddech, N. Liquid organic fertilizer production for growing vegetables under hydroponic condition. *Int. J. Recycling Org. Waste Agriculture* **2019**, *8*, 369–380.
109. Beesigamukama, D., et al. Low-cost technology for recycling agro-industrial waste into nutrient-rich organic fertilizer using black soldier fly. *Waste Manag.* **2021**, *119*, 183–194.
110. Majee, S.; Halder, G.; Mandal, T. Formulating nitrogen-phosphorous-potassium enriched organic manure from solid waste: a novel approach of waste valorization. *Process. Saf. Environ. Prot.* **2019**, *132*, 160–168.
111. Majee, S.; Halder, G.; Krishnaraj, R.; Mandal, T. Development and formulation of an organic fertilizer from industrial and agricultural waste to study the growth of marigold (tagetes) plant. *Int. J. Mathematical, Eng. Manag. Sci.* **2020**, *5* (3), 395.
112. Nogueira, F. G., et al. Recycling of solid waste rich in organic nitrogen from leather industry: mineral nutrition of rice plants. *J. Hazard. Mater.* **2011**, *186* (2-3), 1064–1069.
113. Oliveira, L., Dallago, R.; Nascimento Filho, I. *Process of recycling leather tanning solid residues by chrome extraction and recuperation of the decontaminated.* Patent BR PI, 2004. **1538**.
114. Brown, P.; Saa, S. Biostimulants in agriculture. *Front. plant. Sci.* **2015**, *6*, 671.
115. Kauffman, G. L.; Kneivel, D. P.; Watschke, T. L. Effects of a biostimulant on the heat tolerance associated with photosynthetic capacity, membrane thermostability, and polyphenol production of perennial ryegrass. *Crop. Sci.* **2007**, *47* (1), 261–267.

116. Kumari, M.; Swarupa, P.; Kesari, K. K.; Kumar, A. Microbial inoculants as plant biostimulants: a review on risk status. *Life* **2022**, *13* (1), 12.
117. Li, J., et al. Biopesticide and plant growth-promoting activity in maize distillers' dried grains with solubles. *Ind. Crop. Products* **2023**, *193*, 116175.
118. Torre, L.; Battaglia, V.; Caradonia, F. Legal aspects of the use of plant strengtheners (biostimulants) in Europe. *Bulg. J. Agric. Sci.* **2013**, *19* (6), 1183–1189.
119. Russo, R. O.; Berlyn, G. P. The use of organic biostimulants to help low input sustainable agriculture. *J. Sustain. agriculture* **1991**, *1* (2), 19–42.
120. Filatov, V. Tissue treatment (Doctrine on biogenic stimulators). II. Hypothesis of tissue therapy, or the doctrine on biogenic stimulators. *Priroda* **1951**, *12*, 20–28.
121. Martínez-Viveros, O., et al. Mechanisms and practical considerations involved in plant growth promotion by rhizobacteria. *J. soil. Sci. plant. Nutr.* **2010**, *10* (3), 293–319.
122. Rafiee, H., et al. Application of plant biostimulants as new approach to improve the biological responses of medicinal plants—a critical review. *J. Medicinal Plants* **2016**, *15* (59), 6–39.
123. Rouphael, Y.; Colla, G. *Toward a sustainable agriculture through plant biostimulants: From experimental data to practical applications;* MDPI, 20201461.
124. de Vries, F. T., et al. Harnessing rhizosphere microbiomes for drought-resilient crop production. *Science* **2020**, *368* (6488), 270–274.
125. du Jardin, P.; Xu, L.; Geelen, D. Agricultural functions and action mechanisms of plant biostimulants (PBs) an introduction. *Chem. Biol. plant. biostimulants* **2020**, 1–30.
126. Du Jardin, P. Plant biostimulants: definition, concept, main categories and regulation. *Sci. horticulturae* **2015**, *196*, 3–14.
127. Tolisano, C.; Del Buono, D. Biobased: biostimulants and biogenic nanoparticles enter the scene. *Sci. Total. Environ.* **2023**, *885*, 163912.
128. Kandale, A., et al. Marine algae: an introduction, food value and medicinal uses. *J. Pharm. Res.* **2011**, *4* (1), 219–221.
129. Craigie, J. S. Seaweed extract stimuli in plant science and agriculture. *J. Appl. phycology* **2011**, *23*, 371–393.
130. Godlewska, K.; Michalak, I.; Tuhy, Ł.; Chojnacka, K. Plant growth biostimulants based on different methods of seaweed extraction with water. *BioMed. Res. Int.* **2016**, *2016* (1), 5973760.
131. Mukherjee, A.; Patel, J. Seaweed extract: biostimulator of plant defense and plant productivity. *Int. J. Environ. Sci. Technol.* **2020**, *17* (1), 553–558.
132. Ali, O.; Ramsubhag, A.; Jayaraman, J. Biostimulant properties of seaweed extracts in plants: implications towards sustainable crop production. *Plants* **2021**, *10* (3), 531.
133. Bahuguna, A., et al. *Advance technology for biostimulants in agriculture. New and Future Developments in Microbial Biotechnology and Bioengineering;* Elsevier, 2022393–412.
134. Ugarte, R. A.; Sharp, G.; Moore, B. Changes in the brown seaweed Ascophyllum nodosum (L.) Le Jol. Plant morphology and biomass produced by cutter rake harvests in southern New Brunswick, Canada. In *Eighteenth International Seaweed Symposium: Proceedings of the Eighteenth International Seaweed Symposium*, held in Bergen, Norway, 20–25 June 2004. 2007. Springer.
135. Milton, R. Liquid seaweed as a fertilizer. In *Proc Int Seaweed Symp*. 1964.

136. Eef, B., et al., *Identification of the seaweed biostimulant market (Phase1).* The North Sea Farm Foundation: AD Den Haag, The Netherlands, 2018.
137. El Boukhari, M. E. M.; Barakate, M.; Bouhia, Y.; Lyamlouli, K. Trends in seaweed extract based biostimulants: manufacturing process and beneficial effect on soil-plant systems. *Plants* **2020**, *9* (3), 359.
138. Colla, G., et al. Co-inoculation of *Glomus intraradices*and*Trichoderma atroviride* acts as a biostimulant to promote growth, yield and nutrient uptake of vegetable crops. *J. Sci. Food Agriculture* **2015**, *95* (8), 1706−1715.
139. Ugolini, L., et al. Production of an enzymatic protein hydrolyzate from defatted sunflower seed meal for potential application as a plant biostimulant. *Ind. Crop. Products* **2015**, *75*, 15−23.
140. Lafarga, T.; Rai, D. K.; O'connor, P.; Hayes, M. Generation of bioactive hydrolysates and peptides from bovine hemoglobin with in vitro renin, angiotensin-I-converting enzyme and dipeptidyl peptidase-IV inhibitory activities. *J. Food Biochem.* **2016**, *40* (5), 673−685.
141. McCarthy, A. L.; O'Callaghan, Y. C.; O'Brien, N. M. Protein hydrolysates from agricultural crops—bioactivity and potential for functional food development. *Agriculture* **2013**, *3* (1), 112−130.
142. Moreno-Hernández, J. M., et al. Strategies for production, characterization and application of protein-based biostimulants in agriculture: a review. *Chil. J. Agric. Res.* **2020**, *80* (2), 274−289.
143. Santini, G., et al. Effects of cyanobacterial-based biostimulants on plant growth and development: a case study on basil (*Ocimum basilicum* L.). *J. Appl. Phycology* **2022**, *34* (4), 2063−2073.
144. Priyadarshani, I.; Rath, B. Commercial and industrial applications of micro algae—a review. *J. Algal Biomass Utilization* **2012**, *3* (4), 89−100.
145. Huesemann, M. H.; Benemann, J. R. Biofuels from Microalgae: review of products, processes and potential, with special focus on *Dunaliella* sp. *Alga Dunaliella* **2019**, 445−474.
146. Barone, V., et al. Novel bioprocess for the cultivation of microalgae in hydroponic growing system of tomato plants. *J. Appl. Phycology* **2019**, *31*, 465−470.
147. Acién, F. G., et al. Wastewater treatment using microalgae: how realistic a contribution might it be to significant urban wastewater treatment? *Appl. microbiology Biotechnol.* **2016**, *100*, 9013−9022.
148. Navarro-López, E., et al. Biostimulant potential of Scenedesmus obliquus grown in brewery wastewater. *Molecules* **2020**, *25* (3), 664.
149. Behera, B.; Supraja, K. V.; Paramasivan, B. Integrated microalgal biorefinery for the production and application of biostimulants in circular bioeconomy. *Bioresour. Technol.* **2021**, *339*, 125588.
150. Li, K., et al. Microalgae-based wastewater treatment for nutrients recovery: a review. *Bioresour. Technol.* **2019**, *291*, 121934.
151. Hans, A. L., et al. Microalgal extracts for tomatoes: seed treatment or foliar spray? *Curr. Sci. (00113891)* **2020**, *119* (7).
152. García, J. R.; Fernández, F. A.; Sevilla, J. F. Development of a process for the production of l-amino-acids concentrates from microalgae by enzymatic hydrolysis. *Bioresour. Technol.* **2012**, *112*, 164−170.
153. Ronga, D., et al. Microalgal biostimulants and biofertilisers in crop productions. *Agronomy* **2019**, *9* (4), 192.

154. Santini, G.; Biondi, N.; Rodolfi, L.; Tredici, M. R. Plant biostimulants from cyanobacteria: an emerging strategy to improve yields and sustainability in agriculture. *Plants* **2021**, *10* (4), 643.
155. Prisa, D.; Spagnuolo, D. Plant production with microalgal biostimulants. *Horticulturae* **2023**, *9* (7), 829.
156. De Pascale, S.; Rouphael, Y.; Colla, G. Plant biostimulants: Innovative tool for enhancing plant nutrition in organic farming. *Eur. J. Hortic. Sci.* **2017**, *82* (6), 277–285.
157. Puglisi, I., et al. Biostimulant activity of humic-like substances from agro-industrial waste on Chlorella vulgaris and Scenedesmus quadricauda. *Eur. J. Phycology* **2018**, *53* (3), 433–442.
158. Canellas, L. P., et al. Humic and fulvic acids as biostimulants in horticulture. *Sci. horticulturae* **2015**, *196*, 15–27.
159. Scaglia, B.; Pognani, M.; Adani, F. The anaerobic digestion process capability to produce biostimulant: the case study of the dissolved organic matter (DOM) vs. auxin-like property. *Sci. Total. Environ.* **2017**, *589*, 36–45.
160. Stevenson, F. *Humus Chemistry: Genesis, Composition, Reactions;* John Wiley & Sons, 1994.
161. Berbara, R. L.; García, A. C. *Humic substances and plant defense metabolism. Physiological mechanisms and adaptation strategies in plants under changing environment*, volume 1. Springer, 2013297–319.
162. Dutta, P. K.; Ravikumar, M.; Dutta, J. Chitin and chitosan for versatile applications. *J. Macromol. Science, Part. C: Polym. Rev.* **2002**, *42* (3), 307–354.
163. Prashanth, K. H.; Tharanathan, R. Chitin/chitosan: modifications and their unlimited application potential—an overview. *Trends food Sci. & Technol.* **2007**, *18* (3), 117–131.
164. Stasińska-Jakubas, M.; Hawrylak-Nowak, B. Protective, biostimulating, and eliciting effects of chitosan and its derivatives on crop plants. *Molecules* **2022**, *27* (9), 2801.
165. Shahrajabian, M. H., et al. Sustainable agriculture systems in vegetable production using chitin and chitosan as plant biostimulants. *Biomolecules* **2021**, *11* (6), 819.
166. Koleška, I., et al. Biostimulant prevents yield loss and reduces oxidative damage in tomato plants grown on reduced NPK nutrition. *J. Plant. Interact.* **2017**, *12* (1), 209–218.
167. Philibert, T.; Lee, B. H.; Fabien, N. Current status and new perspectives on chitin and chitosan as functional biopolymers. *Appl. Biochem. Biotechnol.* **2017**, *181*, 1314–1337.
168. Meramo-Hurtado, S. I.; González-Delgado, An. D. Application of techno-economic and sensitivity analyses as decision-making tools for assessing emerging large-scale technologies for production of chitosan-based adsorbents. *ACS omega* **2020**, *5* (28), 17601–17610.
169. Reshad, R. A. I.; Jishan, T. A.; Chowdhury, N. N. *Chitosan and its broad applications: a brief review*. Available at SSRN 3842055, 2021.
170. Yadav, A.; Garg, V. Biotransformation of bakery industry sludge into valuable product using vermicomposting. *Bioresour. Technol.* **2019**, *274*, 512–517.
171. Yang, J.-K.; Shih, L.; Tzeng, Y.-M.; Wang, S.-L. Production and purification of protease from a Bacillus subtilis that can deproteinize crustacean wastes☆. *Enzyme Microb. Technol.* **2000**, *26* (5-6), 406–413.

172. Azmana, M., et al. A review on chitosan and chitosan-based bionanocomposites: promising material for combatting global issues and its applications. *Int. J. Biol. Macromolecules* **2021**, *185*, 832–848.
173. Kowalczyk, D.; Kordowska-Wiater, M.; Nowak, J.; Baraniak, B. Characterization of films based on chitosan lactate and its blends with oxidized starch and gelatin. *Int. J. Biol. Macromolecules* **2015**, *77*, 350–359.
174. Hijri, M. Analysis of a large dataset of mycorrhiza inoculation field trials on potato shows highly significant increases in yield. *Mycorrhiza* **2016**, *26* (3), 209–214.
175. Basiru, S.; Mwanza, H.; Hijri, M. Analysis of arbuscular mycorrhizal fungal inoculant benchmarks. *Microorganisms* **2020**, *9*, 81 s Note: MDPI stays neu-tral with regard to jurisdictional clai-ms in.
176. Philippot, L.; Raaijmakers, J. M.; Lemanceau, P.; Van Der Putten, W. H. Going back to the roots: the microbial ecology of the rhizosphere. *Nat. Rev. microbiology* **2013**, *11* (11), 789–799.
177. Spatafora, J. W., et al. A phylum-level phylogenetic classification of zygomycete fungi based on genome-scale data. *Mycologia* **2016**, *108* (5), 1028–1046.
178. Smith, S. E.; Read, D. The symbionts forming arbuscular mycorrhizas. *Mycorrhizal symbiosis* **2008**, 13–41.
179. Gebrehiwot, H. G., et al. Weed-competitive ability of teff (Eragrostis tef (Zucc.) Trotter) varieties. *Agronomy* **2020**, *10* (1), 108.
180. Adeleke, B. S.; Babalola, O. O.; Glick, B. R. Plant growth-promoting root-colonizing bacterial endophytes. *Rhizosphere* **2021**, *20*, 100433.
181. Yakhin, O. I.; Lubyanov, A. A.; Yakhin, I. A.; Brown, P. H. Biostimulants in plant science: a global perspective. *Front. plant. Sci.* **2017**, *7*, 2049.
182. Berg, G.; Grube, M.; Schloter, M.; Smalla, K. Unraveling the plant microbiome: looking back and future perspectives. *Front. microbiology* **2014**, *5*, 148.
183. Pringle, A., et al. Mycorrhizal symbioses and plant invasions. *Annu. Rev. Ecology, Evolution, Syst.* **2009**, *40* (1), 699–715.
184. Rouphael, Y., et al. Arbuscular mycorrhizal fungi act as biostimulants in horticultural crops. *Sci. Horticulturae* **2015**, *196*, 91–108.
185. Woo, S. L., et al. Trichoderma-based products and their widespread use in agriculture. *Open. Mycology J.* **2014**, *8* (1).
186. Stringlis, I. A., et al. Microbial small molecules—weapons of plant subversion. *Nat. Product. Rep.* **2018**, *35* (5), 410–433.
187. Woo, S. L.; Pepe, O. Microbial consortia: promising probiotics as plant biostimulants for sustainable agriculture. *Front. plant. Sci.* **2018**, *9*, 1801.
188. Viscardi, S., et al. Assessment of plant growth promoting activities and abiotic stress tolerance of Azotobacter chroococcum strains for a potential use in sustainable agriculture. *J. soil. Sci. plant. Nutr.* **2016**, *16* (3), 848–863.
189. Zafar, H.; Javed, R.; Zia, M. Nanotoxicity assessment in plants: an updated overview. *Environ. Sci. Pollut. Res.* **2023**, *30* (41), 93323–93344.
190. García-Gómez, C., et al. Comparative study of the phytotoxicity of ZnO nanoparticles and Zn accumulation in nine crops grown in a calcareous soil and an acidic soil. *Sci. Total. Environ.* **2018**, *644*, 770–780.
191. Rienzie, R.; Adassooriya, N. M. Toxicity of nanomaterials in agriculture and food. *Nanomaterials: Ecotoxicity, safety, public. Percept.* **2018**, 207–234.
192. Du, W., et al. TiO 2 and ZnO nanoparticles negatively affect wheat growth and soil enzyme activities in agricultural soil. *J. Environ. Monit.* **2011**, *13* (4), 822–828.

193. Rafique, R., et al. Dose-dependent physiological responses of Triticum aestivum L. to soil applied TiO2 nanoparticles: alterations in chlorophyll content, H2O2 production, and genotoxicity. *Agriculture, Ecosyst. & Environ.* **2018**, *255*, 95–101.
194. Soran, M.-L., et al. The effect of TiO2 nanoparticles on the composition and ultrastructure of wheat. *Nanomaterials* **2021**, *11* (12), 3413.
195. Abdelsalam, N. R., et al. Genotoxicity assessment of amino zinc nanoparticles in wheat (Triticum aestivum L.) as cytogenetical perspective. *Saudi J. Biol. Sci.* **2022**, *29* (4), 2306–2313.
196. Vannini, C., et al. Phytotoxic and genotoxic effects of silver nanoparticles exposure on germinating wheat seedlings. *J. plant. Physiol.* **2014**, *171* (13), 1142–1148.
197. Feng, L., et al. Synergetic toxicity of silver nanoparticle and glyphosate on wheat (Triticum aestivum L.). *Sci. Total. Environ.* **2021**, *797*, 149200.
198. Dimkpa, C. O., et al. CuO and ZnO nanoparticles: phytotoxicity, metal speciation, and induction of oxidative stress in sand-grown wheat. *J. Nanopart. Res.* **2012**, *14*, 1–15.
199. Lee, W. M.; An, Y. J.; Yoon, H.; Kweon, H. S. Toxicity and bioavailability of copper nanoparticles to the terrestrial plants mung bean (Phaseolus radiatus) and wheat (*Triticum aestivum*): plant agar test for water-insoluble nanoparticles. *Environ. Toxicol. Chemistry: An. Int. J.* **2008**, *27* (9), 1915–1921.
200. Ruffini Castiglione, M.; Giorgetti, L.; Geri, C.; Cremonini, R. The effects of nano-TiO 2 on seed germination, development and mitosis of root tip cells of *Vicia narbonensis* L. and *Zea mays* L. *J. Nanopart. Res.* **2011**, *13*, 2443–2449.
201. Zhao, L., et al. ZnO nanoparticle fate in soil and zinc bioaccumulation in corn plants (*Zea mays*) influenced by alginate. *Environ. Science: Process. & Impacts* **2013**, *15* (1), 260–266.
202. Pokhrel, L. R.; Dubey, B. Evaluation of developmental responses of two crop plants exposed to silver and zinc oxide nanoparticles. *Sci. Total. Environ.* **2013**, *452*, 321–332.
203. Lin, D.; Xing, B. Phytotoxicity of nanoparticles: inhibition of seed germination and root growth. *Environ. Pollut.* **2007**, *150* (2), 243–250.
204. Zhang, R., et al. Phytotoxicity of ZnO nanoparticles and the released Zn (II) ion to corn (*Zea mays* L.) and cucumber (*Cucumis sativus* L.) during germination. *Environ. Sci. Pollut. Res.* **2015**, *22*, 11109–11117.
205. Fayez, K.; El-Deeb, B.; Mostafa, N. Toxicity of biosynthetic silver nanoparticles on the growth, cell ultrastructure and physiological activities of barley plant. *Acta Physiologiae Plant.* **2017**, *39*, 1–13.
206. Lin, D.; Xing, B. Root uptake and phytotoxicity of ZnO nanoparticles. *Environ. Sci. & Technol.* **2008**, *42* (15), 5580–5585.
207. Patlolla, A. K.; Berry, A.; May, L.; Tchounwou, P. B. Genotoxicity of silver nanoparticles in *Vicia faba*: a pilot study on the environmental monitoring of nanoparticles. *Int. J. Environ. Res. public. health* **2012**, *9* (5), 1649–1662.
208. Atha, D. H., et al. Copper oxide nanoparticle mediated DNA damage in terrestrial plant models. *Environ. Sci. & Technol.* **2012**, *46* (3), 1819–1827.
209. Wang, W., et al. Phytotoxicity assessment of copper oxide nanoparticles on the germination, early seedling growth, and physiological responses in *Oryza sativa* L. *Bull. Environ. Contamination Toxicol.* **2020**, *104*, 770–777.
210. Priester, J. H., et al. Soybean susceptibility to manufactured nanomaterials with evidence for food quality and soil fertility interruption. *Proc. Natl Acad. Sci.* **2012**, *109* (37), E2451–E2456.

211. Dh, R.; Jalill, A.; Yousef, A. Comparison the phytotoxicity of TiO2 nanoparticles with bulk particles on Amber 33 variety of rice (*Oryza sativa*) in vitro. *Sch. Acad. J. Biosci.* **2015**, *3* (3), 254–262.
212. Chen, J., et al. Phytotoxicity and bioaccumulation of zinc oxide nanoparticles in rice (*Oryza sativa* L.). *Plant. Physiol. Biochem.* **2018**, *130*, 604–612.
213. Mirzajani, F., et al. Effect of silver nanoparticles on *Oryza sativa* L. and its rhizosphere bacteria. *Ecotoxicol. Environ. Saf.* **2013**, *88*, 48–54.
214. De La Torre-Roche, R., et al. Multiwalled carbon nanotubes and C60 fullerenes differentially impact the accumulation of weathered pesticides in four agricultural plants. *Environ. Sci. & Technol.* **2013**, *47* (21), 12539–12547.
215. Mazumdar, H.; Ahmed, G. Phytotoxicity effect of silver nanoparticles on *Oryza sativa*. *Int. J. ChemTech Res.* **2011**, *3* (3), 1494–1500.
216. Wang, X.; Xie, H.; Wang, P.; Yin, H. Nanoparticles in plants: uptake, transport and physiological activity in leaf and root. *Materials* **2023**, *16* (8), 3097.
217. Rajput, V., et al. Accumulation of nanoparticles in the soil-plant systems and their effects on human health. *Ann. Agric. Sci.* **2020**, *65* (2), 137–143.
218. Marchiol, L. Nanotechnology in agriculture: new opportunities and perspectives. *N. Vis. Plant. Sci.* **2018**, *9* (4), 161.

SECTION

IV

Safety regulations policies and future concerns of biogenic nanomaterials

VI

CHAPTER

15 Nano-omics base plant genetic engineering: a key role in sustainable agriculture

Sagar Mondal[1], Swati Priya[1], Roshni Jha[1], Jutishna Bora[1], Smita Lata[1], Abeer Hashem[2], Elsayed Fathi Abd Allah[3] and Sumira Malik[1]

[1]*Amity Institute of Biotechnology, Amity University Jharkhand, Ranchi, Jharkhand, India*

[2]*Botany and Microbiology Department, College of Science, King Saud University, Riyadh, Saudi Arabia*

[3]*Plant Production Department, College of Food and Agricultural Sciences, King Saud University, Riyadh, Saudi Arabia*

15.1 Introduction

The combination of nanotechnology with omics signifies the seamless fusion of advanced scientific fields. This integration propels us into a phase of heightened understanding and exact manipulation of biological systems. By delving into the intricate realm of molecules in living beings, nano-omics has the capacity to profoundly revolutionize multiple domains such as health, environmental research, and agriculture. This presents opportunities for inventive solutions and significant discoveries.

Nanoscale science and nanotechnology refer to the study of items at the nanometer (nm) scale, where unique features allow for diverse and innovative applications with nanomaterials (NMs). Dimension of less than 100 nm artificially created can take on different forms, including individual particles, fused particles, or clumps of particles. Additionally, these particles can exhibit various forms such as spheres, nearly spherical shapes, tubular structures, or irregular, nonspherical shapes. Additionally, constructed NMs can have either a homogenous or heterogeneous composition. Nanotechnology has significantly advanced research focused on the efficient utilization of agrochemicals. Scientists primarily focus on delivery systems that regulates the discharge of fertilizers in agricultural application.[1,2] The platforms comprise many nanoscale components, including hydrogels, the membranes, fibers, parts, polymers that degrade, lipids, proteins, and minerals. Typically, a common practice is to utilize an active substance (cargo) that has been contained in a carrier using chemical or physical means. These platforms will enable the regulation of cargo's dissolving profiles, hence limiting premature

Biogenic Nanoparticles. DOI: https://doi.org/10.1016/B978-0-443-33819-9.00019-6

physicochemical deterioration. Advancements in farming techniques, such as genetically modified organisms (GMOs), indoor farming, vertical farming, and hydroponic farming, aim to enhance crop production efficiency and precision. When building particle delivery systems for agricultural applications, factors such as particle adhesion, absorption, and features (such as size, chemistry, and geometry) are carefully considered aiming removal of pollutants.

Sensitive plants must develop complex defensive mechanisms to combat environmental disturbances, especially abiotic stresses that reduce plant productivity. The generation and consumption of nourishment are greatly disproportionate as a result of alterations in plant life and the escalating worldwide populace. In order to strike a harmonious equilibrium between crop yield and the need for food, it is crucial to develop effective strategies for acclimatizing plants to challenging environments. Abiotic stresses, including heavy metals (HMs), drought, as well as salt, may have harmful impacts on plants. However, when these stressors are found in low concentrations and for short durations, they can trigger various signaling cascades and pathways in plants. Consequently, the resilience of plant life to future shocks is enhanced.[3] The limitations of these tactics, which activate the plant's defense mechanism in response to surrounding stress, have heightened the necessity to find a powerful and cost-effective alternative to demonstrate the plant's capacity to endure different stressful conditions.[4] Lately, scientists studying the effects of stress on plant physiology and handling of stress have focused heavily on exposing plants to either naturally or artificially produced chemical agents. In order to do this, they are making efforts to develop plant varieties that demonstrate improved resilience to various environmental challenges.[5] The present methodology, known as nanopriming, employs nanoparticles or NPs, as a primer to improve seed germination, development of seedlings, and stress tolerance. As a result, there is an increase in the productivity of plants and the quality of food.[6] Seed nanopriming provides numerous advantages that are associated with various biochemical processes., physiological, and signaling mechanisms in seeds. One of the tasks is to ensure the equilibrium of reactive oxygen species or ROS, and their interplay involving phytohormones.

To facilitate nanopriming and seed germination, which are critical and sensitive processes in the life cycle of plants involving several metabolic activities, it is necessary to overcome environmental stress during this period. The plant will progress to further growth phases only at that point. In order to increase resistance to nonliving environmental factors, seeds can be subjected to pretreatment using either naturally occurring or artificially produced substances to cause incomplete wetness or moderate stress. The method is well recognized as "seed processing".[7] In Abbasi et al.'s (2019) study, *Festuca ovina* seeds were subjected to treatment with silver nanoparticles, also known as Ag NPs, at various levels (25%, 50%, and 75%) in order to improve germination rates and mitigate the adverse effects of drought-related stresses on the seeds. The rate of germination of *Brassica napus L.* seedlings infused with the use of zinc oxide (ZnO) NPs (nanoparticles) exhibited a substantial increase in comparison to hydro-primed as well as unprimed seeds. The germination rate was shown to rise at all levels of ZnO NPs, which was linked to the alleviation of salt stress effects.[7] In addition, Khan et al. (2021) conducted a study which discovered that seedlings infused with encapsulated nanoceria comprising 0.1 mM of polyacrylic acid as well as subjected to salt stress showed enhanced water as well as amylase activity, in addition to higher rates of germination compared to seeds that were not treated. Abou-Zeid and Ismail showed enhanced germination and reduced the negative impact of salt exposure on *Triticum aestivum* seedlings upon AgNPs treatment. The addition of multiwalled carbon nanotubes (MWCNTs) to *Dodonaea viscosa L.* seeds led to a reduction in the average germination time and an enhancement in seed germination even in harsh conditions of drought. Rai et al. (2021) demonstrated that the inclusion of oxide of silicon (SiO2) nanoparticles at a concentration of 15 mg/L in T. aestivum seeds prior to germination, when subjected to drought conditions, enhances both the absorption of water and amylase activity. Furthermore, the application of peelings of mango nanoparticles (nMPs) at a concentration of 1000 mg/L hindered the sprouting of *Zea mays L.* seeds under salt stress. Furthermore, it appears to have a significant effect in reducing the high germination rate.[8] In a study conducted by Park et al. (2016), it was found observed the rate of germination of *Daucus carota L. cv. Mussangochon* seeds decreased when they were treated with a suspension of AgNPs at concentrations of up to 2000 mg/L. In addition, no statistically significant alteration in the germination rate was found. The results unambiguously demonstrate the necessity of investigating the factors and mechanisms that influence seed germination by nanopriming. Extensive research has been carried out on this subject in the past few years, as evidenced by Table 15.1. This review specifically examines the most recent progress in employing nanobionics to alleviate the detrimental impacts of metals such as lead (heavy metals), a lack of moisture and salt stress. It also explores the potential mechanisms by which nanobionics can enhance tolerance to these stressors.

Omics the system of biology involves discovering fundamental rules and trends within living organisms, achieved through analyzing the molecular variations and incorporating this information into intricate models of regulatory systems. The study of various "omics," including genomics, transcriptomics, proteomics, epigenomics, and metabolomics across entire plant systems, enhances comprehension of plant system biology and aids in advancing biotechnology. Omics has become a central area of interest in molecular research. Omics technologies include genomes, transcriptomics, proteomics, as well as metabolomics. Genomics, at first, provided information about the sequencing of genes and proteins. However, the focus has now switched towards understanding their dynamic functions along with interactions. Transcriptomics, proteomics, as well as metabolomics provide insights into the biological functions of gene products. These high-throughput methods greatly enhance the ability to concurrently detect a large number of proteins/genes, enabling the correlation of complex mixes with complicated consequences with gene/protein expression profiles. Omics technologies have a main

Table 15.1 The impact of nanoparticles (NPs) on the growth and development of different plant species.

Crops species	Abiotic stress	Applied NP	Conc. and mode of application	Result	References
Zea mays L.	Drought stress	CuNPs	3.33,4.44 and 5.55 mg/L plant priming	The plant matter, anthocyanin, chlorophyll, as well as carotenoid contents of the grain increased, as well as the quantity of seeds it yielded.	9
T. aestivum	Drought	TiO_2	500,1000 and 2000 mg/kg amended soil	Enhanced stomatal conductivity, transpiration rate, and antioxidative enzyme activity.	10
Gossypium hirsutum L.	Salinity	Poly (acrylic-acid) coated cerium oxide NPs	500 mg/L through seedlings	Salinity stress resulted in an increase in the length, vitality, fresh and dry weight, and root anatomical structure of seedlings.	22

objective of identifying all the products of genes (transcripts, proteins, including metabolites) within a biological sample without specific targeting. Nano-omics is a field that use nanoscale techniques to obtain more accurate data from biological systems. The precision of nanotechnology in manipulating and interacting with biomolecules on the microscale has enhanced researchers' ability to collect more information from biological systems. The combination of nanotechnology along with omics has given rise to a new area of research called nano-omics. This involves the use of nanotechnology to separate and examine chemicals from biological fluids, which may then be used for further omics study and analysis.

15.1.1 Nanomaterial

Nanomaterials, also known as NMs, are substances that can be found in nature or created artificially, with sizes typically falling within the range of 1–100 nanometers. These materials have tiny dimensions and an excellent surface-to-volume ratio, which gives them outstanding chemical and physical properties compared to their larger counterparts. Nanomaterials (NMs) possess unique and flexible physical and chemical properties, making them suitable for use in different fields. In recent times, nanotechnology has gained significant interest in the field of plant science. This interest is driven by the need to create small and effective systems that can improve the germination of seeds, growth, and protect plants from both nonliving and living challenges.

Nanogels (NGs) represent a burgeoning category of nanomaterials attracting increasing attention within the nanotechnology sphere. They exhibit exceptional physicochemical characteristics, including colloidal stability, substantial biomolecule encapsulation capacity (bioconjugation), and responsiveness to various stimuli like pH and temperature. Nanogels (NGs) are hydrogels that consist of nano-sized particles made from synthetic or natural polymers, which are connected through chemical or physical cross-linking. Their composition consists of a substantial amount of water (making about 70%–90% of the overall structure), notable porosity, and remarkable ability to bear heavy loads. Frequently found natural polymers are chitosan, alginate, poly(vinyl alcohol), poly(ethylene oxide), poly(ethyleneimine), poly(vinylpyrrolidone), and poly(N-isopropylacrylamide). Furthermore, it is also possible to create nanocomposites with hybrid architectures that combine both polymeric and nonpolymeric components.

15.2 Applications in plant science

Cutting-edge methodologies in plant science, particularly those pertaining to biosensing, agrochemical delivery, and genetic engineering, represent the forefront of innovation in the field.

1. **Biosensor: Nanomaterials (NMs)** are utilized in the development of biosensors and serve as "sensing materials" in agricultural biotechnology, farming, including the food industry. Multiple categories of nanosensors have been investigated in plant systems, such as plasmonic, fluorescence resonance energy transfer (FRET)-based, carbon-based electrochemical, nanowire, and antibody nanosensors. Preliminary findings indicate that NMs show promise in their ability to identify and measure plant metabolic flux, detect residual pesticides in food, and identify numerous pathogens including bacteria, viruses, and fungus. A new advancement involves the production of a fluorometric optical sensor that uses onion membranes to detect sucrose. This is accomplished by combining invertase-nanogold clusters with plant membranes.

   In addition, single-walled carbon nanotubes (SWNTs) have been employed to monitor nitric oxide in *A. thaliana* using near-infrared fluorescence. In addition, FRET probes that are connected to polystyrene nanoparticles (NPs) have been designed to measure and detect the phytoalexin. Biosensors utilizing nanomaterials (NMs) have significant potential in rapidly detecting and precisely quantifying organisms such as viruses, fungi, and bacteria in plant systems. An approach was developed using fluorescent silica nanoparticles and antibodies to detect *Xanthomonas axonopodis pv. Vesicatoria*, the causative agent of bacterial spot illness in Solanaceae plants. Intelligent nanosensors, like the 4mycosensor, can be used to detect mycotoxins. This particular sensor is an antibody-based assay that is effective in testing ZEA, T-2/HT-2, DON, and FB1/FB2 mycotoxin residues in maize, wheat, oats, and barley.

2. **Agrochemicals**: Nanomaterials (NMs) can be used in soil as nanostructured fertilizers, specifically designed to deliver critical elements such as iron (Fe), manganese (Mn), zinc (Zn), copper (Cu), and molybdenum (Mo) nanoparticles (NPs). Additionally, they can function as improved transportation routes to maximize the absorption and effectiveness of traditional fertilizers, like as nutrients and phosphates. Despite the significant promise of nanofertilizers and NM-enhanced fertilizers in agriculture, their utilization in fertilizer supply remains limited. Hydroxyapatite nanoparticles, when used as phosphorous nanofertilizers, increase the growth rate of soybeans by 33% and the yield of seeds by 20% compared to ordinary P fertilizers. Furthermore, nanofertilizers can be gradually released, which can help preserve soil fertility by minimizing the movement of nutrients into runoff or groundwater. This, in turn, helps to reduce environmental pollution and the harmful consequences of excessive fertilizer use.

   Metallic nanoparticles such as iron oxide, ZnO, TiO_2, and copper have been used directly as nanofertilizers in soil by applying them through irrigation or foliar methods on different plants including mung beans, cucumber, and rape. Moreover, the utilization of multiwalled carbon nanotubes (MWNTs) as additives in soil resulted in a twofold increase in the quantity of flowers and fruits produced by tomato plants. This effect is most likely attributed to the activation of genes and proteins that are essential for the growth and development of plants. Recently, chitosan NPs, which are natural polymers, have been used to release phosphorus, potassium, and nitrogen in wheat by means of foliar uptake in a regulated manner.

3. **Nanomaterials for plant genetic engineering:** The presence of the cell wall poses a challenge to the transportation of exogenous biomolecules into plant cells, which requires the use of new techniques for genetic change, including Agrobacterium-mediated transformation or biolistic strategies. An efficient technique called silicon carbide-mediated transformation has proven to be successful in introducing DNA into several types of calli, including as tobacco, maize, rice, soybean, as well as cotton. Furthermore, pioneering research shown that dsRNA derived from several plant viruses may be incorporated onto nontoxic, biodegradable layered double hydroxide (LDH) clay nanosheets or BioClay. The dsRNAs offer defense against cauliflower mosaic virus (CMV) in tobacco leaves that are sprayed with them. They also provide systemic protection to unsprayed leaves when they are exposed to the virus. This demonstrates the feasibility of delivering genetic material into plant cells without integrating transgenes, in a passive and species-independent manner.

Furthermore, successful genetic modification of cotton plants has been accomplished through the utilization of magnetic nanoparticles (MNPs). Transgenic cotton plants were effectively produced by introducing β-glucuronidase (GUS) reporter gene–MNP complexes into cotton pollen grains using magnetic force, followed by pollination using magnetofected pollen. The exogenous DNA was successfully incorporated into the genome, efficiently expressed, and reliably passed on to the progeny through self-fertilization. Recently, carbon nanotube scaffolds were used to deliver linear and plasmid DNA, as well as siRNA, to external plant tissue. This resulted in strong and temporary expression of Green Fluorescent Protein (GFP) in the leaves of *Nicotiana benthamiana, Eruca sativa, Triticum aestivum, and Gossypium hirsutum*, as well as in *E. sativa* protoplasts.

15.3 Adopting "omics" to study interactions at ENM–plant interface

To maximize the benefits of engineered nanomaterials (ENMs) in the production of nutritious foods, it is crucial to possess a comprehensive understanding of the networks of life at different levels (Yin et al., 2018). Lately, the synthesis of nanomaterials made from engineered materials (ENMs) and plants has shifted from single endpoint tests to more sophisticated, high-throughput system biology techniques called "omics" involving advanced techniques as with higher accuracy as shown in (Fig. 15.1). Omics technologies have gained significance in precision farming and nanotoxicity studies due to the increasing need for a comprehensive comprehension of intricate agronomic characteristics and crops' response to exposure to synthetic nanomaterials (ENMs). This approach facilitates the generation of hypotheses by seeing how biomolecules respond when biological processes are deliberately disrupted using engineered nanomaterials (ENMs). The subsequent stage entails the incorporation of extensive datasets onto routes utilizing advanced bioinformatics techniques.[11,12] Gaining a comprehensive understanding of the basic molecular processes in plants will provide valuable knowledge for creating engineered nanomaterials (ENMs) that serve specific functions. These functions include improving resistance against pests or environmental factors, delivering nutrients or pesticides with precision, releasing agrochemicals in a controlled manner in response to stimuli, and utilizing ENMs for biosensing purposes. Omic techniques exhibit exceptional sensitivity, allowing them to identify, quantify, and distinguish the cellular and molecular alterations that take place in an organism when it is subjected to ionic, nano-, or bulk-form fragments that are important at doses which are considerably lower and more representative of the environment.[12] These biomarkers are linked to observable characteristics or traits. Metabolomic investigations facilitate the identification of precise roles associated with genes or proteins that have not been thoroughly studied, therefore, filling in gaps in our comprehension regarding plant metabolism.[13] In addition, metabolomics does not

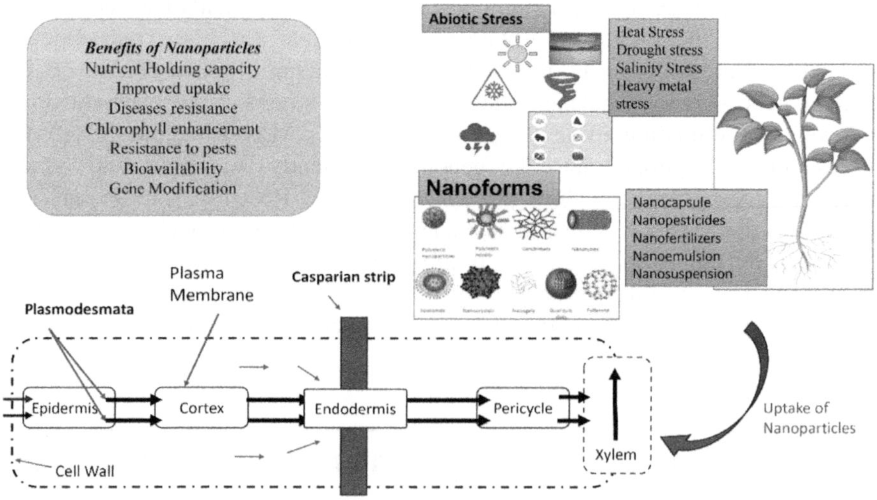

FIGURE 15.1

Schematic representation of potential interactions between NPs and plants resulting from HMs, drought, and salt stress.

depend solely on data obtained from model plant species, which allows it to be easily used to both model along with nonmodel taxa. Hence, to investigate the relationship between ENM and crops, it is recommended to employ a downward causation approach that connects the manifestation of plant characteristics with the metabolome. This technique can enhance the analysis of pathways by complementing proteomic and transcriptome profiles. This review synthesizes the cutting-edge studies in plant metabolomics along with proteomics, with the objective of gaining a more comprehensive comprehension of the interactions among nanomaterials made from engineered materials (ENMs) along with plants. Previous research utilizing alternative omics techniques has been briefly mentioned. The current research investigates integration of various omics datasets can be used to address important mechanistic concerns and promote the wider adoption of nanotechnology in farming (Figs. 15.2 and 15.3).

15.4 Metabolomics mechanisms of ENM–plant interactions

Contact to nanomaterials that have been engineered (ENMs) initiates a sequence of signaling cascade in plant metabolism, resulting in alterations in the synthesis and regulation of cellular metabolites. The exposure was done either acutely or chronically through the roots or leaves of the plants. Table 15.2 contains the

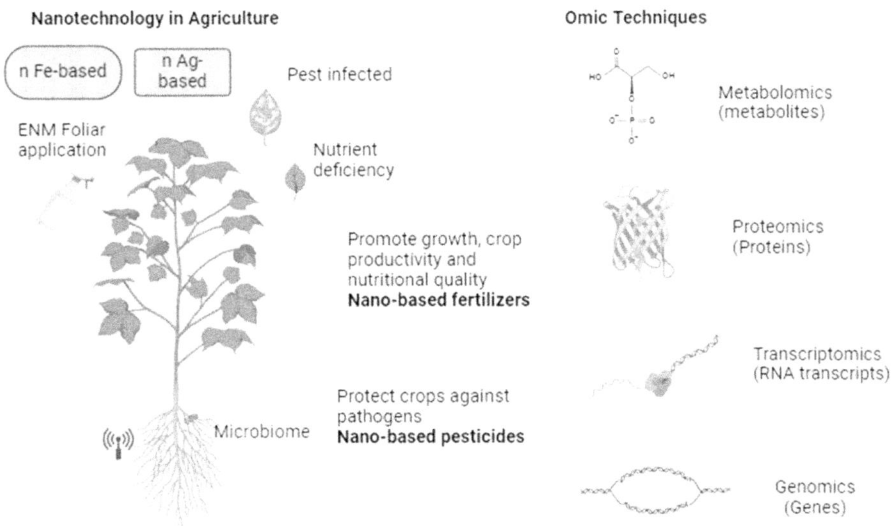

FIGURE 15.2

Utilization of omics techniques in studying the interaction between crop residue and artificial nanomaterials.

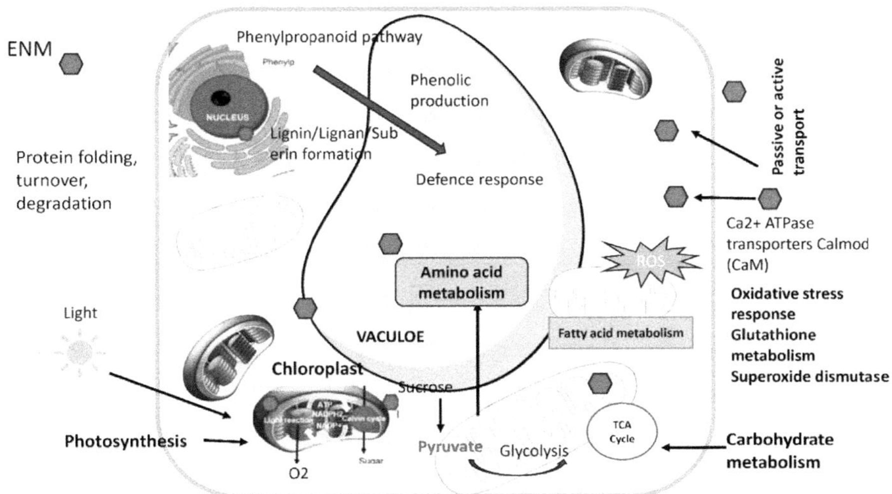

FIGURE 15.3

Integration of omics to understand ENM–responsive biological circuits.

Table 15.2 Overview of metabolomics investigations conducted on plants subjected to engineered nanomaterials (ENMs).

ENM	Size	Plant species and tissue analyzed	Growth media	Exposure route	Exposure concentration	Instrument used	Metabolic pathway affected	References
nAg citrate or PVP coated	NR	*Arabidopsis thaliana* (Columbia-2 ecotype,wild) leaf	Agar with ¼ Hoagland solution	Foliar	1 mg/L	LC QTOF-MS	Fatty acids metabolism	14,15
nFe3O4	30 nm	*Zea mays* leaf, root, soil	Soil	Root	100 mg/kg	GC-MS	Alkaloid biosynthesis glyoxylate metabolism, amino acid metabolism, methane metabolism, pantothenate and CoA biosynthesis, carbon metabolism (TCA cycle, glycolysis, gluconeogenesis, pyruvate metabolism), pyridine metabolism	
nY2O3	30 nm	*Zea mays* seed	Suspension	Seed	10,500 mg/L	1H-NMR	Carbohydrate metabolism, TCA cycle and amino acid metabolism	16
nCdO	7–60 nm	*Hordeum vulgare* leaf, root	Soil	Foliar	2.03 ± 0.45 105 particles per cm3 air	HPLC-LTQ Orbitrap HRMS for polar fractions, GCMS for nonpolar fraction	Amino acid metabolism	17
nCu	40 nm	*Cucumis sativus* root exudate, leaf	Hydroponic	Root	20 mg/L	1H-NMR, GC-TOF-MS	Amino acid metabolism. Fatty acid biosynthesis, Flavonoid biosynthesis	18
nAg	20 nm	*Cucumis sativus* leaf	Soil	Foliar	10, 100 mg/L	GC-MS	Carbohydrate metabolism, TCA cycle, Increased defense response (phenolic compounds, fatty acid, salicylic acid)	19

specific information about these publications. Several studies have employed targeted metabolomics to examine specific sets of metabolites, expanding on the findings of untargeted research.[11]

15.4.1 Copper nanoparticles

Several recent research have investigated the response of several crop species to Cu-based ENMs utilizing untargeted metabolomics. The findings of these investigations are presented in Table 15.2, which includes the works of[18] found that exposing cucumber plants (*Cucumis sativus*) to nCu (10 and 20 mg/L) in the roots for 1 week in hydroponic growth media resulted in alterations in the accumulation of a minimum of 22 leaf metabolites. This innovative metabolomic investigation employed 1 H-NMR to detect elevated concentrations of secondary metabolites linked to defense response and cell signaling, including 4-aminobutyrate (GABA), acetylglucosamine, and phenyl lactate. On the other hand, metabolites related to the metabolism of amino acids [phenylalanine (Phe)/tyrosine (Tyr), tryptophan (Trp), and arginine (Arg)/proline (Pro)], fatty acid production, riboflavin metabolism, and flavonoid production were observed to be inhibited. Cucumber fruits, when grown on soil contaminated with varying levels of copper (200, 400, and 800 mg/kg) until they were fully mature, showed alterations in their metabolite composition, thereby affecting their nutritional value. Two different analytical techniques were used as complementary platforms where NMR discovered just ten metabolites, whereas GC-MS identified a significant 107 molecules. The application of nCu treatments resulted in alterations in carbohydrates, amino acids and fatty acids. The presence of nanoscale copper (nCu) resulted in disturbances in 15 metabolic pathways within the fruits. Of all the modifications, the most significant ones were seen in the galactose metabolism and tricarboxylic acid (TCA) cycle. More precisely, the concentrations of succinate, fumarate, and pyruvate increased, whereas the concentration of citrate decreased. Moreover, the metabolic processes of arginine and proline, glycine/serine/threonine, isoleucine/leucine/valine, phenylalanine/tyrosine/threonine, alpha-linolenic acid, and alkaloids synthesis were affected, suggesting a notable impact on cell membrane and oxidative stress. Kocide-3000 is a readily accessible insecticide composed of nanosheets of micron-sized copper hydroxide (Cu(OH)2). The substance can be applied in the field using either aerial or ground spray techniques. Zhao et al. undertook a study to investigate the metabolomes from lettuce, cucumber, maize, and spinach crops that had been subjected to continuous treatment with Kocide. The administration Kocide to lettuce plants that were 24 days old at dosages of 1050, 1550, and 2100 mg/L. The spraying occurred biweekly over a period of four weeks. Although there was no immediate reaction, prolonged contact led to the buildup of 50 distinct metabolites in the leaves. The buildup had an impact on six metabolic pathways, specifically Gly/Se/Thr metabolism, Ala/Asp/Glu metabolism, TCA cycle, pantothenate as well as CoA biosynthesis, glycolysis or gluconeogenesis, along with pyruvate metabolism. The study conducted by Zhao, Huang, and Keller (2018) revealed that the presence of Kocide resulted in a decrease in the levels of phytol, a consequence of chlorophyll degradation, in spinach leaves. This effect was also observed in maize and cucumber plants. In addition, the Kocide treatment resulted in a decrease in the accumulation of specific amino acids (alanine, valine, leucine, asparagine, methionine, aspartic acid) in spinach leaves. The study conducted a comparison of the metabolic response between 3-week-old cucumber and maize plants that were cultivated in artificial soil showing the enhancement of energy metabolism and facilitating the formation of branched chain amino acids. Amino acids are essential for providing energy in the process of oxidative phosphorylation. Kocide significantly affected the intermediates of the Arg/Pro pathway in cucumber plants. Both plants experienced a reduction in levels of unsaturated fats (particularly linoleic and linolenic acids) as well as a rise in saturated fatty acid (laurate) as a result of exposure to Kocide. This indicates a disturbance in the equilibrium of the plasma membrane.

15.4.2 Carbon-based nanomaterials

Suppression of growth was seen within 7-day young *A. thaliana* plants that were grown on synthetic media with carbon nanodots (C-dots) concentrations of 125 and 1000 mg/L over a period of 7 days. The level of inhibition seen was significantly higher in comparison to both the control with activated-carbon treatments. Metabolite profiling revealed a greater abundance of carbohydrates (melezitose, ribulose-5-phosphate, d-turanose, talose, d-glucopyranoside, inositol) in both the roots and shoots of A. thaliana. These findings indicate that there was an impact on the glucose metabolism and cellular stress on the cellular membrane. The Cdot-exposed A. thaliana tissues displayed a significant buildup of various alkaloids, lignan, carotenoids, flavonoids, amino acids. This observation indicates the initiation of a defensive reaction. Cucumber plants were exposed to C60 fullerols for a continuous period of 3 weeks. The exposure was done by spraying the plants' leaves twice a day for 2 days. The doses used were 1 mg and 2 mg per plant. This exposure led to the breakdown of metabolites in the plasma membrane, notably linolenic and palmitic acid. In contrast, there was a higher buildup of antioxidant metabolites, such as 3-hydroxyflavone, 4-vinylphenol dimer, 1,4 benzenethiol, methyl trans cinnamate, quinin acid, and dehydroascorbic acid.[18]

15.4.3 Additional omics disciplines utilized in nano-enhanced agriculture

By integrating the current developments of genomes and transcriptomics with metabolomics along with proteomics, agricultural growth can be accelerated.[20]

These methodologies can provide significant benefits to the study of plant–microbe interactions. Microbial communities exert a substantial influence on plant growth and productivity through their direct control over soil processes, including soil structure stabilization, enhanced nutrient availability, organic pollutant degradation, CO_2 fixation, and organic matter decomposition. Nevertheless, they exhibit a high degree of sensitivity and vulnerability to the harmful effects of stressors. Pyrosequencing and Illumina-based sequencing, which are next generation sequencing technologies, have successfully addressed the intricacies in the microbial community with more precision compared to traditional methods.[21] Metagenomics allows for the thorough analysis of genome sequences obtained from known and unknown microbes in an ambient sample, without the requirement of isolating every microbe into pure cultures. Engineered nanomaterials (ENMs) can be found in agricultural soil, either deliberately employed as nano-enabled agrochemicals or accidentally established by means of biosolids or irrigation water. This presence has the potential to affect the soil microbial community and, as a result, hinder the cultivation of crops. Studying the soil microbial population using metagenomics can potentially lead to the development of sustainable nanomaterials made from engineered materials (ENMs) that enhance plant defense against pests and boost yield and nutrient content. Metagenomic research offers helpful insights about the functional abilities of the earth's microbial community. However, it does not provide an exact representation of the metabolic processes of individual organisms or the ecosystem as a whole. Analyzing the expression of genes at the RNA level, also known as transcriptomics, can improve the interpretation of genomics. This research can provide vital insights into the advancement of ecologically friendly nanomaterials made from engineered materials (ENMs) for use in agriculture and their impacts on soil and plant populations of microbes. The advancement in the analysis of gene expression by RNA-seq has deepened our understanding of alterations in gene activity in complex samples, which may be further validated through RT-PCR analysis of individual genes. Two publications have extensively analyzed the transcriptome research investigating the effects of nanomaterials that have been engineered (ENMs) on plants. The authors of these reviews investigated the integration of transcriptomics, micro-RNA, as well as proteomics information from plants to discover distinctive biomarkers and impacts associated with nanoparticles.[12] For transcriptome studies, it is essential to have a thorough comprehension of the complete genome sequence and associated functional annotation. Thus, it is more efficient to concentrate on genomes of lower size. *A. thaliana* has the most condensed genome, with a size of 145 Mbp, and it is completely annotated. On the other hand, agricultural species like rice (450 Mbp) along with soybeans (1115 Mbp) possess bigger genomes that have not been fully annotated. Therefore, scientists undertaking worldwide transcriptomic research on how plants interact using the environment have opted to utilize A. thaliana in order to reduce complication.[12] Complementary proteomics and the metabolomics can be used to provide functional annotations for genes that are intriguing in nonmodel species. Ruotolo et al. performed a review and discovered that plants initiate defensive mechanisms when exposed to engineered nanomaterials (ENMs). These pathways entail modifications in the transcriptome and proteome, mostly linked to changes in root structure, phytohormone communication, and stimulation of the system of antioxidants.[12]

15.4.4 Future prospective

Nano-omics combines nanotechnology with genetic sciences like genomics and proteomics.[22] In plant genetic engineering and sustainable farming, nano-omics could be a game-changer. To thoroughly investigate the relationship among nanomaterials and plant cells, it is necessary to incorporate other disciplines and examine the association from numerous perspectives. The utilization of machine learning methodologies, including supervised, semisupervised, as well as unsupervised approaches, assists in the examination of genetic and genomic databases. This process provides vital knowledge for forecasting optimized routes and outcomes when implementing nanotechnology in the field of plant biology. Moreover, the integration of various datasets using deep belief networks, convolutional neural networks, multivariate Poisson machine learning, multilayer perceptron, likely artificial neural systems, or neural networks with radial basis function can improve the accuracy of predicting target genes for nanotechnology-based genetic engineering methods. This integration helps overcome the challenge of dealing with high-dimensional data.

A revolution in agricultural science is imperative to enhance the effectiveness, resilience, and sustainability of the agricultural sector. Nanotechnology will have a crucial impact on the development of intelligent agriculture systems in the field of engineering. These devices will tackle the worldwide problem of insufficient food storage. Nanoparticles (NPs) can effectively deliver micronutrients and offer protection against insects or pathogens by enhancing enzyme and nonenzymatic compound levels alongside genetic resources. In addition, nanomaterials (NMs) can enhance the quality and efficiency of crops, reduce production expenses, and minimize postharvest losses through the use of nanograms, nanofertilizers, nanoinsecticides, nanosensors, nanotags, along with nanochips. Moreover, the use of nanomaterials (NMs) will result in a reduction in the application use sprayed agrochemicals and enhance their efficacy by delivering active components directly to the target areas and minimizing nutrient loss during fertilization.

To sum up, nano-omics in plant genetic engineering could transform sustainable farming by improving crops, saving resources, and protecting the environment. But we need to be careful, ensuring safety, fairness, and ethical use of these powerful tools.

15.5 Conclusion

Nano-omics-based plant genetic engineering is at the forefront of modern agricultural innovation, poised to transform sustainable farming. This approach combines nanotechnology and genomics to precisely modify plant genes, offering numerous benefits crucial for addressing agriculture's challenges. One key advantage of nano-omics-based plant genetic engineering is its ability to target specific genes within plant DNA accurately. Unlike traditional methods that may cause unintended genetic changes, nano-omics allows scientists to edit or insert genes precisely, increasing the success of desired traits without unwanted effects.

Precision is crucial for enhancing characteristics such as resistance to diseases, drought tolerance, along with nutrient content in crops. Through genetic modification of genes associated with these characteristics, scientists can create crops that are resistant to pests and diseases, require reduced amounts of water and fertilizer, and provide enhanced nutritional value. These traits boost agricultural productivity and sustainability. Nano-omics-based genetic engineering also reduces reliance on chemical inputs like pesticides and fertilizers. By creating pest-resistant crops and enhancing nutrient absorption, it minimizes the need for these chemicals, benefiting the environment and reducing health risks. Additionally, this approach helps crops adapt to climate change, ensuring stable yields despite extreme weather. Climate-resilient varieties developed through nano-omics withstand droughts, floods, and heatwaves, maintaining agricultural productivity.

Economically, nano-omics-based genetic engineering offers advantages such as increased yields, better-quality produce, and reduced costs for farmers. Value-added products like nutrient-enriched crops also fetch higher prices, encouraging sustainable farming practices. Despite its benefits, nano-omics-based plant genetic engineering raises ethical concerns like biosafety and equitable technology access. Responsible use involves regulatory frameworks, risk assessments, and stakeholder involvement to ensure safe and fair deployment.

In conclusion, nano-omics-based plant genetic engineering plays a vital role in advancing sustainable agriculture by improving crop traits, promoting resource efficiency, building climate resilience, and creating economic opportunities. It offers a path to a more sustainable food system while preserving the environment and addressing global food challenges.

References

1. Jat, S. K.; Bhattacharya, J.; Sharma, M. K. Nanomaterial based gene delivery: a promising method for plant genome engineering. *J. Mater. Chem. B* **2020**, *8* (19), 4165–4175.
2. Kandhol, N., et al. Nanoparticles as a potential protective agent for arsenic toxicity alleviation in plants. *Environ. Pollut.* **2022**, *300*, 118887.

3. Anwar, K., et al. Elucidating the response of crop plants towards individual, combined and sequentially occurring abiotic stresses. *Int. J. Mol. Sci.* **2021**, *22* (11), 6119.
4. Turgut-Kara, N.; Arikan, B.; Celik, H. Epigenetic memory and priming in plants. *Genetica* **2020**, *148*, 47–54.
5. Sharma, A., et al. Synergistic action of silicon nanoparticles and indole acetic acid in alleviation of chromium (CrVI) toxicity in *Oryza sativa* seedlings. *J. Biotechnol.* **2022**, *343*, 71–82.
6. Shah, T., et al. Seed priming with titanium dioxide nanoparticles enhances seed vigor, leaf water status, and antioxidant enzyme activities in maize (*Zea mays* L.) under salinity stress. *J. King Saud. University-Science* **2021**, *33* (1), 101207.
7. El-Badri, A. M., et al. Modulation of salinity impact on early seedling stage via nano-priming application of zinc oxide on rapeseed (*Brassica napus* L.). *Plant. Physiol. Biochem.* **2021**, *166*, 376–392.
8. Elkhatib, E.; Attia, M. G.; Mahdy, A. M.; Mostafa, R. A. Priming with mango peels nanoparticles enhances seed germination of maize (*Zea mays* L.) under salt stress. *Alex. Sci. Exch. J.* **2019**, *40* (OCTOBER-DECEMBER), 767–780.
9. Van Nguyen, D., et al. Copper nanoparticle application enhances plant growth and grain yield in maize under drought stress conditions. *J. Plant. Growth Regul.* **2022**, *41*, 364–375 5.
10. Faraji, J.; Sepehri, A. Exogenous nitric oxide improves the protective effects of TiO2 nanoparticles on growth, antioxidant system, and photosynthetic performance of wheat seedlings under drought stress. *J. Soil. Sci. Plant. Nutr.* **2020**, *20*, 703–714.
11. Majumdar, S., et al. Proteomic, gene and metabolite characterization reveal the uptake and toxicity mechanisms of cadmium sulfide quantum dots in soybean plants. *Environ. Science: Nano* **2019**, *6* (10), 3010–3026.
12. Ruotolo, R., et al. Plant response to metal-containing engineered nanomaterials: an omics-based perspective. *Environ. Sci. & Technol.* **2018**, *52* (5), 2451–2467.
13. Quanbeck, S. M., et al. Metabolomics as a hypothesis-generating functional genomics tool for the annotation of *Arabidopsis thaliana* genes of "unknown function". *Front. plant. Sci.* **2012**, *3*, 15.
14. Chavez, S. , N. G., et al. Mass spectrometry-based metabolomics to assess uptake of silver nanoparticles by *Arabidopsis thaliana*. *Environ. Science: Nano* **2017**, 1944–1953.
15. Zhao, L., et al. Metabolomics reveals that engineered nanomaterial exposure in soil alters both soil rhizosphere metabolite profiles and maize metabolic pathways. *Environ. Science: Nano* **2019**, *17*, 1716–1727.
16. Gong, C., et al. Responses of seed germination and shoot metabolic profiles of maize (*Zea mays* L.) to Y2O3 nanoparticle stress. *RSC Adv.* **2019**, *9*, 27720–27731.
17. Večeřová, K., et al. Changes of primary and secondary metabolites in barley plants exposed to CdO nanoparticles. *Environ. Pollut.* **2016**, *218*, 207–218.
18. Zhao, L., et al. 1H NMR and GC–MS based metabolomics reveal nano-Cu altered cucumber (*Cucumis sativus*) fruit nutritional supply. *Plant. Physiol. Biochem.* **2017**, *110*, 138–146.
19. Zhang, H., et al. Metabolomics reveals how cucumber (*Cucumis sativus*) reprograms metablites to cope with silver ions and silver nanoparticle-induced oxidative stress. *Environ. Sci. & Technol.* **2018**, *52*, 8016–8026.

20. Gómez-Merino, F. C.; Trejo-Téllez, L. I.; Alarcón, A. Plant and microbe genomics and beyond: potential for developing a novel molecular plant nutrition approach. *Acta Physiologiae Plant.* **2015**, *37*, 1–17.
21. Xu, J.; Wang, N. Where are we going with genomics in plant pathogenic bacteria? *Genomics* **2019**, *111* (4), 729–736.
22. Singh, A; Rajput, V. D.; Agrawal, S., et al. Nanoparticles mediated salt stress resilience: a holistic exploration of physiological, biochemical, and nano-omics approaches. *Reviews Env. Contamination* **2024**, *262*, 19. Available from: https://doi.org/10.1007/s44169-024-00070-4.

CHAPTER

Status for multiomics for crop improvement in world 16

Sagar Mondal[1], Swati Priya[1], Jutishna Bora[1], Smita Lata[1], Abeer Hashem[2], Elsayed Fathi Abd Allah[3] and Sumira Malik[1]

[1]*Amity Institute of Biotechnology, Amity University Jharkhand, Ranchi, Jharkhand, India*

[2]*Botany and Microbiology Department, College of Science, King Saud University, Riyadh, Saudi Arabia*

[3]*Plant Production Department, College of Food and Agricultural Sciences, King Saud University, Riyadh, Saudi Arabia*

16.1 Introduction

Omics is a field that includes various subdisciplines, including genomes, transcriptomics, metabolomics, as well as proteomics, among others. Together, these subdisciplines provide a thorough understanding of the molecular landscape. The primary goal of omics is to acquire a comprehensive comprehension of intricate biological events at a molecular level, hence offering insights that surpass the constraints of conventional approaches. This interdisciplinary approach examines the composition, organization, functioning, and interactions of specific molecules, leading to a more holistic comprehension of biological systems. Omics relies on cooperative high-throughput technologies that allow researchers to quickly gather and analyses large amounts of molecular data.

Nanotechnology has greatly revolutionized omics research by empowering researchers to go more into the domain of the minuscule. This field employs nanoscale interventions to get more precise information from biological systems. Nanotechnology's capacity to manipulate and engage with biomolecules on a small scale has enabled enhanced accuracy, enabling researchers to obtain a larger quantity of data from biological systems. Nanotechnology and omics have combined to create a new field of study known as nanoomics. This discipline focuses on utilizing nanotechnology to extract and examine compounds from biological fluids, with the aim of conducting further omics research and analysis. The incorporation of nanotechnology and omics represents the smooth merging of cutting-edge scientific disciplines. This integration advances us into a period of enhanced comprehension and precise control of biological processes. Exploring the complex world of molecule in living beings, nanoomics offers the potential to greatly transform various fields including health, the study of the environment, and agriculture, offering possibilities for innovative solutions and novel findings. The

Biogenic Nanoparticles. DOI: https://doi.org/10.1016/B978-0-443-33819-9.00020-2

generation and consumption of nourishment are significantly imbalanced due to changes in biodiversity and the increasing global population. To ensure a balance between crop production and food demand, it is imperative to devise efficient strategies for adapting plants to difficult conditions. Abiotic stressors, including heavy metals (HMs), shortages of water and salt, can have deleterious effects on plants. However, if these environmental stresses are present in little quantities and for short durations, they can activate many signaling pathways and systems in plants. Consequently, the resilience of ecosystems to endure future disruptions is enhanced. The constraints of these strategies, which stimulate the plant's defense mechanism in reaction to environmental strain, have increased the need to discover a potent and economical substitute to showcase the plant's ability to withstand various stressful circumstances. Recently, researchers in the area of stress in plant physiology including stress management have placed significant emphasis on subjecting plants to either naturally occurring or synthetically created chemical agents for the purpose of preparation. To accomplish this, they are diligently striving to cultivate plant varieties that exhibit enhanced resilience to diverse environmental challenges. The present technique, referred to as nanopriming, employs nanoparticles or NPs, as a primer to enhance the germination of seeds, seedling growth, and resilience to stress. As a result, this leads to a boost in crop yield and the improvement of food quality. Seed nanopriming offers multiple benefits that are linked to diverse biochemical processes., physiological, and signaling pathways in seeds. These responsibilities entail the control of levels of the reactive oxygen species (ROS) and how they interact with phytohormones.

To guarantee proper nanopriming and germination of seeds, which are critical and sensitive stages throughout the crop's life cycle involving several processes related to metabolism, it is essential to minimize stress from the environment during this period. The plant will advance to subsequent stages of growth only at that juncture. In order to increase resistance to nonliving environmental factors, seeds can be subjected to pretreatment using either natural or manufactured substances to cause partial hydration or moderate stress. The method is widely recognized as "seed preparation." Festuca ovina seeds were subjected to various levels (25%, 50%, and 75%) of silver nanoparticles, also known as Ag NPs, to enhance germination rates and alleviate the negative impact of drought-related stresses on the seeds.[1] The rate of germination of *Brassica napus L.* seeds that had been treated with zinc oxide (ZnO) NPs was significantly higher compared to hydro-primed and unprimed seedlings. At all concentrations of ZnO NPs, the rate of germination was shown to rise. This can be attributed to the reduction of the negative effects of salt stress. Khan et al.[2] performed a study that revealed that seedlings infused with wrapped nanoceria with 0.1 mM polyacrylic acid and exposed to salt stress exhibited improved water and amylase activity, along with greater germination rates in comparison to untreated seeds. Abou-Zeid and Ismail[3] showed that using synthetic AgNPs at a concentration of 1 mg/L improved germination and mitigated the detrimental effects of salt overload on *Triticum aestivum* seedlings. Introducing Multiwalled Carbon Nanotubes (MWCNTs) *to Dodonaea viscosa L.*

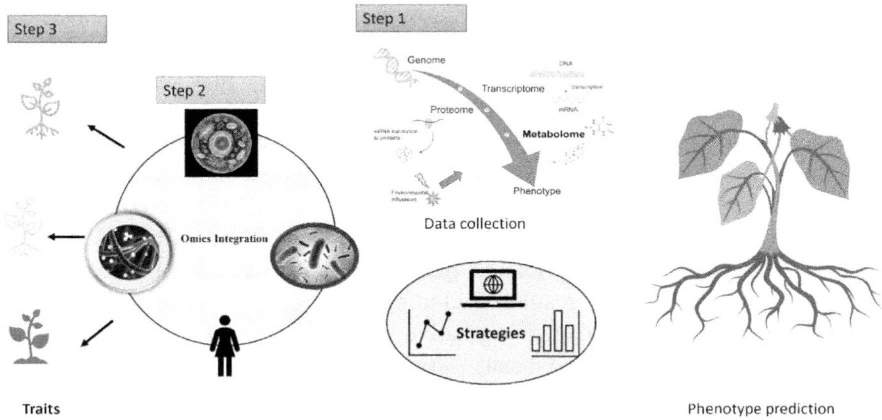

FIGURE 16.1

Multiomics data integration strategies use machine learning approaches to deconstruct complicated agronomic characteristics and forecast phenotypes.

seeds resulted in a decrease in the average time it takes for the seeds to start growing and an improvement in seed germination, even in severe drought conditions. In their study, Rai-Kalal et al.[4] showed that adding silicon oxide (SiO2) nanocrystals to T. aestivum seeds at a concentration of 15 mg/L before germination, under drought conditions, improves both the absorption of water and amylase activity. In addition, the use of mango peel nanoparticles (nMPs) at a concentration of 1000 mg/L impeded the sprouting of *Zea mays L.* seedlings when exposed to salt stress. Moreover, it seems to have a substantial impact on decreasing the heightened germination rate. Park and Ahn[5] conducted a study which revealed that the rate of germination of *Daucus carota L. cv. Mussangochon* seeds reduced when they were exposed to a solution of AgNPs at levels of up to 2000 mg/L. Furthermore, there was no statistically significant change observed in the germination rate. The results unambiguously demonstrate the necessity to investigate the variables and mechanisms that influence seed germination by nanopriming. Extensive research has been carried out on this subject in recent years. This analysis specifically examines the most recent progress in utilizing nanoomics to enhance and cultivate crops (Fig. 16.1).

16.2 Nanotechnology in crop development

The improved flavor and quality of our food, as well as its enhanced shelf life and maximum nutrition, are attributed to nanotechnology. This innovation will revolutionize the field of agriculture. In the future, our food will be designed to possess the capability of detecting the presence of contaminants and substances that cause spoilage.

Nanotechnology is a cutting-edge scientific technology that specifically deals with the manipulation and use of materials at the level of the molecular, namely in the nanoscale range. It is a fast-evolving scientific field that combines multiple disciplines of studies, including agriculture. Several studies have shown the use of Nano-sized particles or nanotechnology to improve the quality of crops. The most prevalent nanoparticles that are routinely used or studied are carbon along with metal-oxide particles. The application of nanoparticles have been discovered to yield several advantageous results, including enhanced germination, elongated roots and shoots, and augmented vegetative biomass of seeds in multiple crops. Some crops, such as soybean, spinach, and peanut, have shown notable enhancements in several physiological parameters, such as photosynthesis and nitrogen metabolism. In a study conducted in 2009, the addition of carbon nanotubes (CNTs) was found to enhance the procedure of germination of seeds in tomato plants. The water absorbing capacity of carbon nanotubes (CNTs) enhanced the seed germination in this sample. TiO_2 nanoparticles have been shown to promote the growth of Spanish. The nanoparticles were responsible for enhancing the activity of Rubisco activase and improving light absorption. According to a research in 2010, ZnO nanoparticles were found to inhibit the germination of maize and rye grass seeds. Additionally, there have been reports indicating that the use of silicon nanoparticles in certain plants has resulted in heightened resilience to diseases and stress. A recent discovery has revealed that the presence of cerium nanoparticles in SWCNTs can triple photosynthetic activity. Exposing seeds to a magnetic field during germination and then using magnetic fluid has been observed to significantly increase the content of nucleic acids.

This occurred as a result of the plant's metabolic renewal mechanism. The application of iron oxide to pumpkins resulted in an observed increase in root elongation. This occurred as a result of the iron dissolving. The genetic consequences of nanoparticles-induced beneficial alterations have been verified by the reduction of oxidative stress on spinach chloroplasts under ultraviolet-B radiation caused by nanotitanium dioxide. Fullerol was transmitted through rice seeds for multiple generations, as observed. Carbon nanotubes were used to alter the genetic expression in potato and tomato by means of nanoparticles.

16.3 Nanoomics technologies in crop improvement

Agriculture has a crucial function in ensuring the provision of nutrition for human beings. Current agricultural practices necessitate enhancement and fortification to meet the escalating food demands of the expanding global population. Gene editing (GE) technology is currently utilized to create plants with enhanced yield, nutrient content, as well as increased resistance to herbicides, insects, along with diseases. Recently, several new GE tools have been introduced, including CRISPR technology using nucleases, which provide a diverse and efficient methodology. The preliminary stages of genetic engineering entail the integration of transgenic crops or CRISPR within plants using targeted gene delivery methods. Nevertheless, genetic engineering approaches do have several limitations, including complex and time-consuming procedures, potential tissue damage, the need to integrate DNA in the genome of the host, and limited efficacy in cell transformation. In order to address these challenges, nanotechnology has arisen as an innovative and contemporary approach. Nanoparticle-mediated gene delivery surpasses conventional biomolecular methods due to its ability to boost the effectiveness of genetic transformation in both temporary and stable changes across different plant species. However, despite the progress made in technological advances, there are still challenges to be addressed in the development of a concise breeding plan for plants. Various analytical techniques are used to increase crop improvement which is shown in Fig. 16.2.

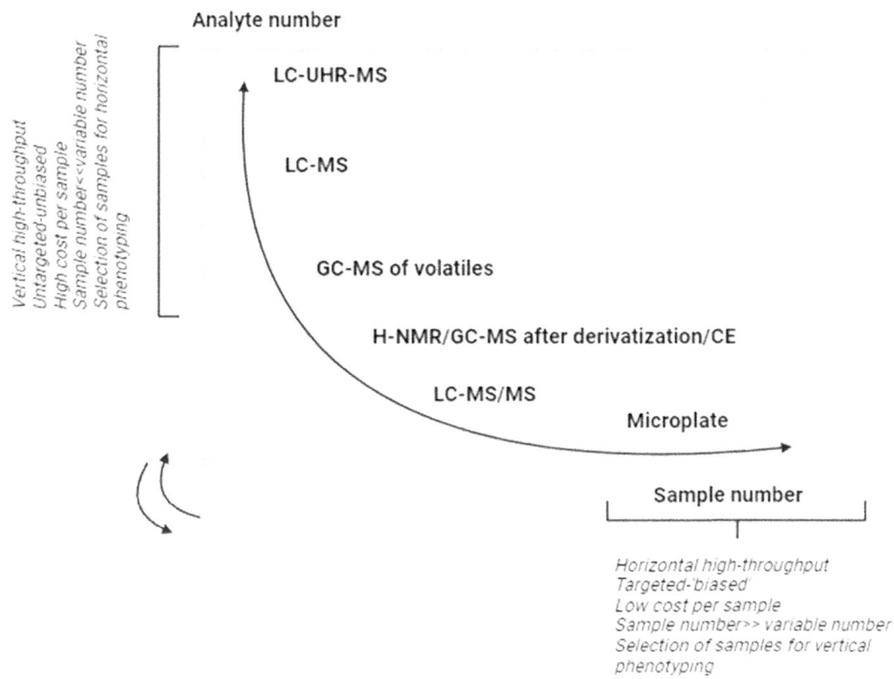

FIGURE 16.2

The combined advantages of efficient horizontal as well as vertical biochemical phenotyping. The expenses and labor requirements are similar for each technique.

16.3.1 Genomic nanotechnologies

Nanotechnology is employed to augment the advancement of genetically modified organisms (GMOs). This entails utilizing nanoparticles (NPs) as carriers to form a compound with biomodifier molecules (such as the CRISPR/Cas system) along with transport them into plant cells. Nanomaterials can be used with existing molecular technologies to tackle issues related to the manufacturing of genetically engineered crops and biotransformation. Nanomaterial engineering is a cutting-edge method used to advance the growth in crops in sustainable agriculture systems. Nanotechnology and nanomaterials (NMs) possess the capacity to alleviate the effects of important demands on energy as well as food manufacturing, while maximizing the utilization of restricted resources such as nutrients a well as water.

Nanobiotechnology methods have enhanced the accuracy of plant breeding by facilitating the selection and transfer of genes, hence lowering the time needed to eliminate undesirable genes and allowing breeders to obtain crucial genes from extensive plant populations. The successful magnetofection of a bioconjugate complex, which includes transgene and nanoparticles, has been recorded in dicot plants. In this case, the utilization of nanotechnologies enables the transportation of DNA without requiring any force or the incorporation of transgenes. This implies that CRISPR procedures, which offer transient expression, can be employed in the majority of countries to accomplish enduring genetic engineering without GMOs. The initial steps to genetic engineering entail the incorporation of transgenes or CRISPR/Cas9 within crops through definite gene transfer techniques. Nanotechnology-based gene transformation transport outperforms conventional biomolecular approaches by greatly enhancing the effectiveness for both temporary (transient) as well as everlasting (stable) genetic alterations in many floral species. Hence, nanotechnology holds the capability to reduce uncertainty and streamline the organization of agriculture management approaches by employing molecular manufacturing techniques as a substitute for traditional advances.

Nanoparticles are now being used to improve and optimize gene editing (GE) by delivering ribonucleoproteins (RNPs), plasmids, as well as in RNA in a precise and effective manner. Nanoparticles (NPs) are commonly employed in mammalian cells to effectively and directly distribute Cas-RNPs to the cytosol and nucleus of various cell types. Research has shown that the utilization of RNP delivery substantially reduces off-target effects as contrasted with plasmid-based CRISPR systems. The presence of the outer cellular layer in plants has impeded the advancement of a comparable mechanism capable of efficiently delivering genetically modified material into mature plants. Consequently, there exists considerable potential to create nanoparticle carriers, including RNA, proteins, as well as DNA, with varied cargo-loading capabilities and appropriate geochemistry to effectively penetrate the cellular walls as well as cellular barriers in the dense tissues of plants, independent of external assistance. A previous study conducted[6] indicated that a number nanoparticle formulations can be naturally taken up by plants, transporting substances like DNA, RNA, even proteins.[7]

In light of the progress in technology, it remains essential to develop crops that can be cultivated rapidly in order to fulfilll the daily dietary needs of the expanding population. This paper presents a plan to accelerate the development of current technologies for crop enhancement through their integration to yield crops that are both economical as well as high-yielding. CRISPR can be used in combination with speed breeding techniques to promote rapid crop improvement. This involves the use of speed editing tactics and nanomaterials. For a more comprehensive understanding of speed breeding.[8] This article provides an overview of nano-based delivery methods and plant genetic engineering approaches, as well as the limitations associated with NP-mediated genetic engineering. It also discusses the concept of rapid crop improvement. Additionally, this document provides a comprehensive description of the future applications and potential obstacles that may arise in the field of nanotechnology. The primary objective is to offer a thorough overview of CRISPR techniques connected to plants, which integrate nanotechnology, and to explore potential future applications for enhancing crops.

16.3.2 Transcriptomic and proteomic nanotechnologies

The plant genome comprises 36,795 proteins, each consisting of an average of 436 amino acid sequences. The average size of this is 27% smaller compared to the proteins from animals.[9] Proteins have a vital function in plant communication and response to stress, as they are directly engaged in cellular homeostasis mechanisms. Characterizing the plant proteome allows for a functional examination of the parts of the genome that have been translated. Hence, proteomics serves as a valuable addition to transcriptome and metabolomic methodologies, enabling a thorough comprehension of the fundamental cellular functions in plants.

Proteomic analysis is an invaluable tool for improving crops, protecting them, and investigating their susceptibility to toxins. It has the ability to identify important regulatory proteins, changes that happen after the process of translation, and the connections among protein in crops when they are subjected to external stimuli. Studying the relationship involving ENMs and plants at the level of protein is crucial for comprehending cellular mechanisms and pinpointing the specific proteins responsible for the response to ENMs. Using time-resolved methods and analysis at the subcellular or organelle levels, it is feasible to identify proteins that react to engineered nanomaterials (ENMs) and play a role in specific biological processes during different stages of plant development.[10] One area which has not been extensively explored is the examination of the protein coronas that develop surrounding nanomaterials made from engineered materials (ENMs) as they encounter plant tissues in different growing settings. These parameters also influence the toxicity and mobility of engineered nanomaterials (ENMs) within plants.

16.3.3 Metabolomic and phenomics nanotechnologies

The plant metabolomics analytical techniques assessed by Steuer in 2007 continue to be in use.[11] Notable progress in the past 5 years has been made in improving the clarity and analysis of spectra (http://www.metabolomicssociety.org/software.html), as well as in creating databases (http://www.metabolomicssociety.org/database.html). Thanks to technological and methodological developments, the variety of analytical substances and compound groups that can be detected in a specimen is constantly growing. However, this expansion sometimes comes at the cost of being able to analyze a larger number of samples due to rising expenses and/or personnel requirements. On the other hand, innovative experimental approaches result in a growing quantity of samples. Therefore, it is necessary to find the optimal balance between the number of analytes (using a vertical high-throughput approach) and the speed at which samples are processed (using a horizontal high-throughput approach), as well as identifying any potential synergies between these approaches. A schematic architecture enabling the development of climate-resilient and nutrient-rich crops, facilitated by integrated multiomics techniques, genetics, advanced plant breeding, gene editing, and computational modeling tools within a systems biology framework is illustrated in Fig. 16.3.

16.4 Role of nanotechnology in the breeding of plants

The existing agronomic practices, formulated during age of sustainability, are currently deemed largely unsustainable given the growing number of people and global warming.[12] Nanobiotechnology provides practical objectives for sustainable agriculture, include efficient prevention of pests as well as proper nutrient utilization, that mitigate the ecological repercussions of food production while decreasing the effects of global warming.[13] Plant nanobiotechnology is an emerging discipline that focuses on the application of synthetic nanomaterials to investigate plant functions.[12,14] At now, nanomaterials (NMs) have been widely employed as an effective means of incorporating biomolecules into plants. These nanoparticles can be modified to regulate the motion and dispersion of biomolecules inside plant organelles and cells.[7]

Various types of nanomaterials (NMs) have been examined, including nanofertilizers, nanopesticides, as well nanobiosensors.[15] Multiple strategies of nitrogen management (NM) can be utilized to enhance crop productivity and soil fertility. Nanofertilizers are classified as either micronutrients or macronutrients and serve as carriers for other chemicals, such as minerals, to enhance nutrient content.[16] found that nanofertilizers have the ability to efficiently encapsulate nutrients within the nanomaterials (NMs). In addition,[17] demonstrated a substantial enhancement in the efficacy of nanofertilizers, ranging from 18% to 29%, when compared to synthetic fertilizers. Metals can enhance transportation pathways, hence improving the absorption and effectiveness of synthetic fertilizers.

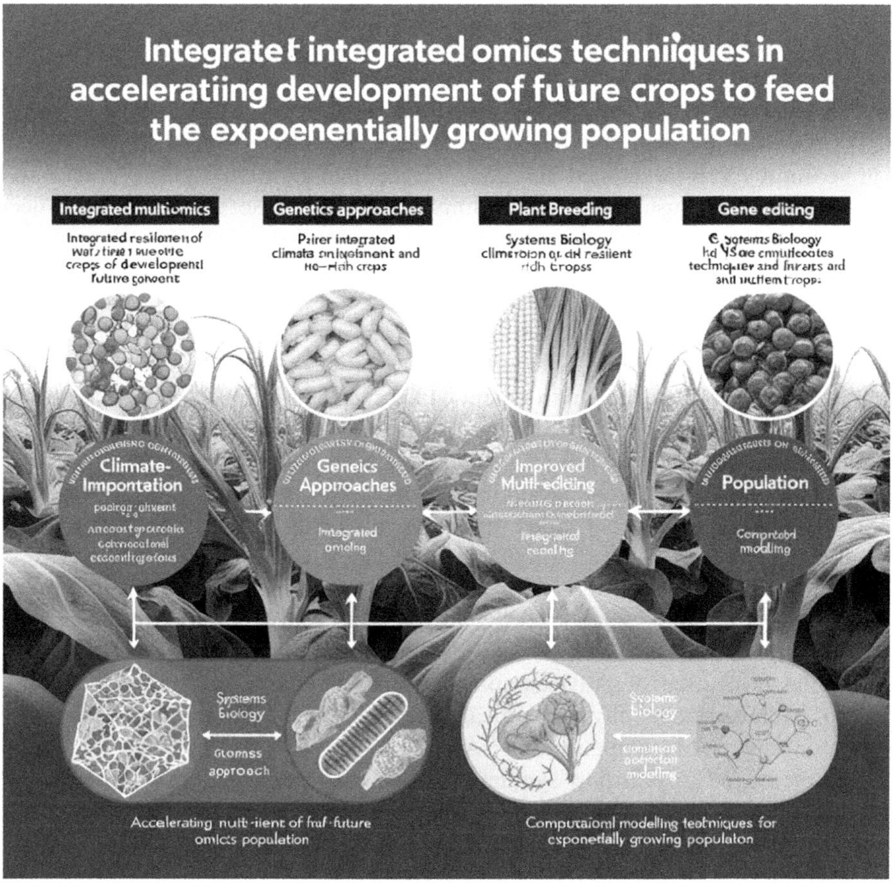

FIGURE 16.3

Schematic portrayal of the role of integrated omics approaches in advancing the production of future crops to feed the world's rapidly rising population. The production of climate-resilient and nutrient-rich crops is aided by integrated multiomics methodologies combined with genetics, enhanced plant breeding, gene editing, and computational modeling tools in a systems biology framework.

Different levels of silver nanoparticles (AgNPs) significantly enhance the rate at which maize, *Citrullus lanatus*, as well as *Cucurbita pepo L.* plants grow from seeds. Although the AgNPs have a mild level of toxicity, they facilitate the germination of seeds.[18] Studies have shown that the germination of lettuce, scientifically known as *Lactuca sativa*, can be enhanced by using titanium dioxide nanoparticles (NM) through electrospraying. Additionally, these nanoparticles have the ability to significantly reduce the need for fertilizers when applied to both the soil and leaves. This not only improves the effectiveness of the fertilizers but also reduces the amount of fertilizer runoff into the environment.[19]

Nanopesticides possess a greater potential capability and can serve as a substitute for synthetic pesticides. By gradually breaking down and precisely releasing active components with suitable nanomaterials, the effectiveness of pest control can be improved for extended durations. Hence, nanopesticides play a crucial role in efficiently and sustainably controlling pests, while also minimizing the need for agrochemicals and thereby mitigating environmental risks. These insecticides demonstrate unique characteristics in comparison to synthetic pesticides, leading to enhanced effectiveness.[17] The solubility of active substances can facilitate the mobility along with degradation of soil-dwelling microorganisms. Moreover, pesticides based on natural products enhance the ability of aluminum to dissolve and provide a lower risk to the environment compared to artificial pesticides.

Typically, nanopesticides exhibit a lower and less frequent release of water and energy compared to synthetic pesticides. Furthermore, they augment the efficacy of pesticides and boost agricultural productivity by minimizing wastage and labor expenditures, hence leading to increased yields and reduced input costs. Natural materials provide a structured antibacterial activity against both viruses and bacteria. Metals serve as vital inorganic nanoparticles with significant pesticidal capabilities. Nanoherbicides, which typically consist of biodegradable polymeric components, can enhance the effectiveness of herbicides. An instance of this is the utilization of poly(-caprolactone) the objective of storing atrazine is due to its remarkable physiochemical characteristics, superior bioaccessibility, along with better biocompatibility.[20] A carbon nanotube (CNT) conveyor system can be employed to transport synthetic substances into hosts, having the goal of minimizing the emission of potentially dangerous substances into the environment that may harm cells from other plants (Table 16.1).

Nanobiosensors are of great importance and are connected to sophisticated sensors capable of detecting different substances at extremely low levels utilizing a physiochemical transducer. Simply put, nanobiosensors have the potential to facilitate plant protection by enabling communication between plants and farmers.[14] This communication can facilitate prompt decision-making to optimize agricultural productivity by effectively managing water, land, fertilizers, and pesticides. Nanobiosensors have a longer lifespan than sensors from earlier generations because they are more stable and sensitive, have faster electron kinetics, and a higher surface-to-volume ratio. Plants have employed many nanosensors, such as plasmonic, fluorescence resonance energy transfer, carbon electrochemical, nanowire, and antibody nanosensors. They have the ability to identify chemicals like urea, glucose, also synthetic pesticides, monitor metabolites, as well as detect different microorganisms or diseases. The utilization of diverse nanomaterials enables accurate control over the timing lag in plant nanotechnology. This might be achieved by employing sophisticated nanomaterials (NMs) and nanoproducts (NPs), that possess the capacity to radically revolutionize the agriculture industry.

Table 16.1 Nanomaterials (NMs) offer numerous benefits for several agronomic crops.

Nanomaterial	Crops	Implementation	Impact	References
Ag	Rice, brown mustard, maize, summer squash, and radish	Plant-nanoparticle interactions	The growth of summer squash and watermelons was increased, while the shoot and root length of brown mustard were also stimulated Additionally, the photosynthetic efficiency of brown mustard was enhanced. However, the growth of maize roots was negatively affected, and the seedling growth of radishes was lowered	21
Au	Arabidopsis, flame lily, barley, rice, and tomato	Interactions between nanoparticles (NPs) in plants and their use in imaging	It does not have a harmful effect on tomato and barley plants. It improves the process of seed germination and the growth of the flame lily plant. It also increases the accumulation of NP (nutrient and protein) in the roots	22
$CaCO_3$	Peanut	Nutrient solution	Increased plant biomass and crop production	23
$Ca_5(PO_4)OH$	Soyabean	Nutrient solution	Enhanced biomass production, growth rate, and crop yield	24

(Continued)

Table 16.1 Nanomaterials (NMs) offer numerous benefits for several agronomic crops. *Continued*

Nanomaterial	Crops	Implementation	Impact	References
Cu	Lettuce, cucumber, mung bean, wheat, and sorghum	Plant–nanoparticle interactions	The nitrogen levels were elevated, resulting in an increase in overall nitrogen, shoot with root length in wheat, mung bean, and sorghum. However, there was a decrease in total biomass, bioaccumulation, and toxicity in these plants. Additionally, the roots of cucumber showed increased NP accumulation and gene deregulation	25
CdSe/ZnS QDs	Onion, Arabidopsis, and alfalfa	Interactions of nanoparticles (NPs) in plants, namely in the context of imaging, fluorescence detection, and nanobiosensors	Biosensors aid in the identification of pathogens, the amplification of reactive oxygen species (ROS), and the reduction of cell and root growth viability	26
CuO	Arabidopsis, rice, wheat, and cucumber	Genetic engineering of plants	Cu augmented the crucial nutrients in plant development, intensified reactive oxygen species (ROS) generation, and diminished shoot and root length	25

(*Continued*)

Table 16.1 Nanomaterials (NMs) offer numerous benefits for several agronomic crops. *Continued*

Nanomaterial	Crops	Implementation	Impact	References
Chitosan	Wheat and tea	Nanofertilizers, nanoherbicides, and plant genetic engineering	Enhanced plant growth, substance that is suitable with living organisms and can naturally break down over time, and ability to inhibit the growth of microorganisms	27
Dendrimer	Bentgrass	Genetic engineering of plants	Endosomal escape is the process of DNA transport that involves the release of DNA from endosomes	28
Fe_3O_4	Soybean, wheat, and maize	Interactions of NPs in plants and nanofertilizers	The soybean exhibited an increased chlorophyll content, the wheat shown enhanced plant height and leaf area, and the maize showed improved visibility of brown spots on its leaves	29
Fullerene	Summer squash, soybean, bitter gourd, poplar, tomato, and maize	Drug delivery in agriculture	The levels of pesticides found in maize, soybean, tomato, and summer squash were reduced. The bitter gourd showed improved growth and higher crop output. Additionally, the poplar tree shown an increased ability to absorb trichloroethylene	30

(*Continued*)

Table 16.1 Nanomaterials (NMs) offer numerous benefits for several agronomic crops. *Continued*

Nanomaterial	Crops	Implementation	Impact	References
Liposomes	Benth and tomato	Transportation of essential nutrients and genetic material	Enhanced transportation of DNA and precise cell localization, along with heightened safeguarding of nucleic acids	31
Mg	Black-eyed pea	Nanofertilizers	Enhanced chlorophyll concentration, together with better stability of the plasma membrane and increased production	32
Mn	Mung bean and chickpea	Interactions between nanoparticles (NPs) and plants	Increased shoot and root length, along with enhanced chlorophyll and carotenoid levels	33
Mo	Chickpea	Interactions between nanoparticles (NPs) and plants	Enhanced antioxidant metabolism resulted in improved nodule number and biomass	34
MSNs	Onion, tobacco, and maize	Plant genetic engineering, delivery of pesticides, and nanofertilizers	Regulation of chemical and nucleic acid release	35
MWCNTs, SWCNTs	Cotton, benth, tobacco, rice, tomato, rocket salad, Arabidopsis, barley, cucumber, ryegrass, rapeseed, and maize	Genetic engineering of plants	Enhanced growth and metabolic activity observed in tobacco plants; Increased germination, development, and blooming observed in tomato plants; Improved	36

(*Continued*)

Table 16.1 Nanomaterials (NMs) offer numerous benefits for several agronomic crops. *Continued*

Nanomaterial	Crops	Implementation	Impact	References
			transport of DNA observed in rocket salad, cotton, and tobacco plants; Increased root growth was observed in cucumber, ryegrass, maize, and rapeseed Additionally, apoptosis and chromatin condensation were observed in rice and Arabidopsis	
SiC whiskers	Cotton	Genetic engineering of plants	Enhanced genetic modification	37
TiO_2	Arabidopsis, rice, and spinach	Nanofertilizers	Spinach exhibits enhanced nitrogen metabolism, leading to better plant growth and enhanced seed germination	38
ZnO	Mung bean, chickpea, onion, Arabidopsis, rapeseed, cucumber, lettuce, ryegrass, rice, radish, and maize	Nanopesticide micronutrient delivery	At higher application rates, the use of this substance resulted in decreased blooming time and yield in onion, while improving plant development. Additionally, it boosted seed germination but inhibited root growth in rapeseed, ryegrass, radish, lettuce, cucumber, and maize	39

(Continued)

Table 16.1 Nanomaterials (NMs) offer numerous benefits for several agronomic crops. *Continued*

Nanomaterial	Crops	Implementation	Impact	References
SiO_2	Arabidopsis	Interactions between nanoparticles (NPs) and plants	Silicon dioxide nanoparticles (SiO_2 NPs) possess the capability to function as a cost-effective and highly efficientefficient, safe, and sustainable alternative for plant disease protection	40

16.5 Challenges and future perspective

Nanotechnology, an interdisciplinary field, is largely acknowledged as the sixth and most complete technology in the current era. The fields of application have been greatly enhanced by the progress in nanoscience and nanotechnology, resulting in a considerable influence on human life. The wide-ranging uses of this technology are noteworthy, as they are recognized in several industries such as medical, textiles, building, energy, food, electronics, cosmetics, defense, and agriculture. The present rate of global research and advancements in nanotechnology is expected to result in a market value of up to 100 billion by the year 2020. Simultaneously, it is widely acknowledged that nanotechnology holds significant potential in the agriculture industry and has the capability to revolutionize it by using nanoscale principles. By 2050, the world's population is projected to reach approximately 9.7 billion, which will be a substantial obstacle in terms of ensuring adequate food and nourishment for all individuals. The primary objective in completing this assignment is to identify methods that operate within the confines of scarce resources and utilize them prudently. Nanotechnology offers superior options for various agricultural aspects in this scenario, enhancing the production and efficiency of agricultural processes. The implementation of advanced nanotechniques has significantly contributed to the growth of the agricultural market, resulting in a growth rate of over 25%. The use of nanotechnologies in the agriculture industry is expected to enhance worldwide economic growth, with an estimated value of approximately 3 trillion USD by 2020. The farming industry has experienced substantial effects from two important developments: the Green Revolution in the 1960s and the current integration of nanotechnology. The heightened adoption of chemical fertilizers and pesticides in agricultural practices during and following the period known as the "Green Revolution" has raised substantial concerns regarding the sustainability of the environment and the risks to human health. The idea of employing environmentally friendly pesticides and fertilizers as a substitute for agrochemicals arose as a method to safeguard the environment and ensure the well-being of both individuals and the ecosystem. Nevertheless, this environmentally conscious method also presents significant concerns regarding the durability, stability in different conditions, the need for high doses to achieve optimal results in certain areas, and their performance in the face of unforeseen environmental changes. Scientists have documented that the use of nano-based formulations can efficiently tackle the issues linked to bio-formulations. In addition, formulations based on nanotechnology have the capacity to address worldwide issues such as the need for food and its availability, the development of crops, the sustainability of the environment, the security of food, and the impact of climate change. The application of nanotechnology in agriculture is being improved by the creation of nanodevices, nanoformulations, including nanoingredients. This encompasses the development of nanoherbicides and insecticides to enhance the control of plants, pests, and weeds, together with nanofertilizers to enhance the delivery of nutrients.[41] created nanobiosensors that can directly detect the water level, and nutritional value in soil at the specific spot. Therefore, it can be concluded that nanotechnology will have a significant impact on the agricultural sector by enhancing crop output while ensuring ecological sustainability, environmental safety, and economic stability. The future opportunities for multi omics agricultural development is shown in Fig. 16.4.

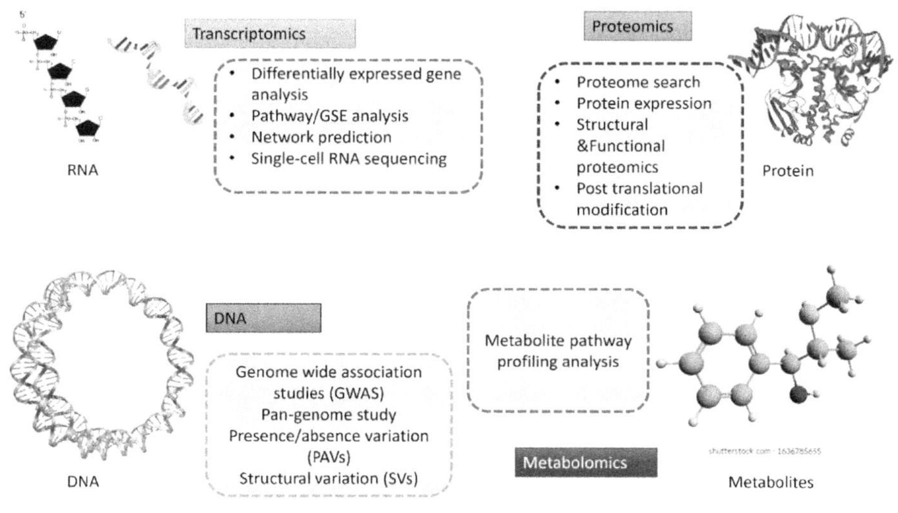

FIGURE 16.4

Future opportunities, constraints, and outlooks for multiomics-based agricultural development. Different multiomics data layers, their relationships, the types of omics features observed in each layer, and the methodologies used to analyse omics data across layers.

16.6 Conclusion

Plant genetic engineering using nanoomics is currently leading the way in modern agricultural innovation and has the potential to revolutionize sustainable farming. This approach integrates nanotechnology and genomics to accurately alter plant genes, providing a multitude of essential advantages for tackling the issues in agriculture. An important benefit of nanoomics-based plant genetic engineering is the precise targeting of specific genes inside plant DNA. Nanoomics enables scientists to accurately edit or insert genes, avoiding unintentional genetic modifications and enhancing the likelihood of desirable features without any undesired consequences.

This level of accuracy is highly advantageous for enhancing characteristics like disease resistance, drought tolerance, and nutritional value in agricultural crops. Through genetic modification of genes associated with these characteristics, scientists can create crops that are resistant to pests and diseases, require reduced amounts of water and fertilizer, and provide enhanced nutritional value. These characteristics enhance the efficiency and long-term viability of agricultural practices. Nanoomics-based genetic engineering diminishes dependence on chemical inputs such as insecticides and fertilizers. Through the development of genetically modified crops that are resistant to pests and have improved nutrient absorption, the need on chemical pesticides is reduced, resulting in environmental benefits and decreased health concerns. Furthermore, this methodology aids in the acclimatization of crops to climate change, guaranteeing consistent harvests in spite of severe weather conditions. Nanoomics has been used to generate climate-resilient cultivars that can endure droughts, floods, and heatwaves, while still sustaining agricultural productivity.

From an economic standpoint, the utilization of nanoomics-based genetic engineering provides benefits such as enhanced crop yields, improved food quality, and decreased expenses for farmers. Value-added products, such as crops that have been fortified with nutrients, can command higher prices, which in turn incentivizes the adoption of sustainable agricultural methods. Although nanoomics-based plant genetic engineering offers advantages, it also gives rise to ethical issues such as biosafety and fair access to technology. Responsible use entails the implementation of regulatory frameworks, conducting risk assessments, and involving stakeholders to guarantee the safe and equitable deployment.

Overall, the utilization of nanoomics in plant genetic engineering is crucial for the progress of sustainable agriculture. It enhances crop characteristics, optimizes resource utilization, strengthens resistance to climate change, and fosters economic growth. It provides a means to achieve a more sustainable food system while simultaneously protecting the environment and tackling global food concerns.

References

1. Abbasi Khalaki, M.; Ghorbani, A.; Dadjou, F. Influence of nano-priming on Festuca ovina seed germination and early seedling traits under drought stress, in laboratory condition. *Ecopersia* **2019**, *7* (3), 133–139.

2. Khan, M. N., et al. Nanoceria seed priming enhanced salt tolerance in rapeseed through modulating ROS homeostasis and α-amylase activities. *J. Nanobiotechnol.* **2021**, *19*, 1–19.
3. Abou-Zeid, H.; Ismail, G. The role of priming with biosynthesized silver nanoparticles in the response of *Triticum aestivum* L to salt stress. *Egypt. J. Bot.* **2018**, *58* (1), 73–85.
4. Rai-Kalal, P.; Tomar, R. S.; Jajoo, A. Seed nanopriming by silicon oxide improves drought stress alleviation potential in wheat plants. *Funct. Plant. Biol.* **2021**, *48* (9), 905–915.
5. Park, S.; Ahn, Y.-J. Multi-walled carbon nanotubes and silver nanoparticles differentially affect seed germination, chlorophyll content, and hydrogen peroxide accumulation in carrot (*Daucus carota* L.). *Biocatalysis Agric. Biotechnol.* **2016**, *8*, 257–262.
6. Burlaka, O.; Pirko, Y. V.; Yemets, A.; Blume, Y. B. Plant genetic transformation using carbon nanotubes for DNA delivery. *Cytology Genet.* **2015**, *49* (6), 349–357.
7. Demirer, G. S., et al. High aspect ratio nanomaterials enable delivery of functional genetic material without DNA integration in mature plants. *Nat. Nanotechnol.* **2019**, *14* (5), 456–464.
8. Watson, A., et al. Speed breeding is a powerful tool to accelerate crop research and breeding. *Nat. Plants* **2018**, *4* (1), 23–29.
9. Ramírez-Sánchez, O.; Pérez-Rodríguez, P.; Delaye, L.; Tiessen, A. Plant proteins are smaller because they are encoded by fewer exons than animal proteins. *Genomics, Proteom. Bioinforma.* **2016**, *14* (6), 357–370.
10. Komatsu, S.; Mock, H.-P.; Yang, P.; Svensson, B. Application of proteomics for improving crop protection/artificial regulation. 2013, Frontiers Media SA. p. 522.
11. Steuer, R.; Morgenthal, K.; Weckwerth, W.; Selbig, J. A gentle guide to the analysis of metabolomic data. *Metabolomics: Methods Protoc.* **2007**, 105–126.
12. Lowry, G. V.; Avellan, A.; Gilbertson, L. M. Opportunities and challenges for nanotechnology in the agri-tech revolution. *Nat. Nanotechnol.* **2019**, *14* (6), 517–522.
13. Hofmann, T., et al. Technology readiness and overcoming barriers to sustainably implement nanotechnology-enabled plant agriculture. *Nat. Food* **2020**, *1* (7), 416–425.
14. Giraldo, J. P.; Wu, H.; Newkirk, G. M.; Kruss, S. Nanobiotechnology approaches for engineering smart plant sensors. *Nat. Nanotechnol.* **2019**, *14* (6), 541–553.
15. Usman, M., et al. Nanotechnology in agriculture: current status, challenges and future opportunities. *Sci. Total. Environ.* **2020**, *721*, 137778.
16. DeRosa, M. C., et al. Nanotechnology in fertilizers. *Nat. Nanotechnol.* **2010**, *5* (2), 91 91.
17. Kah, M.; Kookana, R. S.; Gogos, A.; Bucheli, T. D. A critical evaluation of nanopesticides and nanofertilizers against their conventional analogues. *Nat. Nanotechnol.* **2018**, *13* (8), 677–684.
18. Acharya, P., et al. Nanoparticle-mediated seed priming improves germination, growth, yield, and quality of watermelons (*Citrullus lanatus*) at multi-locations in Texas. *Sci. Rep.* **2020**, *10* (1), 5037.
19. Adisa, I. O., et al. Recent advances in nano-enabled fertilizers and pesticides: a critical review of mechanisms of action. *Environ. Science: Nano* **2019**, *6* (7), 2002–2030.

20. Abigail, E. A.; Chidambaram, R. *Nanotechnology in herbicide resistance*. Nanostructured materials: fabrication to applications. *IntechOpen, Rij.* **2017**, 207–212.
21. Almutairi, Z. M.; Alharbi, A. *Effect of silver nanoparticles on seed germination of crop plants. Int. J. Nucl. Quantum Eng.* **2015**, *9* (6), 689–693.
22. Milewska-Hendel, A., et al. Fate of neutral-charged gold nanoparticles in the roots of the *Hordeum vulgare* L. cultivar Karat. *Sci. Rep.* **2017**, *7*, 3014.
23. Liu, X.-m, et al. Responses of peanut to nano-calcium carbonate. *J. Plant. Nutr. Fertilizers* **2005**, *11* (3), 385–389.
24. Liu, R.; Lal, R. Synthetic apatite nanoparticles as a phosphorus fertilizer for soybean (*Glycine max*). *Sci. Rep.* **2014**, *4* (1), 5686.
25. Mosa, K. A., et al. Copper nanoparticles induced genotoxicty, oxidative stress, and changes in superoxide dismutase (SOD) gene expression in cucumber (*Cucumis sativus*) plants. *Front. Plant. Sci.* **2018**, *9*, 872.
26. Modlitbová, P., et al. Short-term assessment of cadmium toxicity and uptake from different types of Cd-based Quantum Dots in the model plant *Allium cepa* L. *Ecotoxicol. Environ. Saf.* **2018**, *153*, 23–31.
27. Malerba, M.; Cerana, R. Recent advances of chitosan applications in plants. *Polymers* **2018**, *10* (2), 118.
28. Kretzmann, J. A., et al. Synthetically controlling dendrimer flexibility improves delivery of large plasmid DNA. *Chem. Sci.* **2017**, *8* (4), 2923–2930.
29. Fathi, A.; Zahedi, M.; Torabian, S.; Khoshgoftar, A. Response of wheat genotypes to foliar spray of ZnO and Fe_2O_3 nanoparticles under salt stress. *J. Plant. Nutr.* **2017**, *40* (10), 1376–1385.
30. Kole, C., et al. Nanobiotechnology can boost crop production and quality: first evidence from increased plant biomass, fruit yield and phytomedicine content in bitter melon (*Momordica charantia*). *BMC Biotechnol.* **2013**, *13*, 1–10.
31. Karny, A., et al. Therapeutic nanoparticles penetrate leaves and deliver nutrients to agricultural crops. *Sci. Rep.* **2018**, *8* (1), 1–10.
32. Delfani, M.; Baradarn Firouzabadi, M.; Farrokhi, N.; Makarian, H. Some physiological responses of black-eyed pea to iron and magnesium nanofertilizers. *Commun. Soil. Sci. Plant. Anal.* **2014**, *45* (4), 530–540.
33. Pradhan, S., et al. Photochemical modulation of biosafe manganese nanoparticles on Vigna radiata: a detailed molecular, biochemical, and biophysical study. *Environ. Sci. Technol.* **2013**, *47* (22), 13122–13131.
34. Taran, N. Y., et al. The effect of colloidal solution of molybdenum nanoparticles on the microbial composition in rhizosphere of *Cicer arietinum* L. *Nanoscale Res. Lett.* **2014**, *9*, 1–8.
35. Rastogi, A., et al. Application of silicon nanoparticles in agriculture. *3 Biotech.* **2019**, *9*, 1–11.
36. Demirer, G. S., et al. Carbon nanotube–mediated DNA delivery without transgene integration in intact plants. *Nat. Protoc.* **2019**, *14* (10), 2954–2971.
37. Asad, S.; Arshad, M. Silicon carbide whisker-mediated plant transformation. In *Properties and Applications of Silicon Carbide*. 2011, IntechOpen.
38. Liu, J., et al. TiO2 nanoparticles in irrigation water mitigate impacts of aged Ag nanoparticles on soil microorganisms, *Arabidopsis thaliana* plants, and Eisenia fetida earthworms. *Environ. Res.* **2019**, *172*, 202–215.

39. Laware, S.; Raskar, S. Influence of zinc oxide nanoparticles on growth, flowering and seed productivity in onion. *Int. J. Curr. Microbiol. Sci.* **2014**, *3* (7), 874–881.
40. El-Shetehy, M., et al. Silica nanoparticles enhance disease resistance in Arabidopsis plants. *Nat. Nanotechnol.* **2021**, *16* (3), 344–353.
41. Pandey, G. Challenges and future prospects of agri-nanotechnology for sustainable agriculture in India. *Environ. Technol. Innov.* **2018**, *11*, 299–307.

Index

Note: Page numbers followed by "*f*" and "*t*" refer to figures and tables, respectively.

O

Q

R

S

T

9780443338199